# Student Solutions Manual for
### Parks, Musser, Trimpe, Maurer, and
# A Mathematical View
# of Our World

Vikki Maurer

# Student Solutions Manual

for

# A Mathematical
# View of Our World

## Hal Parks
Oregon State University

## Gary Musser
Emeritus, Oregon State University

## Lynn Trimpe
Linn-Benton Community College

## Vikki Maurer
Linn-Benton Community College

## Roger Maurer
Linn-Benton Community College

## Prepared by
## Vikki Maurer
Linn-Benton Community College

BROOKS/COLE
CENGAGE Learning

Australia • Brazil • Japan • Korea • Mexico • Singapore • Spain • United Kingdom • United States

**BROOKS/COLE**
CENGAGE Learning™

**Student Solutions Manual for**
**A Mathematical View of Our World**
Hal Parks, Gary Musser, Lynn Trimpe,
Vikki Maurer, Roger Maurer

For product information and technology assistance, contact us at
**Cengage Learning Customer & Sales Support, 1-800-354-9706**

For permission to use material from this text or product, submit all requests online at **cengage.com/permissions**
Further permissions questions can be emailed to
**permissionrequest@cengage.com**

ISBN-13: 978-0-495-01062-3

ISBN-10: 0-495-01062-6

**Brooks/Cole**
10 Davis Drive
Belmont, CA 94002-3098
USA

Cengage Learning is a leading provider of customized learning solutions with office locations around the globe, including Singapore, the United Kingdom, Australia, Mexico, Brazil, and Japan. Locate your local office at:
**international.cengage.com/region**

Cengage Learning products are represented in Canada by Nelson Education, Ltd.

For your course and learning solutions, visit **academic.cengage.com**

Purchase any of our products at your local college store or at our preferred online store **www.ichapters.com**

Printed in the United States of America
2 3 4 5 6 7 11 10 09 08

# PREFACE

This manual accompanies the text *A Mathematical View of Our World* by Hal Parks, Gary Musser, Lynn Trimpe, Vikki Maurer, and Roger Maurer. For each chapter, there is a summary of key concepts, hints or suggestions for all odd-numbered problems, complete solutions for all odd-numbered problems, and complete solutions for all review problems. Additionally, for selected chapters, there is a prerequisite skill review section and a set of review questions and answers.

Key-Concept Checklist: As you master the ideas presented in a chapter of the textbook, you can check off the corresponding concept in the key-concept checklist located at the beginning of each chapter of the solutions manual.

Prerequisite Skills Review: For several chapters, the authors of the text assume you are proficient in certain prerequisite skills. For these chapters, there is a very brief review of the prerequisite skills along with several practice questions and answers.

Hints: For each odd-numbered problem, there is a hint or suggestion listed. If you have trouble getting started, then consult the hint for that problem.

Solutions: For each odd-numbered problem, there is a complete solution given. You should always attempt a problem on your own before consulting the solution manual. Be sure you read and understand the section in the textbook prior to starting work in the problem set. As you read the text, pay close attention to the detailed examples and study the notes you took in class before consulting the worked-out solution for a problem.

Even if you have solved a problem correctly, you may find the solution in this manual useful, as it may show you a different correct solution or show you a different way to approach the solution to the problem.

Vikki Maurer

**Acknowledgments**
Every effort has been made to ensure that each solution is correct and clearly written. I wish to thank Charisse Hake for her assistance in checking the solutions for accuracy, reading the manual for clarity and completeness, and always paying attention to the details. Thank you to Roger Maurer for his help, support, and feedback. Thank you to John Griffith for his technical support and willingness to always lend a hand.

# CONTENTS

# *Chapter 1:* Numbers in Our Lives

## *Key Concepts*

**By the time you finish studying Section 1.1, you should be able to:**

- ☐ discuss the history and features of the social security number.

- ☐ identify identification numbers by type.

- ☐ explain how the two most common types of transmission errors are made.

- ☐ describe what a check digit is and how it is used.

- ☐ calculate the check digit given a check-digit scheme.

- ☐ determine if a check-digit scheme will catch all single-digit errors.

- ☐ determine if a check-digit scheme will catch all adjacent-transposition errors.

- ☐ discuss the features of the Universal Product Code (UPC).

- ☐ calcluate the check digit for a two-weight scheme such as the UPC check-digit scheme.

**By the time you finish studying Section 1.2, you should be able to:**

- ☐ find the quotient and remainder for a division problem without a calculator.

- ☐ find the quotient and remainder for a division problem using a calculator.

- ☐ use the division algorithm for integers to describe the relationship between the dividend, the divisor, the quotient, and the remainder.

- ☐ explain what it means for two integers to be congruent modulo $m$.

- ☐ verify that two integers are congruent given a certain modulus.

- ☐ use modular arithmetic properties to simplify calculations involving exponents.

- ☐ use modular arithmetic to calculate a check digit.

- ☐ apply the process of casting out nines when a mod 9 check-digit scheme is used.

☐ verify the check digits for U.S. Post Office money order numbers, Euro banknote identification numbers, airline ticket numbers, and vehicle identification numbers.

**By the time you finish studying Section 1.3, you should be able to:**

☐ define a binary code and discuss the basic features of the binary codes introduced in this section.

☐ discuss the advantages of a fixed-length binary code over a variable-length binary code.

☐ translate a message to Morse code or from Morse code.

☐ when working with UPC bar codes, convert a manufacturer number and a product number into a bar code or convert to a number from a bar code.

☐ verify or calculate the check-digit for a UPC bar code.

☐ construct a UPC bar code.

☐ encode or decode messages using the Braille code.

☐ discuss the limitations of some binary codes.

☐ encode or decode messages using ASCII.

☐ encode or decode ZIP + 4 codes or ZIP + 4 + delivery-point codes using the Postnet code.

## Skills Brush-Ups

Some problems in Section 1.2 of Chapter 1 will require that you be comfortable working with whole-number exponents, where whole numbers are 0, 1, 2, 3, etc. There are several rules you should remember.

### EXPONENTS:

Let $a$, $m$, and $n$ be any whole numbers such that $m \neq 0$ and $n \neq 0$.

$$a^m = \underbrace{a \times a \times \cdots \times a}_{m \text{ times}}$$

The whole number $a$ is called the base and $m$ is called the exponent or power. When reading, $3^2$ is read "3 squared" or "3 to the second power", $2^3$ is read "2 cubed" or "2 to the third power", and $a^m$ is read "$a$ to the $m$th" or "$a$ to the $m$th power." The following three rules allow us to write expressions using a single exponent.

(1)  $a^m \times a^n = a^{m+n}$

(2)  $\left(a^m\right)^n = a^{m \times n}$

(3)  $\dfrac{a^m}{a^n} = a^{m-n}$, $a \neq 0$

Also recall that $0^0$ is not defined. If $a \neq 0$, then $a^0 = 1$ and $a^{-m} = \dfrac{1}{a^m}$.

*Examples:*
Rewrite each expression using a single exponent.
(a) $4^2 \times 4^3 = 4^{2+3} = 4^5$
(b) $7 \times 7^5 = 7^1 \times 7^5 = 7^{1+5} = 7^6$
(c) $\left(2^5\right)^3 = 2^{5\times3} = 2^{15}$
(d) $\dfrac{5^9}{5^5} = 5^{9-5} = 5^4$

## *Skills Practice*

1. Rewrite each expression using a single exponent.
   (a) $2^6 \times 2^4$     (b) $6^3 \times 6$     (c) $3^4 \times 3^4$     (d) $9^8 \times 9^5$

2. Rewrite each expression using a single exponent.
   (a) $\left(3^5\right)^6$     (b) $\left(6^2\right)^{100}$     (c) $\left(2^3\right)^{200}$     (d) $\left(9^2\right)^5$

3. Rewrite each expression using a single exponent.
   (a) $\dfrac{6^4}{6^3}$     (b) $\dfrac{3^7}{3}$     (c) $\dfrac{7^5}{7^5}$     (d) $4^0$

*Answers to Skills Practice:*
1. (a) $2^{10}$   (b) $6^4$   (c) $3^8$   (d) $9^{13}$
2. (a) $3^{30}$   (b) $6^{200}$   (c) $2^{600}$   (d) $9^{10}$
3. (a) $6$   (b) $3^6$   (c) $1$   (d) $1$

## *Hints for Odd-Numbered Problems*

**Section 1.1**
1. Refer to Table 1.1.

3. (e) Is there a way to change one digit so that the sum of the first four digits is still 7?

5. (a) Create an identification number and transpose the third and fourth digits. Determine the check digit before and after the transposition.
   (b) How will this single-digit error affect the sum of the first four digits?
   (c) How will this single-digit error affect the sum of the first four digits?
   (d) What digits are possible values for the check digit?

7. What digits are possible values for the check digit in this case?
   (a) Is there a way to change one digit without changing the check digit?
   (b) Think about how the check digit is calculated. Will an adjacent transposition affect the check digit?

9. Find the weighted sum of the first six digits and divide the weighted sum by 7.

11. Many answers are possible.
   (a) If the check digit is 5, then the remainder after the weighted sum is divided by 7 must be 5. The weighted sum is 5 more than a multiple of 7.
   (b) The weighted sum must be 3 more than a multiple of 7.

**(c)** The digits must all be the same. Try using a variable to represent the digit and calculate the weighted sum. If the weighted sum is divided by 7, what is the remainder?

**13. (a)** Compare the numbers. Describe how they differ.
**(b)** Calculate the check digit for the identification number containing the error.

**15.** Calculate the weighted sum of the digits. Divide the weighted sum by 11 and determine the remainder.

**17.** The check digit is correct, so the error must have occurred in one of the other digits. A single-digit error has occurred, so consider how changing one digit can affect the weighted sum. Knowing the check digit is correct tells you something about the weighted sum.

**19.** The check digit is correct. Of the remaining nine digits, two adjacent digits have been transposed. Systematically switch the positions of pairs of digits and calculate the check digit until the correction is found.

**21.** When finding the weighted sum, do not forget to add the digits of any two-digit number that results from multiplying by a weight of two.

**23. (a)** Calculate the weighted sum and simplify as much as possible. What must the missing digit be so that the weighted sum is divisible by 10?

**25. (i)** Refer to Table 1.4. **(iii)** The UPC uses a two-weight check-digit scheme. See Example 1.6 (b).

**27.** Calculate the weighted sum and simplify as much as possible. The weighted sum must be divisible by 10. What must the missing digit be to force the weighted sum to be divisible by 10?

**29.** Think about how the weighted sum is calculated. Is there any way to change one digit in a position that is weighted by 3 and end up with the same check digit? Is there any way to change one digit in a position that is weighted by 1 and end up with the same check digit?

**31. (a)** When a check digit is a remainder after a division, in this case after a division by 9, see if you can change a single digit by a multiple of 9. If you can, then the check digit will be the same.

**33.** Calculate the weighted sum. The check digit is the last digit in the weighted sum, so the check digit is the digit in the one's place of the weighted sum.

**35.** Calculate the check digit of the routing number with the mistake.

**37. (a)** Calculate the weighted sum. The check digit is the last digit in the weighted sum, so the check digit is the digit in the one's place of the weighted sum.
**(b)** What digits are possible for the check digit? A single-digit error will change the weighted sum. By how much would the weighted sum have to change so that the check digit remains the same?

**39.** When there are adjacent letters in the name that would be coded with the same number, use the number one time.

**41. (b)** Many answers are possible. Each name must begin with S. Use the number replacement table to determine which consonants are options with the given numbers. Remember that vowels will all be eliminated before coding.

## Section 1.2

**1. (c)** Check to see if 627 is divisible by 14.

**3.** The statement $m \mid a$ is true if the remainder is zero when $a \div m$.

**5.** See Example 1.9.

7. Every integer that is a multiple of 3 will have a remainder of 0 when divided by 3. Every integer that is 1 more than a multiple of 3 will have a remainder of 1 when divided by 3. Every integer that is 2 more than a multiple of 3 will have a remainder of 2 when divided by 3.

9. See Example 1.11.

11. See Example 1.11.

13. You are supposed to work through each problem two different ways. First, add the integers and then divide the sum by 6 to find the remainder. Then divide each integer by 6 and find the remainders first. Add the remainders and determine congruence mod 6. What do you notice about the answers?

15. You are supposed to work through each problem two different ways. First, multiply the integers and then divide the product by 7 to find the remainder. Then divide each integer by 7 and find the remainders first. Multiply the remainders and determine congruence mod 7. What do you notice about the answers?

17. (b) If $71 \times c$ has a remainder of 4 when divided by 8, then what remainder would $(71 \times c) - 4$ have when divided by 8? Simplify $71 \times c$ first by finding 71 mod 8.

19. See Example 1.13. In each case consider how the expression can be rewritten to take advantage of the modulus. For example, in part (a), determine the value of the exponent so that $3^? \equiv 1$ mod 8. Then rewrite the expression using that fact.

21. (b) Determine the number of days from May 14, 2003 to August 3, 2003.

23. Given any year, it may be a leap year if the year is congruent to 0 mod 4. However, if the year is also congruent to 0 mod 100, then it will not be a leap year in many cases. Be aware that there is an exception to the rule. The exception to the rule means that if the year is congruent to 0 mod 4, congruent to 0 mod 100, and

congruent to 0 mod 400, then it is a leap year.

25. The year 2092 is a leap year. List the possible leap years between 2092 and 2150. Which of these are leap years?

27. (a) What digits are possible remainders when dividing by 7?
   (c) The check digit is 3. Determine the value of the missing digit so that the number represented by the first four digits has a remainder of 3 when divided by 7.

29. The check digit is the remainder when the number represented by the first four digits is divided by the modulus. In this case the remainder is 6. What would the remainder be if $(2439 - 6)$ were divided by the modulus? Is there only one modulus possible?

31. See Example 1.16.

33. (b) Let $d$ stand in for the missing digit. Work through the process to find the mod 9 check digit and simplify as much as possible. Then determine what value for $d$ will make the congruence true.

35. Airline ticket numbers use a mod 7 check-digit scheme. See Example 1.18.

37. Remove the 1Z from the number before calculating the check digit.

39. Assign each letter in the VIN a digit according to Table 1.7. Calculate the check digit leaving the ? in place to represent the missing digit, or use a variable. Simplify as much as possible.

41. Assign each character a value according the given table. Notice that the dash is assigned a value also. The check digit is congruent to the sum of the values of the other characters mod 43. Once you know the value of the check digit, convert it to a character based on the given table.

## Section 1.3

1. First determine how many letters are in the word by identifying the spaces between characters. Remember a single black square represents a dot while three black squares together represent a dash.

3. Each dash will be represented by three black squares together. Each dot will be represented by a single black square. Between all dots and dashes, there must be a white square. Between each character, there must be three white squares together.

5. Write out what would be sent in terms of dots, dashes, and spaces. At what point during transmission, would the code fail to form a letter or a number?

7. Three zeros together separate characters.

9. **(a)** Refer to Table 1.12. Remember that the UPC bar code number representations are fixed-length codes, so there is no need to use any sort of separation between digits.

11. Use Table 1.12.

13. **(a)** The UPC is made up of the item number followed by the manufacturer number and the product number. The check digit is the last digit in the code.
    **(b)** The guard bars are made up of three bars and are located on both ends of the bar code. The center bar pattern is made up of five bars. Be sure to use the manufacturer number codes on the left-hand side of the bar code and the product number codes on the right-hand side of the bar code. See Figures 1.17 and 1.18.

15. Identify the guard bars and the center bars. Each digit is represented by seven bars. The manufacturer number codes must be used on the left-hand bars. The product codes must be used on the right-hand bars.

17. Use the Braille alphabet in Figure 1.20.

19. Use the Braille alphabet in Figure 1.20.

21. Refer to Table 1.14.

23. Refer to Table 1.14.

25. Refer to Table 1.23.

27. In every Postnet representation of a digit, there are exactly two tall bars. Consider each bar and decide if making the bar taller or smaller will result in a correct Postnet representation of a digit.

29. Remember the ZIP + 4 Postnet code has a check digit in the last position. There is also a guard bar at the beginning of the code and one at the end.

31. The check digit must be chosen to make the sum of the digits divisible by 10.

33. The code for the ZIP + 4 digits is visible. The delivery-point code can be found in the address. The code is also missing the check digit.

35. The location of the error will be revealed by the time you have tried to convert each five-bar Postnet code to a digit. The check digit is included in the code, so use it to correct the error.

37. Determine the number of bits sampled for one speaker and double it. It may be helpful to break the problem down into smaller calculations. First, determine how many bits are sampled in each second. Next, determine how many bits are sampled in 4 minutes. Finally, double the number of bits due to there being two speakers.

39. **(a)** From problem 37, we know 338,688,000 bits are sampled in a 4-minute song.
    **(b)** Consider the number of bits sampled in a 4-minute song. Remember that 2 bytes = 16 bits. If the number of bytes is reduced by a factor of 10, then the number of bits is also reduced by a factor of 10.

## Solutions to Odd-Numbered Problems in Section 1.1

**1.** **(a)** It could be a valid Tennessee social security number.
   **(b)** It could be a valid Oregon social security number.
   **(c)** It could not be a valid social security number issued today. The area number 885 is not currently in use.
   **(d)** It could not be a valid social security number issued today. The group number is made up of two zeros. The smallest group number issued is 01.
   **(e)** It could be a valid Massachusetts social security number.
   **(f)** It could not be a valid social security number issued today. The area number 700 was used for the Railroad Board and was discontinued.

**3.** **(a)** The number 3380 is a valid identification number. The sum of the digits is $3 + 3 + 8 + 0 = 14$. Fourteen is divisible by 7.
   **(b)** Select a digit so that the sum of the four digits is divisible by 7. The sum of the three given digits is $8 + 9 + 1 = 18$. If the fourth digit is a 3, then the sum of the digits will be $8 + 9 + 1 + 3 = 21$ which is divisible by 7. The fourth digit would have to be a 3.
   **(c)** If the sum of the four digits is not divisible by 7, then the number will be recognized as invalid.
   **(d)** A single-digit error is made when one of the digits is recorded incorrectly. For the given identification number, every single-digit error will result in a change in the sum of the digits. In most cases, the sum will no longer be a multiple of 7 and will be identified as invalid. Notice, however, for the identification number 2453, if the 2 is mistakenly recorded as a 9, the sum of the digits will still be a multiple of 7 and will not be recognized as invalid.
   **(e)** A single-digit error will not be detected if the correct digit and the incorrect digit differ by 7. For example, if the digit 0 becomes 7, 1 becomes 8, 2 becomes 9, or vice versa, then the sum of the digits will still be a multiple of 7.

**5.** **(a)** Transposing the third and fourth digits will not change the sum of the digits. This error will go undetected.
   **(b)** If the third digit is recorded as 0 rather than 9, then the sum of the digits will change, but the sum will still be divisible by 9. This error will go undetected.
   **(c)** Recording the first digit as an 8 rather than a 3 will change the sum of the five digits by 5. The sum will no longer be divisible by 9. This error will be detected.
   **(d)** The digits 0 through 9 can be used for the first four digits of the code. For the check digit, the digit 9 will never be used. If a code is given with 9 as the check digit, then it will be detected as an error.

**7.** **(a)** Not all single-digit errors will be detected. There are many single-digit errors that can occur in the first four digits that will not be detected. For example, in the first four digits, if an odd digit is replaced by a different odd digit, then the error will go undetected because the sum of the digits will still be even. Also, in the first four digits, if any even digit is replaced by a different even digit, then the error will go undetected because the sum of the digits will still be even. The check digit is chosen as the smallest digit that will make the sum of all of the digits even. Therefore, the check digit can be only a 0 or a 1. If there is a single-digit error in the check digit, it will always be detected.
   **(b)** Not all adjacent-transposition errors will be detected. Switching the order of any two adjacent digits in the first four digits will not change the sum of the digits.

**9.** **(a)** Calculate the weighted sum:
   Weighted sum $= 1(2) + 2(1) + 3(2) + 4(6) + 5(4) + 6(8) = 2 + 2 + 6 + 24 + 20 + 48 = 102$.
   Divide the weighted sum by 7: $\frac{102}{7} = 14\frac{4}{7}$. The remainder is 4. Therefore, the check digit is 4.

**(b)** Calculate the weighted sum:
Weighted sum = $1(9) + 2(7) + 3(7) + 4(4) + 5(2) + 6(5) = 9 + 14 + 21 + 16 + 10 + 30 = 100$.

Divide the weighted sum by 7: $\dfrac{100}{7} = 14\dfrac{2}{7}$. The remainder is 2. Therefore, the check digit is 2.

**(c)** Calculate the weighted sum:
Weighted sum = $1(1) + 2(0) + 3(5) + 4(0) + 5(6) + 6(3) = 1 + 0 + 15 + 0 + 30 + 18 = 64$.

Divide the weighted sum by 7: $\dfrac{64}{7} = 9\dfrac{1}{7}$. The remainder is 1. Therefore, the check digit is 1.

**11.** Many answers are possible.
**(a)** The check digit is the remainder after the weighted sum is divided by 7. We want the check digit to be 5 in this identification number, so we want the weighted sum to be 5 more than a multiple of 7. There are many ways to proceed. We could randomly select the digits in positions two through six. By randomly selecting those digits, we will be able to determine what the first digit, which is weighted by a 1, would have to be to force the weighted sum to be 5 more than a multiple of 7. Let $d$ = the first digit, and suppose digits in positions two through six are 10328. Calculate the weighted sum:
Weighted sum = $1(d) + 2(1) + 3(0) + 4(3) + 5(2) + 6(8) = d + 2 + 0 + 12 + 10 + 48 = 72 + d$.
We know $7 \times 10 = 70$, so if we let $d = 3$, then the weighted sum is $72 + 3 = 75$ which is 5 more than a multiple of 7. One seven-digit identification code with a check digit of 5 is 3103285. You should create an example of your own.
**(b)** Use the same method as in part (a) but be sure not to select any 0 digits in the identification number. This time we want the weighted sum to be 3 more than a multiple of 7. Let $d$ = the first digit, and suppose digits in positions two through six are 48529. Calculate the weighted sum:
Weighted sum = $1(d) + 2(4) + 3(8) + 4(5) + 5(2) + 6(9) = d + 8 + 24 + 20 + 10 + 54 = 116 + d$.
We know $7 \times 17 = 119$, so if we let $d = 6$, then the weighted sum is $116 + 6 = 122$ which is 3 more than a multiple of 7. One seven-digit identification code with no 0 digits and a check digit of 3 is 6485293. You should create an example of your own.
**(c)** Let $d$ represent the digit. The digits must all be the same. The identification number would look like $ddddddd$. The weighted sum = $1(d) + 2(d) + 3(d) + 4(d) + 5(d) + 6(d) = 21d = 7(3d)$. The weighted sum will always be a multiple of 7, so the check digit will be 0 for any choice of $d$. Let $d = 0$, then the number 0000000 is a valid identification number.

**13.** **(a)** A single-digit error was made in the first position. A 9 was recorded rather than a 2.
**(b)** This error will not be detected. The first digit is multiplied by 1 when added to the weighted sum of the digits. The incorrect digit and the correct digit differ by 7, so the weighted sum will still have the same remainder when divided by 7.

**15.** **(a)** The weighted sum = $10(0) + 9(2) + 8(4) + 7(3) + 6(6) + 5(1) + 4(4) + 3(2) + 2(7) = 148$. Divide the weighted sum by 11 and find the remainder: $\dfrac{148}{11} = 13\dfrac{5}{11}$. The remainder when the weighted sum is divided by 11 is 5. The check digit is 5.
**(b)** The weighted sum = $10(3) + 9(9) + 8(2) + 7(3) + 6(9) + 5(2) + 4(2) + 3(0) + 2(6) = 232$. Divide the weighted sum by 11 and find the remainder: $\dfrac{232}{11} = 21\dfrac{1}{11}$. The remainder when the weighted sum is divided by 11 is 1. The check digit is 1.
**(c)** The weighted sum = $10(1) + 9(5) + 8(6) + 7(5) + 6(5) + 5(4) + 4(1) + 3(0) + 2(8) = 208$. Divide the weighted sum by 11 and find the remainder: $\dfrac{208}{11} = 18\dfrac{10}{11}$. The remainder when the weighted sum is divided by 11 is 10. The check digit is X.

**(d)** The weighted sum = 10(0) + 9(1) + 8(3) + 7(6) + 6(3) + 5(9) + 4(4) + 3(4) + 2(4) = 174. Divide the weighted sum by 11 and find the remainder: $\dfrac{174}{11} = 15\dfrac{9}{11}$. The remainder when the weighted sum is divided by 11 is 9. The check digit is 9.

**17. (a)** We know the check digit is correct. The check digit in this case is X, so we know the weighted sum of the first nine correct ISBN digits is 10 more than a multiple of 11. Changing one digit could change the weighted sum significantly. Consider possible changes to the weighted sum.

The largest weight on any digit in the ISBN is a 10 on the first digit. If the digit in that position is changed from a 2 to a 9, then the weighted sum would increase by 70. The next largest weight on a digit is a 9. If the digit in that position is changed from a 0 to a 9, then the weighted sum would increase by 81. We must consider changes in a single digit that could increase the weighted sum by as much as 81. The largest digit currently in the ISBN is an 8 and its largest weight is an 8. If the digit in that position were changed from an 8 to a 0, then the weighted sum would decrease by a total of 64. We must consider changes in a single digit that could decrease the weighted sum by as much as 64.

The current weighted sum = 10(2) + 9(0) + 8(8) + 7(1) + 6(5) + 5(2) + 4(8) + 3(5) + 2(2) = 182. We know there is a mistake in one digit. Correcting the mistake could decrease the weighted sum by as much as 64 or increase the weighted sum by as much as 81. Therefore, the correct weighted sum is between 182 − 64 = 118 and 182 + 81 = 263. We do not need to consider each of those weighted sums. We need only to consider those which are 10 more than a multiple of 11 because we know the original check digit was correct. Consider multiples of 11 such as 11, 22, 33, 44, 55, 66, 77, 88, 99, 110, 121, 132, 143, 154, 165, 176, 187, 198, 209, 220, 231, 242, 253, 264, and 275. Add 10 to each of these and select only those numbers that are between 118 and 263. The list of possible correct weighted sums can be narrowed down to 120, 131, 142, 153, 164, 175, 186, 197, 208, 219, 230, 241, 252, and 263. For each of these possible correct weighted sums, we must decide if a single-digit correction can be made in the ISBN to make the weighted sum one of these values.

| Possible Correct Weighted Sum | Correct Weighted Sum Compared to 182 | Is it possible to change one digit to achieve the correct weighted sum? | Possible Corrected ISBN |
|---|---|---|---|
| 120 | 62 less | No | |
| 131 | 51 less | No | |
| 142 | 40 less | Change the digit weighted by 8 from an 8 to a 3. | 2-0**3**-152852-X |
| 153 | 29 less | No | |
| 164 | 18 less | Change the digit weighted by 6 from a 5 to a 2. | 2-08-1**2**2852-X |
| 175 | 7 less | Change the digit weighted by 7 from a 1 to a 0. | 2-08-**0**52852-X |
| 186 | 4 more | Change the digit weighted by 2 from a 2 to a 4. OR Change the digit weighted by 4 from an 8 to a 9. | 2-08-152854-X  2-08-152**9**52-X |
| 197 | 15 more | Change the digit weighted by 5 from a 2 to a 5. | 2-08-155**5**52-X... 2-08-15**5**852-X |
| 208 | 26 more | No | |
| 219 | 37 more | No | |
| 230 | 48 more | No | |
| 241 | 59 more | No | |
| 252 | 70 more | Change the digit weighted by 10 from a 2 to a 9. | **9**-08-152852-X |
| 263 | 81 more | Change the digit weighted by 9 from a 0 to a 9. | 2-**9**8-152852-X |

**(b)** We know the check digit is correct. The check digit in this case is 0, so we know the weighted sum of the first nine correct ISBN digits is a multiple of 11. Changing one digit could change the weighted sum significantly. Consider possible changes to the weighted sum.

The largest weight on any digit in the ISBN is a 10. If the digit in that position is changed from a 1 to a 9, then the weighted sum would increase by 80. We must consider changes in a single digit that could increase the weighted sum by as much as 80. The largest digit currently in the ISBN is an 8 and its weight is a 9. If the digit in that position were changed from an 8 to a 0, then the weighted sum would decrease by a total of 72. We must consider changes in a single digit that could decrease the weighted sum by as much as 72.

The current weighted sum $= 10(1) + 9(8) + 8(8) + 7(3) + 6(4) + 5(2) + 4(3) + 3(5) + 2(4) = 236$. We know there is a mistake in one digit. Correcting the mistake could decrease the weighted sum by as much as 72 or increase the weighted sum by as much as 80. Therefore, the correct weighted sum is between $236 - 72 = 164$ and $236 + 80 = 316$. We do not need to consider each of those weighted sums. We need only to consider those which are a multiple of 11 because we know the original check digit was correct. Consider multiples of 11 that are between 164 and 316: 165, 176, 187, 198, 209, 220, 231, 242, 253, 264, 275, 286, 297, and 308. For each of these possible correct weighted sums, we must decide if a single-digit correction can be made in the ISBN to make the weighted sum one of these values.

| Possible Correct Weighted Sum | Correct Weighted Sum Compared to 236 | Is it possible to change one digit to achieve the correct weighted sum? | Possible Corrected ISBN |
|---|---|---|---|
| 165 | 71 less | No | |
| 176 | 60 less | No | |
| 187 | 49 less | No | |
| 198 | 38 less | No | |
| 209 | 27 less | Change the digit weighted by 9 from an 8 to a 5. | 1-5̲8-342354-0 |
| 220 | 16 less | Change the digit weighted by 8 from an 8 to a 6. | 1-8̲6-342354-0 |
| 231 | 5 less | Change the digit weighted by 5 from a 2 to a 1. | 1-88-341̲354-0 |
| 242 | 6 more | Change the digit weighted by 2 from a 4 to a 7. OR Change the digit weighted by 3 from a 5 to a 7. OR Change the digit weighted by 6 from a 4 to a 5. | 1-88-342357̲-0  1-88-3423̲7̲4-0  1-88-35̲2354-0 |
| 253 | 17 more | No | |
| 264 | 28 more | Change the digit weighted by 7 from a 3 to a 7. | 1-88-7̲42354-0 |
| 275 | 39 more | No | |
| 286 | 50 more | Change the digit weighted by 10 from a 1 to a 6. | 6̲-88-342354-0 |
| 297 | 61 more | No | |
| 308 | 72 more | No | |

**19. (a)** The check digit is 7, and it is correct. The error must have occurred by switching two of the other adjacent digits. The weighted sum of the incorrect ISBN is $10(0) + 9(3) + 8(1) + 7(0) + 6(1) + 5(1) + 4(6) + 3(9) + 2(0) = 97$. The remainder when 97 is divided by 11 is 9. The remainder is supposed to be 7. Pick adjacent pairs of digits, switch their positions, and see if the weighted sum divided by 11 has a remainder of 7. If the second and third digits are switched, then the weighted sum becomes $10(0) + 9(1) + 8(3) + 7(0) + 6(1) + 5(1) + 4(6) + 3(9) + 2(0) = 95$. The remainder when 95 is divided by 11 is 7. The correct ISBN is 0-13-011690-7.

**(b)** Any error that causes a change in the weighted sum that is a multiple of 5 will not be detected. Therefore, any single-digit error that occurs in the position weighted by 5 or the position weighted by 10 will be undetected. Also, any change in a digit such that the correct digit and the incorrect digit differ by 5 will go undetected.

**21. (a)** Calculate the weighted sum:

$(2\times6) + 0 + (2\times1) + 1 + (2\times9) + 8 + (2\times2) + 6 + (2\times3) + 4 + (2\times5) + 1 + (2\times7) + 1 + (2\times1) + 7$

$(12) + 0 + (2) + 1 + (18) + 8 + (4) + 6 + (6) + 4 + (10) + 1 + (14) + 1 + (2) + 7$

Before finding the weighted sum, add the digits of any two digit product.

$(3) + 0 + (2) + 1 + (9) + 8 + (4) + 6 + (6) + 4 + (1) + 1 + (5) + 1 + (2) + 7$

Weighted sum $= 60$

The weighted sum is divisible by 10, so the credit card number is valid.

**(b)** Let $d$ represent the check digit. Calculate the weighted sum and select the check digit, $d$, to make the weighted sum divisible by 10.

$(2\times6) + 0 + (2\times1) + 1 + (2\times4) + 5 + (2\times3) + 3 + (2\times8) + 9 + (2\times5) + 6 + (2\times8) + 7 + (2\times5) + d$

$(12) + 0 + (2) + 1 + (8) + 5 + (6) + 3 + (16) + 9 + (10) + 6 + (16) + 7 + (10) + d$

Before finding the weighted sum, add the digits of any two digit product.

$(3) + 0 + (2) + 1 + (8) + 5 + (6) + 3 + (7) + 9 + (1) + 6 + (7) + 7 + (1) + d$

Weighted sum $= 66 + d$

Let $d = 4$ so that the weighted sum becomes $66 + 4 = 70$ which is a multiple of 10. The check digit is 4.

**23. (a)** Carry out the process to find the weighted sum and determine what the missing digit must be to have a weighted sum that is a multiple of 10.

$(2\times5) + 5 + (2\times8) + 2 + (2\times1) + 9 + (2\times?) + 4 + (2\times4) + 2 + (2\times3) + 2 + (2\times8) + 6 + (2\times7) + 3$

$(10) + 5 + (16) + 2 + (2) + 9 + (2\times?) + 4 + (8) + 2 + (6) + 2 + (16) + 6 + (14) + 3$

Before finding the weighted sum, add the digits of any two digit product.

$(1) + 5 + (7) + 2 + (2) + 9 + (2\times?) + 4 + (8) + 2 + (6) + 2 + (7) + 6 + (5) + 3$

Weighted sum $= 69 + (2\times?)$

We know the weighted sum must be a multiple of 10. If 1 is added to the weighted sum, then we have $69 + 1 = 70$ which is a multiple of 10. Let the missing number be a 5 because $(2\times5) = (10)$ and the two digits must be added before finding the weighted sum, so a 1 is being added to the weighted sum. The missing digit is a 5.

**(b)** The 9th digit is a 0 and will be multiplied by 2. The 10th digit is a 9. Originally the digits in the 9th and 10th positions will add $(2\times0) + 9 = 9$ to the weighted sum. If the two digits are transposed, then the digit 9 ends up in the 9th position and the digit 0 ends up in the 10th position. In this case, $(2\times9) + 0 = (18) + 0 = 9 + 0 = 9$ is added to the weighted sum. The transposition does not cause a change in the weighted sum, so the error will go undetected.

**25. (a) (i)** The manufacturer number is 51000 so the manufacturer is Campbell's.
   **(ii)** The product number is the group of five digits in positions 7 through 11. The product number is 00011.
   **(iii)** Let $d$ represent the check digit. Calculate the weighted sum and determine the value of $d$ so that the weighted sum is divisible by 10.
   
   Weighted sum $= 3(0 + 1 + 0 + 0 + 0 + 1) + 1(5 + 0 + 0 + 0 + 1 + d)$
   $= 3(2) + 1(6 + d)$
   $= 12 + d$
   
   If we let $d = 8$, then the weighted sum is $12 + 8 = 20$ which is a multiple of 10. The check digit is 8.

**(b) (i)** The manufacturer number is 21000 so the manufacturer is Kraft.
   **(ii)** The product number is the group of five digits in positions 7 through 11. The product number is 65883.

(iii) Let $d$ represent the check digit. Calculate the weighted sum and determine the value of $d$ so that the weighted sum is divisible by 10.

$$\begin{aligned} \text{Weighted sum} &= 3(0 + 1 + 0 + 6 + 8 + 3) + 1(2 + 0 + 0 + 5 + 8 + d) \\ &= 3(18) + 1(15 + d) \\ &= 54 + 15 + d \\ &= 69 + d \end{aligned}$$

If we let $d = 1$, then the weighted sum is $69 + 1 = 70$ which is a multiple of 10. The check digit is 1.

(c) (i) The manufacturer number is 16000 so the manufacturer is General Mills.

(ii) The product number is the group of five digits in positions 7 through 11. The product number is 81120.

(iii) Let $d$ represent the check digit. Calculate the weighted sum and determine the value of $d$ so that the weighted sum is divisible by 10.

$$\begin{aligned} \text{Weighted sum} &= 3(0 + 6 + 0 + 8 + 1 + 0) + 1(1 + 0 + 0 + 1 + 2 + d) \\ &= 3(15) + 1(4 + d) \\ &= 45 + 4 + d \\ &= 49 + d \end{aligned}$$

If we let $d = 1$, then the weighted sum is $49 + 1 = 50$ which is a multiple of 10. The check digit is 1.

27. (a) Calculate the weighted sum and then determine what the missing digit would have to be so that the weighted sum is divisible by 10.

$$\begin{aligned} \text{Weighted sum} &= 3(? + 7 + 0 + 2 + 3 + 0) + 1(4 + 0 + 0 + 1 + 0 + 2) \\ &= 3(? + 12) + 7 \\ &= 3(?) + 36 + 7 \\ &= 3(?) + 43 \end{aligned}$$

Find the missing digit so that $3(?) + 43$ is a multiple of 10. If the missing digit is a 9, then the weighted sum becomes $3(9) + 43 = 27 + 43 = 60$ which is a multiple of 10.

(b) Calculate the weighted sum and then determine what the missing digit would have to be so that the weighted sum is divisible by 10.

$$\begin{aligned} \text{Weighted sum} &= 3(2 + 7 + 3 + 1 + 2 + ?) + 1(9 + 0 + 0 + 8 + 3 + 1) \\ &= 3(15 + ?) + 21 \\ &= 45 + 3(?) + 21 \\ &= 66 + 3(?) \end{aligned}$$

Find the missing digit so that $66 + 3(?)$ is a multiple of 10. If the missing digit is an 8, then the weighted sum becomes $66 + 3(8) = 90$ which is a multiple of 10.

(c) Calculate the weighted sum and then determine what the missing digit would have to be so that the weighted sum is divisible by 10.

$$\begin{aligned} \text{Weighted sum} &= 3(0 + 1 + 0 + 3 + 8 + 1) + 1(2 + 0 + 0 + ? + 5 + 0) \\ &= 3(13) + 1(7 + ?) \\ &= 39 + 7 + ? \\ &= 46 + ? \end{aligned}$$

Find the missing digit so that $46 + ?$ is a multiple of 10. If the missing digit is a 4, then the weighted sum becomes $46 + 4 = 50$ which is a multiple of 10.

29. Consider the two-weight check digit scheme. A change in a digit that is multiplied by 3 will change the weighted sum by a multiple of 3, but the change will be less than 30, so the weighted sum will no longer be divisible by 10 and the error will be detected. A change in a digit that is not multiplied by 3 will change the weighted sum by at most 9 so the error will be detected. All single-digit errors will be detected.

**31. (a)** Not all single-digit errors will be detected. Any change in a single digit that results in an increase or a decrease to the weighted sum by a multiple of 9 will not change the remainder when the weighted sum is divided by 9, so the check digit would be the same. Consider the identification number 00000. If a single-digit error is made in any of the first four digits so that a 0 is mistakenly recorded as a 9, the check digit will still be 0, and the error will go undetected.

**(b)** Many answers are possible. If an identification code contains adjacent pairs of digits that differ by 3, 6, or 9, then transposing them will not change the remainder when the weighted sum is divided by 9. For example, for the identification number 28004, notice the check digit is 4. The first and second digits differ by 6. If they are transposed, the code 82004 is not identified as an error because the check digit is also a 4.

**33. (a)** Calculate the weighted sum and determine the last digit.
Weighted sum $= 7(0) + 3(6) + 9(3) + 7(1) + 3(1) + 9(4) + 7(8) + 3(2) = 153$. The last digit is the digit in the one's place. The check digit is a 3.

**(b)** Calculate the weighted sum and determine the last digit.
Weighted sum $= 7(1) + 3(2) + 9(2) + 7(0) + 3(0) + 9(0) + 7(4) + 3(8) = 83$. The last digit is the digit in the one's place. The check digit is a 3.

**35.** The digit in the 8th position was mistakenly recorded as a 1 rather than a 7. This is a single-digit error. Calculate the weighted sum of the incorrect routing number and determine the check digit.
Weighted sum $= 7(3) + 3(2) + 9(3) + 7(2) + 3(7) + 9(4) + 7(2) + 3(1) = 142$. The last digit is the digit in the one's place. The check digit should be a 2. In the routing number we were given, the check digit was 0, so the error would be detected.

**37. (a)** Calculate the weighted sum: $1(5) + 2(8) + 3(3) + 1(2) + 2(5) + 3(9) + 1(3) + 2(6) + 3(7) + 1(1) = 106$. The last digit in the sum is a 6. The check digit is a 6.

**(b)** The check digit is the last digit in the weighted sum. Any single-digit error in the positions weighted by 2 such that the correct digit and the incorrect digit differ by 5 will cause the weighted sum to differ by 10, so the check digit will be the same.

**(c)** Many adjacent-transposition errors will go undetected. In adjacent positions that are weighted by 1 and 3, if the digits differ by 5, then when they are transposed, the weighted sum will change by 10 and the error will go undetected.

**39. (a) - (d)**

| Surname | Hildebrand | Walczyk | Marr | *Pennington* |
|---|---|---|---|---|
| Leave the first letter alone. Then cross out all occurrences of the letters a, e, i, o, u, y, h, and w. | Hi̸ld̸e̸br̸an̸d | W̸al̸cz̸y̸k | M̸a̸rr | P̸e̸nn̸i̸ngt̸o̸n |
| Cross off the second of any double letters. | Hldbrnd | Wlczk | Mr̸ | Pn̸ngtn |
| Leave the first letter alone, and replace each of the other letters with the appropriate number. | H431653 | W4222 | M6 | *P55235* |

| Use duplicate numbers from adjacent letters in the name only once. | H431653 | W422̸2 | M6 | P55235 |
|---|---|---|---|---|
| Use only the first three numbers or fill in the code with zeros if there are not enough numbers. | H431̸6̸ ̸ ̸ | W422 | M600 | P552̸ ̸ |
| *Soundex coded surname* | *H431* | *W422* | *M600* | *P552* |

**41. (a)**

| Surname | **Smithson** | **Saendogh** | ***Smythes*** |
|---|---|---|---|
| Leave the first letter alone. Then cross out all occurrences of the letters a, e, i, o, u, y, h, and w. | S m̸ t h̸ s ø n | S ǽ n d ø g h̸ | S m̸ t h̸ ǿ s |
| Cross off the second of any double letters. | S m t s n | S n d g | *S m t s* |
| Leave the first letter alone, and replace each of the other letters with the appropriate number. | S5325 | S532 | *S532* |
| Use duplicate numbers from adjacent letters in the name only once. | S5325 | S532 | *S532* |
| Use only the first three numbers or fill in the code with zeros if there are not enough numbers. | S532̸5 | S532 | *S532* |
| *Soundex coded surname* | *S532* | *S532* | *S532* |

Each of the names are coded the same.

**(b)** Many answers are possible.

## Solutions to Odd-Numbered Problems in Section 1.2

1. **(a)** For the given long division problem, we see 14 divides into 627 a total of 44 times with a remainder of 11. The divisor is 14, the dividend is 627, the quotient is 44, and the remainder is 11.
   **(b)** Let $a = 627$, $m = 14$, $q = 44$, and $r = 11$. We have $627 = 14 \times 44 + 11$. Notice that $0 \leq 11 < 14$.
   **(c)** It is not true that $14 | 627$. There was a remainder of 11 when 627 was divided by 14. We would conclude $14 | 627$ only if the remainder after 627 was divided by 14 was 0.

3. In each case, divide and determine if there is a remainder.
   **(a)** True, 3 divides 12 because $12 = 3 \times 4 + 0$. The remainder is zero.
   **(b)** False, 5 does not divide 474 because $474 = 5 \times 94 + 4$. There is a nonzero remainder.

(c) True, 18 divides 1116 because $1116 = 18 \times 62 + 0$. The remainder is zero.

(d) False, 31 does not divide 1458 because $1458 = 31 \times 47 + 1$. There is a nonzero remainder.

5. (a) Given $a = 447$ and $m = 12$, find the values of $q$ and $r$. According to the division algorithm for integers, $a = mq + r$.

$$a = mq + r$$

$$\frac{a}{m} = q + \frac{r}{m}$$

$$\frac{447}{12} = q + \frac{r}{12}$$

$$37.25 = q + \frac{r}{12} \qquad \text{37 is the quotient}$$

$$37.25 = 37 + \frac{r}{12}$$

$$0.25 = \frac{r}{12}$$

$$0.25 \times 12 = r$$

$$3 = r$$

The remainder is 3. The quotient is 37.

(b) Given $a = -887$ and $m = 53$, find the values of $q$ and $r$. According to the division algorithm for integers, $a = mq + r$.

$$a = mq + r$$

$$\frac{a}{m} = q + \frac{r}{m}$$

$$\frac{-887}{53} = q + \frac{r}{53}$$

$$-16.735849 \approx q + \frac{r}{53}$$

The remainder must be nonnegative. Add 17 to both sides of the equation to make the left-hand side of the equation positive and less than 1.

$$-16.735849 + 17 \approx (q + 17) + \frac{r}{53}$$

$$0.264151 \approx (q + 17) + \frac{r}{53}$$

The quotient is $-17$.

$$0.264151 \approx \frac{r}{53}$$

$$0.264151 \times 53 \approx r$$

$$14.000003 \approx r$$

The remainder is 14.

(c) Given $a = 5938$ and $m = 91$, find the values of $q$ and $r$. According to the division algorithm for integers, $a = mq + r$.

$$a = mq + r$$

$$\frac{a}{m} = q + \frac{r}{m}$$

$$\frac{5938}{91} = q + \frac{r}{91}$$

$$65.252747 \approx q + \frac{r}{91} \qquad 65 \text{ is the quotient}$$

$$65.252747 \approx 65 + \frac{r}{91}$$

$$0.252747 \approx \frac{r}{91}$$

$$0.252747 \times 91 \approx r$$

$$22.999977 \approx r$$

The remainder is 23. The quotient is 65.

7.

| Integer | Remainder |
|---|---|
| ..., −9, −6, −3, 0, 3, 6, 9,... | 0 |
| ..., −8, −5, −2, 1, 4, 7, 10,... | 1 |
| ..., −7, −4, −1, 2, 5, 8, 11,... | 2 |

9.  (a) We know $39 - 0 = 39$ and $3 \times 13 = 39$, so $3|(39-0)$ and we can say that 39 is congruent to 0 modulo 3.

(b) We know $29 - 4 = 25$ and $5 \times 5 = 25$, so $5|(29-4)$ and we can say that 29 is congruent to 4 modulo 5.

(c) We know $72 - 18 = 54$ and $6 \times 9 = 54$, so $6|(72-18)$ and we can say that 72 is congruent to 18 modulo 6.

(d) We know $81 - 21 = 60$ and $4 \times 15 = 60$, so $4|(81-21)$ and we can say that 81 is congruent to 21 modulo 4.

11. (a) $-55 \equiv 0 \bmod 4$ is false. We know 4 does not divide $-55 - 0 = -55$.

(b) $64 \equiv -5 \bmod 7$ is false. We know 7 does not divide $64 - (-5) = 64 + 5 = 69$.

(c) $87 \equiv 6 \bmod 13$ is false. We know 13 does not divide $87 - 6 = 81$.

(d) $-24 \equiv -4 \bmod 10$ is true. We know 10 divides $-24 - (-4) = -24 + 4 = -20$ because $10 \times (-2) = -20$.

13. (a) **Method 1:** Find convenient numbers congruent to 15 and 41 first.
$15 \equiv 3 \bmod 6$
$41 \equiv 5 \bmod 6$
$15 + 41 \equiv 3 + 5 = 8 \equiv 2 \bmod 6$
**Method 2:** Add first.
$15 + 41 = 56 \equiv 2 \bmod 6$

(b) **Method 1:** Find convenient numbers congruent to 25 and 58 first.
$25 \equiv 1 \bmod 6$
$58 \equiv 4 \bmod 6$
$25 + 58 \equiv 1 + 4 = 5 \equiv 5 \bmod 6$
**Method 2:** Add first.
$25 + 58 = 83 \equiv 5 \bmod 6$

(c) **Method 1:** Find convenient numbers congruent to 76 and 14 first.
$76 \equiv 4 \bmod 6$
$14 \equiv 2 \bmod 6$
$76 + 14 \equiv 4 + 2 = 6 \equiv 0 \bmod 6$
**Method 2:** Add first.
$76 + 14 = 90 \equiv 0 \bmod 6$

15. (a) **Method 1:** Find convenient numbers congruent to 17 and 11 first.
$17 \equiv 3 \bmod 7$
$11 \equiv 4 \bmod 7$
$17 \times 11 \equiv 3 \times 4 = 12 \equiv 5 \bmod 7$
**Method 2:** Multiply first.
$17 \times 11 = 187 \equiv 5 \bmod 7$

(b) **Method 1:** Find convenient numbers congruent to 55 and 35 first.
$55 \equiv 6 \bmod 7$
$35 \equiv 0 \bmod 7$
$55 \times 35 \equiv 6 \times 0 = 0 \equiv 0 \bmod 7$
**Method 2:** Multiply first.
$55 \times 35 = 1925 \equiv 0 \bmod 7$

(c) **Method 1:** Find convenient numbers congruent to 44 and 29 first.
$44 \equiv 2 \bmod 7$
$29 \equiv 1 \bmod 7$
$44 \times 29 \equiv 2 \times 1 = 2 \equiv 2 \bmod 7$
**Method 2:** Multiply first.
$44 \times 29 = 1276 \equiv 2 \bmod 7$

17. (a) We know $52 \equiv 2 \bmod 5$ and $28 \equiv 3 \bmod 5$, so $52 \times 28 \equiv 2 \times 3 = 6 \equiv 1 \bmod 5$. Thus, $b = 1$.

(b) We know $71 \equiv 7 \bmod 8$, so $71 \times c \equiv 7 \times c$. We want to find the smallest whole-number value for $c$ such that $7 \times c \equiv 4 \bmod 8$. By the definition of congruence modulo $m$, we know $7 \times c \equiv 4 \bmod 8$ if and only if 8 divides $(7 \times c) - 4$. Therefore, find the smallest whole-number value for $c$ so that $(7 \times c) - 4$ is a multiple of 8. Try different whole numbers for $c$ until a multiple of 8 is found.
Let $c = 0$, then $(7 \times 0) - 4 = 0 - 4 = -4$ is not a multiple of 8.
Let $c = 1$, then $(7 \times 1) - 4 = 7 - 4 = 3$ is not a multiple of 8.
Let $c = 2$, then $(7 \times 2) - 4 = 14 - 4 = 10$ is not a multiple of 8.
Let $c = 3$, then $(7 \times 3) - 4 = 21 - 4 = 17$ is not a multiple of 8.
Let $c = 4$, then $(7 \times 4) - 4 = 28 - 4 = 24$ is a multiple of 8.
The smallest whole-number value for $c$ such that $71 \times c \equiv 4 \bmod 8$ is 4.

19. (a)
$$3^2 = 9$$
$$\equiv 1 \bmod 8$$
$$3^{200} = (3^2)^{100}$$
$$\equiv 1^{100}$$
$$\equiv 1 \bmod 8$$

**(b)**

$$2^6 = 64 \equiv 1 \bmod 9$$

$$2^5 = 32 \equiv 5 \bmod 9$$

$$2^{311} = (2^6)^{51}(2^5)$$

$$\equiv (1)^{51}(5)$$

$$\equiv 5 \bmod 9$$

**(c)**

$$5^6 = 15,625 \equiv 1 \bmod 9$$

$$5^4 = 625 \equiv 4 \bmod 9$$

$$5^{142} = (5^6)^{23}(5^4)$$

$$\equiv (1)^{23}(4)$$

$$\equiv 4 \bmod 9$$

**21. (a)** Because today is a Friday and $48 \equiv 6 \bmod 7$, in 48 days it will be 6 days past a Friday. Six days after Friday is Thursday.

**(b)** Determine how many days there are from May 14 to August 3. There are $31 - 14 = 17$ days left in May, 30 days in June, 31 days in July, and 3 days in August for a total of $17 + 30 + 31 + 3 = 81$ days. We know $81 \equiv 4 \bmod 7$. May 14, 2003 was a Wednesday. Add four days to Wednesday. August 3, 2003 falls on a Sunday.

**23. (a)** We know $1980 \equiv 0 \bmod 4$ and $1980 \not\equiv 0 \bmod 100$, so 1980 is a leap year.

**(b)** We know $2023 \not\equiv 0 \bmod 4$, so 2023 is not a leap year.

**(c)** We know $2008 \equiv 0 \bmod 4$ and $2008 \not\equiv 0 \bmod 100$, so 2008 is a leap year.

**(d)** We know $2000 \equiv 0 \bmod 4$ and $2000 \equiv 0 \bmod 100$, so it appears that the year 2000 is not a leap year. However, this year falls under the exception. The year $2000 \equiv 0 \bmod 400$, so it is a leap year.

**(e)** We know $1800 \equiv 0 \bmod 4$ and $1800 \equiv 0 \bmod 100$, so 1800 is not a leap year.

**25.** The person is born in 2092. List the possible leap years through the year 2150: 2096, 2100, 2104, 2108, 2112, 2116, 2120, 2124, 2128, 2132, 2136, 2140, 2144, and 2148. Each of the years given is congruent to 0 mod 4. A year is not a leap year if it is also congruent to 0 mod 100. The year $2100 \equiv 0 \bmod 100$, so 2100 is not a leap year. There are 13 leap years between 2092 and 2150. The person will have celebrated 13 birthdays by the year 2150.

**27. (a)** The possible digits for the check digit are 0, 1, 2, 3, 4, 5, and 6 because these digits are the possible remainders after a division by 7.

**(b)** The check digit for 3964 is the whole number from the set 0, 1, 2, 3, 4, 5, and 6 that is congruent to 3964 mod 7. Divide 3964 by 7. Because $3964 = 7 \times 566 + 2$, we know the remainder is 2 so the check digit is 2.

**(c)** Notice the check digit is 3. Determine the missing digit so that $3 \equiv 41?7 \bmod 7$. The missing digit must make the number 41?7 three more than a multiple of 7. If the missing digit is 4, then $4147 \equiv 3 \bmod 7$.

29. Notice the check digit is 6, so a check digit scheme of mod 7 or greater must have been used. The modulus $m$ is unknown. For modulus $m$, we know $2439 \equiv 6 \bmod m$. The remainder is 6 when 2439 is divided by $m$. In other words, using the division algorithm, we can express the relationship as $2439 = m \times q + 6$. Subtract 6 from both sides of the equation, $2433 = m \times q$. The modulus will be one of the factors of 2433. Factors of 2433 are 1, 3, 811, and 2433. The modulus has to be at least 7, so the modulus could be 811 or 2433.

31. Because a mod 9 check-digit scheme is being used, we can use the process of casting out nines. We have $3872219457 \equiv 3 + 8 + 7 + 2 + 2 + 1 + 9 + 4 + 5 + 7 \bmod 9 = 48 \equiv 3 \bmod 9$. The check digit is 3. Alternately, if the method of crossing out all sums that are multiples of nine, we have the following: $3872219457 \equiv 3 + \cancel{8} + \cancel{7} + \cancel{2} + \cancel{2} + \cancel{1} + \cancel{9} + \cancel{4} + \cancel{5} + \cancel{7} \equiv 3 \bmod 9$.

33. (a) The prefix letter on a Euro banknote is determined by the country of origin. Each prefix letter has a digit equivalent given in Table 1.6. In this case, the country digit is missing. Let $d$ = the country digit. A mod 9 check-digit scheme is used. For Euro banknotes, the entire number, including the check digit must be divisible by 9. Find $d$ so that the sum of all of the digits is congruent to 0 mod 9, that is, find $d$ so that $d + 8 + 3 + 6 + 5 + 4 + 2 + 9 + 7 + 5 + 5 + 4 \equiv 0 \bmod 9$. Simplify the left-hand side of the equation: $d + 8 + 3 + 6 + 5 + 4 + 2 + 9 + 7 + 5 + 5 + 4 = d + 58$. We know that $58 \equiv 4 \bmod 9$, so $d + 58 \equiv d + 4$. If $d = 5$ then $d + 4 = 5 + 4 = 9 \equiv 0 \bmod 9$. There are two countries whose prefix letters are determined by the digit 5: Portugal and Spain.

(b) The check digit is missing, but the prefix letter is P, so the Euro banknote is from the Netherlands which has a digit equivalent of 8. Let $d$ = the check digit. The serial number with the prefix replaced by the digit 8, and check digit included is $82974653389d$. The entire serial number must be divisible by 9. Find $d$ so that the sum of all the digits is congruent to 0 mod 9, that is, find $d$ so that $8 + 2 + 9 + 7 + 4 + 6 + 5 + 3 + 3 + 8 + 9 + d \equiv 0 \bmod 9$. Simplify the left-hand side of the equation: $8 + 2 + 9 + 7 + 4 + 6 + 5 + 3 + 3 + 8 + 9 + d = 64 + d$. We know that $64 \equiv 1 \bmod 9$, so $64 + d \equiv 1 + d$. If $d = 8$, then $1 + d = 1 + 8 = 9 \equiv 0 \bmod 9$. The check digit is 8.

35. (a) The airline ticket uses a mod 7 check-digit scheme. The last digit in the ticket number is a 0. Therefore, the check digit for the given ticket is 0, so if the airline ticket is valid, then the ticket number $0163428775932 \equiv 0 \bmod 7$. When 163428775932 is divided by 7, there is a nonzero remainder, so the correct check digit is not 0, and this ticket number is not valid.

(b) The check digit is the remainder after the ticket number is divided by 7, so the check digit is congruent to $161466117765 \bmod 7$. Use a calculator or long division to divide the ticket number by 7. We see $161466117765 \div 7 \approx 23066588252.1428571$. The quotient is 23066588252 and $7 \times 23066588252 = 161466117764$ which differs from the dividend by 1, so the remainder is 1. Alternately, the remainder is approximately $7 \times 0.1428571 = 0.9999997$. Thus, the remainder is 1.

37. (a) For a UPS tracking number, the check digit is found in the 18th position, and in this case, the check digit is a 4. To see if the tracking number is valid, we remove the check digit and the characters 1Z, and determine whether $591580686047248 \equiv 4 \bmod 7$. Most calculators will not be able to display enough digits, so you may need to use long division. We see $591580686047248 \div 7 \approx 84511526578177.28571428$, so the remainder is approximately $7 \times 0.28571428 = 1.99999996$. The remainder is 2, so the tracking number is not valid.

(b) Remove the characters 1Z and find the remainder when 433698652447694 is divided by 7. Use long division if your calculator does not display enough digits. We see $433698652447694 \div 7 \approx 61956950349670.57142857$. The remainder is approximately $7 \times 0.57142857 = 3.99999999$. The remainder is 4.

39. (a) The VIN is alphanumeric, and a digit is missing in the fifth position. The check digit is located in the ninth position, and in this case, the check digit is 5. The letters in the given VIN must be converted to digits. From Table 1.7, we see U = 4, Y = 8, X = 7, and L = 3. Substitute the digits in for the letters and apply the weights to each of the 16 digits other than the check digit.

Original VIN:    1  9  U  Y  ?  3  2  5  5  X  L  0  0  1  4  3  8

Converted VIN:  1  9  4  8  ?  3  2  5  5  7  3  0  0  1  4  3  8

Weights:        8  7  6  5  4  3  2  10   9  8  7  6  5  4  3  2

Multiply each digit by the appropriate weight and find the sum: $8(1) + 7(9) + 6(4) + 5(8) + 4(?) + 3(3) + 2(2) + 10(5) + 9(7) + 8(3) + 7(0) + 6(0) + 5(1) + 4(4) + 3(3) + 2(8) = 331 + 4(?)$.

The missing digit must be found so that $331 + 4(?) \equiv 5 \bmod 11$. Equivalently we could say the missing digit must be found so that $331 + 4(?) - 5 \equiv 0 \bmod 11$. We know $331 - 5 = 326$ and $326 \equiv 7 \bmod 11$, so $331 + 4(?) - 5 = 326 + 4(?) \equiv 7 + 4(?)$. If the missing digit is a 1, then $7 + 4(?) = 7 + 4(1) = 11 \equiv 0 \bmod 11$. The missing digit is a 1.

(b) The VIN is alphanumeric, and a digit is missing in the fourth position. The check digit is located in the ninth position, and in this case, the check digit is 8. The letters in the given VIN must be converted to digits. From Table 1.7, we see G = 7, R = 9, and H = 8. Substitute the digits in for the letters and apply the weights to each of the 16 digits other than the check digit.

Original VIN:    1  G  3  ?  R  6  4  H  8  2  4  2  3  6  9  5  0

Converted VIN:  1  7  3  ?  9  6  4  8  8  2  4  2  3  6  9  5  0

Weights:        8  7  6  5  4  3  2  10   9  8  7  6  5  4  3  2

Multiply each digit by the appropriate weight and find the sum: $8(1) + 7(7) + 6(3) + 5(?) + 4(9) + 3(6) + 2(4) + 10(8) + 9(2) + 8(4) + 7(2) + 6(3) + 5(6) + 4(9) + 3(5) + 2(0) = 380 + 5(?)$.

The missing digit must be found so that $380 + 5(?) \equiv 8 \bmod 11$. Equivalently we could say the missing digit must be found so that $380 + 5(?) - 8 \equiv 0 \bmod 11$. We know $380 - 8 = 372$ and $372 \equiv 9 \bmod 11$, so $380 + 5(?) - 8 = 372 + 5(?) \equiv 9 + 5(?)$. If the missing digit is a 7, then $9 + 5(?) = 9 + 5(7) = 44 \equiv 0 \bmod 11$. The missing digit is a 7.

**41. (a)** When a check digit is used with a Code 39 number, it is given as the last character. For the identification number provided, the check digit is Z which corresponds to the value 35. Convert the other characters in the identification number to their corresponding values and find the sum. Determine whether the sum of the values is congruent to 35 mod 43. The sum of the values of the characters is $2 + 10 + 36 + 4 + 8 + 2 + 29 + 32 = 123$. We know $123 \equiv 37 \bmod 43$, so we know the check digit should have a value of 37 and the given identification number is not valid.

**(b)** The sum of the values of the characters is $28 + 25 + 2 + 4 + 5 + 3 + 9 + 9 + 22 + 33 + 6 + 5 = 151$. We know $151 \equiv 22 \bmod 43$, so the check digit has a value of 22. From the table, we know the character that has a value of 22 is M. The check digit character is M.

## Solutions to Odd-Numbered Problems in Section 1.3

**1.** When three black squares are together, it represents a dash. A single black square represents a dot. Between any dots or dashes, there must be a white square to represent that the circuit is off for one unit of time. Entire characters are separated by three white squares to represent that the circuit is off for three time units between characters.

**(a)** Notice there are two occurrences of three white squares in a row. This means the word is made up of three letters. The first letter is represented by a dash followed by two dots. Refer to Table 1.9 to see that the letter is a D. The second letter represented by three dashes, so it is an O. The third letter is represented by a single dash, so it is a T. The word is DOT.

**(b)** Notice there are three occurrences of three white squares in a row. This means the word is made up of four letters. The first letter is represented by a dash followed by two dots, so it is a D. The second letter is represented by a dot and a dash, so it is an A. The third letter is represented by three dots, so it is an S. The fourth letter is represented by four dots, so it is an H. The word is DASH.

**3.** Be sure to separate characters using three white squares in a row. Separate dots and dashes with one white square.

5.  Without the needed spacing between characters, the code for the T and the H would appear to be the code for a 6 so the receiver would not notice any problems yet. When the code for E is sent, with no spacing to signify a character has ended after the H was sent, the receiver would know an error has been made.

7.  In this case, the three time unit separation between characters will be represented by three 0s. Each dash is represented by three 1s while each dot is represented by a single 1. All dots and dashes are separated by a single 0.
    (a) Separate the characters first. Then determine which letters are represented by referring to Table 1.11. In this case, the one-word message is WHY.

    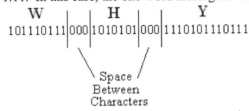

    (b) The Morse code representation for the word CHECK is 11101011101000101010100010001110101110100011101011.

9.  (a) Use the manufacturer number representations in Table 1.12. The number 194 is represented as 001100100010110100011.
    (b) The following figure is the bar code representation of the number 194 using the sequence of 0s and 1s from part (a). A gray border has been placed around the bar code so the left-hand white bar is visible.

11. (a) The sequence of 0s and 1s represents the product number 470.
    (b) The sequence of 0s and 1s represents the product number 528.

13. (a) Let $d$ represent the check digit. The first 11 digits of the UPC identification number are 02100065833. The check digit is the 12th digit selected so that the total of 3 times the sum of the digits in the odd-numbered positions and 1 times the sum of the digits in the even-numbered positions is divisible by 10. The check-digit calculation follows:
    $3(0 + 1 + 0 + 6 + 8 + 3) + 1(2 + 0 + 0 + 5 + 3 + d) = 3(18) + 1(10 + d) = 54 + 10 + d = 64 + d$.
    If we let $d = 6$, then $64 + 6 = 70$ which is divisible by 10.
    (b) Remember to include the guard bars and the center bars in the bar code. A grid at the bottom has been left in place so that the spacing of the bars is clear.

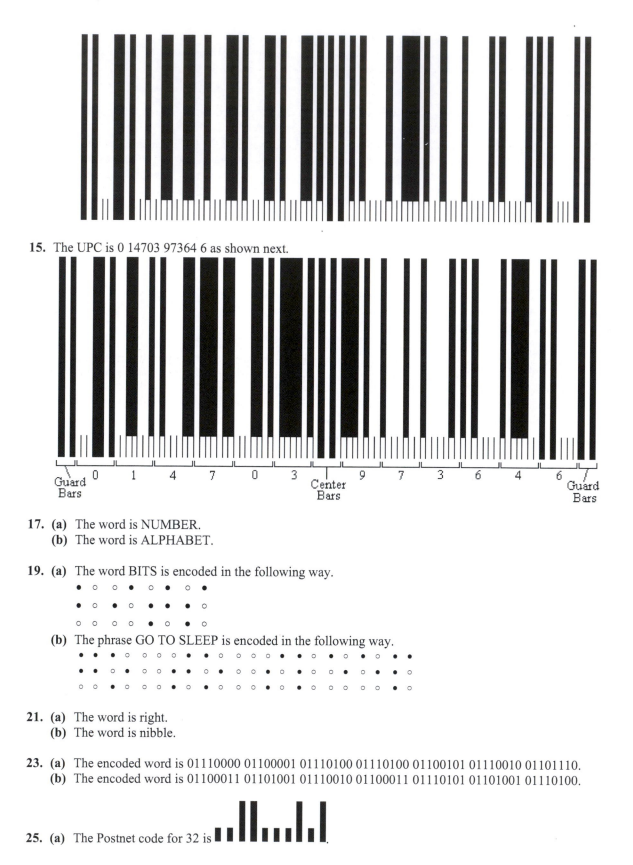

**15.** The UPC is 0 14703 97364 6 as shown next.

**17. (a)** The word is NUMBER.
**(b)** The word is ALPHABET.

**19. (a)** The word BITS is encoded in the following way.

● ○ ○ ● ○ ● ○ ●
● ○ ● ○ ● ● ● ○
○ ○ ○ ○ ● ○ ● ○

**(b)** The phrase GO TO SLEEP is encoded in the following way.

● ● ● ○ ○ ○ ○ ● ● ○ ○ ○ ○ ● ● ● ● ○ ● ○ ● ○ ● ●
● ● ○ ● ○ ○ ● ● ○ ● ○ ○ ● ○ ● ○ ● ○ ● ○ ● ○ ● ○
○ ○ ● ● ○ ○ ● ○ ● ○ ○ ○ ● ○ ● ○ ○ ○ ○ ○ ○ ● ○

**21. (a)** The word is right.
**(b)** The word is nibble.

**23. (a)** The encoded word is 01110000 01100001 01110100 01110100 01100101 01110010 01101110.
**(b)** The encoded word is 01100011 01101001 01110010 01100011 01110101 01101001 01110100.

**25. (a)** The Postnet code for 32 is

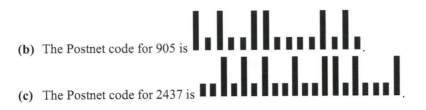

**(b)** The Postnet code for 905 is                                      .

**(c)** The Postnet code for 2437 is                                      .

**27. (a)** A single mistake would be a mistake in a single bar such that the bar is tall when it was supposed to be short or the bar is short when it was supposed to be tall. The given Postnet code does not match any of the digits. If the first tall bar was made short, then the code would represent the digit 6. If the second bar was made short, then the code would represent the digit 9. If the third bar was made short, then the code would represent the digit 0. No correct representations of digits would result from making either of the short bars tall. No Postnet code for a digit has more than two tall bars. Therefore, there are three ways to fix a mistake in the given Postnet code. The corrected Postnet code could be a 6, 9, or 0.

**(b)** The number represented is 368.

**29.** Ignore the single guard bars at both ends of the Postnet code. There are ten digits represented by the code. The last digit is the check digit and is chosen so that the sum of all of the digits is divisible by 10. The ZIP + 4 code is 97322-3232. The check digit is 7.

Guard Bar                                                        Guard Bar

9   7   3   2   2   3   2   3   2   7

Check Digit

**31.** The check digit is selected so that the sum of the digits is divisible by 10. The sum of the digits is $1 + 9 + 8 + 0 + 4 + 0 + 0 + 0 + 1 = 23$. If the check digit is 7, then $23 + 7 = 30$, which is divisible by 10. The following Postnet code includes the check digit and the guard bars.

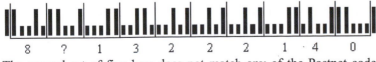

**33.** The delivery-point code consists of the last two digits of the street address. In this case, the delivery-point digits are 22. The check digit is chosen to make the sum of all the digits divisible by 10. The sum of the digits is $2 + 0 + 0 + 8 + 1 + 0 + 0 + 0 + 1 + 2 + 2 = 16$. The check digit must be a 4.

**35.** Remember that there is a guard bar at each end of the Postnet code. The remaining Postnet code is made up of ten digits: the ZIP + 4 and the check digit. Separate the Postnet code into sets of five bars and convert each set of five bars to a digit as shown next:

8   ?   1   3   2   2   2   1   4   0

The second set of five bars does not match any of the Postnet code representations of a digit. The mistake must be in the second set of five bars. The check digit is 0. The check digit is selected so that the sum of the digits is divisible by 10. The sum of the digits is $8 + ? + 1 + 3 + 2 + 2 + 2 + 1 + 4 + 0 = 23$. The missing digit must be a 7. The corrected Postnet code is given next.

```
8   7   1   3   2   2   2   1   4   0
```

**37.** Music samples are taken for both speakers, so double the number of bits sampled for one speaker. First, determine how many bits per second are sampled for a speaker.

$$\frac{44,100 \text{ samples}}{1 \text{ second}} \times \frac{2 \text{ bytes}}{1 \text{ sample}} \times \frac{16 \text{ bits}}{2 \text{ bytes}} = 705,600 \text{ bits per second}$$

We know music is sampled for 4 minutes, so determine how many bits are sampled for 4 minutes for a speaker.

$$\frac{705,600 \text{ bits}}{1 \text{ second}} \times \frac{4 \text{ minutes}}{1} \times \frac{60 \text{ seconds}}{1 \text{ minute}} = 169,344,000 \text{ bits}$$

Finally, we know music is sampled for each speaker, so there is a total of $2 \times 169,344,000 = 338,688,000$ bits sampled.

**39. (a)** We know from problem 37 that a total of 338,688,000 bits are sampled for a 4-minute song. Under ideal conditions a 56K modem will download at a rate of 56Kilobits per second, and a Kilobit is 1024 bits.

$$\frac{338,688,000 \text{ bits}}{1} \times \frac{1 \text{ Kilobit}}{1024 \text{ bits}} \times \frac{1 \text{ second}}{56 \text{ Kilobits}} = 5906.25 \text{ seconds}$$

It would take 5906.25 seconds which is 98 minutes and approximately 26 seconds.

**(b)** For a 4-minute song, there are 338,688,000 bits sampled. The MP3 format reduces the number of bytes by a factor of 10, so the number of bits is reduced by a factor of 10. Determine how many seconds it would take a 56K modem, under ideal conditions, to download that many bits.

$$\frac{338,688,000 \text{ bits}}{10} \times \frac{1 \text{ Kilobit}}{1024 \text{ bits}} \times \frac{1 \text{ second}}{56 \text{ Kilobits}} = 590.625 \text{ seconds}$$

It would take 590.625 seconds which is 9 minutes and approximately 51 seconds.

**(c)** A DSL can download at a maximum rate of 512 Kilobits per second.

$$\frac{338,688,000 \text{ bits}}{1} \times \frac{1 \text{ Kilobit}}{1024 \text{ bits}} \times \frac{1 \text{ second}}{512 \text{ Kilobits}} \approx 646 \text{ seconds}$$

It would take approximately 646 seconds which is 10 minutes and 46 seconds.

## Solutions to Chapter 1 Review Problems

**1. (a)** Social security numbers are issued in a certain pattern. We can tell the order the three social security numbers were issued by looking at the group numbers. The group number is the two-digit number in the fourth and fifth positions of the social security number. See Table 1.2. Of the three group numbers given, 09, 04, and 07, the social security number containing 07 was issued first. The social security number containing 09 was issued next. The most recent social security number in the group is the one with 04 as the group number.

**(b)** See Table 1.1. The area number 434 was issued from the state of Louisiana.

**2. (a)** The identification number consists of a string of digits and letters. It is an alphanumeric identification number.

**(b)** An adjacent-transposition error was made.

**(c)** The digit and letter in positions five and six were transposed.

3. **(a)** The check digit is completely determined once the first five digits are selected. There are ten digits that can be used in each of the first five positions of the number. There are $10 \times 10 \times 10 \times 10 \times 10 = 100{,}000$ valid identification numbers possible.

   **(b)** If 439713 is a valid identification number, then the sum of the six digits must be divisible by 9. The sum of the digits is $4 + 3 + 9 + 7 + 1 + 3 = 27$, and 27 is divisible by 9. The number is valid.

   **(c)** Add the first five digits: $3 + 5 + 0 + 2 + 8 = 18$. The check digit must be selected so that when it is added, the sum of all the digits is divisible by 9. The sum of the first five digits is 18 which is already divisible by 9. The check digit must be 0.

   **(d)** Determine the sum of the four digits that can be read: $1 + 3 + 4 + 4 = 12$. Two digits are missing. The missing digits must be added so that the sum of all the digits is divisible by 9. The sum without the missing digits is 12. If the missing digits add to 6, then the sum of all the digits will be 18 which is divisible by 9. Also, if the missing digits add to 15, then the sum of all the digits will be 27 which is divisible by 9. Consider all possible options and orders of the two digits so that they add to 6 or 15: 06, 60, 15, 51, 24, 42, 33, 69, 96, 78, and 87.

4. **(a)** The first four digits in the identification number must be chosen from the digits 0, 1, 2, 3, 4, 5, and 6. There are 7 choices for each of the first four digits. Once the first four digits have been selected, there is only one choice for the fifth digit. There are $7 \times 7 \times 7 \times 7 = 2401$ different valid identification numbers possible.

   **(b)** The single-digit error must be in one of the first four digits. The check digit is correct. Currently the sum of all the digits is $3 + 6 + 2 + 5 + 3 = 19$. Any single-digit correction must be made so that the sum of all of the digits is divisible by 7. A change in one of the current digits must decrease the original sum of the digits by 5 because $19 - 5 = 14$ which is divisible by 7; must increase the original sum of the digits by 2 because $19 + 2 = 21$ which is divisible by 7; or must increase the original sum of the digits by 9 because $19 + 9 = 28$ which is divisible by 7. For each of the first four digits, decide if the digit can be decreased by 5, increased by 2, and/or increased by 9. The following six corrections are possible. The underlined digit is the one that has been corrected: 5̲6253, 31̲253, 364̲53, or 362̲0̲3.

5. **(a)** The first digit of the UPC is 0, so it is a general grocery product. The manufacturer number is 38000, so the manufacturer is Kellogg. The product number is 21451.

   **(b)** The check digit is selected to make the weighted sum divisible by 10. Multiply the sum of the digits in odd-numbered positions by 3 and add it to the sum of the digits in the even-numbered positions. Let $d =$ the check digit. The UPC is 52100043891$d$. The weighted sum is calculated as $3(5 + 1 + 0 + 4 + 8 + 1) + 1(2 + 0 + 0 + 3 + 9 + d) = 3(19) + 14 + d = 57 + 14 + d = 71 + d$. Let $d = 9$ so that $71 + 9 = 80$ which is divisible by 10.

6. **(a)** Using long division, the quotient is 1491, and the remainder is 1.

   ```
        1491
   17)25348
      17
      ──
       83
       68
      ───
       154
       153
      ───
        18
        17
       ───
         1
   ```

   Using a calculator, $25348 \div 17 \approx 1491.05882353$. The quotient is 1491. We know $1491.05882353 \approx 1491 + \dfrac{r}{17}$ and $0.05882353 \approx \dfrac{r}{17}$. Therefore the remainder is approximately $17 \times 0.05882353 = 1.00000001$. Round the remainder down to 1.

(b) The divisor is 17. The dividend is 25,348. The quotient is 1491. The remainder is 1. Using the division algorithm, we have 25,348 = 17(1491) + 1.

(c) It is not true that 17| 25,348 because we know there is a nonzero remainder when 25,348 is divided by 17.

7. (a) The possible remainders that can result from a division by 11 are 0, 1, 2, 3, 4, 5, 6, 7, 8, 9, and 10.

(b)

| Integer | Remainder |
|---|---|
| ..., −22, −11, 0, 11, 22, ... | 0 |
| ..., −21, −10, 1, 12, 23, ... | 1 |
| ..., −20, −9, 2, 13, 24, ... | 2 |
| ..., −19, −8, 3, 14, 25, ... | 3 |
| ..., −18, −7, 4, 15, 26, ... | 4 |
| ..., −17, −6, 5, 16, 27, ... | 5 |
| ..., −16, −5, 6, 17, 28, ... | 6 |
| ..., −15, −4, 7, 18, 29, ... | 7 |
| ..., −14, −3, 8, 19, 30, ... | 8 |
| ..., −13, −2, 9, 20, 31, ... | 9 |
| ..., −12, −1, 10, 21, 32, ... | 10 |

8. (a) $32 \equiv 2 \bmod 3$ because $32 - 2 = 30$ and $30 = 3 \times 10$, so $3 \mid (32 - 2)$.

(b) $49 \equiv 4 \bmod 5$ because $49 - 4 = 45$ and $45 = 5 \times 9$, so $5 \mid (49 - 4)$.

(c) $114 \equiv 0 \bmod 6$ because $114 - 0 = 114$ and $114 = 6 \times 19$, so $6 \mid (114 - 0)$.

(d) $65 \equiv 1 \bmod 4$ because $65 - 1 = 64$ and $64 = 4 \times 16$, so $4 \mid (65 - 1)$.

9. (a) True, $-83 - 23 = -106 = -53 \times 2$. The difference is a multiple of 2, so $-83 \equiv 23 \bmod 2$.

(b) False, $44 - 15 = 29$, and 29 is not a multiple of 6.

(c) True, $32 - (-13) = 45$, and $45 = 3 \times 15$. The difference is a multiple of 15, so $32 \equiv -13 \bmod 15$.

(d) True, $-10 - (-6) = -4$, and $-4 = 4 \times (-1)$. The difference is a multiple of 4, so $-10 \equiv -6 \bmod 4$.

10. (a) $b = 6$; $6 \equiv 54 \bmod 16$ because $6 - 54 = -48$ and $-48 = 16 \times (-3)$.

(b) $b = 5$; $5 \equiv -93 \bmod 7$ because $5 - (-93) = 98$ and $98 = 7 \times 14$.

(c) $b = 1$; $1 \equiv 4522 \bmod 3$ because $1 - 4522 = -4521$ and $-4521 = 3 \times (-1507)$.

11. **Method 1:** Find convenient numbers congruent to 13 and 22 first.

$13 \equiv 1 \bmod 4$

$22 \equiv 2 \bmod 4$

$13 \times 22 \equiv 1 \times 2 = 2 \equiv 2 \bmod 4$

**Method 2:** Multiply first.

$13 \times 22 = 286 \equiv 2 \bmod 4$

12. Determine what power of 5 is congruent to 1 mod 6. We know $5^2 = 25 \equiv 1 \bmod 6$. Use the fact that the remainder is 1 when $5^2$ is divided by 6 and rewrite the expression using $5^2$.

$5^{200} = (5^2)^{100} \equiv 1^{100} = 1$.

13. (a) Find the weighted sum: $2 + (2 \times 4) + 5 + (2 \times 8) + 0 + (2 \times 1) + 9 + (2 \times 4) + 0 = 2 + 8 + 5 + 16 + 0 + 2 + 9 + 8 + 0 = 50$. The weighted sum is 50 and $50 \equiv 0 \bmod 10$, so the SIN is valid.

(b) Let $d =$ the check digit. Calculate the weighted sum: $9 + (2 \times 0) + 1 + (2 \times 8) + 6 + (2 \times 4) + 3 + (2 \times 5) + d = 9 + 0 + 1 + 16 + 6 + 8 + 3 + 10 + d = 53 + d$. The check digit must be 7 because $53 + 7 = 60$ and $60 \equiv 0 \bmod 10$.

(c) Calculate the weighted sum: $1 + (2 \times 2) + ? + (2 \times 6) + 3 + (2 \times 7) + 0 + (2 \times 9) + 4 = 1 + 4 + ? + 12 + 3 + 14 + 0 + 18 + 4 = 56 + ?$. The missing digit must be 4 because $56 + 4 = 60$ and $60 \equiv 0 \mod 10$.

14. (a) The sum of the digits is $3 + 0 + 1 + 8 + 3 = 15$ and $15 \equiv 6 \mod 9$, so the check digit cannot be 2. The number is not valid.

   (b) The sum of the digits is $7 + 6 + 6 + 2 + 8 = 29$. We know $29 \equiv 2 \mod 9$, so the check digit is 2.

   (c) The check digit is 6. Add the digits: $2 + 7 + 0 + ? + 3 = 12 + ?$. If the missing digit is 3, then $12 + 3 = 15$ and $15 \equiv 6 \mod 9$. Therefore, the missing digit is 3.

15. (a) Find the weighted sum. Add the digits of any two-digit product in the weighted sum.
   $(2\times5) + 2 + (2\times8) + 1 + (2\times0) + 0 + (2\times2) + 3 + (2\times9) + 8 + (2\times9) + 4 + (2\times7) + 2 + (2\times2) =$
   $(10) + 2 + (16) + 1 + (0) + 0 + (4) + 3 + (18) + 8 + (18) + 4 + (5) + 2 + (4) =$
   $(1) + 2 + (7) + 1 + (0) + 0 + (4) + 3 + (9) + 8 + (9) + 4 + (5) + 2 + (4) = 59$. The check digit must be 1 so that $59 + 1 = 60$ and $60 \equiv 0 \mod 10$. The given check digit is 1, so the MasterCard number is valid.

   (b) Find the weighted sum, and let $d$ = the check digit. Add the digits of any two digit product in the weighted sum. $(2\times5) + 5 + (2\times9) + 0 + (2\times2) + 7 + (2\times2) + 3 + (2\times8) + 1 + (2\times0) + 0 + (2\times1) + 3 + (2\times3) + d = (1) + 5 + (9) + 0 + (4) + 7 + (4) + 3 + (7) + 1 + (0) + 0 + (2) + 3 + (6) + d = 52 + d$. The check digit must be 8 so that $52 + 8 = 60$ and $60 \equiv 0 \mod 10$.

   (c) Find the missing digit. Add the digits of any two-digit product in the weighted sum. $(2\times5) + 4 + (2\times?) + 1 + (2\times0) + 0 + (2\times3) + 2 + (2\times1) + 8 + (2\times1) + 6 + (2\times2) + 2 + (2\times3) + 5 = (1) + 4 + (2\times?) + 1 + (0) + 0 + (6) + 2 + (2) + 8 + (2) + 6 + (4) + 2 + (6) + 5 = 49 + (2\times?)$. The missing digit must be 5 because when it is multiplied by 2 and the digits are added, the result is 1.

16. For each part, the check digit is given as a ?.

   (a) The check digit in a VIN is located in position 9. Convert the letters to digits according to Table 1.7. Find the weighted sum of the digits, not including the check digit.
   Original VIN: 2 J 5 G P 3 2 M ? K L 1 5 6 7 3 2
   Converted VIN: 2 1 5 7 7 3 2 4 ? 2 3 1 5 6 7 3 2
   Weights: 8 7 6 5 4 3 2 10 9 8 7 6 5 4 3 2
   Multiply each digit by the appropriate weight and find the sum: $8(2) + 7(1) + 6(5) + 5(7) + 4(7) + 3(3) + 2(2) + 10(4) + 9(2) + 8(3) + 7(1) + 6(5) + 5(6) + 4(7) + 3(3) + 2(2) = 319$.
   The check digit is congruent to 319 mod 11. Because 319 is a multiple of 11, the check digit is 0.

   (b) The check digit is missing, but the prefix letter is R, so the Euro banknote is from Luxembourg which has a digit equivalent of 1. Let $d$ = the check digit. The serial number with the prefix replaced by the digit 1, and check digit included is $11324753448d$. The entire serial number must be divisible by 9. Find $d$ so that the sum of all the digits is congruent to 0 mod 9, that is, find $d$ so that $1 + 1 + 3 + 2 + 4 + 7 + 5 + 3 + 4 + 4 + 8 + d \equiv 0 \mod 9$. Simplify the left-hand side of the equation: $1 + 1 + 3 + 2 + 4 + 7 + 5 + 3 + 4 + 4 + 8 + d = 42 + d$. We know that $45 \equiv 0 \mod 9$, so let the check digit be 3.

17. (a) The translated word is EASY.

   (b) Remember to separate characters by 000. The Morse code representation of the word TEST is 11100010001010101000111.

18. (a) The word is EXAMS.

   (b) The word VACATION is written in Braille as

**19.** The bar code for the UPC 0 23119 45030 6 is given next. A grid at the bottom has been retained so that the spacing of the bars is clear.

**20.** The number represented is 9621.

**21.** The check digit is selected so that the sum of the digits is divisible by 10. The sum of the digits is 9 + 7 + 3 + 5 + 5 + 1 + 8 + 4 + 4 = 46. The check digit is 4.

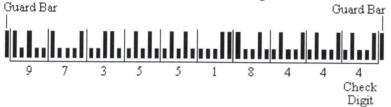

**22.** (a) It is ASCII.
   (b) The word PARITY is 0101 0000 0100 0001 0101 0010 0100 1001 0101 0100 0101 1001.

# *Chapter 2:* Shapes in Our Lives

## *Key Concepts*

**By the time you finish studying Section 2.1, you should be able to:**

☐ identify polygons by name and explain what a polygonal region is.

☐ list two requirements for polygonal regions to form a tiling.

☐ use the fact that the sum of the measures of the vertex angles in a triangle is 180° to determine the sum of the measures of the vertex angles in any polygon.

☐ describe the difference between regular polygons and irregular polygons.

☐ calculate the measure of a vertex angle in any regular polygon.

☐ describe what makes a tiling regular or edge to edge.

☐ explain why there are only three regular, edge-to-edge tilings possible.

☐ explain the difference between a regular tiling and a semiregular tiling.

☐ use vertex figures to determine whether a tiling is semiregular.

☐ create a tiling using polygons.

☐ state the Pythagorean theorem and use it to find the length of a side of a right triangle when the length of the other side and the length of the hypotenuse are given, or use it to find the length of the hypotenuse of a right triangle when the two side lengths are given.

☐ verify a triangle is a right triangle using the converse of the Pythagorean theorem.

**By the time you finish studying Section 2.2, you should be able to:**

☐ identify lines of symmetry in a pattern.

☐ specify the center of rotation and the degrees through which a pattern must be rotated if it has rotation symmetry.

☐ determine whether a pattern has translation symmetry.

☐ combine reflections, rotations, and translations to produce a strip pattern.

☐ use the definition of a reflection to describe the image of a point or a collection of points under a reflection with respect to a given line.

☐ use the definition of a translation to describe the image of a point or a collection of points under a translation determined by a given vector.

☐ use the definition of a rotation to describe the image of a point or a collection of points under a rotation given the center of rotation and a directed angle.

☐ describe a glide reflection.

☐ use the crystallographic classification system to classify the symmetries in a strip pattern.

**By the time you finish studying Section 2.3, you should be able to:**

☐ explain how to generate terms of the Fibonacci sequence.

☐ generate terms of a sequence given a recursion rule and the starting value(s).

☐ list several ways in which the Fibonacci numbers occur in nature.

☐ build a figure using geometric recursion.

☐ define the golden ratio and a golden rectangle.

☐ determine lengths required to make a rectangle a golden rectangle.

## Skills Brush-Ups

Some problems in Chapter 2 will require that you be comfortable using a protractor and plotting points in a coordinate system. You should be familiar with both, but we will review them here so they are fresh in your mind.

### PROTRACTOR:

A protractor is a tool used to measure angles. See the following figure. The protractor is evenly divided into 180 degrees, which is written 180°. Many protractors have no label for 180° or 0°, but a line on the protractor will make it clear where those measurements are located. Place the vertex of the angle you want to measure in the center of the protractor along the line through 0°. There is usually a hole or a cross to indicate where the vertex should be positioned. Position the initial side of the angle to coincide with the line through 0°. Read the degree measure of the angle at the point where the terminal side of the angle intersects the protractor.

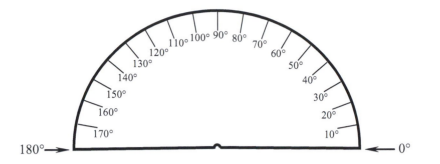

*Example:*    Use a protractor to find the degree measure of the following angle.

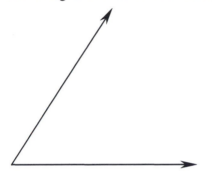

Position the protractor on the angle as shown in the following figure. The measure of the angle is 60°.

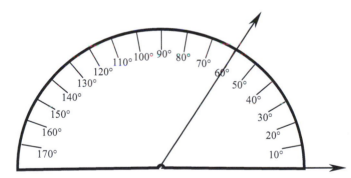

**CARTESIAN COORDINATE SYSTEM:**
The Cartesian coordinate system is formed by perpendicular real-number lines. The point at which the lines intersect is called the origin. The horizontal line is called the $x$-axis; the vertical line, $y$-axis. Each point in the coordinate system can be identified by a pair of points $(x, y)$ called the coordinates of the point. The sign of the $x$-coordinate indicates the position of the point to the right or to the left of the origin. If the $x$-coordinate is positive, then the point is to the right of the origin; if negative, left. The sign of the $y$-coordinate indicates the position of the point above or below the origin. If the $y$-coordinate is positive, then the point is above the origin; if negative, below.

*Example:* Several points are plotted in the following coordinate system. The coordinates of each point are as follows: $A(4, 5)$, $B(-9, 1)$, $C(0, -7)$, and $D(3, -2)$.

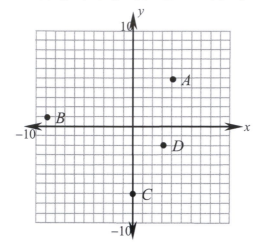

## Skills Practice

1. Use your protractor to find the measure of each angle. Extend the sides of the angle if necessary.

(a)                    (b)                    (c)

2. Give the coordinates of each of the points in the coordinate system.

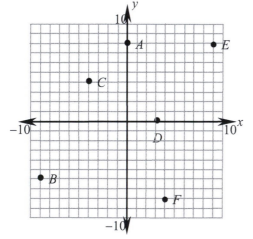

*Answers to Skills Practice:*
1. (a) 28°      (b) 131°      (c) 145°
2. $A(0, 8)$, $B(-9, -6)$, $C(-4, 4)$, $D(3, 0)$, $E(9, 8)$, $F(4, -8)$

## Hints for Odd-Numbered Problems

## Section 2.1

1. (a) See Table 2.1.
   (c) Use the definition of the sum of the measures of the vertex angles in a polygon with $n$ sides.

3. The sum of the measures of the vertex angles in a polygon with $n$ sides is $(n-2)180°$.

5. (a) The measure of a vertex angle in a regular $n$ gon is $\dfrac{(n-2)180°}{n}$.
   (b) One way to solve this problem is to guess possible values for $n$ and check to see if the sum of all but one of the vertex angle measures is $3078°$.

7. Find three toothpicks or pieces of wire that have the same length and arrange them to form a triangle. What do you notice about the measures of the angles? Is there any way to arrange the toothpicks or wire so that you get a different result?

9. You know each polygon is regular, so you can find the measures of a single vertex angle in each. If the measures of the angles surrounding a vertex are added, what must the sum be?

11. Consider the measures of the vertex angles in each of the three regular polygons and notice that $x$ and $y$ are each between the sides of two of the regular polygons.

13. (b) Find the measure of an exterior angle given that the measure of a vertex angle is $144°$.

15. A regular tiling is a tiling composed of regular polygonal regions in which all the polygons are the same size and shape.

17. Any regular tiling must satisfy the condition that the vertex angle divides $360°$.

19. See Example 2.6.

21. Refer to the definition of a semiregular tiling.

23. Select several vertices, draw, and compare the vertex figures. What do you notice?

25. (a) See Figure 2.22.
    (b) See Figure 2.26.

27. See Figure 2.29.

29. (a) Consider each set of conditions. Each condition has a certain angle measure sum or angle congruence requirement as well as side length requirements. Be aware that the hexagon given may be a rotation of the one labeled in the problem.

31. Use the fact that in a right triangle if $a$ and $b$ are the lengths of the sides and $c$ is the length of the hypotenuse, $a^2 + b^2 = c^2$.

33. See Example 2.7.

35. You are given the length of the hypotenuse in each case and a side length, so you are given the value for $c$ and the value for $a$ or $b$.

37. The converse of the Pythagorean theorem states that if $a^2 + b^2 = c^2$, then the triangle with side lengths $a$ and $b$ and hypotenuse length $c$ must be a right triangle.

39. Use the Pythagorean theorem.

41. Draw a diagram and use the Pythagorean theorem.

43. Draw a diagram and use the Pythagorean theorem.

## Section 2.2

1. (b) A complete $360°$ turn does not count as a rotation symmetry.

3. (b) Is there a relationship between the number of sides and the number of lines of reflection symmetry?

5. Reflection symmetry can be vertical, horizontal, or diagonal.

7. (a) The center of rotation is the black dot. See Figure 2.43(c) for an example of a 180° rotation to complete a figure.

9. Refer to Table 2.5.

11. See Figure 2.64 and the discussion related to that figure.

13. (a) and (b) See Figure 2.39 for examples of vertical and horizontal reflection symmetry.
    (c) The center of rotation is the center of the letter.

15. (b) Remember that a 360° rotation does not count as a rotation symmetry.

17. (a) See Figure 2.45.
    (b) See Figure 2.47.
    (c) Specify a vector so that when the entire pattern is translated according to the length of the vector and the direction of the vector the translated pattern falls exactly onto the original pattern.

19. (c) Notice that the rectangles get wider the farther they get from the center.

21. (c) Use a protractor as described in Example 2.9 to locate the coordinates of the image of each point under a reflection with respect to line $l$.

23. See Example 2.9.

25. See Example 2.10.

27. See Example 2.11.

29. (a) The sign of the measure of the angle indicates the direction of rotation. A negative angle measure indicates a clockwise rotation.

31. (b) Compare the coordinates of $\triangle ABC$ to the coordinates of its image after a reflection with respect to the $y$-axis. How are the $x$-coordinates related? How are the $y$-coordinates related?

33. See Figure 2.60.

35. See Figure 2.43 and the discussion in the previous paragraph.

37. See Example 2.10.

39. It may be easier to identify the line of reflection if you identify a translation first.

41. See Figure 2.65 and the discussion.

43. You may find it useful to create your design using a computer program so that exact duplicates are easily made.

## Section 2.3

1. (e) To find the 11th number in the sequence, let $n = 11$ in the rule.

3. (a) Let $n = 5$ in the recursive rule.
   (c) Use the rule to find $a_2$ first, then use $a_2$ to find the value of $a_1$.

5. First find $a_3$. The value of the third term is found by multiplying the previous two terms. The value of each term is generated similarly.

7. Use the rule for the Fibonacci sequence: $f_n = f_{n-1} + f_{n-2}$.

9. (a) Can you tell which Fibonacci numbers will be divisible by 3 just by looking at the term number?

11. Calculate the values of the fractions inside the parentheses first and write down all of the displayed digits or store the values in your calculator's memory. Use all the digits the calculator displays.

13. There are spirals in both directions making individual spirals difficult to see. You might try holding the photo at an arm's length from your eye, or photocopy the photo and enlarge it if you have trouble seeing the spirals.

15. Create a notation so that you can tell which are adults and which are calves. You might use subscripts to denote the age of the calf.

**17. (b)** Compare the numbers in the last two columns. Write an equation based on your observations.

**19. (b)** Leave at least 4 digits. Consider Table 2.11.

**21.** Pay attention to the order of the subtraction when a T comes up.

**23.** Consider each term beginning with the third. What operation must have been carried out on the previous two terms to generate the one under consideration?

**25. (b)** Keep at least four places after the decimal.

**27.** List the sums of the numbers on the diagonals in order from the top diagonal to the bottom.

**29. (a)** Remember $\phi \approx 1.618$. For each triangle divide the length of the long side by the length of the short side.

**(b) – (d)** Substitute the given length into the equation and solve for the unknown length. Use the fact that $\phi = \dfrac{1+\sqrt{5}}{2}$.

**31.** Neither may be a perfect golden rectangle, but determine which is the best approximate golden rectangle.

**33.** The ratio of the side lengths in a golden rectangle is $\phi$. We do not know if the width is the longer side or the shorter side, so there is more than one answer.

**35.** Try measuring using millimeters.

**37.** Measure as carefully as possible for the best results.

**39.** Measure as carefully as possible for the best results.

**41.** Use the numbered and lettered figure provided as a guide and label the given cycle in a similar way. How many sets of numbered waves are there? How many sets of lettered waves are there? How many different Fibonacci numbers can you recognize?

## Solutions to Odd-Numbered Problems in Section 2.1

**1. (a)** There are 8 sides, 8 vertices, and it is called an octagon.
   **(b)** Six triangles are formed.

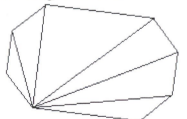

   **(c)** In this case $n = 8$. The sum of the measures of the vertex angles is $(n - 2)180° = (8 - 2)180° = (6)180° = 1080°$.

**3. (a)** A dodecagon has $n = 12$ sides. The sum of the measures of the vertex angles is $(n - 2)180° = (12 - 2)180° = (10)180° = 1800°$.
   **(b)** A decagon has $n = 10$ sides. The sum of the measures of the vertex angles is $(n - 2)180° = (10 - 2)180° = (8)180° = 1440°$.
   **(c)** A 16-gon has $n = 16$ sides. The sum of the measures of the vertex angles is $(n - 2)180° = (16 - 2)180° = (14)180° = 2520°$.
   **(d)** A 24-gon has $n = 24$ sides. The sum of the measures of the vertex angles is $(n - 2)180° = (24 - 2)180° = (22)180° = 3960°$.

**(b)** Yes, it can. The following edge-to-edge semiregular tiling is made up of regular octagons and squares.

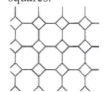

21. Each edge-to-edge tiling consists of two regular polygons. In each tiling, no matter where a vertex figure is drawn, each vertex figure is the same size and shape.
    **(a)** All vertex figures have the following shape or a rotation of the following shape.

    **(b)** All vertex figures have the following shape or a rotation of the following shape.

    **(c)** All vertex figures have the following shape or a rotation of the following shape.

    **(d)** All vertex figures have the following shape or a rotation of the following shape.

23. The tiling is not semiregular because vertex figures are different depending on which vertex is selected.

25. **(a)**

    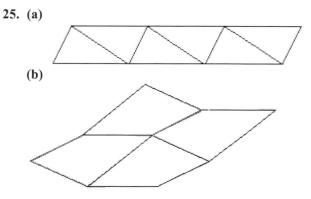

    **(b)**

27.

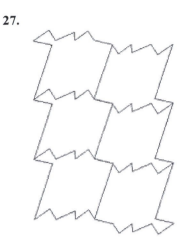

29. (a) If we label the vertices and side lengths as is done in the following figure, notice that three consecutive angle measures add to 360°, 150° + 80° + 130° = 360°, and there are two side lengths that are 3 units in length. Condition (I) is met.

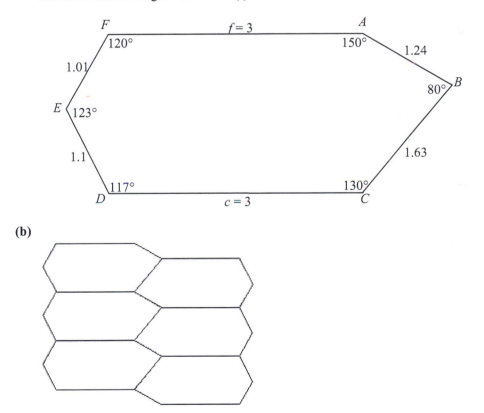

(b)

31. In each case, let $a$ and $b$ represent side lengths and let $c$ represent the length of the hypotenuse. Verify that $a^2 + b^2 = c^2$.

(a) $5^2 + 12^2 = 25 + 144 = 169 = 13^2$

(b) $1^2 + \left(\sqrt{3}\right)^2 = 1 + 3 = 4 = 2^2$

(c) $12^2 + 16^2 = 144 + 256 = 400 = 20^2$

**33.** In each case, let $a$ and $b$ represent the given side lengths. Use the Pythagorean theorem to find $c$.

(a) Let $a = 15$ and $b = 20$.

$$c^2 = 15^2 + 20^2$$
$$= 225 + 400$$
$$c^2 = 625$$
$$c = \sqrt{625}$$

Therefore, $c = \sqrt{625} = 25$.

(b) Let $a = 2$ and $b = 7$.

$$c^2 = 2^2 + 7^2$$
$$= 4 + 49$$
$$c^2 = 53$$
$$c = \sqrt{53}$$

Therefore, $c = \sqrt{53}$.

(c) Let $a = 5$ and $b = 5$.

$$c^2 = 5^2 + 5^2$$
$$= 25 + 25$$
$$c^2 = 50$$
$$c = \sqrt{50}$$

Therefore, $c = \sqrt{50} = \sqrt{(25)(2)} = 5\sqrt{2}$.

**35.** In each case, let $a$ represent the given side length. Let $c$ represent the given hypotenuse. Use the Pythagorean theorem to find $b$.

(a) Let $a = 28$ and $c = 35$.

$$a^2 + b^2 = c^2$$
$$28^2 + b^2 = 35^2$$
$$784 + b^2 = 1225$$
$$b^2 = 441$$
$$b = \sqrt{441}$$

Therefore, $b = \sqrt{441} = 21$.

(b) Let $a = 2$ and $c = 4$.

$$a^2 + b^2 = c^2$$
$$2^2 + b^2 = 4^2$$
$$4 + b^2 = 16$$
$$b^2 = 12$$
$$b = \sqrt{12}$$

Therefore, $b = \sqrt{12} = \sqrt{(4)(3)} = 2\sqrt{3}$.

(c) Let $a = 7$ and $c = \sqrt{113}$.

$$a^2 + b^2 = c^2$$
$$7^2 + b^2 = \sqrt{113}^2$$
$$49 + b^2 = 113$$
$$b^2 = 64$$
$$b = \sqrt{64}$$

Therefore, $b = \sqrt{64} = 8$.

**37.** In each case, the longest length must be the hypotenuse. Let the two shorter lengths be represented by $a$ and $b$. Let the longest length be represented by $c$. Determine whether $a^2 + b^2 = c^2$.

**(a)** Let $a = 10$, $b = 24$, and $c = 26$.
$$a^2 + b^2 = c^2$$
$$10^2 + 24^2 = 26^2$$
$$100 + 576 = 676$$
$$676 = 676$$
This equation is true, so the triangle is a right triangle.

**(b)** Let $a = \sqrt{2}$, $b = \sqrt{3}$, and $c = \sqrt{5}$.
$$a^2 + b^2 = c^2$$
$$\left(\sqrt{2}\right)^2 + \left(\sqrt{3}\right)^2 = \left(\sqrt{5}\right)^2$$
$$2 + 3 = 5$$
$$5 = 5$$
This equation is true, so the triangle is a right triangle.

**(c)** Let $a = 6$, $b = 8$, and $c = 12$.
$$a^2 + b^2 = c^2$$
$$6^2 + 8^2 = 12^2$$
$$36 + 64 = 144$$
$$100 = 144$$
This equation is false, so the triangle is not a right triangle.

**39.** Notice $x$ is the length of the hypotenuse of a right triangle for which the two side lengths are known. Use the Pythagorean theorem to solve for $x$.

**Find the value of $x$:**

Let $a = 7.4$, $b = 8.1$, and $c = x$.
$$a^2 + b^2 = c^2$$
$$7.4^2 + 8.1^2 = x^2$$
$$54.76 + 65.61 = x^2$$
$$120.37 = x^2$$
$$\sqrt{120.37} = x$$

Now that the value for $x$ is known, we can use the Pythagorean theorem to find the value of $y$ which is the length of a side of a right triangle.

**Find the value of $y$**

Let $a = 4$, $b = y$, and $c = \sqrt{120.37}$.

$$a^2 + b^2 = c^2$$

$$4^2 + y^2 = \left(\sqrt{120.37}\right)^2$$

$$16 + y^2 = 120.37$$

$$y^2 = 104.37$$

$$y = \sqrt{104.37}$$

Therefore, $x = \sqrt{120.37} \approx 10.97$ cm and $y = \sqrt{104.37} \approx 10.22$ cm.

**41.** Draw a diagram. A right triangle is formed with a right angle at first base. The side lengths are known to be 90 feet each. The length of the hypotenuse is unknown. Let $c$ represent the length of the hypotenuse.

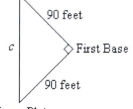

Second Base

90 feet

$c$ ◇ First Base

90 feet

Home Plate

$$a^2 + b^2 = c^2$$

$$90^2 + 90^2 = c^2$$

$$8100 + 8100 = c^2$$

$$16,200 = c^2$$

$$\sqrt{16,200} = c$$

$$127.28 \approx c$$

The catcher must throw the ball approximately 127.28 feet.

**43.** Draw a diagram. The ladder is leaning against a wall that is assumed to be perpendicular to the ground so a right triangle is formed. The length of the hypotenuse is 16 feet and the length of one side is 4 feet. Let $b$ represent the missing side length.

16 feet     $b$

4 feet

$$a^2 + b^2 = c^2$$
$$4^2 + b^2 = 16^2$$
$$16 + b^2 = 256$$
$$b^2 = 240$$
$$b = \sqrt{240}$$
$$b \approx 15.5$$

Therefore, the highest point on the house that the ladder will reach is approximately 15.5 feet.

## Solutions to Odd-Numbered Problems in Section 2.2

1. **(a)** The rectangle has two axes of reflection symmetry.

   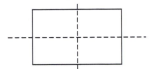

   **(b)** The rectangle has one rotation symmetry. It is a half-turn or a 180° rotation. The center of rotation is the center of the rectangle.

3. **(a)** The regular pentagon (i) has 5 lines of reflection symmetry. The regular hexagon (ii) has 6. The regular octagon (iii) has 8.

   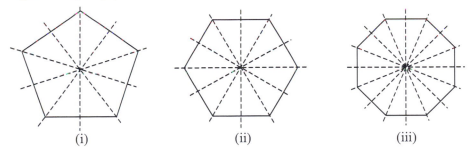

   (i)          (ii)          (iii)

   **(b)** A regular *n*-gon has *n* lines of reflection symmetry.

5. **(a)** The pattern has vertical reflection symmetry only. It would take a complete 360° rotation to yield an identical pattern, so there is no rotation symmetry.
   **(b)** The pattern has two lines of reflection symmetry that pass diagonally through the pattern. There is one 180° or half-turn rotation symmetry through a point in the center of the pattern.
   **(c)** The pattern has vertical reflection symmetry only. It would take a complete 360° rotation to yield an identical pattern, so there is no rotation symmetry.

7. **(a)**

**(c)** There is no translation symmetry. The rectangles get longer the farther away from the center they are.

**21. (a)** The image of any point reflected with respect to the *x*-axis will have the same *x*-coordinate as the original point. The image will be the same distance from the *y*-axis as the original point. Any point on the *x*-axis will be its own image. The images of *A*, *B*, *C*, and *D* are *A′*(8, −3), *B′*(0, 7), *C′* (−3, 0), and *D′*(−2,−5).

**(b)** The image of any point reflected with respect to the *y*-axis will have the same *y*-coordinate as the original point. The image will be the same distance from the *x*-axis as the original point. Any point on the *y*-axis will be its own image. The images of *A*, *B*, *C*, and *D* are *A′* (−8, 3), *B′* (0, −7), *C′* (3, 0), and *D′* (2, 5).

**(c)** The images of *A*, *B*, *C*, and *D* are *A′* (3, 8), *B′* (−7, 0), *C′* (0, −3), and *D′* (5, −2).

**23.** Use a protractor and a ruler or compass as described in Example 2.9.
**(a)** Each point and its image are the same distance from line *l*.

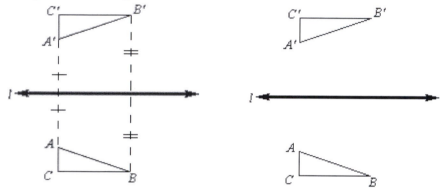

**(b)** The point *A* is on line *l* so *A* and *A′* are the same point.

**25. (a)** The vector *v* moves all the points the same distance and in the same direction. In this case, △*ABC* will be shifted horizontally to the right a distance equal to the length of vector *v*. One vector, $\overrightarrow{CC'}$, has been drawn in.

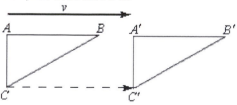

**(b)** The vector *v* moves all the points the same distance and in the same direction. $\overrightarrow{AA'}$, $\overrightarrow{BB'}$, and $\overrightarrow{CC'}$ are drawn in, and each has the same length and orientation as vector *v*.

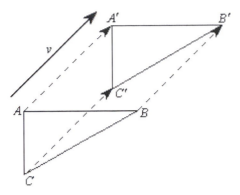

(c) The vector *v* moves all the points the same distance and in the same direction. $\overrightarrow{AA'}$, $\overrightarrow{BB'}$, and $\overrightarrow{CC'}$ are drawn in, and each has the same length and orientation as vector *v*.

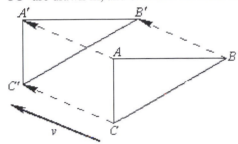

27. Suppose we locate the image of *B* first. Draw $\overline{OB}$. Use your protractor to mark off an angle, using $\overline{OB}$ as the initial side, so that the angle measures 35°. Use your compass to swing an arc that intersects the terminal side of the angle. The intersection point is *B′*. See the left-hand figure. The image of each point will be located in the same manner. See the right-hand figure.

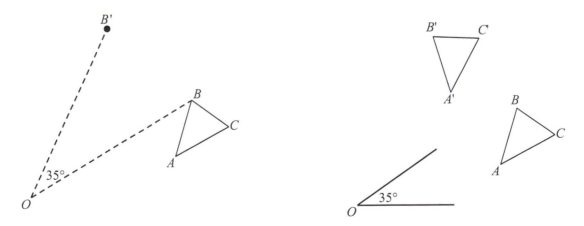

29. (a) The center of rotation is *O*. The −90° rotation is a clockwise rotation of 90°. The image of point *A* has been located in the following figure. The image of each point on the segment is located in the same manner.

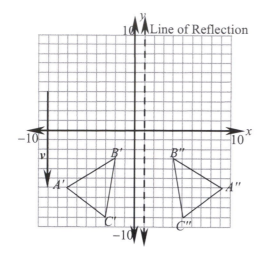

**41. (a)**

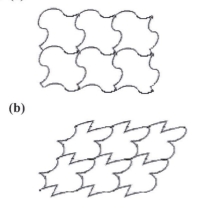

**(b)**

**43.** Answers will vary.

## Solutions to Odd-Numbered Problems in Section 2.3

**1.** **(a)** Give the values of the second and the third terms in the sequence: $a_2 = 4$ and $a_3 = 9$.

**(b)** Find the sum of the fourth term and the sixth term in the sequence: $a_4 + a_6 = 16 + 36 = 52$.

**(c)** If $n = 3$, then $a_{n-2} = a_{3-2} = a_1 = 1$ and $a_{n-1} = a_{3-1} = a_2 = 4$.

**(d)** If $n = 8$, then $a_{n-3} + a_{n-1} = a_{8-3} + a_{8-1} = a_5 + a_7 = 25 + 49 = 74$.

**(e)** If $a_n = n^2$, then $a_{11} = 11^2 = 121$, $a_{15} = 15^2 = 225$, and $a_{20} = 20^2 = 400$.

**3.** **(a)** Let $n = 5$ in the rule: $a_5 = a_4 + 4$.

**(b)** Let $n = 8$, then $a_8 = a_7 + 4 = 35 + 4 = 39$.
Let $n = 9$, then $a_9 = a_8 + 4 = 39 + 4 = 43$.
Let $n = 10$, then $a_{10} = a_9 + 4 = 43 + 4 = 47$.
Let $n = 11$, then $a_{11} = a_{10} + 4 = 47 + 4 = 51$.

**(c)** We know $a_3 = a_2 + 4$, and we know $a_3 = 28$, so $28 = a_2 + 4$. Therefore, $a_2 = 24$.
We know $a_2 = a_1 + 4$, and we know $a_2 = 24$, so $24 = a_1 + 4$. Therefore, $a_1 = 20$.

**(d)** The first ten numbers in the sequence are 7, 11, 15, 19, 23, 27, 31, 35, 39, and 43.

**5.** The first ten numbers in the sequence are 1; 2; 2; 4; 8; 32; 256; 8192; 2,097,152; and 17,179,869,184.

7. The Fibonacci sequence is defined by $f_n = f_{n-1} + f_{n-2}$. We know $f_{27} = f_{26} + f_{25}$ so $f_{26} = f_{27} - f_{25} = 196{,}418 - 75{,}025 = 121{,}393$.

9. (a) List the first 20 numbers in the Fibonacci sequence: 1, 1, 2, 3, 5, 8, 13, 21, 34, 55, 89, 144, 233, 377, 610, 987, 1597, 2584, 4181, and 6765. The Fibonacci numbers that are divisible by 3 are $f_4 = 3$, $f_8 = 21$, $f_{12} = 144$, $f_{16} = 987$, and $f_{20} = 6765$. Every 4th Fibonacci number is divisible by 3.

   (b) The term numbers from the list that are multiples of 4 are 48, 196, and 1000. Therefore, $f_{48}$, $f_{196}$, and $f_{1000}$ are divisible by 3.

11. (a) Let $n = 1$ in Binet's formula:

$$f_1 = \frac{\left(\frac{1+\sqrt{5}}{2}\right)^1 - \left(\frac{1-\sqrt{5}}{2}\right)^1}{\sqrt{5}} = 1$$

Let $n = 2$ in Binet's formula:

$$f_2 = \frac{\left(\frac{1+\sqrt{5}}{2}\right)^2 - \left(\frac{1-\sqrt{5}}{2}\right)^2}{\sqrt{5}} = 1$$

Let $n = 3$ in Binet's formula:

$$f_3 = \frac{\left(\frac{1+\sqrt{5}}{2}\right)^3 - \left(\frac{1-\sqrt{5}}{2}\right)^3}{\sqrt{5}} = 2$$

Let $n = 4$ in Binet's formula:

$$f_4 = \frac{\left(\frac{1+\sqrt{5}}{2}\right)^4 - \left(\frac{1-\sqrt{5}}{2}\right)^4}{\sqrt{5}} = 3$$

   (b) Let $n = 40$ in Binet's formula: $f_{40} = 102{,}334{,}155$.

13. There are a total of 55 spirals marked in red and a total of 89 spirals marked in green. Both are Fibonacci numbers.

15. (a) We begin with one calf in her first year. This calf will not be considered an adult cow until the end of the second year, so for the first two years, we have 1 calf. In the third year, the calf has become an adult cow and produces a female calf. Then that adult cow produces a single female calf every year. It will be helpful to create a notation to keep track of the calves as they mature, so we will let $c_1$ represent a calf in her first year, $c_2$ represent a calf in her second year, and C represent an adult cow. The following table displays the total number of adult cows and total number of calves each year for the first 10 years.

| Year | Adult Cows | Calves | Total Cows and Calves |
|---|---|---|---|
| 1 | 0 | $c_1$ | 1 |
| 2 | 0 | $c_2$ | 1 |
| 3 | C | $c_1$ | 2 |
| 4 | C | $c_1, c_2$ | 3 |
| 5 | C, C | $c_1, c_1, c_2$ | 5 |
| 6 | C, C, C | $c_1, c_1, c_1, c_2, c_2$ | 8 |
| 7 | C, C, C, C, C | $c_1, c_1, c_1, c_1, c_1, c_2, c_2, c_2$ | 13 |
| 8 | C, C, C, C, C, C, C, C | $c_1, c_1, c_1, c_1, c_1, c_1, c_1, c_1,$ $c_2, c_2, c_2, c_2, c_2$ | 21 |

Multiply both sides of the equation by the short side length, divide by $\phi$, and use the fact that

$$\phi = \frac{1+\sqrt{5}}{2}: \text{ short side length} = \frac{1}{\phi} = \left(\frac{1}{\frac{1+\sqrt{5}}{2}}\right) = \frac{2}{1+\sqrt{5}}.$$

The length of the short side is $\dfrac{2}{1+\sqrt{5}} \approx 0.618$ inch.

(d) An isosceles triangle is a golden triangle if $\dfrac{\text{long side length}}{\text{short side length}} = \phi$. Suppose the length of the long

side is 7. Substitute the length of the long side into the equation: $\dfrac{7}{\text{short side length}} = \phi$.

Multiply both sides of the equation by the short side length, divide by $\phi$, and use the fact that

$$\phi = \frac{1+\sqrt{5}}{2}: \text{ short side length} = \frac{7}{\phi} = \left(\frac{7}{\frac{1+\sqrt{5}}{2}}\right) = \frac{14}{1+\sqrt{5}}.$$

The length of the short side is $\dfrac{14}{1+\sqrt{5}} \approx 4.326$ cm.

31. For each rectangle, find $\dfrac{\text{long side length}}{\text{short side length}}$. For the credit card, $\dfrac{\text{long side length}}{\text{short side length}} = \dfrac{86}{54} \approx 1.5926$. For

the Post Alpha-Bits box, $\dfrac{\text{long side length}}{\text{short side length}} = \dfrac{11.75}{8} \approx 1.4688$. The credit card most closely

approximates a golden rectangle. The ratio of the long side to the short side is closer to the golden ratio.

33. We know the rectangle is a golden rectangle, so the ratio of the long side to the short side is the golden ratio. The measure of the short side of the rectangle is given. If the short side measures 11 mm, then $\dfrac{\text{long side}}{11} = \phi$ and long side $= 11 \times \phi \approx 17.8$ mm.

35. The flag that most closely approximates a golden rectangle is the flag of Finland. The flags listed in order of best approximate golden rectangle to least are Finland, Portugal, Bangladesh, Brazil, Australia, and United States of America.

37. Each is approximately the golden ratio.

39. (a) The $\dfrac{\text{length}}{\text{width}}$ will be approximately 1.6 which is close to the golden ratio.

(b) The ratios are all approximately the golden ratio.

**41.** Note the 5 upward (numbered) waves and 3 downward (lettered) waves. Notice $3 + 5 = 8$.

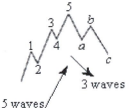

This pattern repeats in the figure.

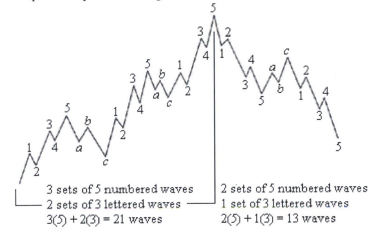

| 3 sets of 5 numbered waves | 2 sets of 5 numbered waves |
| 2 sets of 3 lettered waves | 1 set of 3 lettered waves |
| $3(5) + 2(3) = 21$ waves | $2(5) + 1(3) = 13$ waves |

There are 5 sets of 5 waves and 3 sets of 3 waves for a total of 34 waves. The numbers 3, 5, 8, 13, 21, and 34 are all Fibonacci numbers.

## Solutions to Chapter 2 Review Problems

**1.** **(a)** The polygons are (i), (iv), and (v). Figures (ii) and (iii) are not polygons because they both have curved sides. Figure (vi) is not a polygon because it cannot be traced so that the starting point and ending point are the same.
   **(b)** Convex polygons are (i) and (v).
   **(c)** The regular polygon is (v).

**2.** **(a)** In a polygon with 9 sides, called a nonagon, the sum of the measures of the vertex angles is $(9-2)180° = 1260°$.

   **(b)** In a regular nonagon, the measure of each vertex angle is $\dfrac{(9-2)180°}{9} = \dfrac{1260°}{9} = 140°$.

**3.** The measure of one vertex angle of a regular nonagon is 140° which is not a factor of 360°. A regular tiling of nonagons would have at least three figures around a vertex, but the sum of the measures of the vertex angles of three nonagons around one vertex is $3(140°) = 420°$ which is greater than 360°.

**4.** All vertex figures have the same size and shape.

**5.** **(a)** Yes, it is an edge-to-edge tiling because each triangle shares an entire edge with another triangle.
   **(b)** The tiling would be a regular tiling if the triangles were equilateral and $a = b = c = 60°$.

**(c)**

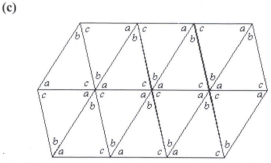

**(d)** When rearranged, the three angles of the triangle come together to form a straight angle which has a measure of 180°.

6.  Let $a = 5$, $b = 10$, and $c$ be the length of the hypotenuse. Then, by the Pythagorean theorem we know $c^2 = a^2 + b^2$, so $c^2 = 5^2 + 10^2 = 25 + 100 = 125$ and $c = \sqrt{125}$ Therefore, the length of the hypotenuse is $\sqrt{125} = 5\sqrt{5}$ units.

7.  For a triangle with side lengths $a$, $b$, and $c$, if $a^2 + b^2 = c^2$, then the triangle is a right triangle. In this case, let $a = 47$, $b = 85$, and $c = 97$. Since $47^2 + 85^2 \neq 97^2$ the triangle is not a right triangle.

8.  Draw a picture. Label the height of the flagpole 40 feet, the shadow 30 feet, and the distance from the top of the flagpole to the end of the shadow $c$. Assume the flagpole makes a right angle with the level ground. Find $c$ using the Pythagorean theorem.

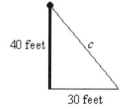

We have $c^2 = a^2 + b^2 = 40^2 + 30^2 = 1600 + 900 = 2500$, so $c = \sqrt{2500} = 50$ feet.

9.  **(a)** The pattern has translation symmetry, and 180° rotation symmetry. It is classified as a $p112$ pattern.
    **(b)** The pattern has translation symmetry only and is classified as a $p111$ pattern.
    **(c)** The pattern has translation symmetry only and is classified as a $p111$ pattern.
    **(d)** The pattern has translation symmetry, vertical reflection symmetry, horizontal reflection symmetry, and 180°rotation symmetry. It is classified as a $pmm2$ pattern.
    **(e)** The pattern has translation symmetry, vertical reflection symmetry, glide reflection symmetry, and 180° rotation symmetry. It is classified as a $pma2$ pattern.

10. The figure has 6 lines of reflection symmetry.

11. The figure has seven rotation symmetries and the center of rotation is in the center of the figure.

**41.** Note the 5 upward (numbered) waves and 3 downward (lettered) waves. Notice 3 + 5 = 8.

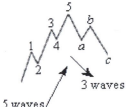

This pattern repeats in the figure.

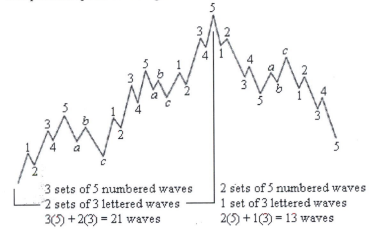

3 sets of 5 numbered waves | 2 sets of 5 numbered waves
2 sets of 3 lettered waves | 1 set of 3 lettered waves
3(5) + 2(3) = 21 waves | 2(5) + 1(3) = 13 waves

There are 5 sets of 5 waves and 3 sets of 3 waves for a total of 34 waves. The numbers 3, 5, 8, 13, 21, and 34 are all Fibonacci numbers.

## Solutions to Chapter 2 Review Problems

1.  **(a)** The polygons are (i), (iv), and (v). Figures (ii) and (iii) are not polygons because they both have curved sides. Figure (vi) is not a polygon because it cannot be traced so that the starting point and ending point are the same.
    **(b)** Convex polygons are (i) and (v).
    **(c)** The regular polygon is (v).

2.  **(a)** In a polygon with 9 sides, called a nonagon, the sum of the measures of the vertex angles is $(9-2)180° = 1260°$.

    **(b)** In a regular nonagon, the measure of each vertex angle is $\dfrac{(9-2)180°}{9} = \dfrac{1260°}{9} = 140°$.

3.  The measure of one vertex angle of a regular nonagon is 140° which is not a factor of 360°. A regular tiling of nonagons would have at least three figures around a vertex, but the sum of the measures of the vertex angles of three nonagons around one vertex is 3(140°) = 420° which is greater than 360°.

4.  All vertex figures have the same size and shape.

5.  **(a)** Yes, it is an edge-to-edge tiling because each triangle shares an entire edge with another triangle.
    **(b)** The tiling would be a regular tiling if the triangles were equilateral and $a = b = c = 60°$.

**(c)**

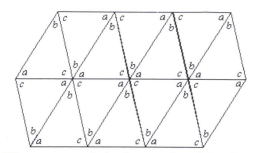

**(d)** When rearranged, the three angles of the triangle come together to form a straight angle which has a measure of 180°.

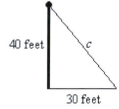

6. Let $a = 5$, $b = 10$, and $c$ be the length of the hypotenuse. Then, by the Pythagorean theorem we know $c^2 = a^2 + b^2$, so $c^2 = 5^2 + 10^2 = 25 + 100 = 125$ and $c = \sqrt{125}$ Therefore, the length of the hypotenuse is $\sqrt{125} = 5\sqrt{5}$ units.

7. For a triangle with side lengths $a$, $b$, and $c$, if $a^2 + b^2 = c^2$, then the triangle is a right triangle. In this case, let $a = 47$, $b = 85$, and $c = 97$. Since $47^2 + 85^2 \neq 97^2$ the triangle is not a right triangle.

8. Draw a picture. Label the height of the flagpole 40 feet, the shadow 30 feet, and the distance from the top of the flagpole to the end of the shadow $c$. Assume the flagpole makes a right angle with the level ground. Find $c$ using the Pythagorean theorem.

We have $c^2 = a^2 + b^2 = 40^2 + 30^2 = 1600 + 900 = 2500$, so $c = \sqrt{2500} = 50$ feet.

9. **(a)** The pattern has translation symmetry, and 180° rotation symmetry. It is classified as a $p112$ pattern.
   **(b)** The pattern has translation symmetry only and is classified as a $p111$ pattern.
   **(c)** The pattern has translation symmetry only and is classified as a $p111$ pattern.
   **(d)** The pattern has translation symmetry, vertical reflection symmetry, horizontal reflection symmetry, and 180° rotation symmetry. It is classified as a $pmm2$ pattern.
   **(e)** The pattern has translation symmetry, vertical reflection symmetry, glide reflection symmetry, and 180° rotation symmetry. It is classified as a $pma2$ pattern.

10. The figure has 6 lines of reflection symmetry.

11. The figure has seven rotation symmetries and the center of rotation is in the center of the figure.

**12.** The coordinates of the vertices are $A = (-9, 0)$, $B = (-4, -8)$, $C = (-8, -9)$, and $D = (-10, -4)$. The image of $ABCD$ after a reflection with respect to the $y$-axis is $A'B'C'D'$ as shown in the following figure. The coordinates of the image of each vertex point are $A' = (9, 0)$, $B' = (4, -8)$, $C' = (8, -9)$, and $D' = (10, -4)$.

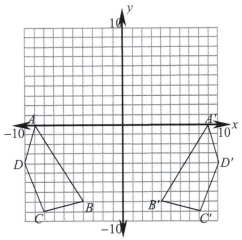

The image of $A'B'C'D'$ after a translation according to vector $v$ will shift each point of $A'B'C'D'$ left 2 units and up 3 units. The coordinates of the vertices of the image are found as follows:

$A'' = (9 - 2, 0 + 3) = (7, 3)$
$B'' = (4 - 2, -8 + 3) = (2, -5)$
$C'' = (8 - 2, -9 + 3) = (6, -6)$
$D'' = (10 - 2, -4 + 3) = (8, -1)$.

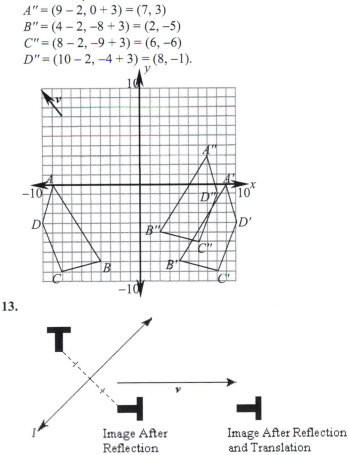

**13.**

Image After Reflection          Image After Reflection and Translation

14. The center of rotation is vertex $A$ and each point in the triangle will be rotated clockwise 90°. The image of $A$ is $A$ under a rotation about point $A$. Locate the image of $B$ then locate the image of $C$.

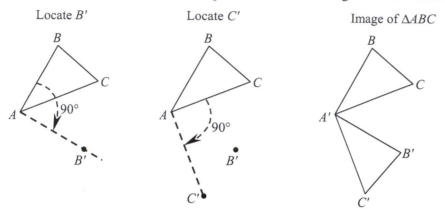

15. Notice $OA = OA'$ and $\angle AOA' = 60°$. The images of $B$ and $C$ are located similarly.

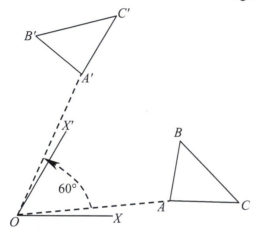

16. **(a)** The third number is 1, the sixth number is 11, and the eighth number is 6.
    **(b)** The second term $a_2$ is 2. The fifth term $a_5$ is –2. The tenth term $a_{10}$ is –9.
    **(c)** $a_{11} = 2a_8 + a_9 - a_{10} = 2(6) + 17 - (-9) = 38$
    $a_{12} = 2a_9 + a_{10} - a_{11} = 2(17) + (-9) - 38 = -13$
    $a_{13} = 2a_{10} + a_{11} - a_{12} = 2(-9) + 38 - (-13) = 33$

17. **(a)** The first 12 numbers in the sequence are 4, 6, 10, 16, 26, 42, 68, 110, 178, 288, 466, and 754.
    **(b)**

| $n$ | $\dfrac{f_{n+1}}{f_n}$ |
|---|---|
| 2 | $\dfrac{f_3}{f_2} = \dfrac{10}{6} \approx 1.6667$ |
| 3 | $\dfrac{f_4}{f_3} = \dfrac{16}{10} = 1.6$ |
| 4 | $\dfrac{f_5}{f_4} = \dfrac{26}{16} = 1.625$ |

| 5 | $\dfrac{f_6}{f_5} = \dfrac{42}{26} \approx 1.6154$ |
|---|---|
| 6 | $\dfrac{f_7}{f_6} = \dfrac{68}{42} \approx 1.6190$ |
| 7 | $\dfrac{f_8}{f_7} = \dfrac{110}{68} \approx 1.6176$ |
| 8 | $\dfrac{f_9}{f_8} = \dfrac{178}{110} \approx 1.6182$ |
| 9 | $\dfrac{f_{10}}{f_9} = \dfrac{288}{178} \approx 1.6180$ |
| 10 | $\dfrac{f_{11}}{f_{10}} = \dfrac{466}{288} \approx 1.6181$ |
| 11 | $\dfrac{f_{12}}{f_{11}} = \dfrac{754}{466} \approx 1.6180$ |

The values of the ratios get closer and closer to the golden ratio, $\dfrac{1+\sqrt{5}}{2} \approx 1.618$.

18.

| $n$ | $f_n$ | $f_{n+1}$ | $1^2 + 1^2 + \ldots + f_n^2$ |
|---|---|---|---|
| 3 | 2 | 3 | $1^2 + 1^2 + 2^2 = 6$ |
| 4 | 3 | 5 | $1^2 + 1^2 + 2^2 + 3^2 = 15$ |
| 5 | 5 | 8 | $1^2 + 1^2 + 2^2 + 3^2 + 5^2 = 40$ |
| 6 | 8 | 13 | $1^2 + 1^2 + 2^2 + 3^2 + 5^2 + 8^2 = 104$ |
| 7 | 13 | 21 | $1^2 + 1^2 + 2^2 + 3^2 + 5^2 + 8^2 + 13^2 = 273$ |
| 8 | 21 | 34 | $1^2 + 1^2 + 2^2 + 3^2 + 5^2 + 8^2 + 13^2 + 21^2 = 714$ |
| 9 | 34 | 55 | $1^2 + 1^2 + 2^2 + 3^2 + 5^2 + 8^2 + 13^2 + 21^2 + 34^2 = 1870$ |
| 10 | 55 | 89 | $1^2 + 1^2 + 2^2 + 3^2 + 5^2 + 8^2 + 13^2 + 21^2 + 34^2 + 55^2 = 4895$ |

In each case, $1^2 + 1^2 + \ldots + f_n^2 = f_n \times f_{n+1}$.

19. (a) Notice the Fibonacci numbers in the sum are even-numbered terms. The sum of the even-numbered terms is 1 less than the value of the next odd-numbered term.
$1 + 3 + 8 + 21 + 55 = 89 - 1$
$1 + 3 + 8 + 21 + 55 + 144 = 233 - 1$

   (b) The number 377 is the 14th Fibonacci number. The 15th Fibonacci number is 610, so the sum is $610 - 1 = 609$.

   (c) The number 987 is the 16th Fibonacci number. The 17th Fibonacci number is 1597, so the sum is $1597 - 1 = 1596$.

20. (a) Let B represent each adult animal and let $b_1$, $b_2$, $b_3$, and $b_4$ represent babies in their first, second, third, and fourth years respectively.

| Year | Adult | Baby | Total |
|---|---|---|---|
| 1 | 0 | $b_1$ | 1 |
| 2 | 0 | $b_2$ | 1 |
| 3 | 0 | $b_3$ | 1 |
| 4 | 0 | $b_4$ | 1 |

 ☐  demonstrate that any of the four voting methods presented in this section can violate the irrelevant-alternatives criterion.

 ☐  describe the method of approval voting.

**By the time you finish studying Section 3.3, you should be able to:**

 ☐  use proper notation to indicate weights for voters and the quota.

 ☐  describe what is meant by a winning coalition and a losing coalition.

 ☐  determine if a motion passes or fails given the quota and weights of the voters.

 ☐  determine the number of coalitions in a weighted voting system.

 ☐  list all possible coalitions and determine which are winning and which are losing coalitions.

 ☐  describe voters who are dummies, dictators, and those who have veto power.

 ☐  determine if a voter is a critical voter.

 ☐  calculate the Banzhaf power index for each voter in a weighted voting system.

## Hints for Odd-Numbered Problems

### Section 3.1

1. **(a)** A majority win requires more than 50% of the votes.
   **(b)** To find the percentages, divide the candidate's vote total by the total number of votes cast and multiply by 100%.

3. A majority win would result from a location receiving more than 50% of the votes. How many votes were cast?

5. How many first-place, second-place, and third-place votes did each consultant receive?

7. **(a)** There are 32 voters. Any player receiving a first-place vote will earn 14 points. The maximum number of points would result from the maximum number of first-place votes.

9. Each first-place vote is worth 3 points. Each second-place vote is worth 2 points. Each third-place vote is worth 1 point.

11. What ranks are possible with 5 candidates and how many points would each rank be awarded? Each of the 30 voters will rank each candidate.

13. This is a modified Borda count method because every player is not ranked, and only those who receive a vote for first, second, and third earn points. There were many players who did not receive any votes at all.

15. **(b)** Remember, the winner using the plurality method is the candidate with the most first-place votes.

17. **(b)** Every location which was ranked lower than the eliminated location moves up in the ranking. Every remaining location will be ranked first or second.

19. **(b)** There are three pairwise comparisons that must be made. Compare Ann to Pat, Ann to Eno, and Eno to Pat.

21. **(b)** The park that is preferred in each comparison is given 1 point.

23. **(a)** The first candidate must be paired with each of the other four candidates. The second candidate does not need to be paired with the first, because that comparison has already been made. For each candidate consider which comparisons have already been made and which still need to be made.
    **(b)** There are four places in which to rank the candidates. Once the first candidate is placed, then how many options are there for the second candidate?

25. The candidate with the largest point total will receive the largest scholarship. The candidate with the smallest point total will receive the smallest scholarship.

27. Only consider the number of first-place votes. Rank candidates according to how many first-place votes they received.

29. Keep track of the order in which candidates are eliminated. The candidates will be ranked in reverse order of elimination.

31. Form all pairwise comparisons and assign points. Rank candidates according to the number of points earned.

33. This modified Borda count method only differs from the Borda count method in that the first-place vote earns 4 points.

35. Use the given table to count the number of first-place through fourth-place finishes for each team.

37. **(a)** Eliminate the site with the most third-place votes first and create a new preference table. Then eliminate the site with the most second-place votes.
    **(b)** This time apply the standard plurality with elimination method in which the site with the fewest first-place votes is eliminated first.

39. **(a)** Count the first-place votes for each senior and rank the seniors from greatest number of first-place votes to least. The second and third ranked senior will be involved in the run-off election so create a preference table with only those two seniors included. The winner will enter a run-off election with the senior who had the most first-place votes originally.

41. In each round, eliminate the candidate with the fewest number of votes and use the percentages given to redistribute the eliminated candidate's votes to the other candidates. When calculating the number of votes to distribute to the other candidates, round to the nearest vote.

## Section 3.2

1. **(a)** Determine the number of voters and how many votes are required to have a majority.
   **(b)** Assign 3 points to every first-place vote, 2 points to every second-place vote, and 1 point for every third-place vote.
   **(c)** Compare the results from part (a) and (b). Are they the same or different?

3. **(a)** Remember, when using the plurality method, count the number of first-place votes received by each proposal.
   **(b)** There are three proposals, so there are three pairwise comparisons.
   **(c)** Compare the results from part (a) and (b). Are they the same or different?

5. Find the result using the plurality method. If the plurality method violates the head-to-head criterion, then one candidate would have to be favored over every other candidate and the plurality method would have to select a different candidate as the winner.

7. If there is a Condorcet candidate, then there is a candidate who is favored over every other candidate. If there is a Condorcet candidate, then how did he or she do in each election?

9. **(b)** Is there a Condorcet candidate?

11. **(b)** Remove C and create a new preference table before determining the winner under the plurality method.
    **(c)** Did the winner from part (a) differ from the winner from part (b)?

13. (c) Did any candidate have a majority? If so, was it the same candidate who won the election using the Borda count method?
    (d) Is there a Condorcet candidate?

15. To show a method fails the irrelevant-alternatives criterion, determine the winner using the method, then eliminate a losing candidate, and apply the method again. If the result of the new election is different, then a violation has occurred.

17. To show a method fails the head-to-head criterion, determine the winner using the method. Also determine if there is a Condorcet candidate. If the winner using the method is not the Condorcet candidate, then there is a violation.

19. (b) Create a new preference table to reflect the changes in preference.
    (c) Which criterion might be violated when voters change how they rank candidates?

21. (d) Which criterion involves a candidate being eliminated? Candidate B was the one that was eliminated. Originally, did candidate B win or lose?

23. (b) If there is a Condorcet program, then it is possible to have a violation of the head-to-head criterion. If you remove a losing candidate from the election and hold the election again, do the results change?
    (c) If a candidate has a majority of first place votes, then it is possible to have a violation of the majority criterion.

25. (c) There are four fairness criteria that could be violated. The majority criterion might be violated if there is a candidate who has a majority of first-place votes. The head-to-head criterion might be violated if there is a Condorcet candidate. The irrelevant-alternatives criterion might be violated if a losing candidate is eliminated and the election is held again. The monotonicity criterion might be violated if voters change their preferences and the election is held again. Explore these possibilities and determine which criterion has been violated.

27. (a) Delegates ranked the candidates. Count the votes in Round 1.
    (b) In the first three rounds, which candidates received the fewest number of votes? Compare the votes in Round 1 to the votes in Round 2. When the first candidate is eliminated in Round 1, who receives another vote in Round 2? What does that tell you about the first-place and second-place ranking of the voter who cast that ballot?
    (c) Compare the total number of voters to the number of votes in each round.

29. The three candidates with the greatest number of votes are the winners.

31. Three candidates will be selected. Which candidates have the top three vote totals?

33. (b) With approval voting voters vote for all options they feel are acceptable.

35. Study the summary of methods and fairness criteria in Table 3.25. Which criteria can be violated by the plurality method? How can you determine if there is a violation?

37. Study the summary of methods and fairness criteria in Table 3.25. Which criteria can be violated by the Borda count method? How can you determine if there is a violation?

## Section 3.3

1. (a) Express the weighted voting system using square brackets as follows: [quota | $W_1$, $W_2$, $W_3$, $W_4$]. Weights are listed from largest to smallest.
   (b) A simple majority is one more than half of the total weight if half of the total weight is a whole number. If half of the total weight is not a whole number, round up to the nearest whole number.

3. A simple majority is one more than half of the total weight if half of the total weight is a whole number. If half of the total weight is not a whole number, round up to the nearest whole number.

5. Find the quota by multiplying the percentage required to pass a measure by the total weight in the system. In each case, round the quota up to the nearest whole number.

7. (c) Consider both the Yes votes and the No votes.

9. Every person on the committee must agree to pass a motion. The quota must be set high enough to reflect this.

11. Add the weights of each voter in the coalition. A coalition is a winning coalition if it has enough votes to pass a motion.

13. Add the weights of each voter in the coalition. A coalition is a winning coalition if it has enough votes to pass a motion.

15. Refer to the theorem for the number of coalitions in a weighted voting system with $n$ voters.

17. Determine the number of coalitions possible for the weighted voting system and list each one along with the weight of the coalition. A coalition is a winning coalition if it has enough votes to pass a motion.

19. (a) Divide the weight of a voter by the total weight in the system and express it as a percent.
    (b) List all of the coalitions and determine which are winning coalitions. For each winning coalition, determine which voter(s) are critical. Divide the total number of times a voter is critical by the total number of times all voters are critical.

21. (a) What does the quota have to be if the verdict must be unanimous?

23. It will be helpful to lay out your work as in Example 3.19.

25. Notice how the power changes as the quota increases.

27. List all of the coalitions and determine which of them are winning coalitions. Systematically determine which voters are critical in each winning coalition.

29. List all of the coalitions. For each calculate the weight and determine if it is greater than the total weight of the system minus the quota.

31. Which voters are permanent members and which are elected?

33. (b) If two voters always vote the same, it is as if there is one voter controlling the votes.

35. Notice how many votes each voter controls and how the power is distributed before and after the shift in votes.

37. Notice how many votes each voter controls compared to the power held by each voter.

## Solutions to Odd-Numbered Problems in Section 3.1

1. (a) Requiring a majority to win means that a candidate must win more than 50% of the votes. The total number of votes cast in the 2002 California election was 7,326,133. A majority of votes would be a vote total greater than 50% of 7,326,133 = $0.5 \times 7,326,133 = 3,663,066.5$. Therefore, any candidate receiving at least 3,663,067 votes would be declared the winner. No candidate in the election received a majority.

**(b)** Calculate the percentage of votes each candidate received by dividing the candidate's vote total by the total number of votes cast and multiply by 100%.

| Candidate | Votes | Percentage of Vote Received |
|---|---|---|
| Gray Davis | 3,469,025 | $\dfrac{3,469,025}{7,326,133} \times 100\% \approx 47.35\%$ |
| Bill Simon | 3,105,477 | $\dfrac{3,105,477}{7,326,133} \times 100\% \approx 42.39\%$ |
| Reinhold S. Gulke | 125,338 | $\dfrac{125,338}{7,326,133} \times 100\% \approx 1.71\%$ |
| Peter Miguel Camejo | 381,700 | $\dfrac{381,700}{7,326,133} \times 100\% \approx 5.21\%$ |
| Gary David Copeland | 158,161 | $\dfrac{158,161}{7,326,133} \times 100\% \approx 2.16\%$ |
| Iris Adam | 86,432 | $\dfrac{86,432}{7,326,133} \times 100\% \approx 1.18\%$ |
| Total | 7,326,133 | 100% |

**(c)** Under the plurality method, Gray Davis won the election with 47.35% of the votes.

3. **(a)** The city council has nine voting members. Fifty percent of nine is 4.5, so a majority of votes would be at least 5 votes.
   **(b)** Read the first-place votes across the row labeled 1st in the table. Davis Ave. received 4 first-place votes. Two votes went to 9th Street. Beca Blvd. is listed in two columns for a total of 3 first-place votes.
   **(c)** Davis Ave. received a plurality of votes. It did not receive a majority, but it received more than any other location. Therefore, Davis Ave. is selected.

5. **(a)** Determine the number of first-place, second-place, and third-place votes each consultant received. Then multiply each first-place vote by 3 points, every second-place vote by 2 points, and every third-place vote by 1 point.

| Consultant | Points For Each Vote Received | | | Total Points |
|---|---|---|---|---|
| | 1st Place Votes 3 Points Each | 2nd Place Votes 2 Points Each | 3rd Place Votes 1 Point Each | |
| Finster | $2 \times 3 = 6$ | $4 \times 2 = 8$ | $1 \times 1 = 1$ | 15 |
| Gorman | $1 \times 3 = 3$ | $2 \times 2 = 4$ | $4 \times 1 = 4$ | 11 |
| Yamada | $4 \times 3 = 12$ | $1 \times 2 = 2$ | $2 \times 1 = 2$ | 16 |

   **(b)** The consultant who receives the most points is selected. In this case, Yamada received 16 points, so under the Borda count method, Yamada is selected.

7. **(a)** There are 32 voters. The maximum number of points a player could receive would result from a player receiving 32 first-place votes worth 14 points each. The maximum number is $32 \times 14 = 448$ points.

**(b)** Each player receives 14 points for a first-place vote, 9 points for a second-place vote, 8 points for a third-place vote,…, and 1 point for a tenth-place vote. The following table contains the point total each player received for each place and the total modified Borda count points received.

| Player | Points For Each Vote Received | | | | | | | | | | Total Points |
|---|---|---|---|---|---|---|---|---|---|---|---|
| | 1st | 2nd | 3rd | 4th | 5th | 6th | 7th | 8th | 9th | 10th | |
| Johnson | 0 | 0 | 40 | 21 | 24 | 20 | 8 | 12 | 2 | 0 | 127 |
| Berkman | 0 | 9 | 56 | 35 | 36 | 25 | 8 | 9 | 2 | 1 | 181 |
| Green | 0 | 0 | 24 | 56 | 24 | 15 | 8 | 9 | 6 | 4 | 146 |
| Bonds | 448 | 0 | 0 | 0 | 0 | 0 | 0 | 0 | 0 | 0 | 448 |
| Kent | 0 | 0 | 24 | 14 | 24 | 40 | 20 | 6 | 6 | 1 | 135 |
| Sosa | 0 | 0 | 0 | 14 | 18 | 5 | 4 | 12 | 8 | 2 | 63 |
| Schilling | 0 | 0 | 0 | 7 | 12 | 15 | 8 | 3 | 6 | 2 | 53 |
| Pujols | 0 | 234 | 32 | 0 | 6 | 0 | 4 | 0 | 0 | 0 | 276 |
| Smoltz | 0 | 9 | 24 | 35 | 12 | 5 | 24 | 6 | 6 | 3 | 124 |
| Guerrero | 0 | 36 | 40 | 21 | 18 | 25 | 8 | 9 | 10 | 1 | 168 |

**(c)** The winner is the player who received the most points. Bonds received 448 points, so he is the winner.

**9.** Each location receives 3 points for a first-place vote, 2 points for a second-place vote, and 1 point for a third-place vote. The following table contains the point total each location received for each place and the total Borda count points received.

| Location | Points For Each Vote Received | | | Total Points |
|---|---|---|---|---|
| | 1st Place Votes 3 Points Each | 2nd Place Votes 2 Points Each | 3rd Place Votes 1 Point Each | |
| Davis Ave. | 4 × 3 = 12 | 1 × 2 = 2 | 4 × 1 = 4 | 18 |
| 9th Street | 2 × 3 = 6 | 6 × 2 = 12 | 1 × 1 = 1 | 19 |
| Beca Blvd. | 3 × 3 = 9 | 2 × 2 = 4 | 4 × 1 = 4 | 17 |

The 9th Street location received 19 points, which was more than any other location, so 9th Street is selected using the Borda count method.

**11.** If there are 5 candidates in an election, then there will be first, second, third, fourth, and fifth-place. A fifth-place vote would earn 1 point. A fourth-place vote would earn 2 points. A third-place vote would earn 3 points. A second-place vote would earn 4 points. A first-place vote would earn 5 points. Each of the 30 voters will rank each candidate, so each voter has control of $1 + 2 + 3 + 4 + 5 = 15$ points. There are a total of $15 \times 30 = 450$ Borda count points possible in this case.

**13. (a)** The following table contains the Borda count total for each player.

| Player | Points For Each Vote Received | | | Total Points |
|---|---|---|---|---|
| | 1st Place Votes 3 Points Each | 2nd Place Votes 2 Points Each | 3rd Place Votes 1 Point Each | |
| Johnson | 108 × 3 = 324 | 130 × 2 = 260 | 142 × 1 = 142 | 726 |
| Dorsey | 122 × 3 = 366 | 89 × 2 = 178 | 99 × 1 = 99 | 643 |
| Palmer | 242 × 3 = 726 | 224 × 2 = 448 | 154 × 1 = 154 | 1328 |
| Griffin | 1 × 3 = 3 | 8 × 2 = 16 | 9 × 1 = 9 | 28 |
| Leftwich | 22 × 3 = 66 | 26 × 2 = 52 | 34 × 1 = 34 | 152 |
| Kingsbury | 6 × 3 = 18 | 2 × 2 = 4 | 11 × 1 = 11 | 33 |

31. Create all possible pairwise comparisons. With three consultants, there will be three comparisons.

| Consultants Compared | Preferred Consultant | Points Awarded |
|---|---|---|
| Finster and Gorman | Finster is preferred 6 to 1 | Finster receives 1 point |
| Finster and Yamada | Yamada is preferred 5 to 2 | Yamada receives 1 point |
| Gorman and Yamada | Yamada is preferred 4 to 3 | Yamada receives 1 point |

Yamada has the most points followed by Finster. Gorman did not receive any points. The consultants are ranked Yamada, Finster, and Gorman.

33.

| Candidate | Points For Each Rank | | | Total Points |
|---|---|---|---|---|
| | 1st Place Votes 4 Points Each | 2nd Place Votes 2 Points Each | 3rd Place Votes 1 Point Each | |
| Peter | $33 \times 4 = 132$ | $68 \times 2 = 136$ | $34 \times 1 = 34$ | 302 |
| Carmen | $53 \times 4 = 212$ | $28 \times 2 = 56$ | $54 \times 1 = 54$ | 322 |
| Shawna | $49 \times 4 = 196$ | $39 \times 2 = 78$ | $47 \times 1 = 47$ | 321 |

Carmen received the most points. Carmen is elected senior class president.

35. Count the number of first-place through fourth-place finishes each team earns in the four events. The Raiders earned 1 first-place, 2 third-place, and 1 fourth-place finish. The Spartans earned 2 first-place and 1 second-place finish. The Titans earned 2 second-place and 1 third-place finish. The Vikings earned 1 first-place, 1 second-place, 1 third-place, and 3 fourth-place finishes.

| Team | Points For Each Rank | | | | Total Points |
|---|---|---|---|---|---|
| | 1st Place Finishes 4 Points Each | 2nd Place Finishes 3 Points Each | 3rd Place Finishe 2 Points Each | 4th Place Finishes 1 Point Each | |
| Raiders | $1 \times 4 = 4$ | $0 \times 3 = 0$ | $2 \times 2 = 4$ | $1 \times 1 = 1$ | 9 |
| Spartans | $2 \times 4 = 8$ | $1 \times 3 = 3$ | $0 \times 2 = 0$ | $0 \times 1 = 0$ | 11 |
| Titans | $0 \times 4 = 0$ | $2 \times 3 = 6$ | $1 \times 2 = 2$ | $0 \times 1 = 0$ | 8 |
| Vikings | $1 \times 4 = 4$ | $1 \times 3 = 3$ | $1 \times 2 = 2$ | $3 \times 1 = 3$ | 12 |

In order of total points, the teams are ranked Vikings, Spartans, Raiders, and Titans.

37. (a) Consider the number of third-place votes for each site. Site A has 8, Site B has 4, and Site C has 3. Site A is the site with the most third-place votes, so eliminate Site A and create a new preference table.

| Ranking | Number of Commissioners | | | | |
|---|---|---|---|---|---|
| | 4 | 4 | 4 | 2 | 1 |
| 1st | C | B | C | B | B |
| 2nd | B | C | B | C | C |

Consider the number of second-place votes for each site. Site B has 8 and Site C has 7. Site B has the most second-place votes, so eliminate Site B. Site C is selected.

(b) Consider the number of first-place votes for each site. Site B has 5, Site A has 6, and Site C has 4. Site C is the site with the fewest first-place votes, so eliminate Site C and create a new preference table.

| Ranking | Number of Commissioners | | | | |
|---|---|---|---|---|---|
| | 4 | 4 | 4 | 2 | 1 |
| 1st | A | B | B | A | B |
| 2nd | B | A | A | B | A |

Consider the number of first-place votes for each site. Site B has 9 and Site A has 6. Site A has the fewest first-place votes, so eliminate Site A. Site B is selected.

**(c)** Apply the Borda count method.

| Site | Total Points For Each Rank | | | Total Points |
|---|---|---|---|---|
| | 1st Place Votes 3 Points Each | 2nd Place Votes 2 Points Each | 3rd Place Votes 1 Point Each | |
| A | $6 \times 3 = 18$ | $1 \times 2 = 2$ | $8 \times 1 = 8$ | 28 |
| B | $5 \times 3 = 15$ | $6 \times 2 = 12$ | $4 \times 1 = 4$ | 31 |
| C | $4 \times 3 = 12$ | $8 \times 2 = 16$ | $3 \times 1 = 3$ | 31 |

In this case, Site B and Site C have the same number of points, so there is a tie.

**(d)** Apply the modified Borda count method.

| Site | Total Points For Each Rank | | | Total Points |
|---|---|---|---|---|
| | 1st Place Votes 4 Points Each | 2nd Place Votes 2 Points Each | 3rd Place Votes 1 Point Each | |
| A | $6 \times 4 = 24$ | $1 \times 2 = 2$ | $8 \times 1 = 8$ | 34 |
| B | $5 \times 4 = 20$ | $6 \times 2 = 12$ | $4 \times 1 = 4$ | 36 |
| C | $4 \times 4 = 16$ | $8 \times 2 = 16$ | $3 \times 1 = 3$ | 35 |

Site B is selected.

**39. (a)** Count the number of first-place votes each senior received. Aaron received 8, Griffey received 6, Bonds received 4, and Ruth received 3. Note the senior with the most first-place votes is Aaron with 8 votes. Use the table to conduct the run-off election between the candidate who received the second largest number of first-place votes and the candidate who received the third largest number of first-place votes. The run-off election will be held between Griffey and Bonds. Eliminate Ruth and Aaron from the preference table. Create a preference table with only Griffey and Bonds for the run-off election.

| Ranking | Number of Players | | | | | | | | | | |
|---|---|---|---|---|---|---|---|---|---|---|---|
| | 4 | 3 | 2 | 2 | 2 | 2 | 2 | 1 | 1 | 1 | 1 |
| 1st | B | B | B | G | G | B | G | B | G | G | G |
| 2nd | G | G | G | B | B | G | B | G | B | B | B |

Bonds is preferred by 12 team members while Griffey is preferred by 9 team members. Bonds wins the run-off election. Now carry out the run-off election between Bonds and Aaron. Eliminate Ruth and Griffey from the original preference table. Create a preference table with only Bonds and Aaron for the run-off election.

| Ranking | Number of Players | | | | | | | | | | |
|---|---|---|---|---|---|---|---|---|---|---|---|
| | 4 | 3 | 2 | 2 | 2 | 2 | 2 | 1 | 1 | 1 | 1 |
| 1st | A | B | B | A | B | A | A | B | A | B | B |
| 2nd | B | A | A | B | A | B | B | A | B | A | A |

Bonds has 10 first-place votes while Aaron has 11 first-place votes. Aaron is the new team captain.

**(b)** Count the number of fourth-place votes for each senior. Ruth has 4, Aaron has 8, Griffey has 4, and Bonds has 5. Aaron has the most fourth-place votes and is eliminated. Create a new preference table.

| Ranking | Number of Players | | | | | | | | | | |
|---|---|---|---|---|---|---|---|---|---|---|---|
| | 4 | 3 | 2 | 2 | 2 | 2 | 2 | 1 | 1 | 1 | 1 |
| 1st | B | B | R | G | G | B | R | B | G | G | R |
| 2nd | G | G | B | R | B | R | G | R | R | R | G |
| 3rd | R | R | G | B | R | G | B | G | B | B | B |

Count the number of third-place votes for each senior. Ruth has 9, Griffey has 5, and Bonds has 7. Ruth has the most third-place votes and is eliminated. Create a new preference table.

| Ranking | Number of Players | | | | | | | | | | |
|---|---|---|---|---|---|---|---|---|---|---|---|
| | 4 | 3 | 2 | 2 | 2 | 2 | 2 | 1 | 1 | 1 | 1 |
| 1st | B | B | B | G | G | B | G | B | G | G | G |
| 2nd | G | G | G | B | B | G | B | G | B | B | B |

Count the number of second-place votes. Griffey has 12 and Bonds has 9. Griffey is eliminated and Bonds is selected as the team captain.

**41.** A majority of votes for a candidate would be more than 50% of $104,787,749 = 0.50 \times 104,787,749 = 52,393,874.5$ votes. Any candidate with at least 52,393,875 votes has a majority. In the first round, Buchanan received the fewest number of votes, so he is eliminated, and his votes are redistributed to the other three candidates. Use the percentages given to redistribute Buchanan's 448,895 votes. Round to the nearest vote.

**Redistribute Buchanan's Votes:**

93.9% of $448,895 = 0.939 \times 448,895 = 421,512.405 \approx 421,512$ votes to Bush

3.9% of $448,895 = 0.039 \times 448,895 = 17,506.905 \approx 17,507$ votes to Gore

2.2% of $448,895 = 0.022 \times 448,895 = 9875.69 \approx 9876$ votes to Nader

**New Vote Totals:**

Gore: $50,999,897 + 17,507 = 51,017,404$

Bush: $50,456,002 + 421,512 = 50,877,514$

Nader: $2,882,955 + 9876 = 2,892,831$

After redistributing Buchanan's votes, none of the three remaining candidates has a majority. Nader has the fewest number of votes, so he is eliminated, and his votes are redistributed to the other two candidates. Use the percentages given to redistribute Nader's 2,892,831 votes. Round to the nearest vote.

**Redistribute Nader's Votes:**

53.1% of $2,892,831 = 0.531 \times 2,892,831 = 1,536,093.261 \approx 1,536,093$ votes to Bush

46.9% of $2,892,831 = 0.469 \times 2,892,831 = 1,356,737.739 \approx 1,356,738$ votes to Gore

**New Vote Totals:**

Gore: $51,017,404 + 1,356,738 = 52,374,142$

Bush: $50,877,514 + 1,536,093 = 52,413,607$

After redistributing Nader's votes, George Bush has a majority of the votes and is declared the winner of the election. He has $\dfrac{52,413,607}{104,787,749} \times 100\% \approx 50.02\%$ of the popular vote.

## *Solutions to Odd-Numbered Problems in Section 3.2*

**1.** **(a)** There are 9 voters in this election. A majority would be at least 5 votes. Candidate A has 5 first-place votes, so candidate A wins this election.

**(b)** The following table contains the point totals if the Borda count method is used.

| Candidate | Points For Each Rank | | | Total Points |
|---|---|---|---|---|
| | 1st Place Votes 3 Points Each | 2nd Place Votes 2 Points Each | 3rd Place Votes 1 Point Each | |
| A | $5 \times 3 = 15$ | $0 \times 2 = 0$ | $4 \times 1 = 4$ | 19 |
| B | $1 \times 3 = 3$ | $4 \times 2 = 8$ | $4 \times 1 = 4$ | 15 |
| C | $3 \times 3 = 9$ | $5 \times 2 = 10$ | $1 \times 1 = 1$ | 20 |

Candidate C has the most points and is the winner.

**(c)** From part (a), we know candidate A was the winner with a majority of votes. The Borda count method produces a different winner. This is a violation of the majority criterion.

3. (a) Proposal A has 4 first-place votes, proposal B has 2 first-place votes, and proposal C has 3 first-place votes. Proposal A wins by a plurality of votes.

   (b) With three proposals, there are three pairwise comparisons to be made.

| Proposals Compared | Preferred Proposal | Points Awarded |
|---|---|---|
| A and B | B is preferred 5 to 4 | B receives 1 point |
| A and C | C is preferred 5 to 4 | C receives 1 point |
| B and C | C is preferred 6 to 3 | C receives 1 point |

   Proposal C has the most points and is selected.

   (c) From part (b) we know that proposal C won every head-to-head comparison, yet from part (a), we know the plurality method did not select C. In this case, the plurality method violates the head-to-head criterion.

5. Considering the first-place votes, candidate A has more votes than any of the other candidates and is the winner by a plurality. The following table contains each pairwise comparison.

| Candidates Compared | Preferred Candidate | Points Awarded |
|---|---|---|
| A and B | B is preferred 6 to 4 | B receives 1 point |
| A and C | C is preferred 6 to 4 | C receives 1 point |
| B and C | B is preferred 7 to 3 | B receives 1 point |

   Notice that B is preferred to both candidates A and C, yet B did not win with the plurality method. This shows that, in this case, the plurality method violates the head-to-head criterion.

7. (a) Consider first-place votes. Jackson has 4, Carter has 3, and Morton has 2. Jackson wins by a plurality of votes.

   (b) With three candidates, there are three pairwise comparisons.

| Candidates Compared | Preferred Candidate | Points Awarded |
|---|---|---|
| J and C | J is preferred 5 to 4 | J receives 1 point |
| J and M | J is preferred 5 to 4 | J receives 1 point |
| C and M | C is preferred 7 to 2 | C receives 1 point |

   Jackson has the most points and is the winner. Notice Jackson is a Condorcet candidate.

   (c)

| Candidate | Points For Each Rank | | | Total Points |
|---|---|---|---|---|
| | 1st Place Votes 3 Points Each | 2nd Place Votes 2 Points Each | 3rd Place Votes 1 Point Each | |
| C | $3 \times 3 = 9$ | $5 \times 2 = 10$ | $1 \times 1 = 1$ | 20 |
| J | $4 \times 3 = 12$ | $2 \times 2 = 4$ | $3 \times 1 = 3$ | 19 |
| M | $2 \times 3 = 6$ | $2 \times 2 = 4$ | $5 \times 1 = 5$ | 15 |

   Carter has the most points and is the winner.

   (d) Notice from part (b) that Jackson is preferred over Carter and Morton. Jackson won in every head-to-head match up. The plurality method from part (a) resulted in Jackson winning the election. However, the Borda count method from part (c) resulted in Carter winning the election. Therefore, the Borda count method violated the head-to-head criterion.

9. (a)

| Candidate | Points For Each Rank | | | | Total Points |
|---|---|---|---|---|---|
| | 1st Place Votes 4 Points Each | 2nd Place Votes 3 Points Each | 3rd Place Votes 2 Points Each | 4th Place Votes 1 Point Each | |
| A | $18 \times 4 = 72$ | $0 \times 3 = 0$ | $7 \times 2 = 14$ | $26 \times 1 = 26$ | 112 |
| B | $7 \times 4 = 28$ | $14 \times 3 = 42$ | $9 \times 2 = 18$ | $21 \times 1 = 21$ | 109 |
| C | $19 \times 4 = 76$ | $3 \times 3 = 9$ | $25 \times 2 = 50$ | $4 \times 1 = 4$ | 139 |
| D | $7 \times 4 = 28$ | $34 \times 3 = 102$ | $10 \times 2 = 20$ | $0 \times 1 = 0$ | 150 |

   Candidate D has the most points and is the winner.

   (b) With four candidates, there are six pairwise comparisons.

| Candidates Compared | Preferred Candidate | Points Awarded |
|---|---|---|
| A and B | B is preferred 30 to 21 | B receives 1 point |
| A and C | C is preferred 29 to 22 | C receives 1 point |
| A and D | D is preferred 33 to 18 | D receives 1 point |
| B and C | C is preferred 40 to 11 | C receives 1 point |
| B and D | D is preferred 34 to 17 | D receives 1 point |
| C and D | D is preferred 32 to 19 | D receives 1 point |

Candidate D has the most points and is the winner. Notice that candidate D is preferred over candidates A, B, and C. Candidate D is the Condorcet candidate.

(c) Candidate A received 18 first-place votes, B received 7, C received 19, and D received 7. Eliminate candidate B or candidate D. They both received only 7 votes. (Note: In this case, it will not matter whether candidate B or D is eliminated first. The result of the election will be the same. You can verify this yourself.)
Eliminate candidate B and create a new preference table.

| Ranking | Number of Voters | | | | | |
|---|---|---|---|---|---|---|
| | 18 | 10 | 9 | 7 | 4 | 3 |
| 1st | A | C | C | D | D | D |
| 2nd | D | D | D | C | A | C |
| 3rd | C | A | A | A | C | A |

Candidate A received 18 first-place votes, C received 19, and D received 14. Eliminate candidate D and create a new preference table.

| Ranking | Number of Voters | | | | | |
|---|---|---|---|---|---|---|
| | 18 | 10 | 9 | 7 | 4 | 3 |
| 1st | A | C | C | C | A | C |
| 2nd | C | A | A | A | C | A |

Candidate C has 29 first-place votes which is a majority of the 51 votes, so C is the winner.

(d) From part (b), we know that candidate D is the Condorcet candidate yet the plurality with elimination method resulted in candidate C winning. In this case, plurality with elimination violated the head-to-head criterion.

11. (a) Candidate A received more first-place votes than either candidate B or C. Candidate A is the winner under the plurality method.

(b) Eliminate candidate C and create a new preference table.

| Ranking | Number of Voters | | |
|---|---|---|---|
| | 3 | 2 | 2 |
| 1st | A | B | B |
| 2nd | B | A | A |

Candidate B received 4 of the 7 votes and is selected by a majority.

(c) From part (a), we know candidate A was originally selected. From part (b) we know that when a losing candidate, in this case candidate C, was removed from the race, candidate B won the election. This is an example of the plurality method violating the irrelevant-alternatives criterion.

13. (a)

| Candidate | Points For Each Rank | | | Total Points |
|---|---|---|---|---|
| | 1st Place Votes 3 Points Each | 2nd Place Votes 2 Points Each | 3rd Place Votes 1 Point Each | |
| A | $2 \times 3 = 6$ | $3 \times 2 = 6$ | $0 \times 1 = 0$ | 12 |
| B | $3 \times 3 = 9$ | $0 \times 2 = 0$ | $2 \times 1 = 2$ | 11 |
| C | $0 \times 3 = 0$ | $2 \times 2 = 4$ | $3 \times 1 = 3$ | 7 |

Candidate A has the most points and is the winner.

(b) Candidate B has more first-place votes than candidates A or C. Candidate B wins the election by a majority of the five votes.

(c) From part (b), we know candidate B won by a majority of votes. However, from part (a), the Borda count method resulted in candidate A being selected as the winner. Therefore the Borda count method violated the majority criterion in this case.

(d) Notice from the preference table that candidate B is preferred over candidates A and C in pairwise comparisons. Candidate B is a Condorcet candidate. However, from part (a) we know the Borda count method resulted in candidate A winning. Therefore, in this case, the Borda count method also violated the head-to-head criterion.

**15.** Find the winner under the Borda count method.

| Candidate | Points For Each Rank | | | Total Points |
|---|---|---|---|---|
| | 1st Place Votes 3 Points Each | 2nd Place Votes 2 Points Each | 3rd Place Votes 1 Point Each | |
| A | $2 \times 3 = 6$ | $3 \times 2 = 6$ | $0 \times 1 = 0$ | 12 |
| B | $3 \times 3 = 9$ | $0 \times 2 = 0$ | $2 \times 1 = 2$ | 11 |
| C | $0 \times 3 = 0$ | $2 \times 2 = 4$ | $3 \times 1 = 3$ | 7 |

Candidate A has the most points and is the winner.
To show that the Borda count method violates the irrelevant-alternatives criterion, we must eliminate a losing candidate, hold the election again, and show that there would be a different winner. Suppose we eliminate candidate C. Create a new preference table.

| Ranking | Number of Voters | |
|---|---|---|
| | 3 | 2 |
| 1st | B | A |
| 2nd | A | B |

Based on the new preference table, find the winner using the Borda count method.

| Candidate | Points For Each Rank | | Total Points |
|---|---|---|---|
| | 1st Place Votes 2 Points Each | 2nd Place Votes 1 Point Each | |
| A | $2 \times 2 = 4$ | $3 \times 1 = 3$ | 7 |
| B | $3 \times 2 = 6$ | $2 \times 1 = 2$ | 8 |

Candidate B has the most points and is the winner.
When a losing candidate was eliminated, and the election was held again, a different winner emerged. In this case, the Borda count method violated the irrelevant-alternatives criterion.

**17.** Find the winner under the plurality with elimination method. Candidate D received the fewest first-place votes, so eliminate candidate D, and create a new preference table.

| Ranking | Number of Voters | | |
|---|---|---|---|
| | 2 | 2 | 1 |
| 1st | A | B | C |
| 2nd | B | C | A |
| 3rd | C | A | B |

A majority of first-place votes would be at least 3 votes. No candidate has a majority, so eliminate candidate C and create a new preference table.

| Ranking | Number of Voters | | |
|---|---|---|---|
| | 2 | 2 | 1 |
| 1st | A | B | A |
| 2nd | B | A | B |

Candidate A now has a majority of first-place votes and is the winner using the plurality with elimination method. Notice, however, that candidate D is preferred in all pairwise comparisons with candidates A, B, and C. Candidate D is the Condorcet candidate. In a case such as this in which there is a Condorcet candidate, yet the method used produces a different winner, we say there is a violation of the head-to-head criterion.

**19. (a)** Consider first-place votes. Candidate A received 6, candidate B received 5, and candidate C received 6. Eliminate candidate B and create a new preference table.

| Ranking | Number of Faculty | | | |
|---------|---|---|---|---|
| | 6 | 5 | 4 | 2 |
| 1st | A | A | C | C |
| 2nd | C | C | A | A |

Candidate A received a majority of votes and is the winner.

**(b)** Create a new preference table to reflect the changes.

| Ranking | Number of Faculty | | | |
|---------|---|---|---|---|
| | 6 | 5 | 4 | 2 |
| 1st | A | B | C | A |
| 2nd | C | A | B | C |
| 3rd | B | C | A | B |

Consider first-place votes. Candidate A received 8, candidate B received 5, and candidate C received 4. Eliminate candidate C and create a new preference table.

| Ranking | Number of Faculty | | | |
|---------|---|---|---|---|
| | 6 | 5 | 4 | 2 |
| 1st | A | B | B | A |
| 2nd | B | A | A | B |

Candidate B has a majority and is the winner.

**(c)** The original election resulted in candidate A winning. When a change was made that benefited candidate A, and the election was held again, then candidate B won. This is a violation of the monotonicity criterion.

**21. (a)** There are nine voters. Candidate A has more first-place votes than either of the other two candidates. Candidate A wins by a plurality.

**(b)** Candidate C had the fewest first-place votes and must be eliminated. Create a new preference table.

| Ranking | Number of Voters | | |
|---------|---|---|---|
| | 4 | 3 | 2 |
| 1st | A | B | A |
| 2nd | B | A | B |

Candidate A has a majority of first-place votes and is the winner.

**(c)** Create a preference table without candidate B. Notice that Candidate C has a majority of first-place votes and wins by both the plurality method and the plurality with elimination method.

| Ranking | Number of Voters | | |
|---------|---|---|---|
| | 4 | 3 | 2 |
| 1st | A | C | C |
| 2nd | C | A | A |

**(d)** Both the plurality method and the plurality with elimination method produced candidate A as a winner when there were three candidates in the race. When candidate B was eliminated, both methods produced candidate C as the winner. This is a violation of the irrelevant-alternatives criterion.

**23. (a)** There are four programs, so there are six pairwise comparisons that can be made. Program A is preferred to B and D but not to C. Program B is not preferred to any other program. Program C is preferred to A and B but not to D. Program D is preferred to B and C but not to A. There is no Condorcet program.

**(b)** Program A received more first-place votes than any other program and is the winner by a plurality. There can be no violation of the head-to-head criterion in this case because there was no Condorcet program. However, if a losing program is removed, such as program B, and the election is held again, then program C has a more first-place votes and wins. This is a violation of the irrelevant-alternatives criterion.

(c)

| Program | Points For Each Rank | | | | Total Points |
|---|---|---|---|---|---|
| | 1st Place Votes 4 Points Each | 2nd Place Votes 3 Points Each | 3rd Place Votes 2 Points Each | 4th Place Votes 1 Point Each | |
| A | $14 \times 4 = 56$ | $6 \times 3 = 18$ | $7 \times 2 = 14$ | $12 \times 1 = 12$ | 100 |
| B | $12 \times 4 = 48$ | $0 \times 3 = 0$ | $0 \times 2 = 0$ | $27 \times 1 = 27$ | 75 |
| C | $6 \times 4 = 24$ | $19 \times 3 = 57$ | $14 \times 2 = 28$ | $0 \times 1 = 0$ | 109 |
| D | $7 \times 4 = 28$ | $14 \times 3 = 42$ | $18 \times 2 = 36$ | $0 \times 1 = 0$ | 106 |

Program C has the most points and is the winner under the Borda count method. In order for there to be a violation of the majority criterion, one program would have to have a majority of first-place votes. No program had a majority, so no violation occurred. There also can be no violation of the head-to-head criterion in this case because there was no Condorcet program.

25. (a)

| Candidate | Points For Each Rank | | | Total Points |
|---|---|---|---|---|
| | 1st Place Votes 3 Points Each | 2nd Place Votes 2 Points Each | 3rd Place Votes 1 Point Each | |
| A | $4 \times 3 = 12$ | $5 \times 2 = 10$ | $3 \times 1 = 3$ | 25 |
| B | $3 \times 3 = 9$ | $4 \times 2 = 8$ | $5 \times 1 = 5$ | 22 |
| C | $5 \times 3 = 15$ | $3 \times 2 = 6$ | $4 \times 1 = 4$ | 25 |

Candidates A and C both have 25 points while candidate B has 22 points. Candidates A and C tie.

(b) From part (a), we know candidate B had the lowest Borda count total. Nanson's method requires that we eliminate candidate B and recalculate the points. A new preference table is included next without candidate B.

| Ranking | Number of Voters | | |
|---|---|---|---|
| | 5 | 4 | 3 |
| 1st | C | A | C |
| 2nd | A | C | A |

Calculate the new point totals.

| Candidate | Points For Each Rank | | Total Points |
|---|---|---|---|
| | 1st Place Votes 2 Points Each | 2nd Place Votes 1 Point Each | |
| A | $4 \times 2 = 8$ | $8 \times 1 = 8$ | 16 |
| C | $8 \times 2 = 16$ | $4 \times 1 = 4$ | 20 |

Candidate C has the most points and is the winner.

(c) From part (b), we know candidate C wins under Nanson's method. Suppose candidate A is eliminated from the race. Create a new preference table and find the winner using Nanson's method.

| Ranking | Number of Voters | | |
|---|---|---|---|
| | 5 | 4 | 3 |
| 1st | C | B | B |
| 2nd | B | C | C |

Determine the number of points each candidate will receive.

| Candidate | Points For Each Rank | | Total Points |
|---|---|---|---|
| | 1st Place Votes 2 Points Each | 2nd Place Votes 1 Point Each | |
| B | $7 \times 2 = 14$ | $5 \times 1 = 5$ | 19 |
| C | $5 \times 2 = 10$ | $7 \times 1 = 7$ | 17 |

Candidate C has fewer points and is eliminated. Candidate B is the winner. This demonstrates that Nanson's method does not always satisfy the irrelevant-alternatives criterion. When three candidates were in the race, candidate C won. When a losing candidate was eliminated, the results changed.

27. **(a)** Add the number of votes in the Round 1 column. There were 1048 delegates.

**(b)** In the first three rounds, Barker, Sundwall, and Lee were eliminated. When Barker was eliminated, one extra vote showed up in Wyatt's Round 2 total. When Sundwall was eliminated, one extra vote showed up in Gross' Round 3 total. When Lee was eliminated, eleven votes had to be redistributed. By examining the Round 4 totals, we can discover who received those eleven votes. One vote went to Jacobs, 3 votes went to Probasco, 1 vote went to Gross, 2 votes went to Wyatt, 1 vote went to McCall, and 3 votes went to Garn. For the 13 ballots in which the votes were redistributed, the first-place and second-place candidates were listed as follows:
Barker, Wyatt (1 ballot)
Sundwall, Gross (1 ballot)
Lee, Jacobs (1 ballot)
Lee, Probasco (3 ballots)
Lee, Gross (1 ballot)
Lee, Wyatt (2 ballots)
Lee, McCall (1 ballot)
Lee, Garn (3 ballots)

**(c)** If every voter ranked every candidate, then every round of voting would have the same number votes distributed to the candidates. Count the total number of votes cast in each round. The vote totals are 1048 for Round 1 through Round 6, so all of the voters ranked six of the candidates. In Round 7, there were only 1043 votes cast. Five voters failed to rank more than six candidates. In Round 9, there were only 1034 votes cast. Fourteen voters failed to rank every candidate.

**(d)** Bishop had the most votes in Round 1 and would have been the winner. If Wyatt was disqualified and those 136 votes went to McCall, then McCall would have 136 + 182 = 318 votes. In this case, McCall would win by a plurality. This demonstrates that the plurality method can violate the irrelevant-alternatives criterion.

29. Count the number of votes received by each candidate. Althaus received a total of 152 votes. Wirt received 218. Medley received 147. Brown received 161. Kofoid received 78. There are three Trustee positions to fill. The candidates with the top three vote totals are Wirt, Brown, and Althaus.

31. Count the number of votes received by each candidate. Deegear received 950 votes, Maio received 793 votes, Soderstrom received 1129 votes, and Vaca received 1247 votes. The three positions will be filled by Vaca, Soderstrom, and Deegear.

33. **(a)** Count the number of votes for each of the main dishes. Pizza has 28 votes; tacos, 49 votes; hamburgers, 37 votes; Chinese food, 66 votes; and sub sandwiches, 53 votes. The two types of food that will be served at the party are Chinese food and sub sandwiches.

**(b)** If the eleven students who voted for all of the options had left their ballots completely blank instead, then each of the totals would decrease by 11, but the same two options would be selected. Blank ballots are valid ballots. They indicate that a voter does not approve of any of the choices.

35. Flavor C has the most first-place votes and wins by a plurality. The plurality method can violate the head-to-head criterion and the irrelevant-alternatives criterion.
To determine if the plurality method violates the head-to-head criterion, examine every pairwise comparison and determine if there is a Condorcet flavor. Notice that flavor B is preferred over every other flavor, that is, B wins in every pairwise comparison yet C is selected by a plurality. The head-to-head criterion has been violated in this case.
To determine if the plurality method violates the irrelevant-alternatives criterion, eliminate a losing flavor and see if the results change. If flavor A is eliminated, flavor B receives ten more votes for first place for a total of 23 votes and a plurality win. Eliminating a losing flavor changed the results. The irrelevant-alternatives criterion has been violated.

**37.** Use the Borda count method to determine the winning flavor.

| Flavor | Points For Each Rank | | | | Total Points |
|---|---|---|---|---|---|
| | 1st Place Votes 4 Points Each | 2nd Place Votes 3 Points Each | 3rd Place Votes 2 Points Each | 4th Place Votes 1 Point Each | |
| A | $10 \times 4 = 40$ | $25 \times 3 = 75$ | $7 \times 2 = 14$ | $8 \times 1 = 8$ | 137 |
| B | $13 \times 4 = 52$ | $17 \times 3 = 51$ | $20 \times 2 = 40$ | $0 \times 1 = 0$ | 143 |
| C | $15 \times 4 = 60$ | $0 \times 3 = 0$ | $10 \times 2 = 20$ | $25 \times 1 = 25$ | 105 |
| D | $12 \times 4 = 48$ | $8 \times 3 = 24$ | $13 \times 2 = 26$ | $17 \times 1 = 17$ | 115 |

Flavor B has the most points and is the winner using the Borda count method. The Borda count method can violate the majority criterion, the head-to-head criterion, and the irrelevant-alternatives criterion. In this case, no flavor had a majority of the first-place votes, so there is no violation of the majority criterion. From problem 35, we know flavor B is the Condorcet flavor, and flavor B won using the Borda count method, so there is no violation of the head-to-head criterion. To determine if the irrelevant-alternatives criterion is violated, a losing flavor must be eliminated. If the Borda count method is applied again, and there is a change in the winner, then there is a violation. In this case there is no violation.

## Solutions to Odd-Numbered Problems in Section 3.3

**1. (a)** The total weight is $3 + 5 + 7 + 12 = 27$. Half of 27 is 13.5. A simple majority is 14 votes.

   **(b)** The total weight is 27, and $\frac{2}{3} \times 27 = 18$. Therefore, 18 votes is the minimum needed to pass a motion.

   **(c)** The weighted voting system would be $[17 \mid 12, 7, 5, 3]$.

**3.** A simple majority is one more than half of the total weight if half the total weight is a whole number. If half of the total weight is not a whole number, then round up to the nearest whole number.

   **(a)** The total weight in the system is $6 + 5 + 5 + 3 + 3 + 2 + 1 + 1 = 26$. Half of the total weight is 13. A simple majority is needed to pass a measure, so the quota is 14. The weighted voting system is expressed as $[14 \mid 6, 5, 5, 3, 3, 2, 1, 1]$.

   **(b)** The total weight in the system is $4 + 8 + 5 + 3 + 2 + 5 + 2 = 29$. Half of the total weight is 14.5. A simple majority is needed to pass a measure, so the quota is 15. The weighted voting system is expressed as $[15 \mid 8, 5, 5, 4, 3, 2, 2]$.

   **(c)** The total weight in the system is $10 + 5 + 5 + 5 + 3 + 3 + 3 = 34$. Half of the total weight is 17. A simple majority is needed to pass a measure, so the quota is 18. The weighted voting system is expressed as $[18 \mid 10, 5, 5, 5, 3, 3, 3]$.

   **(d)** The total weight in the system is $1 + 2 + 2 + 5 + 5 + 7 = 22$. Half of the total weight is 11. A simple majority is needed to pass a measure, so the quota is 12. The weighted voting system is expressed as $[12 \mid 7, 5, 5, 2, 2, 1]$.

**5. (a)** The total weight in the system is $8 + 5 + 5 + 3 + 3 + 2 = 26$ votes. The quota for this weighted voting system is 60% of the total number of votes. The quota is 60% of $26 = 0.60 \times 26 = 15.6$. Round the quota up to 16 votes. The weighted voting system is expressed as $[16 \mid 8, 5, 5, 3, 3, 2]$.

   **(b)** The total weight in the system is $9 + 6 + 5 + 5 + 3 + 2 + 2 = 32$ votes. The quota for this weighted voting system is 67% of the total number of votes. The quota is 67% of $32 = 0.67 \times 32 = 21.44$. Round the quota up to 22 votes. The weighted voting system is expressed as $[22 \mid 9, 6, 5, 5, 3, 2, 2]$.

   **(c)** The total weight in the system is $7 + 5 + 5 + 5 + 3 + 2 = 27$ votes. The quota for this weighted voting system is 75% of the total number of votes. The quota is 75% of $27 = 0.75 \times 27 = 20.25$. Round the quota up to 21 votes. The weighted voting system is expressed as $[21 \mid 7, 5, 5, 5, 3, 2]$.

(d) The total weight in the system is $5 + 5 + 4 + 4 + 3 + 3 = 24$ votes. The quota for this weighted voting system is 60% of the total number of votes. The quota is 60% of $24 = 0.60 \times 24 = 14.4$. Round the quota up to 15 votes. The weighted voting system is expressed as $[15 \mid 5, 5, 4, 4, 3, 3]$.

7. (a) Add the weights of the three voters who voted Yes: $W_1 + W_2 + W_3 = 7 + 6 + 6 = 19$. The quota is 10, so the motion will pass.
   (b) Add the weights of the voters who voted Yes: $W_3 + W_4 = 6 + 5 = 11$. The quota is 10, so the motion will pass.
   (c) The total weight of the system is 24. The quota is set too low. A measure can pass, as it did in part (b), even though two voters with a total weight of 13 voted No.

9. Answers will vary. The quota will have to be the sum of the weights so that all of the voters would have to agree and vote the same way to pass a motion. Suppose each of the six people has 1 vote. The weighted voting system would be given as $[6 \mid 1, 1, 1, 1, 1, 1]$.

11. A coalition is a winning coalition if it has enough votes to pass a measure.
    (a) Add the weights of the voters in the coalition: $W_1 + W_4 + W_6 = 10 + 7 + 4 = 21$. The quota is 21, and the coalition has exactly 21 votes. This is a winning coalition.
    (b) Add the weights of the voters in the coalition: $W_2 + W_3 + W_6 = 8 + 7 + 4 = 19$. The quota is 21, and the coalition has less than 21 votes. This is a losing coalition.
    (c) Add the weights of the voters in the coalition: $W_2 + W_3 + W_4 = 8 + 7 + 7 = 22$. The quota is 21, and the coalition has more than 21 votes. This is a winning coalition.
    (d) Add the weights of the voters in the coalition: $W_3 + W_4 + W_5 + W_6 = 7 + 7 + 4 + 4 = 22$. The quota is 21, and the coalition has more than 21 votes. This is a winning coalition.

13. A coalition is a winning coalition if it has enough votes to pass a measure.
    (a) Add the weights of the voters in the coalition: $W_1 + W_2 + W_3 = 8 + 4 + 3 = 15$. The quota is 15, and the coalition has exactly 15 votes. This is a winning coalition.
    (b) Add the weights of the voters in the coalition: $W_2 + W_3 + W_6 = 4 + 3 + 2 = 9$. The quota is 15, and the coalition has less than 15 votes. This is a losing coalition.
    (c) Add the weights of the voters in the coalition: $W_2 + W_3 + W_4 + W_5 = 4 + 3 + 3 + 2 = 12$. The quota is 15, and the coalition has less than 15 votes. This is a losing coalition.
    (d) Add the weights of the voters in the coalition: $W_2 + W_3 + W_4 + W_5 + W_6 = 4 + 3 + 3 + 2 + 2 = 14$. The quota is 15, and the coalition has less than 15 votes. This is a losing coalition.

15. If there are $n$ voters in a weighted voting system, then there are exactly $2^n - 1$ possible coalitions.
    (a) There are 8 voters, so let $n = 8$. There are $2^8 - 1 = 256 - 1 = 255$ possible coalitions.
    (b) There are 10 voters, so let $n = 10$. There are $2^{10} - 1 = 1024 - 1 = 1023$ possible coalitions.

17. (a) There are 3 voters, so let $n = 3$. There are $2^3 - 1 = 8 - 1 = 7$ possible coalitions. In this system, the quota is 4 votes.

| Coalition | Sum of the Weights | Winning or Losing |
|---|---|---|
| $\{P_1\}$ | 3 | Losing |
| $\{P_2\}$ | 2 | Losing |
| $\{P_3\}$ | 1 | Losing |
| $\{P_1, P_2\}$ | $3 + 2 = 5$ | Winning |
| $\{P_1, P_3\}$ | $3 + 1 = 4$ | Winning |
| $\{P_2, P_3\}$ | $2 + 1 = 3$ | Losing |
| $\{P_1, P_2, P_3\}$ | $3 + 2 + 1 = 6$ | Winning |

**(b)** There are 4 voters, so let $n = 4$. There are $2^4 - 1 = 16 - 1 = 15$ possible coalitions. In this system, the quota is 26 votes.

| Coalition | Sum of the Weights | Winning or Losing |
|---|---|---|
| $\{P_1\}$ | 20 | Losing |
| $\{P_2\}$ | 15 | Losing |
| $\{P_3\}$ | 10 | Losing |
| $\{P_4\}$ | 5 | Losing |
| $\{P_1, P_2\}$ | $20 + 15 = 35$ | Winning |
| $\{P_1, P_3\}$ | $20 + 10 = 30$ | Winning |
| $\{P_1, P_4\}$ | $20 + 5 = 25$ | Losing |
| $\{P_2, P_3\}$ | $15 + 10 = 25$ | Losing |
| $\{P_2, P_4\}$ | $15 + 5 = 20$ | Losing |
| $\{P_3, P_4\}$ | $10 + 5 = 15$ | Losing |
| $\{P_1, P_2, P_3\}$ | $20 + 15 + 10 = 45$ | Winning |
| $\{P_1, P_2, P_4\}$ | $20 + 15 + 5 = 40$ | Winning |
| $\{P_1, P_3, P_4\}$ | $20 + 10 + 5 = 35$ | Winning |
| $\{P_2, P_3, P_4\}$ | $15 + 10 + 5 = 30$ | Winning |
| $\{P_1, P_2, P_3, P_4\}$ | $20 + 15 + 10 + 5 = 50$ | Winning |

**19. (a)** Voter 1 controls 10 out of the 19 votes, so Voter 1 controls $\frac{10}{19} \times 100\% \approx 52.6\%$ of the votes.

Voter 2 controls 5 out of the 19 votes, so Voter 2 controls $\frac{5}{19} \times 100\% \approx 26.3\%$ of the votes.

Voter 3 controls 4 out of the 19 votes, so Voter 3 controls $\frac{4}{19} \times 100\% \approx 21.1\%$ of the votes.

**(b)** List all of the coalitions and the weights, and determine which of them are winning coalitions. There are 3 voters so there are $2^n - 1 = 2^3 - 1 = 8 - 1 = 7$ possible coalitions.

| Coalition | Sum of the Weights | Winning or Losing |
|---|---|---|
| $\{P_1\}$ | 10 | Losing |
| $\{P_2\}$ | 5 | Losing |
| $\{P_3\}$ | 4 | Losing |
| $\{P_1, P_2\}$ | $10 + 5 = 15$ | Losing |
| $\{P_1, P_3\}$ | $10 + 4 = 14$ | Losing |
| $\{P_2, P_3\}$ | $5 + 4 = 9$ | Losing |
| $\{P_1, P_2, P_3\}$ | $10 + 5 + 4 = 19$ | Winning |

There is one winning coalition, $\{P_1, P_2, P_3\}$. If any one of the three voters leaves the coalition, it becomes a losing coalition. Therefore, each voter is a critical voter once in this coalition. Therefore, the total Banzhaf power in the system is 3. To find the Banzhaf power index for each voter, divide the number of times a voter was a critical voter by the total Banzhaf power in the system. In this case, each voter was critical one time, so they each have the same Banzhaf power index of $\frac{1}{3}$. The Banzhaf power index may be expressed as a fraction or a percent. We can also

say that each voter has a Banzhaf power index of $33\frac{1}{3}\%$.

(c) The percentages from part (a) are different for each voter. They each control a different number of votes; however, they each have the same Banzhaf power index. The weight of a voter is not a good measure of the voter's power.

21. (a) Prior to 1999, murder convictions required a unanimous vote. All 12 jurors had to agree. Each of the 12 jurors had one vote so the quota was 12. The weighted voting system would be given as [12 | 1, 1, 1, 1, 1, 1, 1, 1, 1, 1, 1, 1].

   (b) There is only one winning coalition. It is the coalition containing all twelve of the jurors.

   (c) Every one of the jurors has veto power. Any one of the jurors can decide to vote No while all of the eleven other jurors vote Yes, and the jury will not be able to produce a conviction.

   (d) There is one winning coalition. It is the coalition containing all twelve of the jurors. Each juror is a critical voter once. The total Banzhaf power is 12. Each juror's Banzhaf power index is $\frac{1}{12}$.

23. (a) List all of the coalitions and determine which are winning coalitions. The quota is 8.

| Coalition | Sum of the Weights | Winning or Losing |
|---|---|---|
| $\{P_1\}$ | 5 | Losing |
| $\{P_2\}$ | 4 | Losing |
| $\{P_3\}$ | 3 | Losing |
| $\{P_1, P_2\}$ | $5 + 4 = 9$ | Winning |
| $\{P_1, P_3\}$ | $5 + 3 = 8$ | Winning |
| $\{P_2, P_3\}$ | $4 + 3 = 7$ | Losing |
| $\{P_1, P_2, P_3\}$ | $5 + 4 + 3 = 12$ | Winning |

There are three winning coalitions. Consider each of the winning coalitions and determine which voters are critical voters in each.

| Winning Coalition | Coalition Without One Voter | Weight of New Coalition | Winning or Losing | Conclusion |
|---|---|---|---|---|
| $\{P_1, P_2\}$ | $\{P_1\}$ | 5 | Losing | $P_2$ is Critical |
|  | $\{P_2\}$ | 4 | Losing | $P_1$ is Critical |
| $\{P_1, P_3\}$ | $\{P_1\}$ | 5 | Losing | $P_3$ is Critical |
|  | $\{P_3\}$ | 3 | Losing | $P_1$ is Critical |
| $\{P_1, P_2, P_3\}$ | $\{P_1, P_2\}$ | $5 + 4 = 9$ | Winning |  |
|  | $\{P_1, P_3\}$ | $5 + 3 = 8$ | Winning |  |
|  | $\{P_2, P_3\}$ | $4 + 3 = 7$ | Losing | $P_1$ is Critical |

We see from the table that $P_1$ is a critical voter 3 times, $P_2$ is a critical voter 1 time, and $P_3$ is a critical voter 1 time. The total Banzhaf power in the system is $3 + 1 + 1 = 5$. To find the Banzhaf power index for each voter, divide the number of times the voter was critical by the total Banzhaf power.

| Voter | Banzhaf Power Index |
|-------|---------------------|
| $P_1$ | $\dfrac{3}{5}$ or 60% |
| $P_2$ | $\dfrac{1}{5}$ or 20% |
| $P_3$ | $\dfrac{1}{5}$ or 20% |

(i) There are no dictators in this voting system. (ii) Each voter is a critical voter at least once, so there are no dummies. (iii) $P_1$ has veto power.

**(b)** List all of the coalitions and determine which are winning coalitions. The quota is 25.

| Coalition | Sum of the Weights | Winning or Losing |
|-----------|--------------------|-------------------|
| $\{P_1\}$ | 14 | Losing |
| $\{P_2\}$ | 13 | Losing |
| $\{P_3\}$ | 12 | Losing |
| $\{P_4\}$ | 8 | Losing |
| $\{P_1, P_2\}$ | $14 + 13 = 27$ | Winning |
| $\{P_1, P_3\}$ | $14 + 12 = 26$ | Winning |
| $\{P_1, P_4\}$ | $14 + 8 = 22$ | Losing |
| $\{P_2, P_3\}$ | $13 + 12 = 25$ | Winning |
| $\{P_2, P_4\}$ | $13 + 8 = 21$ | Losing |
| $\{P_3, P_4\}$ | $12 + 8 = 20$ | Losing |
| $\{P_1, P_2, P_3\}$ | $14 + 13 + 12 = 39$ | Winning |
| $\{P_1, P_2, P_4\}$ | $14 + 13 + 8 = 35$ | Winning |
| $\{P_1, P_3, P_4\}$ | $14 + 12 + 8 = 34$ | Winning |
| $\{P_2, P_3, P_4\}$ | $13 + 12 + 8 = 33$ | Winning |
| $\{P_1, P_2, P_3, P_4\}$ | $14 + 13 + 12 + 8 = 47$ | Winning |

There are eight winning coalitions. Consider each of the winning coalitions and determine which voters are critical voters in each.

| Winning Coalition | Coalition Without One Voter | Weight of New Coalition | Winning or Losing | Conclusion |
|-------------------|------------------------------|--------------------------|-------------------|------------|
| $\{P_1, P_2\}$ | $\{P_2\}$ | 13 | Losing | $P_1$ is Critical |
|                | $\{P_1\}$ | 14 | Losing | $P_2$ is Critical |
| $\{P_1, P_3\}$ | $\{P_3\}$ | 12 | Losing | $P_1$ is Critical |
|                | $\{P_1\}$ | 14 | Losing | $P_3$ is Critical |
| $\{P_2, P_3\}$ | $\{P_3\}$ | 12 | Losing | $P_2$ is Critical |
|                | $\{P_2\}$ | 13 | Losing | $P_3$ is Critical |
| $\{P_1, P_2, P_3\}$ | $\{P_2, P_3\}$ | $13 + 12 = 25$ | Winning | |
|                     | $\{P_1, P_3\}$ | $14 + 12 = 26$ | Winning | |
|                     | $\{P_1, P_2\}$ | $14 + 13 = 27$ | Winning | |
| $\{P_1, P_2, P_4\}$ | $\{P_2, P_4\}$ | $13 + 8 = 21$ | Losing | $P_1$ is Critical |
|                     | $\{P_1, P_4\}$ | $14 + 8 = 22$ | Losing | $P_2$ is Critical |
|                     | $\{P_1, P_2\}$ | $14 + 13 = 27$ | Winning | |

| | | | | |
|---|---|---|---|---|
| {$P_1, P_3, P_4$} | {$P_3, P_4$} | 12 + 8 = 20 | Losing | $P_1$ is Critical |
| | {$P_1, P_4$} | 14 + 8 = 22 | Losing | $P_3$ is Critical |
| | {$P_1, P_3$} | 14 + 12 = 26 | Winning | |
| {$P_2, P_3, P_4$} | {$P_3, P_4$} | 12 + 8 = 20 | Losing | $P_2$ is Critical |
| | {$P_2, P_4$} | 13 + 8 = 21 | Losing | $P_3$ is Critical |
| | {$P_2, P_3$} | 13 + 12 = 25 | Winning | |
| {$P_1, P_2, P_3, P_4$} | {$P_2, P_3, P_4$} | 13 + 12 + 8 = 33 | Winning | |
| | {$P_1, P_3, P_4$} | 14 + 12 + 8 = 34 | Winning | |
| | {$P_1, P_2, P_4$} | 14 + 13 + 8 = 35 | Winning | |
| | {$P_1, P_2, P_3$} | 14 + 13 + 12 = 39 | Winning | |

The power of each voter is the number of times the voter is critical. $P_1$ is critical 4 times, $P_2$ is critical 4 times, and $P_3$ is critical 4 times. The total Banzhaf power in the system is $4 + 4 + 4 = 12$. To find the Banzhaf power index for each voter, divide the number of times the voter was critical by the total Banzhaf power.

| Voter | Banzhaf Power Index |
|---|---|
| $P_1$ | $\dfrac{4}{12} = \dfrac{1}{3}$ or 33.3% |
| $P_2$ | $\dfrac{4}{12} = \dfrac{1}{3}$ or 33.3% |
| $P_3$ | $\dfrac{4}{12} = \dfrac{1}{3}$ or 33.3% |
| $P_4$ | $\dfrac{0}{12} = 0$ or 0% |

(i) There are no dictators. (ii) $P_4$ is never a critical voter so $P_4$ is a dummy. (iii) Every coalition of three voters is a winning coalition, so there is no voter with veto power.

(c) List all of the coalitions and determine which are winning coalitions. The quota is 7.

| Coalition | Sum of the Weights | Winning or Losing |
|---|---|---|
| {$P_1$} | 7 | Winning |
| {$P_2$} | 2 | Losing |
| {$P_3$} | 2 | Losing |
| {$P_4$} | 2 | Losing |
| {$P_1, P_2$} | 7 + 2 = 9 | Winning |
| {$P_1, P_3$} | 7 + 2 = 9 | Winning |
| {$P_1, P_4$} | 7 + 2 = 9 | Winning |
| {$P_2, P_3$} | 2 + 2 = 4 | Losing |
| {$P_2, P_4$} | 2 + 2 = 4 | Losing |
| {$P_3, P_4$} | 2 + 2 = 4 | Losing |
| {$P_1, P_2, P_3$} | 7 + 2 + 2 = 11 | Winning |
| {$P_1, P_2, P_4$} | 7 + 2 + 2 = 11 | Winning |
| {$P_1, P_3, P_4$} | 7 + 2 + 2 = 11 | Winning |
| {$P_2, P_3, P_4$} | 2 + 2 + 2 = 6 | Losing |
| {$P_1, P_2, P_3, P_4$} | 7 + 2 + 2 + 2 = 13 | Winning |

There are eight winning coalitions. Consider each of the winning coalitions and determine which voters are critical voters in each.

| Winning Coalition | Coalition Without One Voter | Weight of New Coalition | Winning or Losing | Conclusion |
|---|---|---|---|---|
| $\{P_1\}$ | { } | 0 | Losing | $P_1$ is Critical |
| $\{P_1, P_2\}$ | $\{P_2\}$ | 2 | Losing | $P_1$ is Critical |
|  | $\{P_1\}$ | 7 | Winning |  |
| $\{P_1, P_3\}$ | $\{P_3\}$ | 2 | Losing | $P_1$ is Critical |
|  | $\{P_1\}$ | 7 | Winning |  |
| $\{P_1, P_4\}$ | $\{P_4\}$ | 2 | Losing | $P_1$ is Critical |
|  | $\{P_1\}$ | 7 | Winning |  |
| $\{P_1, P_2, P_3\}$ | $\{P_2, P_3\}$ | $2 + 2 = 4$ | Losing | $P_1$ is Critical |
|  | $\{P_1, P_3\}$ | $7 + 2 = 9$ | Winning |  |
|  | $\{P_1, P_2\}$ | $7 + 2 = 9$ | Winning |  |
| $\{P_1, P_2, P_4\}$ | $\{P_2, P_4\}$ | $2 + 2 = 4$ | Losing | $P_1$ is Critical |
|  | $\{P_1, P_4\}$ | $7 + 2 = 9$ | Winning |  |
|  | $\{P_1, P_2\}$ | $7 + 2 = 9$ | Winning |  |
| $\{P_1, P_3, P_4\}$ | $\{P_3, P_4\}$ | $2 + 2 = 4$ | Losing | $P_1$ is Critical |
|  | $\{P_1, P_4\}$ | $7 + 2 = 9$ | Winning |  |
|  | $\{P_1, P_3\}$ | $7 + 2 = 9$ | Winning |  |
| $\{P_1, P_2, P_3, P_4\}$ | $\{P_2, P_3, P_4\}$ | $2 + 2 + 2 = 6$ | Losing | $P_1$ is Critical |
|  | $\{P_1, P_3, P_4\}$ | $7 + 2 + 2 = 11$ | Winning |  |
|  | $\{P_1, P_2, P_4\}$ | $7 + 2 + 2 = 11$ | Winning |  |
|  | $\{P_1, P_2, P_3\}$ | $7 + 2 + 2 = 11$ | Winning |  |

The power of each voter is the number of times the voter is critical. $P_1$ is critical 8 times, and no other voter is ever a critical voter. The total Banzhaf power in the system is 8. To find the Banzhaf power index for each voter, divide the number of times the voter was critical by the total Banzhaf power.

| Voter | Banzhaf Power Index |
|---|---|
| $P_1$ | $\dfrac{8}{8} = 1$ or 100% |
| $P_2$ | $\dfrac{0}{8} = 0$ or 0% |
| $P_3$ | $\dfrac{0}{8} = 0$ or 0% |
| $P_4$ | $\dfrac{0}{8} = 0$ or 0% |

(i) $P_1$ has all of the power in the system and is a dictator. (ii) If a system contains a dictator, all of the other voters are automatically dummies. $P_2$, $P_3$, and $P_4$ are dummies. (iii) Every dictator has veto power, so $P_1$ has veto power.

**25.** Create a table of all of the coalitions and the weights. The table of coalitions will be used for each part of the problem. In parts (a) through (e), the quotas are given, so the quota in each part will determine whether a coalition is a winning or a losing coalition.

| Coalition | Sum of the Weights | Winning or Losing |
|---|---|---|
| $\{P_1\}$ | 5 | |
| $\{P_2\}$ | 3 | |
| $\{P_3\}$ | 1 | |
| $\{P_1, P_2\}$ | $5 + 3 = 8$ | |
| $\{P_1, P_3\}$ | $5 + 1 = 6$ | |
| $\{P_2, P_3\}$ | $3 + 1 = 4$ | |
| $\{P_1, P_2, P_3\}$ | $5 + 3 + 1 = 9$ | |

**(a)** The quota is 5.

| Coalition | Sum of the Weights | Winning or Losing |
|---|---|---|
| $\{P_1\}$ | 5 | Winning |
| $\{P_2\}$ | 3 | Losing |
| $\{P_3\}$ | 1 | Losing |
| $\{P_1, P_2\}$ | $5 + 3 = 8$ | Winning |
| $\{P_1, P_3\}$ | $5 + 1 = 6$ | Winning |
| $\{P_2, P_3\}$ | $3 + 1 = 4$ | Losing |
| $\{P_1, P_2, P_3\}$ | $5 + 3 + 1 = 9$ | Winning |

There are four winning coalitions. Consider each of the winning coalitions and determine which voters are critical voters in each.

| Winning Coalition | Coalition Without One Voter | Weight of New Coalition | Winning or Losing | Conclusion |
|---|---|---|---|---|
| $\{P_1\}$ | $\{\ \}$ | 0 | Losing | $P_1$ is Critical |
| $\{P_1, P_2\}$ | $\{P_2\}$ | 3 | Losing | $P_1$ is Critical |
| | $\{P_1\}$ | 5 | Winning | |
| $\{P_1, P_3\}$ | $\{P_3\}$ | 1 | Losing | $P_1$ is Critical |
| | $\{P_1\}$ | 5 | Winning | |
| $\{P_1, P_2, P_3\}$ | $\{P_2, P_3\}$ | $3 + 1 = 4$ | Losing | $P_1$ is Critical |
| | $\{P_1, P_3\}$ | $5 + 1 = 6$ | Winning | |
| | $\{P_1, P_2\}$ | $5 + 3 = 8$ | Winning | |

The power of each voter is the number of times the voter is critical. $P_1$ is a critical voter 4 times, and no other voter is ever a critical voter. The total Banzhaf power in the system is 4. To find the Banzhaf power index for each voter, divide the number of times the voter was critical by the total Banzhaf power.

| Voter | Banzhaf Power Index |
|---|---|
| $P_1$ | $\dfrac{4}{4} = 1$ or 100% |
| $P_2$ | $\dfrac{0}{4}$ or 0% |
| $P_3$ | $\dfrac{0}{4}$ or 0% |

Notice $P_1$ is a dictator and the other two voters are dummies.

**(b)** The quota is 6.

| Coalition | Sum of the Weights | Winning or Losing |
|---|---|---|
| $\{P_1\}$ | 5 | Losing |
| $\{P_2\}$ | 3 | Losing |
| $\{P_3\}$ | 1 | Losing |
| $\{P_1, P_2\}$ | 5 + 3 = 8 | Winning |
| $\{P_1, P_3\}$ | 5 + 1 = 6 | Winning |
| $\{P_2, P_3\}$ | 3 + 1 = 4 | Losing |
| $\{P_1, P_2, P_3\}$ | 5 + 3 + 1 = 9 | Winning |

There are three winning coalitions. Consider each of the winning coalitions and determine which voters are critical voters in each.

| Winning Coalition | Coalition Without One Voter | Weight of New Coalition | Winning or Losing | Conclusion |
|---|---|---|---|---|
| $\{P_1, P_2\}$ | $\{P_2\}$ | 3 | Losing | $P_1$ is Critical |
|  | $\{P_1\}$ | 5 | Losing | $P_2$ is Critical |
| $\{P_1, P_3\}$ | $\{P_3\}$ | 1 | Losing | $P_1$ is Critical |
|  | $\{P_1\}$ | 5 | Losing | $P_3$ is Critical |
| $\{P_1, P_2, P_3\}$ | $\{P_2, P_3\}$ | 3 + 1 = 4 | Losing | $P_1$ is Critical |
|  | $\{P_1, P_3\}$ | 5 + 1 = 6 | Winning |  |
|  | $\{P_1, P_2\}$ | 5 + 3 = 8 | Winning |  |

The power of each voter is the number of times the voter is critical. $P_1$ is a critical voter 3 times, $P_2$ is a critical voter 1 time, $P_3$ is a critical voter 1 time. The total Banzhaf power in the system is 5. To find the Banzhaf power index for each voter, divide the number of times the voter was critical by the total Banzhaf power.

| Voter | Banzhaf Power Index |
|---|---|
| $P_1$ | $\frac{3}{5}$ or 60% |
| $P_2$ | $\frac{1}{5}$ or 20% |
| $P_3$ | $\frac{1}{5}$ or 20% |

**(c)** The quota is 7.

| Coalition | Sum of the Weights | Winning or Losing |
|---|---|---|
| $\{P_1\}$ | 5 | Losing |
| $\{P_2\}$ | 3 | Losing |
| $\{P_3\}$ | 1 | Losing |
| $\{P_1, P_2\}$ | 5 + 3 = 8 | Winning |
| $\{P_1, P_3\}$ | 5 + 1 = 6 | Losing |
| $\{P_2, P_3\}$ | 3 + 1 = 4 | Losing |
| $\{P_1, P_2, P_3\}$ | 5 + 3 + 1 = 9 | Winning |

There are two winning coalitions. Consider each of the winning coalitions and determine which voters are critical voters in each.

| Winning Coalition | Coalition Without One Voter | Weight of New Coalition | Winning or Losing | Conclusion |
|---|---|---|---|---|
| $\{P_1, P_2\}$ | $\{P_2\}$ | 3 | Losing | $P_1$ is Critical |
| | $\{P_1\}$ | 5 | Losing | $P_2$ is Critical |
| $\{P_1, P_2, P_3\}$ | $\{P_2, P_3\}$ | $3 + 1 = 4$ | Losing | $P_1$ is Critical |
| | $\{P_1, P_3\}$ | $5 + 1 = 6$ | Losing | $P_2$ is Critical |
| | $\{P_1, P_2\}$ | $5 + 3 = 8$ | Winning | |

The power of each voter is the number of times the voter is critical. $P_1$ is a critical voter 2 times and $P_2$ is a critical voter 2 times. The total Banzhaf power in the system is 4. To find the Banzhaf power index for each voter, divide the number of times the voter was critical by the total Banzhaf power.

| Voter | Banzhaf Power Index |
|---|---|
| $P_1$ | $\dfrac{2}{4} = \dfrac{1}{2}$ or 50% |
| $P_2$ | $\dfrac{2}{4} = \dfrac{1}{2}$ or 50% |
| $P_3$ | $\dfrac{0}{4} = 0$ or 0% |

Notice that $P_3$ has no power and is a dummy.

(d) The quota is 8.

| Coalition | Sum of the Weights | Winning or Losing |
|---|---|---|
| $\{P_1\}$ | 5 | Losing |
| $\{P_2\}$ | 3 | Losing |
| $\{P_3\}$ | 1 | Losing |
| $\{P_1, P_2\}$ | $5 + 3 = 8$ | Winning |
| $\{P_1, P_3\}$ | $5 + 1 = 6$ | Losing |
| $\{P_2, P_3\}$ | $3 + 1 = 4$ | Losing |
| $\{P_1, P_2, P_3\}$ | $5 + 3 + 1 = 9$ | Winning |

There are two winning coalitions. Consider each of the winning coalitions and determine which voters are critical voters in each.

| Winning Coalition | Coalition Without One Voter | Weight of New Coalition | Winning or Losing | Conclusion |
|---|---|---|---|---|
| $\{P_1, P_2\}$ | $\{P_2\}$ | 3 | Losing | $P_1$ is Critical |
| | $\{P_1\}$ | 5 | Losing | $P_2$ is Critical |
| $\{P_1, P_2, P_3\}$ | $\{P_2, P_3\}$ | $3 + 1 = 4$ | Losing | $P_1$ is Critical |
| | $\{P_1, P_3\}$ | $5 + 1 = 6$ | Losing | $P_2$ is Critical |
| | $\{P_1, P_2\}$ | $5 + 3 = 8$ | Winning | |

The power of each voter is the number of times the voter is critical. $P_1$ is a critical voter 2 times and $P_2$ is a critical voter 2 times. The total Banzhaf power in the system is 4. To find the Banzhaf power index for each voter, divide the number of times the voter was critical by the total Banzhaf power.

| Voter | Banzhaf Power Index |
|-------|---------------------|
| $P_1$ | $\dfrac{2}{4} = \dfrac{1}{2}$ or 50% |
| $P_2$ | $\dfrac{2}{4} = \dfrac{1}{2}$ or 50% |
| $P_3$ | $\dfrac{0}{4} = 0$ or 0% |

Notice that $P_3$ has no power and is a dummy.

**(e)** The quota is 9.

| Coalition | Sum of the Weights | Winning or Losing |
|-----------|--------------------|-------------------|
| $\{P_1\}$ | 5 | Losing |
| $\{P_2\}$ | 3 | Losing |
| $\{P_3\}$ | 1 | Losing |
| $\{P_1, P_2\}$ | $5 + 3 = 8$ | Losing |
| $\{P_1, P_3\}$ | $5 + 1 = 6$ | Losing |
| $\{P_2, P_3\}$ | $3 + 1 = 4$ | Losing |
| $\{P_1, P_2, P_3\}$ | $5 + 3 + 1 = 9$ | Winning |

Consider the winning coalition and determine which voters are critical voters.

| Winning Coalition | Coalition Without One Voter | Weight of New Coalition | Winning or Losing | Conclusion |
|-------------------|------------------------------|-------------------------|-------------------|------------|
| $\{P_1, P_2, P_3\}$ | $\{P_2, P_3\}$ | $3 + 1 = 4$ | Losing | $P_1$ is Critical |
| | $\{P_1, P_3\}$ | $5 + 1 = 6$ | Losing | $P_2$ is Critical |
| | $\{P_1, P_2\}$ | $5 + 3 = 8$ | Losing | $P_3$ is Critical |

The power of each voter is the number of times the voter is critical. $P_1$ is a critical voter 1 time, $P_2$ is a critical voter 1 time, and $P_3$ is a critical voter 1 time. The total Banzhaf power in the system is 3. To find the Banzhaf power index for each voter, divide the number of times the voter was critical by the total Banzhaf power.

| Voter | Banzhaf Power Index |
|-------|---------------------|
| $P_1$ | $\dfrac{1}{3}$ or $33\dfrac{1}{3}\%$ |
| $P_2$ | $\dfrac{1}{3}$ or $33\dfrac{1}{3}\%$ |
| $P_3$ | $\dfrac{1}{3}$ or $33\dfrac{1}{3}\%$ |

Power is shared equally.

**27. (a)** List the coalitions and determine which are winning coalitions. The quota is 4.

| Coalition | Sum of the Weights | Winning or Losing |
|-----------|--------------------|-------------------|
| $\{P_1\}$ | 3 | Losing |
| $\{P_2\}$ | 2 | Losing |
| $\{P_3\}$ | 1 | Losing |
| $\{P_1, P_2\}$ | $3 + 2 = 5$ | Winning |
| $\{P_1, P_3\}$ | $3 + 1 = 4$ | Winning |
| $\{P_2, P_3\}$ | $2 + 1 = 3$ | Losing |
| $\{P_1, P_2, P_3\}$ | $3 + 2 + 1 = 6$ | Winning |

Both $\{P_1, P_2\}$ and $\{P_1, P_3\}$ are minimal coalitions because if either member leaves the coalition it will become a losing coalition. The winning coalition $\{P_1, P_2, P_3\}$ is not minimal because $P_3$ can leave and it will still be a winning coalition.

(b) List the coalitions and determine which are winning coalitions. The quota is 6.

| Coalition | Sum of the Weights | Winning or Losing |
|---|---|---|
| $\{P_1\}$ | 4 | Losing |
| $\{P_2\}$ | 3 | Losing |
| $\{P_3\}$ | 2 | Losing |
| $\{P_4\}$ | 1 | Losing |
| $\{P_1, P_2\}$ | $4 + 3 = 7$ | Winning |
| $\{P_1, P_3\}$ | $4 + 2 = 6$ | Winning |
| $\{P_1, P_4\}$ | $4 + 1 = 5$ | Losing |
| $\{P_2, P_3\}$ | $3 + 2 = 5$ | Losing |
| $\{P_2, P_4\}$ | $3 + 1 = 4$ | Losing |
| $\{P_3, P_4\}$ | $2 + 1 = 3$ | Losing |
| $\{P_1, P_2, P_3\}$ | $4 + 3 + 2 = 9$ | Winning |
| $\{P_1, P_2, P_4\}$ | $4 + 3 + 1 = 8$ | Winning |
| $\{P_1, P_3, P_4\}$ | $4 + 2 + 1 = 7$ | Winning |
| $\{P_2, P_3, P_4\}$ | $3 + 2 + 1 = 6$ | Winning |
| $\{P_1, P_2, P_3, P_4\}$ | $4 + 3 + 2 + 1 = 10$ | Winning |

$\{P_1, P_2\}$, $\{P_1, P_3\}$, and $\{P_2, P_3, P_4\}$ are minimal coalitions because if any one member leaves the coalition it will become a losing coalition.

29. (a) The total weight of the system is $5 + 4 + 2 = 11$. The quota is 7. The difference is $11 - 7 = 4$. Any coalition that is a blocking coalition will have a weight that is larger than 4. Consider the table of coalitions and the weights from problem 28 (a). The following coalitions are blocking coalitions: $\{P_1\}$, $\{P_1, P_2\}$, $\{P_1, P_3\}$, $\{P_2, P_3\}$, and $\{P_1, P_2, P_3\}$. Whenever these voters get together, they can block the passage of any motion.

(b) The total weight of the system is $20 + 15 + 10 + 5 = 50$. The quota is 26. The difference is $50 - 26 = 24$. Any coalition that is a blocking coalition will have a weight that is larger than 24. Consider the table of coalitions and the weights from problem 28 (b). The following coalitions are blocking coalitions: $\{P_1, P_2\}$, $\{P_1, P_3\}$, $\{P_1, P_4\}$, $\{P_2, P_3\}$, $\{P_1, P_2, P_3\}$, $\{P_1, P_2, P_4\}$, $\{P_1, P_3, P_4\}$, $\{P_2, P_3, P_4\}$, and $\{P_1, P_2, P_3, P_4\}$. Whenever these voters get together, they can block the passage of any motion.

31. (a) The United States is a permanent member. Because the United States voted against the resolution, it failed. Every permanent member has veto power.

(b) The three countries that voted against the resolution were all elected members. The resolution passed.

**33. (a)** List the coalitions and determine which are winning coalitions. The quota is 6.

| Coalition | Sum of the Weights | Winning or Losing |
|---|---|---|
| $\{P_1\}$ | 3 | Losing |
| $\{P_2\}$ | 2 | Losing |
| $\{P_3\}$ | 2 | Losing |
| $\{P_4\}$ | 2 | Losing |
| $\{P_1, P_2\}$ | $3 + 2 = 5$ | Losing |
| $\{P_1, P_3\}$ | $3 + 2 = 5$ | Losing |
| $\{P_1, P_4\}$ | $3 + 2 = 5$ | Losing |
| $\{P_2, P_3\}$ | $2 + 2 = 4$ | Losing |
| $\{P_2, P_4\}$ | $2 + 2 = 4$ | Losing |
| $\{P_3, P_4\}$ | $2 + 2 = 4$ | Losing |
| $\{P_1, P_2, P_3\}$ | $3 + 2 + 2 = 7$ | Winning |
| $\{P_1, P_2, P_4\}$ | $3 + 2 + 2 = 7$ | Winning |
| $\{P_1, P_3, P_4\}$ | $3 + 2 + 2 = 7$ | Winning |
| $\{P_2, P_3, P_4\}$ | $2 + 2 + 2 = 6$ | Winning |
| $\{P_1, P_2, P_3, P_4\}$ | $3 + 2 + 2 + 2 = 9$ | Winning |

There are five winning coalitions. Consider each of the winning coalitions and determine which voters are critical voters.

| Winning Coalition | Coalition Without One Voter | Weight of New Coalition | Winning or Losing | Conclusion |
|---|---|---|---|---|
| $\{P_1, P_2, P_3\}$ | $\{P_2, P_3\}$ | $2 + 2 = 4$ | Losing | $P_1$ is Critical |
| | $\{P_1, P_3\}$ | $3 + 2 = 5$ | Losing | $P_2$ is Critical |
| | $\{P_1, P_2\}$ | $3 + 2 = 5$ | Losing | $P_3$ is Critical |
| $\{P_1, P_2, P_4\}$ | $\{P_2, P_4\}$ | $2 + 2 = 4$ | Losing | $P_1$ is Critical |
| | $\{P_1, P_4\}$ | $3 + 2 = 5$ | Losing | $P_2$ is Critical |
| | $\{P_1, P_2\}$ | $3 + 2 = 5$ | Losing | $P_4$ is Critical |
| $\{P_1, P_3, P_4\}$ | $\{P_3, P_4\}$ | $2 + 2 = 4$ | Losing | $P_1$ is Critical |
| | $\{P_1, P_4\}$ | $3 + 2 = 5$ | Losing | $P_3$ is Critical |
| | $\{P_1, P_3\}$ | $3 + 2 = 5$ | Losing | $P_4$ is Critical |
| $\{P_2, P_3, P_4\}$ | $\{P_3, P_4\}$ | $2 + 2 = 4$ | Losing | $P_2$ is Critical |
| | $\{P_2, P_4\}$ | $2 + 2 = 4$ | Losing | $P_3$ is Critical |
| | $\{P_2, P_3\}$ | $2 + 2 = 4$ | Losing | $P_4$ is Critical |
| $\{P_1, P_2, P_3, P_4\}$ | $\{P_2, P_3, P_4\}$ | $2 + 2 + 2 = 6$ | Winning | |
| | $\{P_1, P_3, P_4\}$ | $3 + 2 + 2 = 7$ | Winning | |
| | $\{P_1, P_2, P_4\}$ | $3 + 2 + 2 = 7$ | Winning | |
| | $\{P_1, P_2, P_3\}$ | $3 + 2 + 2 = 7$ | Winning | |

The power of each voter is the number of times the voter is critical. $P_1$ is a critical voter 3 times, $P_2$ is a critical voter 3 times, $P_3$ is a critical voter 3 times, and $P_4$ is a critical voter 3 times. The total Banzhaf power in the system is 12. To find the Banzhaf power index for each voter, divide the number of times the voter was critical by the total Banzhaf power.

| Voter | Banzhaf Power Index |
|-------|---------------------|
| $P_1$ | $\dfrac{3}{12} = \dfrac{1}{4}$ or 25% |
| $P_2$ | $\dfrac{3}{12} = \dfrac{1}{4}$ or 25% |
| $P_3$ | $\dfrac{3}{12} = \dfrac{1}{4}$ or 25% |
| $P_4$ | $\dfrac{3}{12} = \dfrac{1}{4}$ or 25% |

**(b)** If $P_1$ and $P_2$ always vote the same way, then together, they control a block of 5 votes. Represent the voting system as [6 | 5, 2, 2].

List the coalitions and determine which are winning coalitions. The quota is 6.

| Coalition | Sum of the Weights | Winning or Losing |
|-----------|--------------------|--------------------|
| $\{P_1\}$ | 5 | Losing |
| $\{P_2\}$ | 2 | Losing |
| $\{P_3\}$ | 2 | Losing |
| $\{P_1, P_2\}$ | $5 + 2 = 7$ | Winning |
| $\{P_1, P_3\}$ | $5 + 2 = 7$ | Winning |
| $\{P_2, P_3\}$ | $2 + 2 = 4$ | Losing |
| $\{P_1, P_2, P_3\}$ | $5 + 2 + 2 = 9$ | Winning |

There are three winning coalitions. Consider each of the winning coalitions and determine which voters are critical voters.

| Winning Coalition | Coalition Without One Voter | Weight of New Coalition | Winning or Losing | Conclusion |
|-------------------|-----------------------------|-------------------------|--------------------|------------|
| $\{P_1, P_2\}$ | $\{P_2\}$ | 2 | Losing | $P_1$ is Critical |
|                | $\{P_1\}$ | 5 | Losing | $P_2$ is Critical |
| $\{P_1, P_3\}$ | $\{P_3\}$ | 2 | Losing | $P_1$ is Critical |
|                | $\{P_1\}$ | 5 | Losing | $P_3$ is Critical |
| $\{P_1, P_2, P_3\}$ | $\{P_2, P_3\}$ | $2 + 2 = 4$ | Losing | $P_1$ is Critical |
|                     | $\{P_1, P_3\}$ | $5 + 2 = 7$ | Winning | |
|                     | $\{P_1, P_2\}$ | $5 + 2 = 7$ | Winning | |

The power of each voter is the number of times the voter is critical. $P_1$ is a critical voter 3 times, $P_2$ is a critical voter 1 time, and $P_3$ is a critical voter 1 time. The total Banzhaf power in the system is 5. To find the Banzhaf power index for each voter, divide the number of times the voter was critical by the total Banzhaf power.

| Voter | Banzhaf Power Index |
|-------|---------------------|
| $P_1$ | $\dfrac{3}{5}$ or 60% |
| $P_2$ | $\dfrac{1}{5}$ or 20% |
| $P_3$ | $\dfrac{1}{5}$ or 20% |

**35.** The change from part (a) to part (b) is that the first voter loses a vote and the third voter gains a vote. Compare the Banzhaf power index for each voter before and after the change.

**(a)** List the coalitions and determine which are winning coalitions. The quota is 11. There are eight winning coalitions. Consider each of the winning coalitions and determine which voters are critical voters. The power of each voter is the number of times the voter is critical. $P_1$ is a critical voter 6 times, $P_2$ is a critical voter 2 times, $P_3$ is a critical voter 2 times, and $P_4$ is a critical voter 2 times. The total Banzhaf power in the system is 12. To find the Banzhaf power index for each voter, divide the number of times the voter was critical by the total Banzhaf power.

| Voter | Banzhaf Power Index |
|---|---|
| $P_1$ | $\frac{6}{12}=\frac{1}{2}$ or 50% |
| $P_2$ | $\frac{2}{12}=\frac{1}{6}$ or 16.7% |
| $P_3$ | $\frac{2}{12}=\frac{1}{6}$ or 16.7% |
| $P_4$ | $\frac{2}{12}=\frac{1}{6}$ or 16.7% |

$P_1$ has half of the power and the other three voters are equally powerful.

**(b)** List the coalitions and determine which are winning coalitions. The quota is 11. There are eight winning coalitions. Consider each of the winning coalitions and determine which voters are critical voters. The power of each voter is the number of times the voter is critical. $P_1$ is a critical voter 4 times, $P_2$ is a critical voter 4 times, $P_3$ is a critical voter 4 times, and $P_4$ is a critical voter 0 times. The total Banzhaf power in the system is 12. To find the Banzhaf power index for each voter, divide the number of times the voter was critical by the total Banzhaf power.

| Voter | Banzhaf Power Index |
|---|---|
| $P_1$ | $\frac{4}{12}=\frac{1}{3}$ or 33.3% |
| $P_2$ | $\frac{4}{12}=\frac{1}{3}$ or 33.3% |
| $P_3$ | $\frac{4}{12}=\frac{1}{3}$ or 33.3% |
| $P_4$ | $\frac{0}{12}=0$ or 0% |

After the change, $P_4$ is a dummy and the other three voters have equal power.

**37. (a)** List the coalitions and determine which are winning coalitions. The quota is 51. There are four winning coalitions. Consider each of the winning coalitions and determine which voters are critical voters. The power of each voter is the number of times the voter is critical. $P_1$ is a critical voter 2 times, $P_2$ is a critical voter 2 times, and $P_3$ is a critical voter 2 times. The total Banzhaf power in the system is 6. To find the Banzhaf power index for each voter, divide the number of times the voter was critical by the total Banzhaf power.

| Voter | Banzhaf Power Index |
|---|---|
| $P_1$ | $\frac{2}{6}=\frac{1}{3}$ or 33.3% |
| $P_2$ | $\frac{2}{6}=\frac{1}{3}$ or 33.3% |
| $P_3$ | $\frac{2}{6}=\frac{1}{3}$ or 33.3% |

Most of the votes are controlled by the first two voters, yet the three voters share the power equally.

(b) List the coalitions and determine which are winning coalitions. The quota is 51. There are.eight winning coalitions. Consider each of the winning coalitions and determine which voters are critical voters. The power of each voter is the number of times the voter is critical. $P_1$ is a critical voter 6 times, $P_2$ is a critical voter 2 times, $P_3$ is a critical voter 2 times, and $P_4$ is a critical voter 2 times. The total Banzhaf power in the system is 12. To find the Banzhaf power index for each voter, divide the number of times the voter was critical by the total Banzhaf power.

| Voter | Banzhaf Power Index |
|-------|---------------------|
| $P_1$ | $\dfrac{6}{12}=\dfrac{1}{2}$ or 50% |
| $P_2$ | $\dfrac{2}{12}=\dfrac{1}{6}$ or 16.7% |
| $P_3$ | $\dfrac{2}{12}=\dfrac{1}{6}$ or 16.7% |
| $P_4$ | $\dfrac{2}{12}=\dfrac{1}{6}$ or 16.7% |

Most of the votes are controlled by the first two voters. $P_1$ controls only four more votes than $P_2$, yet $P_1$ controls half of the power. Despite the fact that $P_2$ controls 39 more votes than $P_3$ and 41 more votes than $P_4$, the second, third and fourth voters have an equal amount of power.

## Solutions to Chapter 3 Review Problems

1.  Anne received the greatest number of first-place votes, so she is the winner.

2.  Assign 3 points to every first-place vote, 2 points to every second-place vote, and 1 point to every third-place vote.

| Candidate | Points For Each Rank | | | Total Points |
|-----------|----------------------|---|---|--------------|
| | 1st Place Votes 3 Points Each | 2nd Place Votes 2 Points Each | 3rd Place Votes 1 Point Each | |
| Anne | $6 \times 3 = 18$ | $2 \times 2 = 4$ | $4 \times 1 = 4$ | 26 |
| Brad | $4 \times 3 = 12$ | $7 \times 2 = 14$ | $1 \times 1 = 1$ | 27 |
| Carlisse | $2 \times 3 = 6$ | $3 \times 2 = 6$ | $7 \times 1 = 7$ | 19 |

Brad has the most points, so he is the winner.

3.  Initially, we know that Carlisse has the fewest first-place votes so she should be eliminated. The problem is that there is no way to tell how the voters who ranked Carlisse first would rank Anne or Brad. This is not a preferential ballot. The plurality with elimination method cannot be applied.

4.  Preferences are not given, so a pairwise comparison cannot be made between candidates.

5.  Assign 3 points to every first-place vote, 2 points to every second-place vote, and 1 point to every third-place vote. Candidate A is the winner.

| Candidate | Points For Each Vote Received | | | Total Points |
|-----------|-------------------------------|---|---|--------------|
| | 1st Place Votes 3 Points Each | 2nd Place Votes 2 Points Each | 3rd Place Votes 1 Point Each | |
| A | $5 \times 3 = 15$ | $4 \times 2 = 8$ | $2 \times 1 = 2$ | 25 |
| B | $2 \times 3 = 6$ | $5 \times 2 = 10$ | $4 \times 1 = 4$ | 20 |
| C | $4 \times 3 = 12$ | $2 \times 2 = 4$ | $5 \times 1 = 5$ | 21 |

6. With three candidates, there are three pairwise comparisons to be made.

| Candidates Compared | Preferred Candidate | Points Awarded |
|---|---|---|
| A and B | A is preferred 9 to 2 | A receives 1 point |
| A and C | C is preferred 6 to 5 | C receives 1 point |
| B and C | B is preferred 7 to 4 | B receives 1 point |

There is a three-way tie.

7. (a) There are 11 voters. A majority would be 6 votes. The candidate with the fewest first-place votes is candidate B. Eliminate B and create a new preference table.

| Ranking | Number of Voters | | |
|---|---|---|---|
| | 5 | 4 | 2 |
| 1st | A | C | C |
| 2nd | C | A | A |

Candidate A has 5 first-place votes while candidate C has 6 first-place votes. C has a majority of votes and is the winner.

   (b) The candidate with the most third-place votes is candidate C. Eliminate candidate C and create a new preference table.

| Ranking | Number of Voters | | |
|---|---|---|---|
| | 5 | 4 | 2 |
| 1st | A | A | B |
| 2nd | B | B | A |

Candidate A has 9 first-place votes which is a majority. Candidate A is the winner.

8. (a) Create a new preference table to reflect the changed preferences.

| Ranking | Number of Voters | | |
|---|---|---|---|
| | 3 | 4 | 4 |
| 1st | A | C | B |
| 2nd | B | A | C |
| 3rd | C | B | A |

The candidate with the fewest first-place votes is candidate A. Eliminate A and create a new preference table.

| Ranking | Number of Voters | | |
|---|---|---|---|
| | 5 | 4 | 2 |
| 1st | B | C | B |
| 2nd | C | B | C |

Candidate B has 7 first-place votes which is a majority. Candidate B wins the election.

   (b) When plurality with elimination was applied to the original preference table, candidate C was the winner. After two voters changed their preferences, candidate B became the winner.

   (c) Monotonicity was not violated in this case because two voters not only improved C's rank, but they also improved B's rank enough so that B was not eliminated first.

9. **(a)** Assign 3 points to every first-place vote, 2 points to every second-place vote, and 1 point to every third-place vote.

| Project | Points For Each Vote Received | | | Total Points |
|---|---|---|---|---|
| | 1st Place Votes 3 Points Each | 2nd Place Votes 2 Points Each | 3rd Place Votes 1 Point Each | |
| G | $8 \times 3 = 24$ | $4 \times 2 = 8$ | $8 \times 1 = 8$ | 40 |
| L | $7 \times 3 = 21$ | $13 \times 2 = 26$ | $0 \times 1 = 0$ | 47 |
| U | $5 \times 3 = 15$ | $3 \times 2 = 6$ | $12 \times 1 = 12$ | 33 |

The library project is selected.

**(b)** Form all pairwise comparisons.

| Projects Compared | Preferred Projects | Points Awarded |
|---|---|---|
| G and L | L is preferred 12 to 8 | L receives 1 point |
| G and U | G is preferred 12 to 8 | G receives 1 point |
| L and U | L is preferred 15 to 5 | L receives 1 point |

The library project is selected.

**(c)** The gymnasium received the most first-place votes. The gymnasium project is selected.

**(d)** The student union project received the fewest first-place votes so it is eliminated. Create a new preference table.

| Ranking | Number of Regents | | | |
|---|---|---|---|---|
| | 8 | 4 | 3 | 5 |
| 1st | G | L | L | L |
| 2nd | L | G | G | G |

The library project has 12 first-place votes. There were only 20 voters, so the library project has a majority. The library project is selected.

10. The only voting method we have studied that might violate the majority criterion is the Borda count method. Compare the results from problem 9 (a) and (c). Notice the plurality method selected the gymnasium project while the Borda count method selected the library project. This is not a violation of the majority criterion because the gymnasium project was selected by a plurality rather than a majority.

11. The three voting methods we have studied that might violate the head-to-head criterion are the plurality method, the Borda count method, and the plurality with elimination method. The head-to-head criterion can be violated if there is a Condorcet project. In this case, the library project is preferred to the gymnasium project and to the student union project, so the library project is a Condorcet project and should be selected. If a method selects a different project, then it will violate the head-to-head criterion. We see from problem 9 (c), that the plurality method selected the gymnasium project. In this case, the plurality method violates the head-to-head criterion. The other two methods, Borda count and plurality with elimination, both selected the library so there was no violation from either of those methods.

12. It is possible for each of the four voting methods to violate the irrelevant-alternatives criterion. Consider the results of those methods from problem 9.
**Borda count method:**
Originally, the Borda count method selected the library project. To see if this method will violate the irrelevant-alternatives criterion, eliminate a losing project and determine the winner.
First, eliminate the student union project and create a new preference table.

| Ranking | Number of Regents | | | |
|---|---|---|---|---|
| | 8 | 4 | 3 | 5 |
| 1st | G | L | L | L |
| 2nd | L | G | G | G |

The library project has a majority of first-place votes and is again selected. This time, eliminate the gymnasium project and create a new preference table.

| Ranking | Number of Regents | | | |
|---|---|---|---|---|
| | 8 | 4 | 3 | 5 |
| 1st | L | L | L | U |
| 2nd | U | U | U | L |

The library project again has a majority of first-place votes and is selected.
The Borda count method does not violate the irrelevant-alternatives criterion in this case.

**Pairwise-comparison method:**
Originally, the pairwise-comparison method selected the library project. To see if this method will violate the irrelevant-alternatives criterion, eliminate a losing project and determine the winner. Study the preference tables created by eliminating the student union project or the gymnasium project. Notice that no matter which losing project is eliminated, the library will still be preferred over the other project. The library project will still be selected no matter which losing alternative is eliminated, so this method does not violate the irrelevant-alternatives criterion.

**Plurality method:**
Originally, the plurality method selected the gymnasium as the winning project. Study the preference table with the student union eliminated. The library has 12 first-place votes compared to 8 first-place votes for the gymnasium. When the student union project is eliminated, the library becomes the winning project. This is a violation of the irrelevant-alternatives criterion.

**Plurality with elimination method:**
Originally, the plurality with elimination method selected the library project as the winner. Study the preference tables created by eliminating either the student union project or the gymnasium project. In both cases, the library project has a majority of the votes and is the winning project. There is no violation.

13. The only method we have studied that can violate the monotonicity criterion is the plurality with elimination method. From problem 9 (d), we know that originally, this method selected the library project as the winner. Consider what would happen if any voter who might change his or her mind switched the library project with a project ranked one position higher than the library. There are twenty voters. If four voters move the library up, then the library will have a majority and will be the winner. The library is ranked second by the 13 voters who ranked G or U first. No matter how the preferences are changed and no matter which order candidates G and U are eliminated, the library is still the winner. There is no violation of the monotonicity criterion in this case.

14. **(a)**

| Candidate | Points For Each Rank | | | | Total Points |
|---|---|---|---|---|---|
| | 1st Place Votes 4 Points Each | 2nd Place Votes 3 Points Each | 3rd Place Votes 2 Points Each | 4th Place Votes 1 Point Each | |
| A | 2 × 4 = 8 | 0 × 3 = 0 | 1 × 2 = 2 | 2 × 1 = 2 | 12 |
| B | 2 × 4 = 8 | 0 × 3 = 0 | 2 × 2 = 4 | 1 × 1 = 1 | 13 |
| C | 1 × 4 = 4 | 0 × 3 = 0 | 2 × 2 = 4 | 2 × 1 = 2 | 10 |
| D | 0 × 4 = 0 | 5 × 3 = 15 | 0 × 2 = 0 | 0 × 1 = 0 | 15 |

Candidate D has the most points and is the winner using the Borda count method.

**(b)** Change the point values and find the winner of the election.

| Candidate | Points For Each Rank | | | | Total Points |
|---|---|---|---|---|---|
| | 1st Place Votes 6 Points Each | 2nd Place Votes 3 Points Each | 3rd Place Votes 2 Points Each | 4th Place Votes 1 Point Each | |
| A | $2 \times 6 = 12$ | $0 \times 3 = 0$ | $1 \times 2 = 2$ | $2 \times 1 = 2$ | 16 |
| B | $2 \times 6 = 12$ | $0 \times 3 = 0$ | $2 \times 2 = 4$ | $1 \times 1 = 1$ | 17 |
| C | $1 \times 6 = 6$ | $0 \times 3 = 0$ | $2 \times 2 = 4$ | $2 \times 1 = 2$ | 12 |
| D | $0 \times 6 = 0$ | $5 \times 3 = 15$ | $0 \times 2 = 0$ | $0 \times 1 = 0$ | 15 |

Candidate B has the most points and is the winner using this modified Borda count method.

(c) Candidate D would be eliminated first. Create a new preference table.

| Ranking | Number of Members | | |
|---|---|---|---|
| | 2 | 2 | 1 |
| 1st | A | B | C |
| 2nd | B | C | A |
| 3rd | C | A | B |

No candidate has a majority of first-place votes. Candidate C would be eliminated next. Create a new preference table.

| Ranking | Number of Members | | |
|---|---|---|---|
| | 2 | 2 | 1 |
| 1st | A | B | A |
| 2nd | B | A | B |

Candidate A has a majority of first-place votes and is the winner.

(d) With four candidates, there are six pairwise comparisons.

| Candidates Compared | Preferred Candidate | Points Awarded |
|---|---|---|
| A and B | A is preferred 3 to 2 | A receives 1 point |
| A and C | C is preferred 3 to 2 | C receives 1 point |
| A and D | D is preferred 3 to 2 | D receives 1 point |
| B and C | B is preferred 4 to 1 | B receives 1 point |
| B and D | D is preferred 3 to 2 | D receives 1 point |
| C and D | D is preferred 4 to 1 | D receives 1 point |

Candidate D has 3 points and is the winner.

(e) Candidate D is preferred in every head-to-head comparison and is the Condorcet candidate. The modified Borda count method selected B and the plurality with elimination method selected A. Both of those methods violated the head-to-head criterion in this case.

15. (a) Count the votes for each candidate. Matto received 8 votes. Benton received 9 votes. Newiski received 6 votes. Yao received 6 votes. Benton received the most votes and is the winner.

(b) If there were two representative positions, then the candidates receiving the most votes would be selected. In this case, the two representatives would be Benton and Matto.

16. (a) Using the plurality method based on the number of A grades, the student who earned 8 A grades would win the grant.

(b) Assign 4 points to every A grade, 3 points to every B grade, and 2 points to every C grade. The first student would have $(8 \times 4) + (4 \times 2) = 32 + 8 = 40$ points. The other student would have $(7 \times 4) + (5 \times 3) = 28 + 15 = 43$ points. The grant would go to the student who earned fewer A grades.

(c) Answers will vary. The Borda count method may be the most appropriate method to use in this case. It uses all of the information rather than only the A grades.

17. (a) Candidate C has 5 first-place ranks while candidate A has 4. Candidate C wins under the plurality method. There were nine voters, so C won by a majority.

**(b)**

| Candidate | Points For Each Vote Received | | | Total Points |
|---|---|---|---|---|
| | 1st Place Votes 3 Points Each | 2nd Place Votes 2 Points Each | 3rd Place Votes 1 Point Each | |
| A | $4 \times 3 = 12$ | $5 \times 2 = 10$ | $0 \times 1 = 0$ | 22 |
| B | $0 \times 3 = 0$ | $4 \times 2 = 8$ | $5 \times 1 = 5$ | 13 |
| C | $5 \times 3 = 15$ | $0 \times 2 = 0$ | $4 \times 1 = 4$ | 19 |

Candidate A has the most points and is the winner using the Borda count method.

**(c)** Despite the fact that candidate C had a majority of first-place votes, candidate A won using the Borda count method. In this case, the Borda count method violated the majority criterion.

**18. (a)** Add the weights of the voters. $W_1 + W_2 + W_3 + W_4 + W_5 = 11 + 6 + 4 + 2 + 1 = 24$. The quota is $\frac{3}{4} \times 24 = 18$ votes.

**(b)** There are ten two-person coalitions. From part (a), we know the quota is 18.

| Coalition | Sum of the Weights | Winning or Losing |
|---|---|---|
| $\{P_1, P_2\}$ | $11 + 6 = 17$ | Losing |
| $\{P_1, P_3\}$ | $11 + 4 = 15$ | Losing |
| $\{P_1, P_4\}$ | $11 + 2 = 13$ | Losing |
| $\{P_1, P_5\}$ | $11 + 1 = 12$ | Losing |
| $\{P_2, P_3\}$ | $6 + 4 = 10$ | Losing |
| $\{P_2, P_4\}$ | $6 + 2 = 8$ | Losing |
| $\{P_2, P_5\}$ | $6 + 1 = 7$ | Losing |
| $\{P_3, P_4\}$ | $4 + 2 = 6$ | Losing |
| $\{P_3, P_5\}$ | $4 + 1 = 5$ | Losing |
| $\{P_4, P_5\}$ | $2 + 1 = 3$ | Losing |

There are no two-person winning coalitions.

**(c)** The quota is 18 votes.

| Coalition | Sum of the Weights | Winning or Losing |
|---|---|---|
| $\{P_1, P_2, P_3, P_4\}$ | $11 + 6 + 4 + 2 = 23$ | Winning |
| $\{P_1, P_2, P_3, P_5\}$ | $11 + 6 + 4 + 1 = 22$ | Winning |
| $\{P_2, P_3, P_4, P_5\}$ | $6 + 4 + 2 + 1 = 13$ | Losing |
| $\{P_1, P_3, P_4, P_5\}$ | $11 + 4 + 2 + 1 = 18$ | Winning |
| $\{P_1, P_2, P_4, P_5\}$ | $11 + 6 + 2 + 1 = 20$ | Winning |

There are four winning four-person coalitions. A voter is critical in a winning coalition if the coalition changes from a winning coalition to a losing coalition when the voter leaves the coalition. Consider each winning coalition and the quota of 18 votes. Determine which voters could change the coalition from winning to losing if they were to remove their votes from the coalition. The following table contains each winning coalition of four voters and lists the critical voters for each.

| Winning Coalition With Four Voters | Coalition Without One Voter | Sum of the Weights | Winning or Losing | Conclusion |
|---|---|---|---|---|
| $\{P_1,P_2,P_3,P_4\}$ | $\{P_2,P_3,P_4\}$ | $6+4+2=12$ | Losing | $P_1$ is Critical |
| | $\{P_1,P_3,P_4\}$ | $11+4+2=17$ | Losing | $P_2$ is Critical |
| | $\{P_1,P_2,P_4\}$ | $11+6+2=19$ | Winning | |
| | $\{P_1,P_2,P_3\}$ | $11+6+4=21$ | Winning | |
| $\{P_1,P_2,P_3,P_5\}$ | $\{P_2,P_3,P_5\}$ | $6+4+1=11$ | Losing | $P_1$ is Critical |
| | $\{P_1,P_3,P_5\}$ | $11+4+1=16$ | Losing | $P_2$ is Critical |
| | $\{P_1,P_2,P_5\}$ | $11+6+1=18$ | Winning | |
| | $\{P_1,P_2,P_3\}$ | $11+6+4=21$ | Winning | |
| $\{P_1,P_3,P_4,P_5\}$ | $\{P_3,P_4,P_5\}$ | $4+2+1=7$ | Losing | $P_1$ is Critical |
| | $\{P_1,P_4,P_5\}$ | $11+2+1=14$ | Losing | $P_3$ is Critical |
| | $\{P_1,P_3,P_5\}$ | $11+4+1=16$ | Losing | $P_4$ is Critical |
| | $\{P_1,P_3,P_4\}$ | $11+4+2=17$ | Losing | $P_5$ is Critical |
| $\{P_1,P_2,P_4,P_5\}$ | $\{P_2,P_4,P_5\}$ | $6+2+1=9$ | Losing | $P_1$ is Critical |
| | $\{P_1,P_4,P_5\}$ | $11+2+1=14$ | Losing | $P_2$ is Critical |
| | $\{P_1,P_2,P_5\}$ | $11+6+1=18$ | Winning | |
| | $\{P_1,P_2,P_4\}$ | $11+6+2=19$ | Winning | |

(d) $P_5$ is not a dummy because $P_5$ is a critical voter in one of the four-voter winning coalitions from part (c).

19. (a) The largest possible quota would result from requiring a unanimous decision of the voters. Add all of the weights. The largest quota is $13+4+3+2+1=23$.

(b) The smallest acceptable quota would be the quota that results from requiring a simple majority. Half of the total weight of the weighted voting system is 11.5. A simple majority requirement would set the quota at 12 votes.

(c) A dictator is a voter who can pass a motion alone by voting for it or defeat a motion alone by not voting for it. The first voter controls 13 votes. The quota would have to be set to 14 so that the first voter would not be able to pass a motion alone.

(d) There are 5 voters. Let $n = 5$. The number of coalitions possible for this voting system is $2^n - 1 = 2^5 - 1 = 32 - 1 = 31$.

20. (a) There are 3 voters, so let $n = 3$. There are $2^3 - 1 = 8 - 1 = 7$ possible coalitions. In this system, the quota is 8 votes.

| Coalition | Sum of the Weights | Winning or Losing |
|---|---|---|
| $\{P_1\}$ | 5 | Losing |
| $\{P_2\}$ | 4 | Losing |
| $\{P_3\}$ | 3 | Losing |
| $\{P_1, P_2\}$ | $5+4=9$ | Winning |
| $\{P_1, P_3\}$ | $5+3=8$ | Winning |
| $\{P_2, P_3\}$ | $4+3=7$ | Losing |
| $\{P_1,P_2,P_3\}$ | $5+4+3=12$ | Winning |

**(b)** There are 4 voters, so let $n = 4$. There are $2^4 - 1 = 16 - 1 = 15$ possible coalitions. In this system, the quota is 6 votes.

| Coalition | Sum of the Weights | Winning or Losing |
|---|---|---|
| $\{P_1\}$ | 4 | Losing |
| $\{P_2\}$ | 3 | Losing |
| $\{P_3\}$ | 2 | Losing |
| $\{P_4\}$ | 1 | Losing |
| $\{P_1, P_2\}$ | $4 + 3 = 7$ | Winning |
| $\{P_1, P_3\}$ | $4 + 2 = 6$ | Winning |
| $\{P_1, P_4\}$ | $4 + 1 = 5$ | Losing |
| $\{P_2, P_3\}$ | $3 + 2 = 5$ | Losing |
| $\{P_2, P_4\}$ | $3 + 1 = 4$ | Losing |
| $\{P_3, P_4\}$ | $2 + 1 = 3$ | Losing |
| $\{P_1, P_2, P_3\}$ | $4 + 3 + 2 = 9$ | Winning |
| $\{P_1, P_2, P_4\}$ | $4 + 3 + 1 = 8$ | Winning |
| $\{P_1, P_3, P_4\}$ | $4 + 2 + 1 = 7$ | Winning |
| $\{P_2, P_3, P_4\}$ | $3 + 2 + 1 = 6$ | Winning |
| $\{P_1, P_2, P_3, P_4\}$ | $4 + 3 + 2 + 1 = 10$ | Winning |

**21. (a)** (i) No voter is a dictator because no voter can pass a motion alone. (ii) $P_3$ is a dummy because $P_3$ has no influence on the outcome of a vote. That is, $P_3$ is never a critical voter. It does not matter if $P_3$ is a part of any two-voter coalition. The coalition will still be a losing coalition. It does not matter if $P_3$ is part of a three-voter coalition. The coalition will still be a winning coalition. (iii) $P_1$ and $P_2$ are both voters with veto power. When either $P_1$ or $P_2$ is in a coalition with $P_3$, the coalition is a losing coalition. When the third voter joins the coalition, it becomes a winning coalition. Both $P_1$ and $P_2$ have the power to prevent a motion from passing even though it is supported by the other two voters.

**(b)** (i) The first voter has a weight of 25, and the quota is 25. Even if the three other voters form a coalition, it will be a losing coalition. For a motion to pass, the first voter must vote for it. Therefore, $P_1$ is a dictator. (ii) Because $P_1$ is a dictator, the other three voters are dummies. (iii) Every dictator has veto power, so $P_1$ has veto power, but no other voter has veto power.

**(c)** (i) There are no dictators. No voter has the power to pass or defeat a motion alone. (ii) $P_4$ and $P_5$ are both dummies. Neither is ever a critical voter. (iii) None of the voters have veto power. A voter has veto power if that voter can prevent the rest of the voters from passing a motion. Together, any four of the five voters can pass a motion no matter how the fifth voter votes.

**22.** List the coalitions and determine which are winning coalitions. The quota is 51. There are eight winning coalitions. Consider each of the winning coalitions and determine which voters are critical voters. The power of each voter is the number of times the voter is critical. $P_1$ is a critical voter 4 times, $P_2$ is a critical voter 4 times, $P_3$ is a critical voter 4 times, and $P_4$ is a critical voter 0 times. The total Banzhaf power in the system is 12. To find the Banzhaf power index for each voter, divide the number of times the voter was critical by the total Banzhaf power.

| Voter | Banzhaf Power Index |
|-------|---------------------|
| $P_1$ | $\dfrac{4}{12} = \dfrac{1}{3}$ |
| $P_2$ | $\dfrac{4}{12} = \dfrac{1}{3}$ |
| $P_3$ | $\dfrac{4}{12} = \dfrac{1}{3}$ |
| $P_4$ | $\dfrac{0}{12} = 0$ |

**23.** List the coalitions and determine which are winning coalitions. The quota is 18. There are sixteen winning coalitions. Consider each of the winning coalitions and determine which voters are critical voters. The power of each voter is the number of times the voter is critical. $P_1$ is a critical voter 10 times, $P_2$ is a critical voter 6 times, $P_3$ is a critical voter 6 times, $P_4$ is a critical voter 2 times, and $P_5$ is a critical voter 2 times. The total Banzhaf power in the system is 26. To find the Banzhaf power index for each voter, divide the number of times the voter was critical by the total Banzhaf power.

| Voter | Banzhaf Power Index |
|-------|---------------------|
| $P_1$ | $\dfrac{10}{26} = \dfrac{5}{13}$ |
| $P_2$ | $\dfrac{6}{26} = \dfrac{3}{13}$ |
| $P_3$ | $\dfrac{6}{26} = \dfrac{3}{13}$ |
| $P_4$ | $\dfrac{2}{26} = \dfrac{1}{13}$ |
| $P_5$ | $\dfrac{2}{26} = \dfrac{1}{13}$ |

**24.** It is possible to have a dummy in a three-voter weighted voting system that has a majority rule and no dictator. However, the quota would have to be a supermajority. Consider a weighted voting system in which $P_3$ is a dummy: [6 | 4, 4, 1]. In this case, there are two winning coalitions: $\{P_1, P_2\}$ and $\{P_1, P_2, P_3\}$. Notice that if $P_3$ leaves the three-voter winning coalition, it remains a winning coalition as it should because $P_3$ is a dummy. Now, consider a quota for this weighted voting system that is a simple majority rather than a supermajority. A simple majority is one more than half of the total weight of the system when half of the total weight is a whole number. In this case, half of the total weight is $\dfrac{4+4+1}{2} = \dfrac{9}{2} = 4.5$. Round the quota up to the nearest whole number. Thus, the quota is 5 if a simple majority is required. The weighted voting system is written [5 | 4, 4, 1]. It is impossible now for $P_3$ to be a dummy with a quota equal to the simple majority. $P_3$ is part of two winning two-voter coalitions, $\{P_1, P_3\}$ and $\{P_2, P_3\}$, and if $P_3$ leaves both coalitions, they both become losing coalitions. In this case, $P_3$ is a critical voter and has power. $P_3$ cannot be a dummy.

# *Chapter 4:* Fair Division

## *Key Concepts*

**By the time you finish studying Section 4.1, you should be able to:**

☐ explain the difference between a discrete and a continuous fair-division problem.

☐ explain what it means for a player to receive a fair share in a fair-division problem.

☐ be able to determine whether a division is proportional.

☐ explain and be able to carry out the divide-and-choose method for two players.

☐ explain and be able to carry out the divide-and-choose method for three players.

☐ explain and be able to carry out the last-diminisher method for three or more players.

**By the time you finish studying Section 4.2, you should be able to:**

☐ explain and be able to carry out the method of sealed bids.

☐ explain and be able to carry out the method of points.

☐ explain and carry out the adjusted-winner procedure.

**By the time you finish studying Section 4.3, you should be able to:**

☐ define what it means for a division to be envy-free.

☐ carry out the continuous envy-free division method for three players.

☐ verify that a division is proportional.

☐ verify that a division is envy-free.

## *Skills Brush-Ups*

Problems in Chapter 4 will require that you be comfortable working with ratios and fractions.

### *RATIO:*

A ratio is an ordered pair of numbers. We will express the ratio of the number $a$ to the number $b$ as $a$ to $b$ or as $a:b$. A ratio may also be written as a fraction $\frac{a}{b}$ in which case $b \neq 0$. Ratios can give a part-to-part comparison or a part-to-whole comparison.

*Example:* Suppose there are 14 women in a class and 13 men. The ratio of women to men is 14 to 13, 14:13, or $\frac{14}{13}$. This ratio gives a part-to-part comparison because women make up part of the class and men make up part of the class. By knowing the ratio of women to men, we can express the fraction of women in the class as $\frac{14}{14+13} = \frac{14}{27}$, and the fraction of men in the class as $\frac{13}{14+13} = \frac{13}{27}$. Both of these ratios give a part-to-whole comparison because the number of women **or** men is compared to the total number of women **and** men.

## FRACTIONS:

A fraction is a number written as $\frac{a}{b}$ where $a$ and $b$ are whole numbers and $b \neq 0$. The number $a$ is called the numerator and $b$ is called the denominator.

### Addition and Subtraction of Fractions with Common Denominators

Let $\frac{a}{b}$ and $\frac{c}{b}$ be fractions, then $\frac{a}{b} + \frac{c}{b} = \frac{a+c}{b}$ and $\frac{a}{b} - \frac{c}{b} = \frac{a-c}{b}$.

You can add or subtract fractions only if they have the same denominator. When the denominator is the same, add or subtract the numerators.

*Examples:* (a) $\frac{3}{7} + \frac{2}{7} = \frac{3+2}{7} = \frac{5}{7}$

(b) $\frac{7}{9} - \frac{5}{9} = \frac{7-5}{9} = \frac{2}{9}$

### Addition and Subtraction of Fractions with Unlike Denominators

Let $\frac{a}{b}$ and $\frac{c}{d}$ be fractions, then

$$\frac{a}{b} + \frac{c}{d} = \frac{a \times d}{b \times d} + \frac{b \times c}{b \times d} = \frac{a \times d + b \times c}{b \times d} \text{ and } \frac{a}{b} - \frac{c}{d} = \frac{a \times d}{b \times d} - \frac{b \times c}{b \times d} = \frac{a \times d - b \times c}{b \times d}.$$

When the denominators of fractions are different, create equivalent fractions with common denominators and then add or subtract the numerators.

*Examples:* (a) $\frac{3}{7} + \frac{2}{5} = \frac{3 \times 5}{7 \times 5} + \frac{7 \times 2}{7 \times 5} = \frac{15}{35} + \frac{14}{35} = \frac{29}{35}$

(b) $2\frac{1}{4} + 3\frac{2}{5} = \frac{9}{4} + \frac{17}{5} = \frac{9 \times 5}{4 \times 5} + \frac{4 \times 17}{4 \times 5} = \frac{45}{20} + \frac{68}{20} = \frac{113}{20} = 5\frac{13}{20}$

(c) $\frac{7}{9} - \frac{5}{8} = \frac{7 \times 8}{9 \times 8} - \frac{9 \times 5}{9 \times 8} = \frac{56}{72} - \frac{45}{72} = \frac{11}{72}$

### Improper Fractions and Mixed Numbers

When the numerator of a fraction is larger than the denominator, we call the expression an improper fraction. An example of an improper fraction is $\frac{12}{5}$. A combination of a whole number and a fraction is called a mixed number. An examples of a mixed number is $2\frac{1}{3}$. We can write improper fractions as mixed numbers and vice versa.

*Example:* (a) $\dfrac{12}{5} = \dfrac{10}{5} + \dfrac{2}{5} = 2 + \dfrac{2}{5} = 2\dfrac{2}{5}$

(b) $2\dfrac{1}{3} = 2 + \dfrac{1}{3} = \dfrac{2}{1} + \dfrac{1}{3} = \dfrac{2\times3}{1\times3} + \dfrac{1}{3} = \dfrac{6}{3} + \dfrac{1}{3} = \dfrac{6+1}{3} = \dfrac{7}{3}$

### Multiplication of Fractions.

Let $\dfrac{a}{b}$ and $\dfrac{c}{d}$ be fractions, then $\dfrac{a}{b} \times \dfrac{c}{d} = \dfrac{a\times c}{b\times d}$.

Fractions do not need to have common denominators when they are multiplied.

*Examples:* (a) $\dfrac{8}{13} \times \dfrac{2}{3} = \dfrac{8\times2}{13\times3} = \dfrac{16}{39}$

(b) $\dfrac{4}{5} \times \dfrac{1}{6} = \dfrac{4\times1}{5\times6} = \dfrac{4}{30} = \dfrac{2}{15}$

(c) $3 \times 2\dfrac{4}{5} = \dfrac{3}{1} \times \dfrac{14}{5} = \dfrac{3\times14}{1\times5} = \dfrac{42}{5} = 8\dfrac{2}{5}$

**Note:** Express fractions in simplest form and express improper fractions as mixed numbers.

## Skills Practice

1. Suppose you prefer to drink coffee rather than tea 3 times out of every 5.
   (a) What fraction of the time do you drink coffee?
   (b) What fraction of the time do you drink tea?
   (c) What is your coffee to tea preference ratio?
   (d) What is your tea to coffee preference ratio?

2. Add and simplify.
   (a) $\dfrac{8}{15} + \dfrac{4}{15}$   (b) $\dfrac{5}{11} + \dfrac{2}{3}$   (c) $5\dfrac{2}{9} + 7\dfrac{1}{6}$   (d) $4 + 7\dfrac{1}{3} + 2\dfrac{1}{2}$

3. Multiply and simplify.
   (a) $\dfrac{5}{1} \times \dfrac{4}{7}$   (b) $\dfrac{8}{9} \times \dfrac{4}{7}$   (c) $4 \times 2\dfrac{5}{6}$   (d) $\left(\dfrac{5}{13} \times 12\right) + \left(\dfrac{8}{13} \times 17\right)$

*Answers to Skills Practice:*

1. (a) $\dfrac{3}{5}$   (b) $\dfrac{2}{5}$   (c) 3:2   (d) 2:3

2. (a) $\dfrac{4}{5}$   (b) $1\dfrac{4}{33}$   (c) $12\dfrac{7}{18}$   (d) $13\dfrac{5}{6}$

3. (a) $2\dfrac{6}{7}$   (b) $\dfrac{32}{63}$   (c) $11\dfrac{1}{3}$   (d) $15\dfrac{1}{13}$

## Hints for Odd-Numbered Problems

### Section 4.1

1. In a continuous problem the items can be divided into pieces without a loss in value.

3. (d) How does Madeline, the divider, value the pieces of the division? How does Graham value the piece he received compared to the piece Madeline received?

5. (c) A preference ratio of 2 to 1 means that, out of a total value of 3, the cheddar cheese is preferred 2 out of 3 times.
   (d) The cheddar half is valued more than the Monterey jack half. What dollar amount would be placed on the cheddar half? There are 32 ounces of cheddar cheese. The wedge contains 20 ounces. The value of the wedge is found as follows:

$$\frac{\text{oz in cheddar wedge}}{\text{total cheddar oz}} \times \text{value of cheddar}$$

   (e) This time wedges are given in terms of the degree measure of the angle. In an entire half, there are 180°. The value of a wedge is found as follows:

$$\frac{\text{degrees in wedge}}{180°} \times \text{dollar value}$$

7. (a) All the value will be placed on the laundry soap.
   (c) If the total dollar value is divided into four pieces, then three of them are associated with the laundry soap. Thus, the laundry soap is $\frac{3}{4}$ of the dollar value of the package while the fabric softener is $\frac{1}{4}$ of the value.

9. Patrick can assign 3 points to every daytime minute and 1 point to every evening minute.

11. A 5-to-1 preference ratio means the vanilla frosted half of the cake is $\frac{5}{6}$ the value of the entire cake.

13. Kurt does not value vanilla frosting at all. All of the value is associated with the chocolate frosting.

15. Select a fraction of one type of cake and calculate its monetary value. The remaining dollar amount must come from the other type of cake.

17. To Kurt, the vanilla half has no value. Any amount of the vanilla side may be used or none need be used. All of the value must come from the chocolate side.

19. (d) Each division divides the ice cream into two portions. Both portions in a division must be valued equally by Jeanne.

21. (e) Max will select the portion in a division which he values the most, so calculate the value of each portion in each division.

23. (a) There are many ways this can be done. Assign points to each hour based on Casey's preferences and determine the total value of the combined hours. The two schedules Casey must create have to both be valued at half of the total value of the combined hours.

25. If a plan is acceptable to only one person, then give them that plan.

27. (b) With a 5-to-3-to-1 preference ratio, the weekday set of 1000 minutes is $\frac{5}{9}$ of the total value of the plan. What fraction of the total value would the weeknight and weekend minutes be?
    (d) Calculate the values of the plans for each of the two choosers. Determine which plans each person finds acceptable.

29. **(a) – (c)** Assign points to each foot of beach and determine the value of all 1800 feet of beach for each person. A fair share will be one-third of the total value.
   **(d)** Determine the value Lanie and Elsa place on each division. Which divisions do each woman find acceptable?

31. **(a)** The divider must create portions that are equal in value.
   **(b)** Determine each player's fair share requirement. Which portions are acceptable to each chooser?

33. **(b) – (d)** Find the total point value assigned to all of the desserts for one woman. A fair share to her will be a dessert that is at least one-third of the total.
   **(e)** Determine which desserts are acceptable to each chooser based on their fair share requirements.

35. Begin allocating slices according to the most restrictive player.

37. The last person to trim the portion created by the divider is the person who keeps the portion.

39. **(b)** One player has been given a piece of pie at the end of round 1, so in the second round, there are only three players.

41. **(b)** The same player cuts a piece of cake in each round until that player keeps the piece and is no longer a part of the process.

43. There are 15 days available at each time share location.
   **(d)** Once a player receives a package in part (c), find the number of days remaining at each location. For the person who is the divider, calculate the value that person would place on the total remaining days and determine that person's fair share requirement. There are many ways for the divider to create two packages of equal value.

Try to create a different solution than the one given.

## Section 4.2

1. Work through the method of sealed bids to see who keeps the necklace and see how much money is paid for the necklace.

3. Work through the method of sealed bids to see who keeps the necklace and see how much money is paid for the necklace.

5. Cheryl does not want the necklace so her bid must be less than Luann's. Consider the result from problem 1.

7. If you knew another person's bid, and you really wanted the item, how could you make a bid to guarantee that you get it without spending more than necessary? If you did not want the item, how could you make a bid to guarantee that you get as much compensation as possible?

9. **(b)** See Example 4.5.

11. Round dollar amounts to the nearest penny.

13. **(b)** Only the highest bid for each card goes into the compensation fund.
   **(c)** Each boy receives one-third of their total bids from the compensation fund.

15. **(a)** Study Table 4.17 for an example of how to list the six arrangements.

17. **(a)** There are six possible arrangements of friends and seats.
   **(b)** How do we interpret the smallest point value in each arrangement? By selecting the arrangement for which the smallest point value is the largest, what has been accomplished?

19. **(c)** A proportional division would result in each couple receiving a fair share. With four couples participating, a fair share would be a boat worth at least one-fourth of the 100 points.

21. There are six possible arrangements of chores to kids.

**23.** A proportional division would result in each player receiving a fair share.

**25.** See Step 2 of the adjusted-winner procedure for two players.

**27. (a)** See Step 2 of the adjusted-winner procedure for two players.
**(b)** This is Step 3 of the adjusted-winner procedure.
**(c)** Once the item has been moved in part (b), recalculate the players' point totals and determine which case applies from Step 4 of the adjusted-winner procedure.

**29.** Step 1 of the adjusted-winner procedure has been completed. Begin with Step 2.

## Section 4.3

**1. (a)** Assign points to each ounce of ice cream.
**(c)** When there are two players, a proportional division will result in each player feeling they have received at least half.
**(d)** Envy results from one player valuing another player's portion more than their own.

**3.** How does the divider create the two portions?

**5. (a)** Assign points to each degree in a slice of pizza.
**(f)** How much value does Renn place on each serving?

**7.** Consider what happens in the divide-and-choose method for three players when the allocation of portions is determined easily because each chooser finds a different portion acceptable compared to what can happen if two players have to use the divide-and-choose method for two players to finish the division.

**9. (c)** Each woman's fair share requirement was determined in part (a).
**(d)** Does each woman feel she received a fair share?
**(e)** Does any woman feel the portion given to another woman is worth more than the one she received?

**11.** Consider what options are possible in the last-diminisher method when two remaining players have to use the divide-and-choose method for two players to finish the division.

**13.** When there are three players, and they each allocate 100 points, what would be a fair share of points to each player?

**15. (a)** Assign points to each ounce of juice according to the preference ratio.
**(b)** There are many ways in which the trimming can be done. Determine how many points must be trimmed and then decide how to trim ounces. Can you trim whole ounces, or must you trim fractions of ounces?

**17. (a)** Assign points to each ounce of ice cream according to the preference ratio.
**(b)** There are many ways in which the trimming can be done. Determine how many points must be trimmed and then decide how to trim ounces. Can you trim whole ounces, or must you trim fractions of ounces?

**19. (a)** Using Lucy's preference ratio, determine how much value she places on all of the ice cream first.
**(d)** Lucy was the divider, and Porter was the trimmer and trimmed a serving. According to Step 4 of the continuous envy-free division method for three players, who will select a serving first?
**(e)** The excess must be shared. See Step 1 of Part 2 of the continuous envy-free division method for three players.
**(g)** Verify that each player feels they received a fair share. Verify that each player values their own serving at least as much as the servings of the other players.

**21.** There are many ways to achieve a proportional, envy-free division. Follow the steps outlined in the section for the continuous envy-free division method for three players.

**23.** There are many ways to achieve a proportional, envy-free division. Follow the steps outlined in the section for the continuous envy-free division method for three players.

25. (a) There are many ways to achieve a proportional, envy-free division. Follow the steps outlined in the section for the continuous envy-free division method for three players.

(b) Each player must feel they received a fair share, or one-third of the total value of the hours.

(c) Each player must value their portion at least as much as they value the portions of the other players.

## Solutions to Odd-Numbered Problems in Section 4.1

1. (a) The land could be split into pieces without loss of value, but the car and necklace could not. This is a mixed fair-division problem.
   (b) Cutting an antique painting into pieces would destroy its value. Paintings cannot be subdivided, so this is a discrete fair-division problem.
   (c) A dog, a cat, and a hamster cannot be divided into pieces. This is a discrete fair-division problem.
   (d) A peanut butter pie may be divided into pieces without loss of value. This is a continuous fair-division problem.

3. (a) Graham prefers chocolate over maple, so he would select the half that is completely chocolate.
   (b) Both portions are identical. Graham would be equally happy with either half.
   (c) Graham would select the half with more chocolate.
   (d) Because Graham prefers chocolate over maple, he would feel like he received more than a fair share. Madeline was the divider so she created portions that she felt were equally valuable.

5. (a) If you like cheddar cheese twice as much as Monterey jack cheese, then you have a cheddar cheese to Monterey jack preference ratio of 2 to 1.
   (b) The cheddar cheese is $\frac{1}{2}$ of the cheese wheel.
   (c) We want to create a fraction that compares the value of the cheddar cheese to the value of the cheese wheel. Because you prefer cheddar cheese 2 to 1 over Monterey jack cheese, it will be convenient to think of the entire cheese wheel as having a value of 3 (using the preference ratio, we see 2 + 1 = 3). Then the value of the cheddar cheese is 2 out of 3 and the value of the Monterey Jack cheese is 1 out of 3.

   Fraction: $\dfrac{\text{value of cheddar cheese}}{\text{value of cheese wheel}} = \dfrac{2}{3}$

   (d) The entire cheese wheel was purchased for $15. The 32 ounces of cheddar cheese makes up $\frac{2}{3}$ of the value of the cheese. Therefore, the 32 ounces of cheddar cheese has a dollar value of $\frac{2}{3} \times \$15 = \$10$. If you cut off a wedge containing 20 ounces of cheddar cheese, then it will have a dollar value that is a fraction of the $10 as calculated next.

Dollar value of 20 ounces of cheddar cheese

$$= \frac{\text{ounces in cheddar cheese wedge}}{\text{total ounces of cheddar cheese}} \times \text{dollar value of cheddar cheese}$$

$$= \frac{20}{32} \times \$10$$

$$= \frac{200}{32}$$

$$= \$6.25.$$

(e) **Wedge I** contains $\frac{50°}{180°} = \frac{5}{18}$ of the cheddar cheese. From part (d), we know the cheddar cheese

is valued at $10, so wedge I has a dollar value of $\frac{5}{18} \times \$10 = \frac{\$50}{18} \approx \$2.78.$

**Wedge III** contains $\frac{15°}{180°} = \frac{1}{12}$ of the cheddar cheese and $\frac{120°}{180°} = \frac{2}{3}$ of the Monterey jack

cheese. We know the cheddar cheese is valued at $10. The entire cheese was $15, so the
Monterey jack cheese is valued at $15 − $10 = $5. Wedge III has a dollar value of

$$\frac{1}{12} \times \$10 + \frac{2}{3} \times \$5 = \frac{\$10}{12} + \frac{\$10}{3} = \frac{\$10}{12} + \frac{\$40}{12} = \frac{\$50}{12} \approx \$4.17.$$

7. (a) Donna never uses fabric softener, so it would have no value to her. Donna's fabric softener to
laundry soap preference ratio is 0 to 1. All of the dollar value would be associated with the
laundry soap, so she would place a value of $0 on the fabric softener and a value of $6.40 on the
laundry soap.

(b) Pierce values both items equally, so his fabric softener to laundry soap preference ratio is 1 to 1.

The fabric softener is $\frac{1}{2}$ of the value, and the laundry soap is $\frac{1}{2}$ of the value. Therefore, the

dollar value Pierce would place on each item is $\frac{1}{2} \times \$6.40 = \$3.20.$

(c) Because John values laundry soap 3 to 1 over fabric softener, he probably uses laundry soap 3
times more often than he uses fabric softener. A preference ratio of 3 to 1 means the laundry soap

represents $\frac{3}{4}$ of the value while the fabric softener represents $\frac{1}{4}$ of the value. Therefore, John

would place a dollar value of $\frac{3}{4} \times \$6.40 = \frac{\$19.20}{4} = \$4.80$ on the laundry soap, and he would

place a dollar value of $\frac{1}{4} \times \$6.40 = \frac{\$6.40}{4} = \$1.60$ on the fabric softener.

(d) Because Bethany values fabric softener 4 to 1 over laundry soap, she probably uses fabric softener
4 times more often than she uses laundry soap. Or perhaps she adds four times as much fabric
softener as laundry soap for a load of clothes. A preference ratio of 4 to 1 means the fabric

softener represents $\frac{4}{5}$ of the value while the laundry soap represents $\frac{1}{5}$ of the value. Therefore,

Bethany would place a dollar value of $\frac{4}{5} \times \$6.40 = \frac{\$25.60}{5} = \$5.12$ on the fabric softener, and she

would place a dollar value of $\frac{1}{5} \times \$6.40 = \frac{\$6.40}{5} = \$1.28$ on the laundry soap.

9. Patrick can assign points to the minutes. One way to do this is to assign 3 points to every daytime minute and 1 point to every evening minute for a total of $(500 \times 3) + (500 \times 1) = 1500 + 500 = 2000$ points. The entire package of cell-phone minutes has a value of 2000 points, so a fair share of minutes according to Patrick would be a share worth $\dfrac{2000}{2} = 1000$ points. For each plan, find the point value according to Patrick.

| Plan | Phone 1 | Phone 2 |
|------|---------|---------|
| Plan A | $(300 \times 3) + (150 \times 1) = 1050$ points | $(200 \times 3) + (350 \times 1) = 950$ points |
| Plan B | $(400 \times 3) + (100 \times 1) = 1300$ points | $(100 \times 3) + (400 \times 1) = 700$ points |
| Plan C | $(300 \times 3) + (100 \times 1) = 1000$ points | $(200 \times 3) + (400 \times 1) = 1000$ points |
| Plan D | $(100 \times 3) + (500 \times 1) = 800$ points | $(400 \times 3) + (0 \times 1) = 1200$ points |

Plan C offers a fair division according to Patrick.

11. Brian paid $9.00 for the entire cake. It is frosted with half vanilla and half chocolate.

(a) Brian's vanilla to chocolate preference ratio is 5 to 1. The vanilla portion is worth $\dfrac{5}{6}$ of the value of the cake. He would place a value of $\dfrac{5}{6} \times \$9 = \dfrac{\$45}{6} = \$7.50$ on the vanilla half of the cake.

(b) The chocolate portion is worth $\dfrac{1}{6}$ of the value of the cake. He would place a value of $\dfrac{1}{6} \times \$9 = \dfrac{\$9}{6} = \$1.50$ on the chocolate half of the cake.

(c) Monetary Value of portion = (fraction of chocolate × $1.50) + (fraction of vanilla × $7.50)

| Portion | Monetary Value |
|---------|----------------|
| I | $(1 \times \$1.50) + \left(\dfrac{1}{2} \times \$7.50\right) = \$5.25$ |
| II | $\left(\dfrac{1}{2} \times \$1.50\right) + (1 \times \$7.50) = \$8.25$ |
| III | $\left(\dfrac{1}{2} \times \$1.50\right) + \left(\dfrac{1}{2} \times \$7.50\right) = \$4.50$ |
| IV | $\left(\dfrac{3}{4} \times \$1.50\right) + \left(\dfrac{1}{3} \times \$7.50\right) \approx \$3.63$ |

13. Kurt paid $9.00 for the entire cake. It is frosted with half vanilla and half chocolate.
   (a) Kurt's vanilla to chocolate preference ratio is 0 to 1. The vanilla portion has no value to Kurt. He would place a value of $0 \times \$9 = \$0.00$ on the vanilla half of the cake.
   (b) The chocolate portion is worth the entire amount he paid. He would place a value of $1 \times \$9 = \$9.00$ on the chocolate half of the cake.
   (c) Monetary Value of portion = (fraction of chocolate × $9.00) + (fraction of vanilla × $0.00)

| Portion | Monetary Value |
|---------|----------------|
| I | $(1 \times \$9.00) + \left(\dfrac{1}{2} \times \$0.00\right) = \$9.00$ |
| II | $\left(\dfrac{1}{2} \times \$9.00\right) + (1 \times \$0.00) = \$4.50$ |
| III | $\left(\dfrac{1}{2} \times \$9.00\right) + \left(\dfrac{1}{2} \times \$0.00\right) = \$4.50$ |

| IV | $\left(\dfrac{3}{4} \times \$9.00\right) + \left(\dfrac{1}{3} \times \$0.00\right) = \$6.75$ |
|---|---|

**15.** The new portions can be created in different ways. Your answers may vary. Many solutions are possible.

**(a)** Create a portion that is worth $3.50 to Brian. From problem 11, we know that to Brian, the vanilla half is worth $7.50, and the chocolate half is worth $1.50. Suppose we create a portion that uses $\dfrac{1}{3}$ of the vanilla half. Then it has a value of $\dfrac{1}{3} \times \$7.50 = \dfrac{\$7.50}{3} = \$2.50$. The remaining $1.00 must come from the chocolate half of the cake. What fraction of the chocolate half should we take so that it has a value of $1.00? Set up an equation.

$$(\text{Fraction of Chocolate Half}) \times \$1.50 = \$1.00$$

$$\begin{aligned} \text{Fraction of Chocolate Half} &= \frac{\$1.00}{\$1.50} \\ &= \frac{100}{150} \\ &= \frac{2}{3} \end{aligned}$$

Therefore, use $\dfrac{1}{3}$ of the vanilla half and $\dfrac{2}{3}$ of the chocolate half.

**(b)** Create a portion that is worth $2.50 to Brian. The vanilla half is worth $7.50, and the chocolate half is worth $1.50. Suppose we create a portion that uses $\dfrac{1}{5}$ of the vanilla half, then it has a value of $\dfrac{1}{5} \times \$7.50 = \dfrac{\$7.50}{5} = \$1.50$. The remaining $1.00 must come from the chocolate half of the cake. What fraction of the chocolate half should we take so that it has a value of $1.00? Set up an equation.

$$(\text{Fraction of Chocolate Half}) \times \$1.50 = \$1.00$$

$$\begin{aligned} \text{Fraction of Chocolate Half} &= \frac{\$1.00}{\$1.50} \\ &= \frac{100}{150} \\ &= \frac{2}{3} \end{aligned}$$

Therefore, use $\dfrac{1}{5}$ of the vanilla half and $\dfrac{2}{3}$ of the chocolate half.

**(c)** Create a portion that is worth $6.75 to Brian. The vanilla half is worth $7.50, and the chocolate half is worth $1.50. Suppose we create a portion that uses $\dfrac{4}{5}$ of the vanilla half. Then it has a value of $\dfrac{4}{5} \times \$7.50 = \dfrac{\$30.00}{5} = \$6.00$. The remaining $0.75 of value must come from the chocolate half of the cake. What fraction of the chocolate half should we take so that it has a value of $0.75? Set up an equation.

$$(\text{Fraction of Chocolate Half}) \times \$1.50 = \$0.75$$

$$\text{Fraction of Chocolate Half} = \frac{\$0.75}{\$1.50}$$

$$= \frac{75}{150}$$

$$= \frac{1}{2}$$

Therefore, use $\frac{4}{5}$ of the vanilla half and $\frac{1}{2}$ of the chocolate half.

**17.** The new portions can be created in different ways. Your answers may vary. Many solutions are possible.

**(a)** Create a portion that is worth $3.50 to Kurt. From problem 13, we know that to Kurt, the vanilla half is worth $0.00, and the chocolate half is worth $9.00. It does not matter how much of the vanilla half is used, it will not add any value to the final portion. The entire $3.50 must come from the chocolate half of the cake. What fraction of the chocolate half should we take so that it has a value of $3.50? Set up an equation.

$$(\text{Fraction of Chocolate Half}) \times \$9.00 = \$3.50$$

$$\text{Fraction of Chocolate Half} = \frac{\$3.50}{\$9.00}$$

$$= \frac{350}{900}$$

$$= \frac{7}{18}$$

Therefore, use any fraction of the vanilla half and $\frac{7}{18}$ of the chocolate half.

**(b)** Create a portion that is worth $2.50 to Kurt. From problem 13, we know that to Kurt, the vanilla half is worth $0.00, and the chocolate half is worth $9.00. It does not matter how much of the vanilla half is used, it will not add any value to the final portion. The entire $2.50 must come from the chocolate half of the cake. What fraction of the chocolate half should we take so that it has a value of $2.50? Set up an equation.

$$(\text{Fraction of Chocolate Half}) \times \$9.00 = \$2.50$$

$$\text{Fraction of Chocolate Half} = \frac{\$2.50}{\$9.00}$$

$$= \frac{250}{900}$$

$$= \frac{5}{18}$$

Therefore, use any fraction of the vanilla half and $\frac{5}{18}$ of the chocolate half.

**(c)** Create a portion that is worth $6.75 to Kurt. From problem 13, we know that to Kurt, the vanilla half is worth $0.00, and the chocolate half is worth $9.00. It does not matter how much of the vanilla half is used, it will not add any value to the final portion. The entire $6.75 must come from the chocolate half of the cake. What fraction of the chocolate half should we take so that it has a value of $6.75? Set up an equation.

$$(\text{Fraction of Chocolate Half}) \times \$9.00 = \$6.75$$
$$\text{Fraction of Chocolate Half} = \frac{\$6.75}{\$9.00}$$
$$= \frac{675}{900}$$
$$= \frac{3}{4}$$

Therefore, use any fraction of the vanilla half and $\frac{3}{4}$ of the chocolate half.

19. (a) Jeanne could assign 2 points to each ounce of vanilla ice cream. (Many other values could be assigned to an ounce of vanilla ice cream as long as she assigns twice as many points to every ounce of vanilla ice cream as she does to every ounce of chocolate ice cream.)
   (b) Jeanne could assign 1 point to every ounce of chocolate ice cream.
   (c) There are 5 ounces of vanilla ice cream and 2 ounces of chocolate ice cream, so the bowl of ice cream will have a total value of (5 ounces of vanilla × 2 points per ounce) + (2 ounces of chocolate × 1 point per ounce) = 10 points + 2 points = 12 points.
   (d) A fair share to Jeanne would be worth half of the total found in part (c). Any division that is fair would be worth $\frac{12}{2} = 6$ points. Many answers are possible.

   **Division I:** If 2.5 ounces of vanilla are used, then 2 × 2.5 = 5 points are accounted for. One ounce of chocolate ice cream is needed because every ounce of chocolate ice cream is worth 1 point. Both portions would contain 2.5 ounces of vanilla and 1 ounce of chocolate ice cream.
   **Division II:** If 2 ounces of vanilla are used, then 2 × 2 = 4 points are accounted for. The remaining 2 points must come from the chocolate portion. There are only two ounces of chocolate ice cream. Each has been assigned 1 point, so both must be selected. Use 2 ounces of vanilla ice cream and 2 ounces of chocolate ice cream in one portion. The rest of the ice cream would go into the other portion. In the other portion, use 3 ounces of vanilla ice cream.

21. (a) Marcus could assign 5 points to each ounce of ice cream. (Many other values could be assigned to an ounce of ice cream as long as he maintains a 5-to-2 ice cream to fudge cake point ratio.)
   (b) Marcus could assign 2 points to every ounce of fudge cake.
   (c) There are 6 ounces of fudge cake and 4 ounces of ice cream, so the dessert will have a total value of (6 ounces of fudge cake × 2 points per ounce) + (4 ounces of ice cream × 5 points per ounce) = 12 points + 20 points = 32 points.
   (d) A fair share to Marcus would be worth half of the total found in part (c). Any division that is fair would be worth $\frac{32}{2} = 16$ points. Many answers are possible.

   **Division I:** If 3 ounces of fudge cake are used, then 3 × 2 = 6 points are accounted for. The remaining 10 points must come from the ice cream.
   ounces of ice cream × 5 points per ounce = 10 points
   $$\text{ounces of ice cream} = \frac{10}{5}$$
   $$\text{ounces of ice cream} = 2$$
   Use 3 ounces of fudge cake and 2 ounces of ice cream in one portion. The other portion would contain 3 ounces of fudge cake and 2 ounces of ice cream.
   **Division II:** If 4 ounces of fudge cake are used, then 4 × 2 = 8 points are accounted for. The remaining 8 points must come from the ice cream.

ounces of ice cream $\times$ 5 points per ounce = 8 points

$$\text{ounces of ice cream} = \frac{8}{5}$$

ounces of ice cream = 1.6

Use 4 ounces of fudge cake and 1.6 ounces of ice cream in one portion. The other portion would contain 2 ounces of fudge cake and 2.4 ounces of ice cream.

(e) Max prefers fudge cake 3 to 1 over ice cream, so he could assign 1 point to every ounce of ice cream and 3 points to every ounce of fudge cake. The entire dessert is worth (6 ounces fudge cake $\times$ 3 points per ounce) + (4 ounces ice cream $\times$ 1 point per ounce) = 18 points + 4 points = 22 points. A fair share to Max would be worth $\frac{22}{2} = 11$ points. For each division in part (d) determine the value Max would place on each portion, and which portion Max would select.

**Division I:**
Both portions contain 3 ounces of fudge cake and 2 ounces of ice cream, so each is worth (3 $\times$ 3) + (2 $\times$ 1) = 9 + 2 = 11 points. Max would be happy with either portion.

**Division II:**
The 4 ounces of fudge cake and 1.6 ounces of ice cream is worth (4 $\times$ 3) + (1.6 $\times$ 1) = 12 + 1.6 = 13.6 points.
The 2 ounces of fudge cake and 2.4 ounces of ice cream is worth (2 $\times$ 3) + (2.4 $\times$ 1) = 6 + 2.4 = 8.4 points.
Max would prefer the portion containing 4 ounces of fudge cake and 1.6 ounces of ice cream because it is worth more points. It is the more valuable portion to Max.

23. (a) Casey is the divider. Casey must create a division of the tutor's time that is fair according to his own values. He prefers morning hours to afternoon hours in a 3-to-1 ratio. Casey could assign 1 point to every afternoon hour and 3 points to every morning hour. The tutor's time is worth a total of (3 morning hours $\times$ 3 points per hour) + (4 afternoon hours $\times$ 1 point per hour) = 9 points + 4 points = 13 points. A fair share to Casey would be valued at $\frac{13}{2} = 6.5$ points. There are many divisions of the tutor's time that would be fair according to Casey. Your answer may vary. One way Casey could divide the tutor's time would be to schedule 2 morning hours and 0.5 afternoon hours for one student for a total of (2 $\times$ 3) + (0.5 $\times$ 1) = 6 + 0.5 = 6.5 points. The remaining hours, 1 morning hour and 3.5 afternoon hours, would be scheduled for the other student. Notice that Casey would value this schedule at (1 $\times$ 3) + (3.5 $\times$ 1) = 3 + 3.5 = 6.5 points also.

Tran will select the schedule that he values the most. He prefers afternoon hours to morning hours in a 2-to-1 ratio. If he assigns 1 point to every morning hour and 2 points to every afternoon hour, then determine the value of each schedule.
The 2 morning hours and 0.5 afternoon hours are worth (2 $\times$ 1) + (0.5 $\times$ 2) = 2 + 1 = 3 points.
The 1 morning hour and 3.5 afternoon hours are worth (1 $\times$ 1) + (3.5 $\times$ 2) = 1 + 7 = 8 points.
Tran would pick the schedule with 1 morning hour and 3.5 afternoon hours because he values it the most.

(b) Tran is the divider. Tran must create a division of the tutor's time that is fair according to his own values. He prefers afternoon hours to morning hours in a 2-to-1 ratio. Tran could assign 1 point to every morning hour and 2 points to every afternoon hour. The tutor's time is worth a total of (3 morning hours $\times$ 1 point per hour) + (4 afternoon hours $\times$ 2 points per hour) = 3 points + 8 points = 11 points. A fair share to Tran would be valued at $\frac{11}{2} = 5.5$ points. There are many divisions of the tutor's time that would be fair according to Tran. Your answer may vary.

One way Tran could divide the tutor's time would be to schedule 0.5 morning hours and 2.5 afternoon hours for one student for a total of $(0.5 \times 1) + (2.5 \times 2) = 0.5 + 5 = 5.5$ points. The remaining hours, 2.5 morning hours and 1.5 afternoon hours, would be scheduled for the other student. Notice that Tran would value this schedule at $(2.5 \times 1) + (1.5 \times 2) = 2.5 + 3 = 5.5$ points also.

Casey will select the schedule that he values the most. If Casey assigns 3 points to each morning hour and 1 point to each afternoon hour, then determine the value of each schedule.
The 0.5 morning hours and 2.5 afternoon hours are worth $(0.5 \times 3) + (2.5 \times 1) = 1.5 + 2.5 = 4$ points.
The 2.5 morning hours and 1.5 afternoon hours are worth $(2.5 \times 3) + (1.5 \times 1) = 7.5 + 1.5 = 9$ points.
Casey would pick the schedule with 2.5 morning hours and 1.5 afternoon hours because he values it the most.

25. (a) Arnel created three plans that, to him, were all equally valuable, so he would be content to receive any of the three plans. Marc will accept only plan A, so give plan A to Marc. Paula approved of both plans A and C, but we just gave plan A to Marc, so give plan C to Paula. Plan B will be given to Arnel.

(b) Paula has listed only plan B as acceptable, so give plan B to Paula. Marc has listed only plan C as acceptable, so give plan C to Marc. Give plan A to Arnel.

(c) Paula and Marc cannot both receive plan B. Arnel created three plans that are all acceptable to him, so give him plan A or plan C. The remaining two plans should be dissolved. Now that Arnel has a plan he finds acceptable, we no longer have three people to worry about. Paula and Marc can start fresh and divide the remaining computer time using the divide-and-choose method for two players.

27. (a) Duane values weekday, weeknight, and weekend minutes equally, so his preference ratio is 1 to 1 to 1. To Duane, the weekday minutes are $\frac{1}{3}$ of the value of the package, so they have a monetary value of $\frac{1}{3} \times \$24.30 = \frac{\$24.30}{3} = \$8.10$. The weeknight minutes are $\frac{1}{3}$ of the value of the package, so they have a monetary value of $\frac{1}{3} \times \$24.30 = \frac{\$24.30}{3} = \$8.10$. The weekend minutes are $\frac{1}{3}$ of the value of the package, so they have a monetary value of $\frac{1}{3} \times \$24.30 = \frac{\$24.30}{3} = \$8.10$. The total package is valued at $\$24.30$, so a fair share to Duane would be a package valued at $\frac{1}{3} \times \$24.30 = \frac{\$24.30}{3} = \$8.10$.

(b) Doreen's weekday to weeknight to weekend preference ratio is 5 to 3 to 1. To Doreen, the weekday minutes are $\frac{5}{9}$ of the value of the package, so they have a monetary value of $\frac{5}{9} \times \$24.30 = \frac{\$121.50}{9} = \$13.50$. The weeknight minutes are $\frac{3}{9} = \frac{1}{3}$ of the value of the package, so they have a monetary value of $\frac{1}{3} \times \$24.30 = \frac{\$24.30}{3} = \$8.10$. The weekend minutes are $\frac{1}{9}$ of the value of the package, so they have a monetary value of $\frac{1}{9} \times \$24.30 = \frac{\$24.30}{9} = \$2.70$. The total package is valued at \$24.30, so a fair share to Doreen would be a package valued at $\frac{1}{3} \times \$24.30 = \frac{\$24.30}{3} = \$8.10$.

(c) Kaylee's weekday to weeknight to weekend preference ratio is 2 to 3 to 1. To Kaylee, the weekday minutes are $\frac{2}{6} = \frac{1}{3}$ of the value of the package, so they have a monetary value of $\frac{1}{3} \times \$24.30 = \frac{\$24.30}{3} = \$8.10$. The weeknight minutes are $\frac{3}{6} = \frac{1}{2}$ of the value of the package, so they have a monetary value of $\frac{1}{2} \times \$24.30 = \frac{\$24.30}{2} = \$12.15$. The weekend minutes are $\frac{1}{6}$ of the value of the package, so they have a monetary value of $\frac{1}{6} \times \$24.30 = \frac{\$24.30}{6} = \$4.05$. The total package is valued at \$24.30, so a fair share to Kaylee would be a package valued at $\frac{1}{3} \times \$24.30 = \frac{\$24.30}{3} = \$8.10$.

(d) In order to carry out the divide-and-choose method for three players, we must determine which plans are acceptable to the two choosers. (We already know the three plans are acceptable to the divider.) For Doreen and Kaylee, determine the monetary value of each plan.

**Doreen's Preferences:**
From part (b), we know the monetary value Doreen would place on each set of 1000 minutes.

| Plan | Monetary Value of Plan: $\left(\frac{\text{weekday}}{1000} \times \$13.50\right) + \left(\frac{\text{weeknight}}{1000} \times \$8.10\right) + \left(\frac{\text{weekend}}{1000} \times \$2.70\right)$ | Acceptable? (Worth at least \$8.10?) |
|---|---|---|
| Plan I | $\left(\frac{500}{1000} \times \$13.50\right) + \left(\frac{200}{1000} \times \$8.10\right) + \left(\frac{300}{1000} \times \$2.70\right) = \$9.18$ | Yes |
| Plan II | $\left(\frac{100}{1000} \times \$13.50\right) + \left(\frac{700}{1000} \times \$8.10\right) + \left(\frac{200}{1000} \times \$2.70\right) = \$7.56$ | No |
| Plan III | $\left(\frac{400}{1000} \times \$13.50\right) + \left(\frac{100}{1000} \times \$8.10\right) + \left(\frac{500}{1000} \times \$2.70\right) = \$7.56$ | No |

**Kaylee's Preferences:**
From part (c), we know the monetary value Kaylee would place on each set of 1000 minutes.

| Plan | Monetary Value of Plan: $\left(\frac{\text{weekday}}{1000} \times \$8.10\right) + \left(\frac{\text{weeknight}}{1000} \times \$12.15\right) + \left(\frac{\text{weekend}}{1000} \times \$4.05\right)$ | Acceptable? (Worth at least \$8.10?) |
|---|---|---|
| Plan I | $\left(\frac{500}{1000} \times \$8.10\right) + \left(\frac{200}{1000} \times \$12.15\right) + \left(\frac{300}{1000} \times \$4.05\right) \approx \$7.70$ | No |
| Plan II | $\left(\frac{100}{1000} \times \$8.10\right) + \left(\frac{700}{1000} \times \$12.15\right) + \left(\frac{200}{1000} \times \$4.05\right) \approx \$10.13$ | Yes |
| Plan III | $\left(\frac{400}{1000} \times \$8.10\right) + \left(\frac{100}{1000} \times \$12.15\right) + \left(\frac{500}{1000} \times \$4.05\right) = \$6.48$ | No |

Doreen finds only plan I acceptable, so give plan I to Doreen. Kaylee finds only plan II acceptable, so give plan II to Kaylee. Give plan III to Duane.

**29. (a)** Sondra values each area equally, so her sandy beach to rocky beach to grassy beach preference ratio is 1 to 1 to 1. Sondra could assign 1 point to each foot of sandy beach, 1 point to each foot of rocky beach, and 1 point to each foot of grassy beach. To Sondra, the total value of the beach is (600 feet sandy × 1 point per foot) + (600 feet rocky × 1 point per foot) + (600 feet grassy × 1 point per foot) = 600 points + 600 points + 600 points = 1800 points. A fair share to Sondra would be $\frac{1}{3} \times 1800 = \frac{1800}{3} = 600$ points.

**(b)** Lanie's sandy beach to rocky beach to grassy beach preference ratio is 2 to 1 to 1. Lanie could assign 2 points to each foot of sandy beach, 1 point to each foot of rocky beach, and 1 point to each foot of grassy beach. To Lanie, the total value of the beach is (600 feet sandy × 2 points per foot) + (600 feet rocky × 1 point per foot) + (600 feet grassy × 1 point per foot) = 1200 points + 600 points + 600 points = 2400 points. A fair share to Lanie would be $\frac{1}{3} \times 2400 = \frac{2400}{3} = 800$ points.

**(c)** Elsa's sandy beach to rocky beach to grassy beach preference ratio is 1 to 3 to 2. Elsa could assign 1 point to each foot of sandy beach, 3 points to each foot of rocky beach, and 2 points to each foot of grassy beach. To Elsa, the total value of the beach is (600 feet sandy × 1 point per foot) + (600 feet rocky × 3 points per foot) + (600 feet grassy × 2 points per foot) = 600 points + 1800 points + 1200 points = 3600 points. A fair share to Elsa would be $\frac{1}{3} \times 3600 = \frac{3600}{3} = 1200$ points.

**(d)** Sondra thinks it is fair to give each person a different type of beach. In order to carry out the divide-and-choose method for three players, we must determine which beaches are acceptable to the two choosers. (We already know the three beaches are acceptable to the divider.) For Lanie and Elsa, determine the point value of each beach.

**Lanie's Preferences:**
From part (b), we know the point value Lanie would place on each type of beach.

| Division | Point Value of Plan:<br>(feet sandy × 2) + (feet rocky × 1) + (feet grassy × 1) | Acceptable?<br>(Worth at least 800?) |
|---|---|---|
| 600 feet sandy beach | (600 × 2) + (0 × 1) + (0 × 1) = 1200 | Yes |
| 600 feet rocky beach | (0 × 2) + (600 × 1) + (0 × 1) = 600 | No |
| 600 feet grassy beach | (0 × 2) + (0 × 1) + (600 × 1) = 600 | No |

**Elsa's Preferences:**
From part (c), we know the point value Elsa would place on each type of beach.

| Division | Point Value of Plan:<br>(feet sandy × 1) + (feet rocky × 3) + (feet grassy × 2) | Acceptable?<br>(Worth at least 1200?) |
|---|---|---|
| 600 feet sandy beach | (600 × 1) + (0 × 3) + (0 × 2) = 600 | No |
| 600 feet rocky beach | (0 × 1) + (600 × 3) + (0 × 2) = 1800 | Yes |
| 600 feet grassy beach | (0 × 1) + (0 × 3) + (600 × 2) = 1200 | Yes |

Lanie found only the sandy beach acceptable, so give Lanie the 600 feet of sandy beach. Elsa found both the rocky and grassy beaches acceptable. Give Elsa the 600 feet of rocky beach. Give the 600 feet of grassy beach to Sondra. (It would also be fine to give Lanie the sandy beach, Elsa the grassy beach, and Sondra the rocky beach.)

(e) Lanie is the divider and created three options. In order to carry out the divide-and-choose method for three players, we must determine which options are acceptable to the two choosers. (We already know the three options are acceptable to the divider.) For Sondra and Elsa, determine the point value of each option.

**Sondra's Preferences:**
From part (a), we know the point value Sondra would place on each type of beach.

| Option | Point Value of Plan: (feet sandy × 1) + (feet rocky × 1) + (feet grassy × 1) | Acceptable? (Worth at least 600?) |
|---|---|---|
| I | (100 × 1) + (300 × 1) + (300 × 1) = 700 | Yes |
| II | (250 × 1) + (100 × 1) + (200 × 1) = 550 | No |
| III | (250 × 1) + (200 × 1) + (100 × 1) = 550 | No |

**Elsa's Preferences:**
From part (c), we know the point value Elsa would place on each type of beach.

| Option | Point Value of Plan: (feet sandy × 1) + (feet rocky × 3) + (feet grassy × 2) | Acceptable? (Worth at least 1200?) |
|---|---|---|
| I | (100 × 1) + (300 × 3) + (300 × 2) = 1600 | Yes |
| II | (250 × 1) + (100 × 3) + (200 × 2) = 950 | No |
| III | (250 × 1) + (200 × 3) + (100 × 2) = 1050 | No |

Both Sondra and Elsa found that option I is the only acceptable option. Give option II or option III to Lanie. Recombine the remaining feet of beach and let Sondra and Elsa divide it up using the divide-and-choose method for two players.

31. (a) Suzanne was the divider. She is the only player who values each portion equally. The divider is the one who creates portions of equal value.

(b) For Suzanne, each portion is fair.

For Eric, the total value of the three portions is 17 + 10 + 30 = 57. A fair share to Eric is $\frac{57}{3} = 19$ points. Eric finds only portion 3 fair.

For Janet, the total value of the three portions is 20 + 10 + 10 = 40. A fair share to Janet is $\frac{40}{3} \approx 13.33$ points. Janet finds only portion 1 fair.

(c) Give portion 1 to Janet, portion 2 to Suzanne, and portion 3 to Eric.

33. (a) Sharon could assign 1 point to each ounce of pudding, 2 points to each ounce of cobbler, and 3 points to each ounce of cheesecake.
Ally could assign 1 point to each ounce of pudding, 1 point to each ounce of cobbler, and 3 points to each ounce of cheesecake.
Bev could assign 1 point to each ounce of pudding, 1 point to each ounce of cobbler, and 1 point to each ounce of cheesecake.

| Dessert | Sharon's Point Value | Ally's Point Value | Bev's Point Value |
|---|---|---|---|
| 12 oz pudding | 12 oz × 1 point per oz = 12 points | 12 oz × 1 point per oz = 12 points | 12 oz × 1 point per oz = 12 points |
| 12 oz cobbler | 12 oz × 2 points per oz = 24 points | 12 oz × 1 point per oz = 12 points | 12 oz × 1 point per oz = 12 points |
| 12 oz cheesecake | 12 oz × 3 points per oz = 36 points | 12 oz × 3 points per oz = 36 points | 12 oz × 1 point per oz = 12 points |

(b) For Sharon, the total point value she places on all of the desserts is $12 + 24 + 36 = 72$ points. A fair share for Sharon would be $\frac{72}{3} = 24$ points.

(c) For Ally, the total point value she places on all of the desserts is $12 + 12 + 36 = 60$ points. A fair share for Ally would be $\frac{60}{3} = 20$ points.

(d) For Bev, the total point value she places on all of the desserts is $12 + 12 + 12 = 36$ points. A fair share for Bev would be $\frac{36}{3} = 12$ points.

(e) Bev is the divider and finds each dessert acceptable according to her values.
Sharon will accept a dessert if its point value is at least 24 points. She finds the cobbler and the cheesecake acceptable.
Ally will accept a dessert if its point value is at least 20 points. She finds only the cheesecake acceptable, so she must be given the cheesecake.
There is one way to allocate the desserts. Give Ally the cheesecake. Give Sharon the cobbler. Give Bev the pudding.

35. Notice that player 4 finds only slice C acceptable. Give slice C to player 4.
Player 3 is willing to accept both slice B and slice C. Because slice C has been given to player 4, give slice B to player 3.
Player 2 is willing to accept slice A and slice B. Because slice B has been given to player 3, give slice A to player 2.
Player 1, the divider, gets slice D.

37. (a) Tom moved the divider and Tara made no change, so the portion of the yard Tom adjusted goes to him. Tia and Tara will use the divide-and-choose method for two players to divide the remaining yard.

(b) Neither Tom nor Tara made changes, so Tia keeps the portion of the yard. Tom and Tara will use the divide-and-choose method for two players to divide the remaining yard.

(c) Tara was the last-diminisher, so she keeps the portion of the yard. Tom and Tia will use the divide-and-choose method for two players to divide the remaining yard.

39. (a) Player 3 was the last one to trim the pie, so player 3 keeps the piece at the end of the first round.

(b) Player 3 already has a piece of pie, so in the second round, there are three players left. When player 1 cuts a piece of pie, it will be valued at one-third of the remaining pie.

(c) Player 4 keeps the piece since player 4 trimmed it last.

(d) Player 1 and player 2 are left. They can use the divide-and-choose method for two players to equitably divide the remaining pie.

41. (a) After round 1, players 1, 2, 3, 4, and 6 are left.

(b) Player 1 cut the piece of cake in round 3. Player 1 will cut the piece of cake in every round until player 1 is the one who keeps the piece of cake.

(c) In round 1, player 5 keeps the slice because that player trimmed it last. In round 2, player 3 keeps the slice because that player trimmed it last. In round 3, player 1 keeps the slice because player 1 cut the slice and no player trimmed it. In round 4, player 6 keeps the slice because that player trimmed it last.

(d) Player 2 and player 4 are the last two remaining players. They can use the divide-and-choose method for two players to divide the remaining cake.

43. (a) Galen could assign 2 points to each day in the mountains, 1 point to each day at the ocean, and 2 points to each day in the woods. Galen would assign the following point value to each 15-day vacation:

Mountains: 15 days × 2 points per day = 30 points
Ocean: 15 days × 1 point per day = 15 points
Woods: 15 days × 2 point per day = 30 points
The total point value for all the vacation days is 30 + 15 + 30 = 75. A fair share of vacation days

for Galen would be valued at $\dfrac{75}{3} = 25$ points.

Leland could assign 3 points to each day in the mountains, 1 point to each day at the ocean, and 1 point to each day in the woods. Leland would assign the following point value to each 15-day vacation:
Mountains: 15 days × 3 points per day = 45 points
Ocean: 15 days × 1 point per day = 15 points
Woods: 15 days × 1 point per day = 15 points
The total point value for all the vacation days is 45 + 15 + 15 = 75. A fair share of vacation days

for Leland would be valued at $\dfrac{75}{3} = 25$ points.

Sandie could assign 1 point to each day in the mountains, 3 points to each day at the ocean, and 2 points to each day in the woods. Sandie would assign the following point value to each 15-day vacation:
Mountains: 15 days × 1 point per day = 15 points
Ocean: 15 days × 3 points per day = 45 points
Woods: 15 days × 2 point per day = 30 points
The total point value for all the vacation days is 15 + 45 + 30 = 90. A fair share of vacation days

for Sandie would be valued at $\dfrac{90}{3} = 30$ points.

(b) Galen would assign (5 × 2) + (3 × 1) + (6 × 2) = 10 + 3 + 12 = 25 points to the package he created, so it represents a fair share to him.

(c) Leland will place a value on the package created by Galen. Leland assigns (5 × 3) + (3 × 1) + (6 × 1) = 15 + 3 + 6 = 24 points to the package. A fair share according to Leland is 25 points, so Leland will not trim this package. It goes to Sandie next.

Sandie will place a value on the package created by Galen. Sandie assigns (5 × 1) + (3 × 3) + (6 × 2) = 5 + 9 + 12 = 26 points to the package. A fair share according to Sandie is 30 points, so Sandie will not trim this package.

Because neither Leland nor Sandie trimmed the package, Galen keeps the package.

(d) Many answers are possible. One possible solution is provided next.

There are 10 days left in the mountains, 12 days left at the ocean, and 9 days left in the woods for Leland and Sandie to divide using the divide-and-choose method for two players. Suppose Sandie is the divider. Sandie would assign the following point value to the remaining vacation days:
10 days in the Mountain: 10 days × 1 point per day = 10 points
12 days at the Ocean: 12 days × 3 point per day = 36 points
9 days in the Woods: 9 days × 2 point per day = 18 points
The total point value for the remaining days is 10 + 36 + 18 = 64 points. A fair share of vacation

days for Sandie would be valued at $\dfrac{64}{2} = 32$ points.

One fair division according to Sandie's values would be to create the following two packages:
Package I: 5 days in the mountains, 7 days at the ocean, and 3 days in the woods
Package II: 5 days in the mountains, 5 days at the ocean, and 6 days in the woods

Leland now must judge the value of each package and select the one he values the most.
Package I: (5 × 3) + (7 × 1) + (3 × 1) = 15 + 7 + 3 = 25 points
Package II: (5 × 3) + (5 × 1) + (6 × 1) = 15 + 5 + 6 = 26 points
Leland will select package II leaving package I for Sandie.

# Solutions to Odd-Numbered Problems in Section 4.2

1.  Luann is the highest bidder so she will get to keep the necklace and will put $1100 into the compensation fund. Cheryl is the low bidder so she will not keep the necklace and will put $0 into the compensation fund. The compensation fund contains a total of $1100. From the compensation fund, Luann will receive half of her bid back which is $550. Cheryl will also receive half of her bid which is $10. The compensation fund still contains $1100 − $550 − $10 = $540. The remaining $540 will be split between the women giving each woman an additional $270. Let's summarize what happened in a table:

| Player | Total Paid To Compensation Fund | Total Received From Compensation Fund | Total Dollars Paid (−) or Received (+) | Outcome |
|---|---|---|---|---|
| Luann | $1100 | $550 + $270 = $820 | −$1100 + $820 = −$280 | Received necklace worth $1000 Paid $280 to Cheryl |
| Cheryl | $0 | $10 + $270 = $280 | +$280 | Received $280 from Luann |

Luann kept the $1000 necklace and paid only $280 to Cheryl. Cheryl did not keep the necklace, and received only $280 in compensation which is approximately one quarter of what the necklace is worth. If Cheryl had made a reasonable bid, her compensation would have been greater. Cheryl decreased her own compensation by making a very low bid.

3.  Luann is the highest bidder so she will get to keep the necklace and will put $1100 into the compensation fund. Cheryl is the low bidder so she will not keep the necklace and will put $0 into the compensation fund. The compensation fund contains a total of $1100. From the compensation fund, Luann will receive half of her bid back which is $550. Cheryl will also receive half of her bid which is $450. The compensation fund still contains $1100 − $550 − $450 = $100. The remaining $100 will be split between the women giving each woman an additional $50. Let's summarize what happened in a table:

| Player | Total Paid To Compensation Fund | Total Received From Compensation Fund | Total Dollars Paid (−) or Received (+) | Outcome |
|---|---|---|---|---|
| Luann | $1100 | $550 + $50 = $600 | −$1100 + $600 = −$500 | Received necklace worth $1000 Paid $500 to Cheryl |
| Cheryl | $0 | $450 + $50 = $500 | +$500 | Received $500 from Luann |

Luann kept the $1000 necklace and paid $500 to Cheryl. Cheryl did not keep the necklace, and received $500 in compensation which is exactly half of what the necklace is worth. Cheryl's reasonable bid increased her compensation compared to problem 1.

5.  If Cheryl does not want the necklace, then she must make a bid that is less than Luann's. We know, from problem 1, that a very low bid will put Cheryl at a big disadvantage because she will receive less in compensation than if she made a reasonable bid. Therefore, Cheryl should make a bid that is only slightly less than Luann's so that her compensation is maximized. Luann's bid is $1100, so Cheryl should bid one penny less. Cheryl should bid $1099.99.

7.  Suppose you really want a certain item. If you knew someone else's bid, then you could outbid them by a penny. This would guarantee that you keep the item without spending more than is necessary. On the other hand, suppose you did not really want an item. If you knew someone else's bid, then you could underbid them by a penny. This would guarantee that they keep the item while you get paid as much as possible for the item. If no one knows anyone else's bid, no one can take advantage of the other person.

9. Kyle believes the DVD player is worth $110, and Kayla believes it is worth $98, so assume those are the bids they make and carry out the method of sealed bids. Kyle made the highest bid, so Kyle will place $110 into the compensation fund. Kyle and Kayla will each receive half of their bid from the compensation fund, so Kyle will receive $55 and Kayla will receive $49. The compensation fund still contains $110 − $55 − $49 = $6 to be split evenly between them, so each of them will receive an additional $3. Kyle receives the DVD player and pays a total of $52 to Kayla. Kayla receives a total of $52.

   (a) Kyle's bid was higher than Kayla's, so he should keep the DVD player.

   (b) They should consider the value of the DVD player to be $\dfrac{\$110+\$98}{2}=\dfrac{\$208}{2}=\$104.$

   (c) Kyle received the DVD player. Kyle is happy because, to him, the DVD player is worth more than the $104 they established. He felt it was worth $110 yet he only paid $52 to Kayla which is less than half of his value of the DVD player.

   (d) Kayla received $52 from Kyle for a DVD player that, to her, is worth less than the $104 they established. She felt it was worth $98 yet she was paid $52 which is more than half of her value of the DVD player.

11. Joann made the highest bid so she will keep the family farm and will place $1,300,000 into the compensation fund. Each sibling will receive one-third of their bid from the compensation fund. Round all values to the nearest penny.

   Joann will receive $\dfrac{\$1,300,000}{3}\approx\$433,333.33$ back from the compensation fund.

   Tom will receive $\dfrac{\$1,250,000}{3}\approx\$416,666.67$ from the compensation fund.

   Betty will receive $\dfrac{\$950,000}{3}\approx\$316,666.67$ from the compensation fund.

   The compensation fund still contains $1,300,000 − $433,333.33 − $416,666.67 − $316,666.67 = $133,333.33 which will be split evenly between them. Each of them will receive an additional $\dfrac{\$1,333,333.33}{3}\approx\$44,444.44$ .

   Joann receives the farm. She paid a total of $1,300,000 − $433,333.33 − $44,444.44 = $822,222.23 for the farm.
   Tom receives $416,666.67 + $44,444.44 = $461,111.11 and no property.
   Betty receives $316,666.67 + $44,444.44 = $361,111.11 and no property.

13. (a) Pete was the highest bidder for Common Card A. His bid was $0.50.
   Donald was the highest bidder for Common Card B. His bid was $0.55.
   Alex was the highest bidder for Rare Card A and Rare Card B. His bids were $1.35 and $1.60 respectively.

   (b) The highest bidder for each card must place his bid into the compensation fund, so a total of $0.50 + $0.55 + $1.35 + $1.60 = $4.00 is put into the compensation fund.

   (c) Each boy will receive one-third of his bid on each item from the compensation fund.

   Pete bid a total of $3.75 so he will receive $\dfrac{\$3.75}{3}=\$1.25$ from the compensation fund.

   Donald bid a total of $3.45 so he will receive $\dfrac{\$3.45}{3}=\$1.15$ from the compensation fund.

   Alex bid a total of $3.72 so he will receive $\dfrac{\$3.72}{3}=\$1.24$ from the compensation fund.

The compensation fund still contains $4.00 - $1.25 - $1.15 - $1.24 = $0.36 which will be split evenly between them. Each of them will receive an additional $\frac{\$0.36}{3} \approx \$0.12$ .

Therefore, the method of sealed bids yields the following result:

Pete keeps Common Card A. He received $1.25 + $0.12 = $1.37 from the compensation fund and paid $0.50, so he received a total of $1.37 - $0.50 = $0.87.

Donald keeps Common Card B. He received $1.15 + $0.12 = $1.27 from the compensation fund and paid $0.55, so he received a total of $1.27 - $0.55 = $0.72.

Alex keeps Rare Card A and Rare Card B. He received $1.24 + $0.12 = $1.36 from the compensation fund and paid a total of $2.95, so he paid a net amount of $2.95 - $1.36 = $1.59.

**15. (a)** There are six possible arrangements of teachers to rooms.

| All Possible Assignments of Teachers to Rooms | | | Smallest Point Value |
|---|---|---|---|
| Teacher A, room 1 30 points | Teacher B, room 2 32 points | Teacher C, room 3 37 points | 30 points |
| Teacher A, room 1 30 points | Teacher B, room 3 20 points | Teacher C, room 2 12 points | 12 points |
| Teacher A, room 2 50 points | Teacher B, room 1 48 points | Teacher C, room 3 37 points | 37 points |
| Teacher A, room 2 50 points | Teacher B, room 3 20 points | Teacher C, room 1 51 points | 20 points |
| Teacher A, room 3 20 points | Teacher B, room 1 48 points | Teacher C, room 2 12 points | 12 points |
| *Teacher A, room 3 20 points* | *Teacher B, room 2 32 points* | *Teacher C, room 1 51 points* | 20 points |

**(b)** Look at the "Smallest Point Value" column in the table in part (a). The largest of these is 37, and there is only one arrangement of teachers to rooms so that the smallest point value is 37. Therefore, there is one arrangement for which the smallest point value is as large as possible.

**(c)** Assign room 2 to Teacher A, room 1 to Teacher B, and room 3 to Teacher C.

**17. (a)** There are six possible arrangements of people to seats as given in the following table:

| All Possible Assignments of People to Seats | | | Smallest Point Value |
|---|---|---|---|
| Senji, front 20 points | Marta, middle 44 points | Cole, back 37 points | 20 points |
| Senji, front 20 points | Marta, back 35 points | Cole, middle 52 points | 20 points |
| Senji, middle 60 points | Marta, front 21 points | Cole, back 37 points | 21 points |
| Senji, middle 60 points | Marta, back 35 points | Cole, front 11 points | 11 points |
| Senji, back 20 points | Marta, front 21 points | Cole, middle 52 points | 20 points |
| *Senji, back 20 points* | *Marta, middle 44 points* | *Cole, front 11 points* | 11 points |

There is only one arrangement of people to seats such that the smallest point value is as large as possible. It is the arrangement that has a smallest point value of 21. Therefore, assign Senji the middle seat, Marta the front seat, and Cole the back seat.

**(b)** The numbers in the "Smallest Point Value" column in part (a) measure the satisfaction of the person who is the least happy with their seat for each assignment. By selecting the arrangement for which that smallest point value is the largest, we have selected the arrangement that most pleases the person who is the least happy. Any other arrangement would mean the person who is the least happy is even more unhappy.

19. (a) There are twenty four possible arrangements of couples to boats as given in the following table:

| All Possible Arrangements of Couples to Boats | | | | Smallest Point Value |
|---|---|---|---|---|
| A, canoe 25 points | B, Paddle 35 points | C, speed 24 points | D, raft 19 points | 19 points |
| A, canoe 25 points | B, Paddle 35 points | C, raft 32 points | D, speed 24 points | 24 points |
| A, canoe 25 points | B, speed 31 points | C, Paddle 34 points | D, raft 19 points | 19 points |
| A, canoe 25 points | B, speed 31 points | C, raft 32 points | D, Paddle 25 points | 25 points |
| A, canoe 25 points | B, raft 24 points | C, Paddle 34 points | D, speed 24 points | 24 points |
| A, canoe 25 points | B, raft 24 points | C, speed 24 points | D, Paddle 25 points | 24 points |
| A, Paddle 30 points | B, canoe 10 points | C, speed 24 points | D, raft 19 points | 10 points |
| A, Paddle 30 points | B, canoe 10 points | C, raft 32 points | D, speed 24 points | 10 points |
| A, Paddle 30 points | B, speed 31 points | C, canoe 10 points | D, raft 19 points | 10 points |
| A, Paddle 30 points | B, speed 31 points | C, raft 32 points | D, canoe 32 points | 30 points |
| A, Paddle 30 points | B, raft 24 points | C, canoe 10 points | D, speed 24 points | 10 points |
| A, Paddle 30 points | B, raft 24 points | C, speed 24 points | D, canoe 32 points | 24 points |
| A, speed 35 points | B, canoe 10 points | C, Paddle 34 points | D, raft 19 points | 10 points |
| A, speed 35 points | B, canoe 10 points | C, raft 32 points | D, Paddle 25 points | 10 points |
| A, speed 35 points | B, Paddle 35 points | C, canoe 10 points | D, raft 19 points | 10 points |
| A, speed 35 points | B, Paddle 35 points | C, raft 32 points | D, canoe 32 points | 32 points |
| A, speed 35 points | B, raft 24 points | C, canoe 10 points | D, Paddle 25 points | 25 points |
| A, speed 35 points | B, raft 24 points | C, Paddle 34 points | D, canoe 32 points | 24 points |
| A, raft 10 points | B, canoe 10 points | C, Paddle 34 points | D, speed 24 points | 10 points |
| A, raft 10 points | B, canoe 10 points | C, speed 24 points | D, Paddle 25 points | 10 points |
| A, raft 10 points | B, Paddle 35 points | C, canoe 24 points | D, speed 24 points | 10 points |
| A, raft 10 points | B, Paddle 35 points | C, speed 24 points | D, canoe 32 points | 10 points |
| A, raft 10 points | B, speed 31 points | C, canoe 10 points | D, Paddle 25 points | 10 points |
| A, raft 10 points | B, speed 31 points | C, Paddle 34 points | D, canoe 32 points | 10 points |

There is one arrangement for which the smallest point value is as large as possible. The largest point value in the column of smallest point values is 32. Assign the speedboat to couple A, the paddleboat to couple B, the raft to couple C, and the canoe to couple D.

(b) Couple A received their first choice, the speedboat valued at 35 points.
Couple B received their first choice, the paddleboat valued at 35 points.
Couple C received their second choice, the raft valued at 32 points.
Couple D received their first choice, the canoe valued at 32 points.

(c) A division that results in every couple receiving a fair share is called proportional. There are four couples, so a fair share to a couple would be a boat that is worth at least $\frac{1}{4}$ of the point total or at least 25 points. From part (b), we see three of the four couples received their first choice and one couple received their second choice. All of the boat assignments, however, resulted in each couple receiving a boat that, to them, was valued at more than 25 points. Therefore, the division is proportional.

**21.** There are six arrangements of kids to chores as given in the following table:

| All Possible Assignments of Kids to Chores | | | Smallest Point Value |
|---|---|---|---|
| Alex, mop floor 15 points | Suzanne, windows 25 points | James, scrub toilets 10 points | 10 points |
| Alex, mop floor 15 points | Suzanne, scrub toilets 15 points | James, windows 40 points | 15 points |
| Alex, windows 80 points | Suzanne, mop floor 60 points | James, scrub toilets 10 points | 10 points |
| Alex, windows 80 points | Suzanne, scrub toilets 15 points | James, mop floor 50 points | 15 points |
| Alex, scrub toilets 5 points | Suzanne, mop floor 60 points | James, windows 40 points | 5 points |
| Alex, scrub toilets 5 points | Suzanne, windows 25 points | James, mop floor 50 points | 5 points |

The largest of the smallest point values is 15. There are two assignments with a smallest point value of 15. Those arrangements are listed in the following table along with the middle point values.

| Two Assignments of Kids to Chores | | | Middle Point Value |
|---|---|---|---|
| Alex, mop floor 15 points | Suzanne, scrub toilets 15 points | James, windows 40 points | 15 points |
| Alex, windows 80 points | Suzanne, scrub toilets 15 points | James, mop floor 50 points | 50 points |

There is exactly one arrangement for which the middle number is as large as possible. Alex will wash the windows, Suzanne will scrub the toilets, and James will mop the floor.

**23.** With two players, each will receive half of their bid and an equal share of the surplus, so each player will receive at least half. With three players, each will receive a third of their bid and a share of the surplus, so each player will receive at least one third. As long as each player makes an honest bid, each player will always receive property or money that is worth, to them, more than a fair share. Honest bids will always result in a proportional division.

**25.** (a) The player who assigned the greatest number of points to an item is tentatively assigned ownership of the item. Initially, Qing is given the lamp and the wool rug, while Xu is given the armoire as shown in the following table.

| Qing | | Xu | |
|---|---|---|---|
| Lamp | 25 | Armoire | 60 |
| Wool rug | 35 | | |
| Total | 60 | Total | 60 |

**(b)** Notice, from part (a), that the players' point totals are the same so the adjusted-winner procedure ends and Qing keeps the lamp and the wool rug, while Xu keeps the armoire.

**27. (a)** The player who assigned the greatest number of points to an item is tentatively assigned ownership of the item. Initially, Joel is given the painting and the piano, while Jim is given the desk as shown in the following table.

| Joel | | Jim | |
|---|---|---|---|
| Painting | 20 | Desk | 50 |
| Piano | 45 | | |
| Total | 65 | Total | 50 |

After the initial assignment, Joel has more points than Jim.

**(b)** Ratio of points assigned to the painting:

$$\frac{\text{number of points assigned by Joel}}{\text{number of points assigned by Jim}} = \frac{20}{15} \approx 1.33$$

Ratio of points assigned to the piano:

$$\frac{\text{number of points assigned by Joel}}{\text{number of points assigned by Jim}} = \frac{45}{35} \approx 1.29$$

The ratio of points assigned to the piano is the smallest, so move the piano from Joel to Jim.

**(c)** After moving the piano from Joel to Jim, the point totals are given in the following table:

| Joel | | Jim | |
|---|---|---|---|
| Painting | 20 | Desk | 50 |
| | | Piano | 35 |
| Total | 20 | Total | 85 |

We are in Case C of the adjusted-winner procedure because Jim now has more points than Joel. In order to make the point totals equal, we must move a fraction of the piano back to Joel.

Let $q$ = the fraction of the piano to be returned to Joel.

$T_X = 20$ (Joel's point total not including the piano.)

$T_Y = 50$ (Jim's point total not including the piano.)

$P_X = 45$ (The number of points Joel assigned to the piano.)

$P_Y = 35$ (The number of points Jim assigned to the piano.)

Then, $q = \dfrac{T_Y - T_X + P_Y}{P_X + P_Y} = \dfrac{50 - 20 + 35}{45 + 35} = \dfrac{65}{80} = \dfrac{13}{16} = 0.8125.$

Joel will keep the painting and 81.25% ownership of the piano. Jim will keep the desk and 18.75% ownership of the piano.

**29.** The player who assigned the greatest number of points to an item is tentatively assigned ownership of the item. Initially, Holmes is given the finger print kit, the carrying case, and the hat while Watson is given the magnifying glass and cape as shown in the following table.

| Holmes | | Watson | |
|---|---|---|---|
| Finger-print kit | 25 | Magnifying glass | 17 |
| Carrying case | 20 | Cape | 40 |
| Hat | 18 | | |
| Total | 63 | Total | 57 |

After the initial assignment, Holmes has more points than Watson. In Step 3 of the adjusted-winner procedure, we must select the item currently assigned to Holmes for which the ratio of points assigned by Holmes to points assigned by Watson is the smallest.

Ratio of points assigned to the finger print kit:

$$\frac{\text{number of points assigned by Holmes}}{\text{number of points assigned by Watson}} = \frac{25}{24} \approx 1.04$$

Ratio of points assigned to the carrying case:
$$\frac{\text{number of points assigned by Holmes}}{\text{number of points assigned by Watson}} = \frac{20}{15} \approx 1.33$$

Ratio of points assigned to the hat:
$$\frac{\text{number of points assigned by Holmes}}{\text{number of points assigned by Watson}} = \frac{18}{4} = 4.5$$

The ratio of points assigned to the finger print kit is the smallest, so move the finger print kit from Holmes to Watson. In Step 4, reexamine the point totals. After moving the finger print kit from Holmes to Watson, the point totals are given in the following table:

| Holmes | | Watson | |
|---|---|---|---|
| Carrying case | 20 | Magnifying glass | 17 |
| Hat | 18 | Cape | 40 |
| | | Finger-print kit | 24 |
| Total | 38 | Total | 81 |

We are in Case C of the adjusted-winner procedure because Watson now has more points than Holmes. In order to make the point totals equal, we must move a fraction of the finger print kit back to Holmes. Let $q$ = the fraction of the finger print kit to be returned to Holmes.

$T_X$ = 38 (Holmes' point total not including the finger print kit.)
$T_Y$ = 57 (Watson's point total not including the finger print kit.)
$P_X$ = 25 (The number of points Holmes assigned to the finger print kit.)
$P_Y$ = 24 (The number of points Watson assigned to the finger print kit.)

Then, $q = \dfrac{T_Y - T_X + P_Y}{P_X + P_Y} = \dfrac{57 - 38 + 24}{25 + 24} = \dfrac{43}{49} \approx 0.8776.$

Holmes will keep the carrying case, the hat, and 87.76% ownership of the finger print kit. Watson will keep the magnifying glass, the cape, and 12.24% ownership of the finger print kit.

## Solutions to Odd-Numbered Problems in Section 4.3

1. **(a)** Alaina would place a value of 8 points on each bowl of ice cream.
   (2 points per oz of mint × 3 oz) + (1 point per oz of fudge × 2 oz) = 8 points
   (2 points per oz of mint × 2 oz) + (1 point per oz of fudge × 4 oz) = 8 points

   **(b)** Marie's Values:
   (3 points per oz of mint × 3 oz) + (2 point per oz of fudge × 2 oz) = 13 points
   (3 points per oz of mint × 2 oz) + (2 point per oz of fudge × 4 oz) = 14 points
   Marie would choose the bowl containing 2 ounces of mint chip and 4 ounces of fudge swirl.

   **(c)** A proportional division for two players is one in which both players feel the portion they received is worth at least $\frac{1}{2}$. Alaina was the divider so she values each bowl of ice cream equally. To her, both bowls are worth exactly $\frac{1}{2}$. Marie values the total amount of ice cream at (3 points per oz of mint × 5 oz) + (2 point per oz of fudge × 6 oz) = 27 points. To Marie, a fair share would be a portion worth $\frac{27}{2} = 13.5$ points. The bowl of ice cream Marie selected is worth 14 points to her,

so she also feels she received a portion worth at least $\frac{1}{2}$. This is a proportional division.

(d) A division resulting in envy would be a division in which one player values another player's portion more than they value their own. Alaina was the divider so she values her portion and Marie's portion equally. Alaina does not envy Marie's portion. Marie selected the portion that, to her, was worth more than a fair share. She feels Alaina's portion is worth only 13 points, which is less than a fair share. Marie does not envy Alaina's portion. This division is envy-free.

3. The divider always creates two portions that he or she feels are each worth exactly half, so the divider will always receive a share that is worth half of the total value.

5. (a) **Joseph's Values:**
   (3 points per degree of pepperoni × 180°) + (1 point per degree of olive × 180°) = 720 points. A fair share to Joseph is a portion worth at least $\frac{720}{3} = 240$ points.

   **Paul's Values:**
   (4 points per degree of pepperoni × 180°) + (1 point per degree of olive × 180°) = 900 points. A fair share to Paul is a portion worth at least $\frac{900}{3} = 300$ points.

   **Renn's Values:**
   (2 points per degree of pepperoni × 180°) + (1 point per degree of olive × 180°) = 540 points. A fair share to Renn is a portion worth at least $\frac{540}{3} = 180$ points.

   (b) Renn is the divider. Determine the value of each serving according to Renn's values.
   Serving I: (2 points per degree of pepperoni × 90°) + (1 point per degree of olive × 0°) = 180 points.
   Serving II: (2 points per degree of pepperoni × 50°) + (1 point per degree of olive × 80°) = 180 points.
   Serving III: (2 points per degree of pepperoni × 40°) + (1 point per degree of olive × 100°) = 180 points.
   Renn values each serving at 180 points.

   (c) Determine the value of each serving according to Joseph's values.
   Serving I: (3 points per degree of pepperoni × 90°) + (1 point per degree of olive × 0°) = 270 points.
   Serving II: (3 points per degree of pepperoni × 50°) + (1 point per degree of olive × 80°) = 230 points.
   Serving III: (3 points per degree of pepperoni × 40°) + (1 point per degree of olive × 100°) = 220 points.
   From part (a), we know Joseph will find a serving acceptable if it is worth at least 240 points. Joseph finds only Serving I acceptable.

   (d) Determine the value of each serving according to Paul's values.
   Serving I: (4 points per degree of pepperoni × 90°) + (1 point per degree of olive × 0°) = 360 points.
   Serving II: (4 points per degree of pepperoni × 50°) + (1 point per degree of olive × 80°) = 280 points.
   Serving III: (4 points per degree of pepperoni × 40°) + (1 point per degree of olive × 100°) = 260 points.
   From part (a), we know Paul will find a serving acceptable if it is worth at least 300 points. Paul finds only Serving I acceptable.

(e) Both Joseph and Paul find only Serving I acceptable. Renn can be given either Serving II or Serving III. Suppose Renn gets Serving II. Renn feels he received a serving worth exactly one third. Joseph and Paul will use the divide-and-choose method for two players to divide the remaining pizza. Each of them feels Renn received less than one third, so they will be dividing what they believe to be more than two-thirds of the entire pizza. Whoever is the divider will receive a portion worth exactly half of two thirds which is $\frac{1}{2} \times \frac{2}{3} = \frac{1}{3}$ of the pizza. The chooser will receive a portion they feel is worth at least half of two thirds which is at least $\frac{1}{2} \times \frac{2}{3} = \frac{1}{3}$ of the pizza. Therefore, this division is proportional.

(f) Determine Paul's value of each serving.

Serving A: (4 points per degree of pepperoni × 60°) + (1 point per degree of olive × 65°) = 305 points.

Serving B: (4 points per degree of pepperoni × 70°) + (1 point per degree of olive × 35°) = 315 points.

Paul would select Serving B because he values it more than he values Serving A. Joseph will receive Serving A.

This division is not envy-free. Renn was the original divider and has been watching Paul and Joseph divide the remaining pizza. Determine Renn's value of each serving.

Serving A: (2 points per degree of pepperoni × 60°) + (1 point per degree of olive × 65°) = 185 points.

Serving B: (2 points per degree of pepperoni × 70°) + (1 point per degree of olive × 35°) = 175 points.

From part (a), we know Renn feels a fair share is a portion worth 180 points. He received a portion worth exactly 180 points, but now he values Serving A more than he values his own. Renn envy's Joseph's serving.

7. The divide-and-choose method for three players can lead to a fair and envy-free division if the pieces can be allocated without recombining two portions and dividing them using the divide-and-choose method for two players, and the two choosers each feel a single (and different) piece is acceptable. For example, suppose the divider divides a cake into three portions he or she feels are each worth exactly one-third of the cake. If player 2 thinks only portion II is fair, and player 3 thinks only portion III is fair, then giving portion I to player 1, portion II to player 2, and portion III to player 3 results in a fair and envy-free division.

9. (a) **Melody's Values:**

(2 points per oz pie × 24 oz) + (1 point per oz cheesecake × 12 oz) = 60 points. A fair share to Melody is a portion worth at least $\frac{60}{3} = 20$ points.

**June's Values:**

(2 points per oz pie × 24 oz) + (1 point per oz cheesecake × 12 oz) = 60 points. A fair share to June is a portion worth at least $\frac{60}{3} = 20$ points.

**Rose's Values:**

(3 points per oz pie × 24 oz) + (2 point per oz cheesecake × 12 oz) = 96 points. A fair share to Rose is a portion worth at least $\frac{96}{3} = 32$ points.

(b) Melody's serving is worth (2 points per oz pie × 9 oz) + (1 point per oz cheesecake × 2 oz) = 20 points. From part (a), we know Melody feels a portion worth 20 points is a fair share. Melody created a portion that she feels is worth a fair share.

(c) June values the serving Melody created at (2 points per oz pie × 9 oz) + (1 point per oz cheesecake × 2 oz) = 20 points. From part (a), we know June feels a portion worth 20 points is a fair share. June feels this portion is fair, so she will not trim it.

Rose values the serving Melody created at (3 points per oz pie × 9 oz) + (2 points per oz cheesecake × 2 oz) = 31 points. From part (a), we know Rose feels a portion worth 32 points is a fair share. Rose feels this portion is less than a fair share, so she will not trim it.

Melody should keep the piece she created. The remaining desserts will be divided by June and Rose using the divide-and-choose method for two players.

(d) Determine the womens' values of the servings they received.

From part (b), Melody values her serving at 20 points. She has received a fair share.

Rose values her serving at (3 points per oz pie × 7 oz) + (2 points per oz cheesecake × 6 oz) = 33 points. A fair share to Rose is 32 points, so Rose has received more than a fair share.

June values her serving at (2 points per oz pie × 8 oz) + (1 point per oz cheesecake × 4 oz) = 20 points. A fair share to June is 20 points, so June has received a fair share.

Each woman received at least a fair share, so the division is proportional.

(e) To determine if the division from part (d) is envy-free, we must calculate the point value each woman would place on the two portions they did not receive.

**Melody's Values of Servings II and III:**

Serving II: (2 points per oz pie × 7 oz) + (1 point per oz cheesecake × 6 oz) = 20 points.

Serving III: (2 points per oz pie × 8 oz) + (1 point per oz cheesecake × 4 oz) = 20 points.

Melody values all the servings at 20 points. She has no envy for the servings of the other women.

**Rose's Values of Servings I and III:**

Serving I: (3 points per oz pie × 9 oz) + (2 points per oz cheesecake × 2 oz) = 31 points.

Serving III: (3 points per oz pie × 8 oz) + (2 points per oz cheesecake × 4 oz) = 32 points.

Rose values her own serving at 33 points and the servings of Melody and June at 31 points and 32 points respectively. She has a fair share and does not value the servings of the other women more than her own. Rose has no envy.

**June's Values of Servings I and II:**

Serving I: (2 points per oz pie × 9 oz) + (1 point per oz cheesecake × 2 oz) = 20 points.

Serving II: (2 points per oz pie × 7 oz) + (1 point per oz cheesecake × 6 oz) = 20 points.

June values all the servings at 20 points. She has no envy for the servings of the other women.

11. The last-diminisher method can lead to a fair and envy-free division. Consider the three player case. The first player creates a portion that is, to her, worth exactly one third. If no player trims the piece, then player 1 keeps it. The remaining two players finish the division using the divide-and-choose method for two players. As long as player 1 does not value either piece in the division created by player 2 and player 3 more than her own, then the division will be fair and envy-free.

13. With three players, a proportional division using the method of points would result in every player receiving a portion worth at least $\frac{100}{3} = 33\frac{1}{3}$ points.

(a) Each player received an item worth more than $33\frac{1}{3}$ points. The division is proportional.

(b) Each player received an item worth more than $33\frac{1}{3}$ points. The division is proportional.

(c) In this case, player A received an item worth 10 points which is much less than a fair share. This division is not proportional.

**15. (a)** The following table contains the point value you would assign to each serving.

| Serving | Fruit Juice Point Value 2 Points Per Ounce | Club Soda Point Value 3 Points Per Ounce | Total Point Value |
|---|---|---|---|
| Serving I | $2 \times 8 = 16$ | $3 \times 6 = 18$ | 34 |
| Serving II | $2 \times 9 = 18$ | $3 \times 4 = 12$ | 30 |
| Serving III | $2 \times 10 = 20$ | $3 \times 2 = 6$ | 26 |

Serving I has the greatest value, and Serving II has the second greatest value.

**(b)** There are many ways the trimming can be done. Your answers may be different. You must trim 4 points from Serving I so that its value is the same as that of Serving II.

Each ounce of fruit juice is worth 2 points, so you could trim 2 ounces of fruit juice from Serving I because 2 points per oz $\times$ 2 oz = 4 points. Each ounce of club soda is worth 3 points. You could trim $\dfrac{1}{2}$ ounce of fruit juice and 1 ounce of club soda because

$$\left(2 \text{ points per oz } \times \frac{1}{2}\text{oz}\right) + (3 \text{ points per oz } \times 1 \text{ oz}) = \left(\frac{2}{1} \times \frac{1}{2}\right) + 3 = 1 + 3 = 4 \text{ points .}$$

**17. (a)** The following table contains the point value you would assign to each serving.

| Serving | Chocolate Point Value 1 Point Per Ounce | Strawberry Point Value 3 Points Per Ounce | Vanilla Point Value 1 Point Per Ounce | Total Point Value |
|---|---|---|---|---|
| Serving I | $1 \times 6 = 6$ | $3 \times 2 = 6$ | $1 \times 3 = 3$ | 15 |
| Serving II | $1 \times 2 = 2$ | $3 \times 1 = 3$ | $1 \times 5 = 5$ | 10 |
| Serving III | $1 \times 1 = 1$ | $3 \times 3 = 9$ | $1 \times 4 = 4$ | 14 |

Serving I has the greatest value, and Serving III has the second greatest value.

**(b)** There are many ways the trimming can be done. Your answers may be different. You must trim 1 point from Serving I so that its value is the same as that of Serving III.

Each ounce of chocolate is worth 1 point, so you could trim 1 ounce of chocolate because 1 point per oz $\times$ 1 oz = 1 point.

Each ounce of strawberry is worth 3 points. You could trim $\dfrac{1}{3}$ ounce of strawberry because

$$3 \text{ points per oz } \times \frac{1}{3}\text{oz} = \frac{3}{1} \times \frac{1}{3} = \frac{3}{3} = 1 \text{ point.}$$

Each ounce of vanilla is worth 1 point, so you could trim 1 ounce of vanilla because 1 point per oz $\times$ 1 oz = 1 point.

**19. (a)** Lucy prefers chocolate 3 to 1 over strawberry. If she assigns 1 point to each ounce of strawberry and 3 points to each ounce of chocolate, then the ice cream is valued at (3 points per ounce $\times$ 15 ounces) + (1 point per ounce $\times$ 15 ounces) = 60 points. A fair share to Lucy would be a portion worth $\dfrac{60}{3} = 20$ points. The following table contains the point value Lucy would assign to each serving.

| Serving | Chocolate Point Value 3 Points Per Ounce | Strawberry Point Value 1 Point Per Ounce | Total Point Value |
|---|---|---|---|
| Serving I | $3 \times 4 = 12$ | $1 \times 8 = 8$ | 20 |
| Serving II | $3 \times 6 = 18$ | $1 \times 2 = 2$ | 20 |
| Serving III | $3 \times 5 = 15$ | $1 \times 5 = 5$ | 20 |

Lucy values each serving at 20 points.

**(b)** The following table contains the point value Porter would assign to each serving.

| Serving | Chocolate Point Value 1 Point Per Ounce | Strawberry Point Value 2 Points Per Ounce | Total Point Value |
|---|---|---|---|
| Serving I | $1 \times 4 = 4$ | $2 \times 8 = 16$ | 20 |
| Serving II | $1 \times 6 = 6$ | $2 \times 2 = 4$ | 10 |
| Serving III | $1 \times 5 = 5$ | $2 \times 5 = 10$ | 15 |

Porter feels Serving I has the greatest value, and Serving III has the second greatest value.

**(c)** If 1 ounce of chocolate and 2 ounces of strawberry are trimmed from Serving I, then Serving I is made up of 3 ounces of chocolate and 6 ounces of strawberry. Show that Porter values the trimmed Serving I the same as Serving III. Porter would value the trimmed Serving I at (1 point per ounce × 3 ounces) + (2 points per ounce × 6 ounces) = 15 points. Therefore, to Porter, the trimmed Serving I and Serving III are both valued at 15 points.

Alternately, Porter could have trimmed 3 ounces of chocolate and 1 ounce of strawberry from Serving I to trim the 5 points, however, there are many ways this can be done.

**(d)** Lucy was the divider, and Porter was the trimmer, so Arnold will select the serving he feels is the most valuable. Then Porter will select the serving he trimmed if it is available. Lucy will receive the last remaining serving. The following table contains the point value Arnold would assign to each serving.

| Serving | Chocolate Point Value 3 Points Per Ounce | Strawberry Point Value 2 Points Per Ounce | Total Point Value |
|---|---|---|---|
| Serving I | $3 \times 3 = 9$ | $2 \times 6 = 12$ | 21 |
| Serving II | $3 \times 6 = 18$ | $2 \times 2 = 4$ | 22 |
| Serving III | $3 \times 5 = 15$ | $2 \times 5 = 10$ | 25 |

Arnold values Serving III the most, so he will select Serving III. The trimmed serving is available, so Porter will receive Serving I because he trimmed it. Lucy will receive Serving II.

**(e)** The excess must be distributed, so there needs to be a new divider. The player who received the trimmed piece, Porter, becomes the new chooser, and Arnold becomes the new divider.

**(f)** The following table contains the point values Porter would assign to the three servings created from the excess ice cream.

| Serving | Chocolate Point Value 1 Point Per Ounce | Strawberry Point Value 2 Points Per Ounce | Total Point Value |
|---|---|---|---|
| Serving A | $1 \times \dfrac{7}{9} = \dfrac{7}{9}$ | $2 \times 0 = 0$ | $\dfrac{7}{9}$ |
| Serving B | $1 \times \dfrac{2}{9} = \dfrac{2}{9}$ | $2 \times \dfrac{5}{6} = \dfrac{2}{1} \times \dfrac{5}{6} = \dfrac{10}{6} = \dfrac{5}{3}$ | $\dfrac{2}{9} + \dfrac{5}{3} = \dfrac{2}{9} + \dfrac{15}{9} = \dfrac{17}{9} = 1\dfrac{8}{9}$ |
| Serving C | $1 \times 0 = 0$ | $2 \times 1\dfrac{1}{6} = \dfrac{2}{1} \times \dfrac{7}{6} = \dfrac{14}{6} = \dfrac{7}{3}$ | $\dfrac{7}{3} = 2\dfrac{1}{3}$ |

Porter will select Serving C. Next, Lucy will select the serving she values most. The following table contains the point values Lucy would assign to Servings A and B.

| Serving | Chocolate Point Value 3 Points Per Ounce | Strawberry Point Value 1 Point Per Ounce | Total Point Value |
|---|---|---|---|
| Serving A | $3 \times \dfrac{7}{9} = \dfrac{3}{1} \times \dfrac{7}{9} = \dfrac{21}{9} = \dfrac{7}{3}$ | $1 \times 0 = 0$ | $\dfrac{7}{3} = 2\dfrac{1}{3}$ |
| Serving B | $3 \times \dfrac{2}{9} = \dfrac{3}{1} \times \dfrac{2}{9} = \dfrac{6}{9} = \dfrac{2}{3}$ | $1 \times \dfrac{5}{6} = \dfrac{5}{6}$ | $\dfrac{2}{3} + \dfrac{5}{6} = \dfrac{4}{6} + \dfrac{5}{6} = \dfrac{9}{6} = \dfrac{3}{2} = 1\dfrac{1}{2}$ |

Lucy will select Serving A. Arnold will receive Serving B.

**(g)** Add the servings obtained from part (d) to the servings obtained from the excess in part (f).

Arnold received $5\frac{2}{9}$ ounces of chocolate and $5\frac{5}{6}$ ounces of strawberry.

Porter received 3 ounces of chocolate and $7\frac{1}{6}$ ounces of strawberry.

Lucy received $6\frac{7}{9}$ ounces of chocolate and 2 ounces of strawberry.

In order to demonstrate that this division is both proportional and envy-free, we must show that each player feels they received at least one-third of the ice cream and that they do not value any one else's serving more than their own. From part (a), we know a fair share to Lucy is 20 points. Arnold feels all of the ice cream is worth (3 pts per oz × 15 oz) + (2 pts per oz × 15 oz) = 75 points, so a fair share to Arnold is 25 points. Porter feels all the ice cream is worth (1 pt per oz × 15 oz) + (2 pts per oz × 15 oz) = 45 points, so a fair share to Porter is 15 points.

The following table gives the point values each player would give to their own serving and the servings of the other two players.

| Servings | Arnold's Value | Porter's Value | Lucy's Value |
|---|---|---|---|
| Arnold's Serving <br> $5\frac{2}{9}$ oz chocolate <br> $5\frac{5}{6}$ oz strawberry | $27\frac{1}{3}$ | $16\frac{8}{9}$ | $21\frac{1}{2}$ |
| Porter's Serving <br> 3 oz chocolate <br> $7\frac{1}{6}$ oz strawberry | $23\frac{1}{3}$ | $17\frac{1}{3}$ | $16\frac{1}{6}$ |
| Lucy's Serving <br> $6\frac{7}{9}$ oz chocolate <br> 2 oz strawberry | $24\frac{1}{3}$ | $10\frac{7}{9}$ | $22\frac{1}{3}$ |

All three players feel they received a fair share, so the division is proportional. Each player values their own serving more than they value the servings of the other two players, so the division is envy-free.

**21.** Follow the step-by-step method for finding a proportional, envy-free division as described in the text. There are many possible correct solutions. One possible solution will be provided, but you should try to create a different one.

**PART I (Distribute the majority of the minutes.)**

**STEP 1:** Duane is the first divider. He must divide the minutes into three plans that he considers to be of equal value. Duane values all of the minutes at (2 pts per weekday min × 1000 min) + (3 pts per weeknight min × 1000 min) + (1 pt per weekend min × 1000 min) = 6000 points. A fair share to Duane would be a plan worth $\frac{6000}{3} = 2000$ points. Duane could divide the minutes into the following three plans:

Plan I: 250 Weekday, 400 Weeknight, and 300 Weekend
Plan II: 350 Weekday, 300 Weeknight, and 400 Weekend
Plan III: 400 Weekday, 300 Weeknight, and 300 Weekend

**STEP 2:** Kaylee is the trimmer and will evaluate the three plans. Her value of each plan is given in the following table.

| Plan | Weekday Point Value 2 Points Per Day | Weeknight Point Value 3 Points Per Day | Weekend Point Value 4 Points Per Day | Total Point Value |
|---|---|---|---|---|
| Plan I | $2 \times 250 = 500$ | $3 \times 400 = 1200$ | $4 \times 300 = 1200$ | 2900 |
| Plan II | $2 \times 350 = 700$ | $3 \times 300 = 900$ | $4 \times 400 = 1600$ | 3200 |
| Plan III | $2 \times 400 = 800$ | $3 \times 300 = 900$ | $4 \times 300 = 1200$ | 2900 |

Kaylee feels that plan II has the greatest value and must be trimmed to have a point value of 2900.

**STEP 3:** One way Kaylee can trim plan II down is to trim 150 weekday minutes. These excess minutes are set aside.

**STEP 4:** Doreen is the first chooser and will select the plan she feels has the greatest value. The following table contains the value Doreen places on each plan where plan II is the trimmed plan.

| Plan | Weekday Point Value 5 Points Per Day | Weeknight Point Value 3 Points Per Day | Weekend Point Value 1 Point Per Day | Total Point Value |
|---|---|---|---|---|
| Plan I | $5 \times 250 = 1250$ | $3 \times 400 = 1200$ | $1 \times 300 = 300$ | 2750 |
| Plan II | $5 \times 200 = 1000$ | $3 \times 300 = 900$ | $1 \times 400 = 400$ | 2300 |
| Plan III | $5 \times 400 = 2000$ | $3 \times 300 = 900$ | $1 \times 300 = 300$ | 3200 |

Doreen will select plan III.

**STEP 5:** Kaylee, the trimmer, will get plan II because she trimmed that plan and it is still available.

**STEP 6:** Duane will get plan I because it is the remaining plan.

### PART 2 (Share the Excess.)

**STEP 1:** Doreen becomes the second divider, and Kaylee becomes the second chooser.

**STEP 2:** Doreen must divide the excess 150 weekday minutes into three portions of equal value. She can do this by splitting the 150 weekday minutes into three portions each containing 50 weekday minutes.

**STEPS 3 - 5:** Kaylee, the second chooser, will value each of the portions containing the excess minutes the same, because they are identical, so it does not matter which portion she takes. Doreen, Duane, and Kaylee each will receive 50 additional weekday minutes.

The final proportional, envy-free division gives Duane 300 weekday minutes, 400 weeknight minutes, and 300 weekend minutes. Doreen has 450 weekday minutes, 300 weeknight minutes, and 300 weekend minutes. Kaylee has 250 weekday minutes, 300 weeknight minutes, and 400 weekend minutes.

23. Follow the step-by-step method for finding a proportional, envy-free division as described in the text. There are many possible correct solutions. One possible solution will be provided, but you should try to create a different one.

### PART I (Distribute the majority of the linear feet of beach.)

**STEP 1:** Sondra is the first divider. She must divide the beach into three portions that she considers to be of equal value. Sondra values all of the beach at (1 pt per sandy foot × 600 min) + (2 pts per rocky foot × 600 min) + (1 pt per grassy foot × 600 min) = 2400 points. A fair share to Sondra would be a portion worth $\frac{2400}{3} = 800$ points. Sondra could divide the beach into the following three portions:

Portion I: 100 feet sandy, 300 feet rocky, and 100 feet grassy
Portion II: 200 feet sandy, 200 feet rocky, and 200 feet grassy
Portion III: 300 feet sandy, 100 feet rocky, and 300 feet grassy

**STEP 2:** Lanie is the trimmer and will evaluate the three portions. Her value of each portion is given in the following table.

| Portion | Sandy Point Value 4 Points Per Foot | Rocky Point Value 2 Points Per Foot | Grassy Point Value 1 Point Per Foot | Total Point Value |
|---------|-------------------------------------|-------------------------------------|-------------------------------------|-------------------|
| Portion I | $4 \times 100 = 400$ | $2 \times 300 = 600$ | $1 \times 100 = 100$ | 1100 |
| Portion II | $4 \times 200 = 800$ | $2 \times 200 = 400$ | $1 \times 200 = 200$ | 1400 |
| Portion III | $4 \times 300 = 1200$ | $2 \times 100 = 200$ | $1 \times 300 = 300$ | 1700 |

Lanie feels that portion III has the greatest value and must be trimmed to have a point value of 1400.

**STEP 3:** One way Lanie can trim portion III down is to trim 300 feet of grassy beach. These excess feet are set aside.

**STEP 4:** Elsa is the first chooser and will select the portion she feels has the greatest value. The following table contains the value Elsa places on each portion where portion III is the trimmed portion.

| Portion | Sandy Point Value 1 Point Per Foot | Rocky Point Value 3 Points Per Foot | Grassy Point Value 2 Points Per Foot | Total Point Value |
|---------|-------------------------------------|-------------------------------------|-------------------------------------|-------------------|
| Portion I | $1 \times 100 = 100$ | $3 \times 300 = 900$ | $2 \times 100 = 200$ | 1200 |
| Portion II | $1 \times 200 = 200$ | $3 \times 200 = 600$ | $2 \times 200 = 400$ | 1200 |
| Portion III | $1 \times 300 = 300$ | $3 \times 100 = 300$ | $2 \times 0 = 0$ | 600 |

Elsa will select either portion I or II. Suppose she selects portion I.

**STEP 5:** Lanie, the trimmer, will get portion III because she trimmed that plan and it is still available.

**STEP 6:** Sondra will get portion II because it is the remaining portion.

**PART 2 (Share the Excess.)**

**STEP 1:** Elsa becomes the second divider, and Lanie becomes the second chooser.

**STEP 2:** Elsa must divide the excess 300 feet of grassy beach into three portions of equal value. She can do this by splitting the 300 feet into three equal portions of 100 feet of grassy beach each.

**STEPS 3 - 5:** Lanie, the second chooser, values each portion the same because they are identical. Each person will receive 100 feet of grassy beach.

The final proportional, envy-free division gives Lanie 300 feet of sandy beach, 100 feet of rocky beach, and 100 feet of grassy beach. Sondra gets 200 feet of sandy beach, 200 feet of rocky beach, and 300 feet of grassy beach. Elsa gets 100 feet of sandy beach, 300 feet of rocky beach, and 200 feet of grassy beach.

25. (a) Follow the step-by-step method for finding a proportional, envy-free division as described in the text. There are many possible correct solutions. One possible solution will be provided, but you should try to find a different solution.

   **PART I (Distribute the majority of the hours.)**

   **STEP 1:** Tom is the first divider. He must divide the hours into three plans that he considers to be of equal value. Tom values all of the hours at (1 pt per restaurant hour × 30 hours) + (2 pts per video hour × 30 hours) + (3 pts per auto parts hour × 27 hours) = 171 points. A fair share to Tom would be a plan worth $\frac{171}{3} = 57$ points. Tom could divide the hours into the following three plans:

   Plan I: 7 restaurant hours, 10 video store hours, and 10 auto parts store hours
   Plan II: 15 restaurant hours, 12 video store hours, and 6 auto parts store hours
   Plan III: 8 restaurant hours, 8 video store hours, and 11 auto parts store hours

   **STEP 2:** Peter is the trimmer and will evaluate the three plans. His value of each plan is given in the following table.

| Plan | Restaurant Value 1 Point Per Hour | Video Value 1 Point Per Hour | Auto Part Value 1 Point Per Hour | Total Point Value |
|---|---|---|---|---|
| Plan I | $1 \times 7 = 7$ | $1 \times 10 = 10$ | $1 \times 10 = 10$ | 27 |
| Plan II | $1 \times 15 = 15$ | $1 \times 12 = 12$ | $1 \times 6 = 6$ | 33 |
| Plan III | $1 \times 8 = 8$ | $1 \times 8 = 8$ | $1 \times 11 = 11$ | 27 |

Peter feels that plan II has the greatest value and must be trimmed to have a point value of 27.

**STEP 3:** One way Peter can trim plan II down is to trim 6 hours of time working at the auto parts store. These excess hours are set aside.

**STEP 4:** Jack is the first chooser and will select the plan he feels has the greatest value. The following table contains the value Jack places on each plan where plan II is the trimmed plan.

| Plan | Restaurant Value 2 Points Per Hour | Video Value 2 Points Per Hour | Auto Part Value 1 Point Per Hour | Total Point Value |
|---|---|---|---|---|
| Plan I | $2 \times 7 = 14$ | $2 \times 10 = 20$ | $1 \times 10 = 10$ | 44 |
| Plan II | $2 \times 15 = 30$ | $2 \times 12 = 24$ | $1 \times 0 = 0$ | 54 |
| Plan III | $2 \times 8 = 16$ | $2 \times 8 = 16$ | $1 \times 11 = 11$ | 43 |

Jack will select plan II.

**STEP 5:** Peter, the trimmer, will get plan I because plan II is no longer available. (He could have also received plan III because he valued plans I and III equally.)

**STEP 6:** Tom will get plan III because it is the remaining plan.

**PART 2 (Share the Excess.)**

**STEP 1:** Peter becomes the second divider, and Jack becomes the second chooser.

**STEP 2:** Peter must divide the excess 6 hours of time working at the auto parts store into three portions of equal value. He can do this by splitting the excess time into portions that each contain 2 hours.

**STEP 3:** Jack, the second chooser, will value each portion equally because they are identical. Each player will receive an additional 2 hours working at the auto parts store.

The final proportional, envy-free division gives Tom 8 hours at the fast food restaurant, 8 hours at the video store, and 13 hours at the auto parts store. Jack gets 15 hours at the fast food restaurant, 12 hours at the video store, and 2 hours at the auto parts store. Peter gets 7 hours at the fast food restaurant, 10 hours at the video store, and 12 hours at the auto parts store.

(b) From part (a), we know that Tom values all the hours at 171 points and feels a fair share is a portion worth 57 points. Tom values his plan at (1 pt per restaurant hour × 8 hours) + (2 pts per video hour × 8 hours) + (3 pts per auto parts hour × 13 hours) = 63 points.

Jack values all of the hours at (2 pts per restaurant hour × 30 hours) + (2 pts per video hour × 30 hours) + (1 pt per auto parts hour × 27 hours) = 147 points. A fair share to Jack would be a plan worth $\frac{147}{3} = 49$ points. Jack values his plan at (2 pts per restaurant hour × 15 hours) + (2 pts per video hour × 12 hours) + (1 pt per auto parts hour × 2 hours) = 56 points.

Peter values all of the hours at (1 pt per restaurant hour × 30 hours) + (1 pt per video hour × 30 hours) + (1 pt per auto parts hour × 27 hours) = 87 points. A fair share to Peter would be a plan worth $\frac{87}{3} = 29$ points. Peter values his plan at (1 pt per restaurant hour × 7 hours) + (1 pt per video hour × 10 hours) + (1 pt per auto parts hour × 12 hours) = 29 points. Therefore, each player feels they received a fair share. The division is proportional.

(c) The following table contains the values each player places on their own plan and the plans of the other players.

| Plan | Tom's Value | Jack's Value | Peter's Value |
|---|---|---|---|
| Tom's Plan<br>8 Restaurant hours<br>8 Video hours<br>13 Auto part hours | 63 | 45 | 29 |
| Jack's Plan<br>15 Restaurant hours<br>12 Video hours<br>2 Auto part hours | 45 | 56 | 29 |
| Peter's Plan<br>7 Restaurant hours<br>10 Video hours<br>12 Auto part hours | 63 | 46 | 29 |

Each player values their own plan at least as much as they value the plans of the other players. This is an envy-free division.

## Solutions to Chapter 4 Review Problems

1. (a) The total number of ounces is 128 ounces + 72 ounces = 200 ounces. The milk is $\frac{128}{200}=\frac{16}{25}$ of the total, and the soda is $\frac{72}{200}=\frac{9}{25}$ of the total.

(b) Pete's milk to soda preference ratio is 3 to 1, so three times out of four he would select milk over soda.

(c) Based on Pete's preference ratio, he might assign 1 point to every ounce of soda and 3 points to every ounce of milk. The total value of milk and soda would be (3 points per oz × 128 oz) + (1 point per oz × 72 oz) = 456 points. Because the total cost was $4.56, every point is worth $\frac{\$4.56}{456\text{points}}=\$0.01$. The total value of the soda is 1 × 72 = 72 points, so its value in dollars would be $0.01 × 72 = $0.72. The milk must be worth $4.56 – $0.72 = $3.84.

(d) Six ounces of milk would be worth 3 × 6 = 18 points, so its value in dollars would be $0.01 × 18 = $0.18. Twelve ounces of soda would be worth 1 × 12 = 12 points, so its value in dollars would be $0.01 × 12 = $0.12. The total value in dollars of 6 ounces of milk and 12 ounces of soda is $0.18 + $0.12 = $0.30.

(e) Answers will vary, and many answers are possible. From part (c), we know the total value Pete places on the milk and soda is 456 points. When Pete creates a fair division, each portion must be worth $\frac{456}{2}=228$ points.

**Division I:**
Suppose Pete creates a portion containing 60 ounces of milk. To Pete, it is worth 3 × 60 = 180 points out of the 228 points. The remaining 228 – 180 = 48 ounces must come from the soda. Each ounce of soda is worth 1 point, so 48 ounces of soda would have to be used.
Portion I: 60 ounces of milk and 48 ounces of soda
Portion II: 68 ounces of milk and 24 ounces of soda

**Division II:**
Suppose Pete creates a portion containing 70 ounces of milk. To Pete, it is worth 3 × 70 = 210 points out of the 228 points. The remaining 228 – 210 = 18 ounces must come from the soda. Each ounce of soda is worth 1 point, so 18 ounces of soda would have to be used.
Portion I: 70 ounces of milk and 18 ounces of soda
Portion II: 58 ounces of milk and 54 ounces of soda

**(f)** If Russell prefers soda to milk in a ratio of 5 to 4, he might assign 4 points to every ounce of soda and 5 points to every ounce of milk. Calculate the point value Russell would place on each portion in each division from part (e).

**Division I:**
Russell's value of portion I is (5 points per oz × 60 oz) + (4 points per oz × 48 oz) = 492 points.
Russell's value of portion II is (5 points per oz × 68 oz) + (4 points per oz × 24 oz) = 436 points.
Russell would choose portion I.

**Division II:**
Russell's value of portion I is (5 points per oz × 70 oz) + (4 points per oz × 18 oz) = 422 points.
Russell's value of portion II is (5 points per oz × 58 oz) + (4 points per oz × 54 oz) = 506 points.
Russell would choose portion II.

2. **(a)** Maxine might assign 2 points to every mile of running and 5 points to every mile of biking. To Maxine, the entire competition is worth (5 points per mile × 70 miles) + (2 points per mile × 6 miles) = 362 points.

**(b)** From part (a), Maxine would value the entire competition to be worth 362 points, so a fair share would be a portion worth $\frac{362}{2} = 181$ points. A portion made up of 31 miles of biking and 1 mile of running would be valued at (5 points per mile × 31 miles) + (2 points per mile × 1 mile) = 157 points. Maxine would feel this was less than a fair share.

**(c)** Mary might assign 3 points to every mile of running and 1 point to every mile of biking. Calculate the value Mary would place on each option:
Mary's value of option I is (3 points per mile × 4 miles) + (1 point per mile × 34.6 miles) = 46.6 points.
Mary's value of option II is (3 points per mile × 2 miles) + (1 point per mile × 35.4 miles) = 41.4 points.
Mary would choose option I because she values it more than option II.

**(d)** We must determine what a fair share is to Mary. To Mary, the entire competition is worth (1 point per mile × 70 miles) + (3 points per mile × 6 miles) = 88 points. A fair share to Mary would be a portion worth $\frac{88}{2} = 44$ points.

If Mary created a portion that included 29 miles of biking, then she would assign (1 point per mile × 29 miles) = 29 points to the biking. The remaining 44 − 29 = 15 points must come from the running. Mary assigns 3 points for every mile of running so (3 points per mile × the number of miles running) = 15. Therefore, there should be $\frac{15}{3} = 5$ miles of running.

Option I: 29 miles of biking and 5 miles of running
Option II: 41 miles of biking and 1 mile of running

Maxine will choose the option she values the most.
Maxine's value of option I is (5 points per mile × 29 miles) + (2 points per mile × 5 miles) = 155 points.
Maxine's value of option II is (5 points per mile × 41 miles) + (2 points per mile × 1 mile) = 207 points.
Maxine will choose option II.

3. (a) Melinda does not value cheese and salami equally or she would have created slices that are each half of the pizza. She prefers cheese to salami because, to her, a smaller slice of cheese has the same value as all of the salami and a small portion of the cheese. Could her cheese to salami preference ratio be 2 to 1? Calculate the value of each slice and see if they are equal.

   Melinda's value of slice I: (2 points per degree of cheese × 150°) = 300 points

   Melinda's value of slice II: (2 points per degree of cheese × 30°) + (1 point per degree of salami × 180°) = 240 points.

   Therefore, she must not have a cheese to salami preference ratio of 2 to 1. Could she have a cheese to salami preference ratio of 3 to 2? Calculate the value of each slice and see if they are equal.

   Melinda's value of slice I: (3 points per degree of cheese × 150°) = 450 points

   Melinda's value of slice II: (3 points per degree of cheese × 30°) + (2 point per degree of salami × 180°) = 450 points.

   Melinda has a cheese to salami preference ratio of 3 to 2.

   (b) Therese prefers cheese to salami in a 4-to-3 preference ratio. She might assign 3 points for every degree of salami and 4 points for every degree of cheese.

   Therese's value of slice I: (4 points per degree of cheese × 150°) = 600 points

   Therese's value of slice II: (4 points per degree of cheese × 30°) + (3 point per degree of salami × 180°) = 660 points.

   Therefore, Therese would select slice II because she values it more.

4. Stan is the divider. Dawn and Tyler judge whether the parcels are acceptable according to their own values.

   (a) Dawn should get parcel 1 because she feels only parcel 1 is acceptable. Tyler should get parcel 3 because he feels only parcel 3 is acceptable. Stan should get parcel 2. He was the divider, so he will be happy with any of the three parcels.

   (b) Tyler should get parcel 2 because he feels only parcel 2 is acceptable. Dawn would be happy with either parcel 2 or parcel 3, but because parcel 2 has been given to Tyler, Dawn should get parcel 3. Stan should get parcel 1. He was the divider, so he will be happy with any of the three parcels.

   (c) Dawn and Tyler both want the same parcel. Stan should get either parcel 1 or parcel 2 because neither Dawn nor Tyler felt those parcels were acceptable. Once Stan has his parcel, he is out of the process. The remaining two parcels should be recombined, and Dawn and Tyler should use the divide-and-choose method for two players to divide the land fairly between them.

5. (a) The division from problem 4 (a) is an envy-free division. Dawn and Tyler each received the only parcels they felt were acceptable. Dawn will not envy the other parcels because she felt they were both worth less than a fair share. Tyler will not envy the other parcels because he felt they were both worth less than a fair share. Stan was the divider so he values each parcel equally.

   (b) The division from problem 4 (b) could lead to one of the players feeling envy. Stan will not feel envy because he was the divider and values each parcel equally. Tyler felt only parcel 2 was acceptable, and he received parcel 2. He did not feel either parcel 1 or parcel 3 was acceptable so he would not envy anyone else for keeping those parcels. Dawn is the one who might experience envy. When a player in the divide-and-choose method for three players labels a parcel "acceptable", that means the parcel is worth at least one-third of the entire 30 acres of land. It is possible that Dawn valued parcel 2 more than she valued parcel 3 even though they were both labeled as acceptable. Dawn could envy Tyler in this case. While this division was fair, it may not be envy-free.

   (c) The division from problem 4 (c) could lead to envy. Stan will keep either parcel 1 or 2 because neither of the other two players thought they were acceptable. The remaining parcels will be divided using the divide-and-choose method for two players. Once that division is made, it is possible that Stan will value one of the new parcels more than his own. While the division is fair, it may not be envy-free.

6. Emma made the highest bid so she will keep the painting and will place $835.00 into the compensation fund. Each sister will receive half of her bid from the compensation fund.

Emma will receive $\dfrac{\$835.00}{2} = \$417.50$ back from the compensation fund.

Else will receive $\dfrac{\$780.00}{2} = \$390.00$ from the compensation fund.

The compensation fund still contains $835.00 – $417.50 – $390.00 = $27.50 which will be split evenly

between them. Each of them will receive an additional $\dfrac{\$27.50}{2} = \$13.75$.

Emma receives the painting. She paid a total of $835.00 – $417.50 – $13.75 = $403.75 for the painting.
Else receives $390.00 + $13.75 = $403.75.

7. (a) Matt's value of the race is (1 point per mile running × 16 miles) + (4 points per mile biking × 80 miles) + (1 point per mile swimming × 9 miles) = 345 points. A fair share to Matt would be a portion worth $\dfrac{345}{3} = 115$ points.

Jory's value of the race is (2 points per mile running × 16 miles) + (2 points per mile biking × 80 miles) + (1 point per mile swimming × 9 miles) = 201 points. A fair share to Jory would be a portion worth $\dfrac{201}{3} = 67$ points.

Sart's value of the race is (3 points per mile running × 16 miles) + (1 point per mile biking × 80 miles) + (4 point per mile swimming × 9 miles) = 164 points. A fair share to Sart would be a portion worth $\dfrac{164}{3} = 54\dfrac{2}{3}$ points.

   (b) Determine how each player values the 3-mile run, 19-mile ride, and 4-mile swim and compare the point values to what each player would consider to be a fair share from part (a).
   Matt's value of the portion is (1 point per mile running × 3 miles) + (4 points per mile biking × 19 miles) + (1 point per mile swimming × 4 miles) = 83 points. It is less than a fair share according to Matt.
   Jory's value of the portion is (2 points per mile running × 3 miles) + (2 points per mile biking × 19 miles) + (1 point per mile swimming × 4 miles) = 48 points. It is less than a fair share according to Jory.
   Sart's value of the portion is (3 points per mile running × 3 miles) + (1 point per mile biking × 19 miles) + (4 point per mile swimming × 4 miles) = 44 points. It is less than a fair share according to Sart.

8. (a) The divider will be the person who values each option equally.
   **Matt's Values:**
   Option I: (1 point per mile running × 6 miles) + (4 points per mile biking × 27 miles ) + (1 point per mile swimming × 1 miles) = 115 points.
   Option II: (1 point per mile running × 4 miles) + (4 points per mile biking × 26 miles ) + (1 point per mile swimming × 7 miles) = 115 points.
   Option III: (1 point per mile running × 6 miles) + (4 points per mile biking × 27 miles ) + (1 point per mile swimming × 1 miles) = 115 points.

   **Jory's Values:**
   Option I: (2 points per mile running × 6 miles) + (2 points per mile biking × 27 miles ) + (1 point per mile swimming × 1 miles) = 67 points.

Option II: (2 points per mile running × 4 miles) + (2 points per mile biking × 26 miles ) + (1 point per mile swimming × 7 miles) = 67 points.

Option III: (2 points per mile running × 6 miles) + (2 points per mile biking × 27 miles ) + (1 point per mile swimming × 1 miles) = 67 points.

**Sart's Values:**

Option I: (3 points per mile running × 6 miles) + (1 point per mile biking × 27 miles ) + (4 points per mile swimming × 1 miles) = 49 points.

Option II: (3 points per mile running × 4 miles) + (1 point per mile biking × 26 miles ) + (4 points per mile swimming × 7 miles) = 66 points.

Option III: (3 points per mile running × 6 miles) + (1 point per mile biking × 27 miles ) + (4 points per mile swimming × 1 miles) = 49 points.

To Sart, only option II is acceptable.

Both Matt and Jory value each option equally, so either of them could have been the divider.

**(b)** Suppose Matt was the divider. We know from problem 7 (a) how many points each player believes makes up a fair share. From part (a), we see that Jory finds all three options acceptable, but Sart finds only option II acceptable. Give option II to Sart, option I to Matt, and option III to Jory. Another perfectly acceptable way to allocate the options would be to give option II to Sart, option III to Matt, and option I to Jory. The results would be the same if Jory had been the divider.

9. Yes, the division is envy-free. A player will experience envy if that player feels another player receives an option more valuable than his own. Sart received the only option he felt was acceptable. He does not envy the options given to the other two players. Both Matt and Jory felt all three options were equally acceptable, so they feel no envy.

10. **(a)** Holly made the first cut, and neither of the other players trimmed it, so Holly keeps the portion. The remaining players should use the divide-and-choose method for two players to fairly divide the remaining candy bar.

**(b)** Holly made the first cut, and then Hanna trimmed it. Hanna keeps the trimmed portion. The trimmings are recombined with the rest of the candy bar. The remaining players should use the divide-and-choose method for two players to fairly divide the remaining candy bar.

**(c)** Holly made the first cut, and then Harvey trimmed it. Harvey keeps the trimmed portion. The trimmings are recombined with the rest of the candy bar. The remaining players should use the divide-and-choose method for two players to fairly divide the remaining candy bar.

**(d)** Holly made the first cut, Harvey trimmed it, then Hanna trimmed it again. Hanna keeps the trimmed portion because she was the last diminisher. The trimmings are recombined with the rest of the candy bar. The remaining players should use the divide-and-choose method for two players to fairly divide the remaining candy bar.

11. **(a)** Holly received a portion she felt was fair because she was the original divider. Neither Harvey nor Hanna trimmed Holly's portion, so they felt her portion was less than a fair share. Holly kept her portion and left the process. Then Harvey and Hanna used the divide-and-choose method for two players to divide the remaining candy bar. Once they created two portions they both believed were fair, it is possible for Holly to feel one of the portions is more than a fair share. Only Holly could experience envy in this case.

**(b)** Hanna received a portion she felt was fair because she trimmed the portion created by Holly. Once Hanna trimmed it, Holly felt it was less than a fair share. Harvey felt the original portion was less than a fair share, so the trimmed portion was also less than a fair share. Hanna kept her portion and left the process. Then Harvey and Holly used the divide-and-choose method for two players to divide the remaining candy bar. Once they created two portions they both believed were fair, it is possible for Hanna to feel one of the portions is more than a fair share. Only Hanna could experience envy in this case.

(c) Harvey received a portion he felt was fair because he trimmed the portion created by Holly. Once Harvey trimmed it, Holly felt it was less than a fair share. Hanna felt the original portion was less than a fair share, so the trimmed portion was also less than a fair share. Harvey kept his portion and left the process. Then Hanna and Holly used the divide-and-choose method for two players to divide the remaining candy bar. Once they created two portions they both believed were fair, it is possible for Harvey to feel one of the portions is more than a fair share. Only Harvey could experience envy in this case.

(d) Hanna received a portion she felt was fair because she trimmed the portion created by Holly. Once Hanna trimmed it, Holly felt it was less than a fair share. Harvey trimmed the original piece first, so after Hanna trimmed it again, he felt it was less than a fair share. Hanna kept her portion and left the process. Then Harvey and Holly used the divide-and-choose method for two players to divide the remaining candy bar. Once they created two portions they both believed were fair, it is possible for Hanna to feel one of the portions is more than a fair share. Only Hanna could experience envy in this case.

12. (a) For Donna to keep the slice, she would have to trim the slice cut by Abe, and then Eric could not have trimmed it further. Donna would have to be the last diminisher.

(b) In this case, Cathy was the last diminisher so she keeps the piece. Abe will cut the piece in the second round and will continue to cut the pieces in each round until he is the one to keep a piece and he leaves the process.

(c) In the second round, there are four players left. The player to cut the piece will feel the piece is valued at one-fourth of the remaining cake.

(d) In round 1, Abe cut the piece, and Donna, who was the last diminisher, kept the piece.
In round 2, Abe cut the piece, and Eric, who was the last diminisher, kept the piece.
In round 3, Abe cut the piece, and no one trimmed the piece, so Abe kept the piece.
In round 4, Beth cut the piece, and Cathy, who was the last diminisher, kept the piece.

13. (a) **Harmon's Values:**
(2 pts per $mi^2$ east × 30 $mi^2$) + (1 pt per $mi^2$ west × 60 $mi^2$) + (1 pt per $mi^2$ south × 12 $mi^2$) = 132 points. A fair share to Harmon would be a portion worth $\dfrac{132}{3} = 44$ points.

**Darnell's Values:**
(1 pt per $mi^2$ east × 30 $mi^2$) + (3 pts per $mi^2$ west × 60 $mi^2$) + (1 pt per $mi^2$ south × 12 $mi^2$) = 222 points. A fair share to Darnell would be a portion worth $\dfrac{222}{3} = 74$ points.

**Parker's Values:**
(3 pts per $mi^2$ east × 30 $mi^2$) + (2 pts per $mi^2$ west × 60 $mi^2$) + (1 pt per $mi^2$ south × 12 $mi^2$) = 222 points. A fair share to Parker would be a portion worth $\dfrac{222}{3} = 74$ points.

(b) **Harmon's Value of the Package:**
(2 pts per $mi^2$ east × 14 $mi^2$) + (1 pt per $mi^2$ west × 12 $mi^2$) + (1 pt per $mi^2$ south × 4 $mi^2$) = 44 points. Harmon feels a fair share is worth 44 points, so this package does represent a fair share.

(c) Evaluate Darnell's value of the package first. Then evaluate Parker's value of the package.
**Darnell's Value of the Package:**
(1 pt per $mi^2$ east × 14 $mi^2$) + (3 pts per $mi^2$ west × 12 $mi^2$) + (1 pt per $mi^2$ south × 4 $mi^2$) = 54 points. From part (a), we know Darnell feels a fair share is worth at least 74 points, so this package does not represent a fair share and will not be trimmed.
**Parker's Value of the Package:**
(3 pts per $mi^2$ east × 14 $mi^2$) + (2 pts per $mi^2$ west × 12 $mi^2$) + (1 pt per $mi^2$ south × 4 $mi^2$) = 70 points. From part (a), we know Parker feels a fair share is worth at least 74 points, so this package does not represent a fair share and will not be trimmed.
Because neither Darnell nor Parker trimmed the package, Harmon will keep the package.

(d) The remaining land is made up of 16 square miles on the east side, 48 square miles on the west side, and 8 square miles on the south side of the mountain. There are many ways Darnell and Parker could fairly divide the remaining land. For example, they could create two packages that both contain 8 square miles on the east side, 24 square miles on the west side, and 4 square miles on the south side.

14. (a) Maxine was the highest bidder for the antique chair, the patio set, and the treadmill. Dugan was the highest bidder for the plasma television and the rototiller.

(b) The highest bidder for each item will place their bid into the compensation fund. The compensation fund contains $290 + $4500 + $345 + $560 + $490 = $6185.

(c) Each player will receive half of the total amount they bid on all items from the compensation fund. Maxine's bids totaled $290 + $4200 + $345 + $560 + $350 = $5745. She will receive $\frac{\$5745}{2} = \$2872.50$. Dugan's bids totaled $245 + $4500 + $250 + $475 + $490 = $5960. He will receive $\frac{\$5960}{2} = \$2980$. The compensation fund still contains a total of $6185 − $2872.50 − $2980 = $332.50 to be distributed equally between the two. Each of them will receive an additional $\frac{\$332.50}{2} = \$166.25$. Maxine receives the antique chair, the patio set, and the treadmill. She paid a total of $290 + $345 + $560 = $1195 to the compensation fund and received a total of $2872.50 + 166.25 = $3038.75 from the compensation fund. Her net cash amount received was $3038.75 − $1195 = $1843.75.

Dugan receives the plasma television and the rototiller. He paid a total of $4500 + $490 = $4990 to the compensation fund and received a total of $2980 + 166.25 = $3146.25 from the compensation fund. His net cash amount paid was $4990 − $3146.25 = $1843.75.

15. The highest bidder on each item will keep the item and place their bid for the item into the compensation fund. In the following table, the highest bid for each item is marked, each of the roommates' bid totals is given, and one-fourth of each roommates bid total is given.

| Items | Allen's Bids | Miguel's Bids | Carl's Bids | Xie's Bids |
|---|---|---|---|---|
| Couch | $35 | $40 | $10 | **$50** High |
| Basketball hoop | $15 | **$25** High | $15 | $20 |
| Weight set | $100 | $110 | **$125** High | $90 |
| Pitching machine | **$45** High | $25 | $30 | $35 |
| Trampoline | **$120** High | $100 | $85 | $95 |
| Total | $315 | $300 | $265 | $290 |
| $\frac{\text{Total}}{4}$ | $\frac{\$315}{4} = \$78.75$ | $\frac{\$300}{4} = \$75$ | $\frac{\$265}{4} = \$66.25$ | $\frac{\$290}{4} = \$72.50$ |

The compensation fund contains a total of $45 + $120 + $25 + $125 + $50 = $365. Each roommate will keep the item for which they had the highest bid and will receive one-fourth of their total bids from the compensation fund. These dollar amounts are listed in the table. The compensation fund still contains $365 − $78.75 − $75 − $66.25 − $72.50 = $72.50 which will be divided evenly between them.

Each roommate will receive an additional $\frac{\$72.50}{4} = \$18.125$.

Allen received the pitching machine and the trampoline. He paid a total of $120 + $45 − $78.75 − $18.125 ≈ $68.13.

Miguel received the basketball hoop. He received a total of $75 + $18.125 − $25 ≈ $68.13.

Carl received weight set. He paid a total of $125 − $66.25 − $18.125 ≈ $40.63.

Xie received the couch. He received a total of $72.50 + $18.125 − $50 ≈ $40.63.

**16. (a)** There are six possible arrangements of kids to rooms as given in the following table:

| All Possible Assignments of Kids to Rooms | | | Smallest Point Value |
|---|---|---|---|
| Audrey, room 1 10 points | Bruce, room 2 55 points | Carol, room 3 30 points | 10 points |
| Audrey, room 1 10 points | Bruce, room 3 25 points | Carol, room 2 40 points | 10 points |
| Audrey, room 2 60 points | Bruce, room 1 20 points | Carol, room 3 30 points | 20 points |
| Audrey, room 2 60 points | Bruce, room 3 25 points | Carol, room 1 30 points | 25 points |
| Audrey, room 3 30 points | Bruce, room 1 20 points | Carol, room 2 40 points | 20 points |
| Audrey, room 3 30 points | Bruce, room 2 55 points | Carol, room 1 30 points | 30 points |

There is one arrangement of kids to rooms such that the smallest point value is as large as possible. It is the arrangement that has a smallest point value of 30.

**(b)** Assign room 1 to Carol, room 2 to Bruce, and room 3 to Audrey.

**17.** There are six possible arrangements of kids to rooms as given in the following table:

| All Possible Assignments of Kids to Rooms | | | Smallest Point Value |
|---|---|---|---|
| Audrey, room 1 10 points | Bruce, room 2 55 points | Carol, room 3 40 points | 10 points |
| Audrey, room 1 10 points | Bruce, room 3 25 points | Carol, room 2 40 points | 10 points |
| Audrey, room 2 60 points | Bruce, room 1 20 points | Carol, room 3 40 points | 20 points |
| Audrey, room 2 60 points | Bruce, room 3 25 points | Carol, room 1 20 points | 20 points |
| Audrey, room 3 30 points | Bruce, room 1 20 points | Carol, room 2 40 points | 20 points |
| Audrey, room 3 30 points | Bruce, room 2 55 points | Carol, room 1 20 points | 20 points |

There are four arrangements of kids to rooms such that the smallest point value is as large as possible. They are the arrangements that have a smallest point value of 20. The following table contains those arrangements along with the middle point value noted.

| Four Assignments of Kids to Rooms | | | Middle Point Value |
|---|---|---|---|
| Audrey, room 2 60 points | Bruce, room 1 20 points | Carol, room 3 40 points | 40 points |
| Audrey, room 2 60 points | Bruce, room 3 25 points | Carol, room 1 20 points | 25 points |
| Audrey, room 3 30 points | Bruce, room 1 20 points | Carol, room 2 40 points | 30 points |
| Audrey, room 3 30 points | Bruce, room 2 55 points | Carol, room 1 20 points | 30 points |

There is one arrangement for which the middle value is as large as possible. It is the arrangement with a middle value of 40 points. Assign room 1 to Bruce, room 2 to Audrey, and room 3 to Carol.

**18.** The player who assigned the greatest number of points to an item is tentatively assigned ownership of the item. Initially, Ann is given the recliner and the tractor while Gerald is given the piano and the computer as shown in the following table.

| Ann | | Gerald | |
|---|---|---|---|
| Recliner | 30 | Piano | 30 |
| Tractor | 35 | Computer | 25 |
| Total | 65 | Total | 55 |

After the initial assignment, Ann has more points than Gerald. In Step 3 of the adjusted-winner procedure, we must select the item currently assigned to Ann for which the ratio of points assigned by Ann to points assigned by Gerald is the smallest.

Ratio of points assigned to the recliner:
$$\frac{\text{number of points assigned by Ann}}{\text{number of points assigned by Gerald}} = \frac{30}{25} = 1.2$$

Ratio of points assigned to the tractor:
$$\frac{\text{number of points assigned by Ann}}{\text{number of points assigned by Gerald}} = \frac{35}{20} = 1.75$$

The ratio of points assigned to the recliner is the smallest, so move the recliner from Ann to Gerald. In Step 4, reexamine the point totals. After moving the recliner from Ann to Gerald, the point totals are given in the following table:

| Ann | | Gerald | |
|---|---|---|---|
| Tractor | 35 | Piano | 30 |
| | | Computer | 25 |
| | | Recliner | 25 |
| Total | 35 | Total | 80 |

We are in Case C of the adjusted-winner procedure because Gerald now has more points than Ann. In order to make the point totals equal, we must move a fraction of the recliner back to Ann.

Let $q$ = the fraction of the recliner to be returned to Ann.

$T_X = 35$ (Ann's point total not including the recliner.)
$T_Y = 55$ (Gerald's point total not including the recliner.)
$P_X = 30$ (The number of points Ann assigned to the recliner.)
$P_Y = 25$ (The number of points Gerald assigned to the recliner.)

Then, $q = \dfrac{T_Y - T_X + P_Y}{P_X + P_Y} = \dfrac{55 - 35 + 25}{30 + 25} = \dfrac{45}{55} \approx 0.8182.$

Ann will keep the tractor and 81.82% ownership of the recliner. Gerald will keep the piano, the computer, and 18.18% ownership of the recliner.

**19.** The player who assigned the greatest number of points to an item is tentatively assigned ownership of the item. Initially, Gordon is given the iron, rocking chair, and file cabinet while Diana is given the vacuum, dresser, and printer as shown in the following table.

| Gordon | | Diana | |
|---|---|---|---|
| Iron | 5 | Vacuum | 22 |
| Rocking chair | 14 | Dresser | 35 |
| File cabinet | 20 | Printer | 20 |
| Total | 39 | Total | 77 |

After the initial assignment, Diana has more points than Gordon. In Step 3 of the adjusted-winner procedure, we must select the item currently assigned to Diana for which the ratio of points assigned by Diana to points assigned by Gordon is the smallest.

Ratio of points assigned to the vacuum:

$$\frac{\text{number of points assigned by Diana}}{\text{number of points assigned by Gordon}} = \frac{22}{18} \approx 1.22$$

Ratio of points assigned to the dresser:

$$\frac{\text{number of points assigned by Diana}}{\text{number of points assigned by Gordon}} = \frac{35}{28} = 1.25$$

Ratio of points assigned to the printer:

$$\frac{\text{number of points assigned by Diana}}{\text{number of points assigned by Gordon}} = \frac{20}{15} \approx 1.33$$

The ratio of points assigned to the vacuum is the smallest, so move the vacuum from Diana to Gordon. In Step 4, reexamine the point totals. After moving the vacuum from Diana to Gordon, the point totals are given in the following table:

| Gordon | | Diana | |
|---|---|---|---|
| Iron | 5 | Dresser | 35 |
| Rocking chair | 14 | Printer | 20 |
| File cabinet | 20 | | |
| Vacuum | 18 | | |
| Total | 57 | Total | 55 |

We are in Case C of the adjusted-winner procedure because Gordon now has more points than Diana. In order to make the point totals equal, we must move a fraction of the vacuum back to Diana.

Let $q$ = the fraction of the vacuum to be returned to Diana.

$T_X = 55$ (Diana's point total not including the vacuum.)
$T_Y = 39$ (Gordon's point total not including the vacuum.)
$P_X = 22$ (The number of points Diana assigned to the vacuum.)
$P_Y = 18$ (The number of points Gordon assigned to the vacuum.)

Then, $q = \dfrac{T_Y - T_X + P_Y}{P_X + P_Y} = \dfrac{39 - 55 + 18}{22 + 18} = \dfrac{2}{40} = 0.05.$

Diana will keep the dresser, printer, and 5% ownership of the vacuum. Gordon will keep the iron, rocking chair, file cabinet, and 95% ownership of the vacuum.

method for finding a proportional, envy-free division as described in Section ...ible correct solutions. One possible solution will be provided.

...ajority of the desserts.)

...ivider. She must divide the desserts into three portions that she considers to ...lues all of the desserts at (2 pts per oz pudding × 12 oz) + (1 pt per oz ... ...  per oz cheesecake × 12 oz) = 72 points. A fair share to Ally would be a

... $\frac{..}{3} = 24$ points. Ally could divide the desserts into the following three portions:

Portion I: 3 oz pudding, 0 oz cobbler, 6 oz cheesecake
Portion II: 4 oz pudding, 4 oz cobbler, 4 oz cheesecake
Portion III: 5 oz pudding, 8 oz cobbler, 2 oz cheesecake

**STEP 2:** Bev is the trimmer and will evaluate the three portions. Her value of each portion is given in the following table.

| Portion | Pudding Point Value 1 Point Per Ounce | Cobbler Point Value 1 Point Per Ounce | Cheesecake Point Value 2 Points Per Ounce | Total Point Value |
|---------|---------|---------|---------|---------|
| Portion I | $1 \times 3 = 3$ | $1 \times 0 = 0$ | $2 \times 6 = 12$ | 15 |
| Portion II | $1 \times 4 = 4$ | $1 \times 4 = 4$ | $2 \times 4 = 8$ | 16 |
| Portion III | $1 \times 5 = 5$ | $1 \times 8 = 8$ | $2 \times 2 = 4$ | 17 |

Bev feels that portion III has the greatest value and must be trimmed to have a point value of 16.

**STEP 3:** One way Bev can trim portion III down is to trim 1 oz of cobbler. The excess cobbler is set aside.

**STEP 4:** Sharon is the first chooser and will select the portion she feels has the greatest value. The following table contains the value Sharon places on each portion where portion III is the trimmed portion.

| Portion | Pudding Point Value 1 Point Per Ounce | Cobbler Point Value 2 Points Per Ounce | Cheesecake Point Value 5 Points Per Ounce | Total Point Value |
|---------|---------|---------|---------|---------|
| Portion I | $1 \times 3 = 3$ | $2 \times 0 = 0$ | $5 \times 6 = 30$ | 33 |
| Portion II | $1 \times 4 = 4$ | $2 \times 4 = 8$ | $5 \times 4 = 20$ | 32 |
| Portion III | $1 \times 5 = 5$ | $2 \times 7 = 14$ | $5 \times 2 = 10$ | 29 |

Sharon will select portion I.

**STEP 5:** Bev, the trimmer, will get portion III because she trimmed that portion.
**STEP 6:** Ally will get portion II because it is the remaining portion.

**PART 2 (Share the Excess.)**

**STEP 1:** Sharon becomes the second divider, and Bev becomes the second chooser.

**STEP 2:** Sharon must divide the excess 1 oz of cobbler into three portions of equal value. She can do this by splitting the cobbler into three portions each containing $\frac{1}{3}$ oz of cobbler.

**STEP 3:** Bev, the second chooser, will value each portion equally because they are identical. Each player will receive an additional $\frac{1}{3}$ oz of cobbler.

The final proportional, envy-free division gives Sharon 3 oz of pudding, $\frac{1}{3}$ oz of cobbler, and 6 oz of cheesecake. Bev gets 5 oz of pudding, $7\frac{1}{3}$ oz of cobbler, and 2 oz of cheesecake. Ally gets 4 oz of pudding, $4\frac{1}{3}$ oz of cobbler, and 4 oz of cheesecake.

**21.** A proportional division is one in which each player feels they received a fair share [...] woman must feel she received at least one-third of the desserts. Use the results from [...] From problem 20, we know a fair share to Ally is a portion worth at least 24 points. [...] portion at (2 pts per oz pudding × 4 oz) + (1 pt per oz cobbler × $4\frac{1}{3}$ oz) + (3 pts per oz c[...]

oz) = $24\frac{1}{3}$ points.

Sharon values all of the desserts at (1 pt per oz pudding × 12 oz) + (2 pts per oz cobbler × 12 oz[...] pts per oz cheesecake × 12 oz) = 96 points. A fair share to Sharon would be a portion worth $\frac{96}{3} = $ [...] points. Sharon values her portion at (1 pt per oz pudding × 3 oz) + (2 pts per oz cobbler × $\frac{1}{3}$ oz) + (5[...] pts per oz cheesecake × 6 oz) = $33\frac{2}{3}$ points.

Bev values all of the desserts at (1 pt per oz pudding × 12 oz) + (1 pt per oz cobbler × 12 oz) + (2 pts per oz cheesecake × 12 oz) = 48 points. A fair share to Bev would be a portion worth $\frac{48}{3} = 16$ points.

Bev values her portion at (1 pt per oz pudding × 5 oz) + (1 pt per oz cobbler × $7\frac{1}{3}$ oz) + (2 pts per oz cheesecake × 2 oz) = $16\frac{1}{3}$ points. Therefore, each woman feels she received a portion that was fair. The division is proportional.

**22.** The following table contains the values each woman places on their own portions from problem 20 and the portions of the other players.

| Portion | Sharon's Value | Bev's Value | Ally's Value |
|---|---|---|---|
| Sharon's Portion<br>3 oz Pudding<br>$\frac{1}{3}$ oz Cobbler<br>6 oz Cheesecake | $33\frac{2}{3}$ | $15\frac{1}{3}$ | $24\frac{1}{3}$ |
| Bev's Portion<br>5 oz Pudding<br>$7\frac{1}{3}$ oz Cobbler<br>2 oz Cheesecake | $29\frac{2}{3}$ | $16\frac{1}{3}$ | $23\frac{1}{3}$ |
| Ally's Portion<br>4 oz Pudding<br>$4\frac{1}{3}$ oz Cobbler<br>4 oz Cheesecake | $32\frac{2}{3}$ | $16\frac{1}{3}$ | $24\frac{1}{3}$ |

Each woman values her own portion at least as much as she values the portions of the other women. This is an envy-free division.

In this case, each
problem 20.
Ally values her
eesecake × 4

+ (5

32

# *Chapter 5:* Apportionment

## *Key Concepts*

**By the time you finish studying Section 5.1, you should be able to:**

- ☐ discuss the apportionment problem.

- ☐ calculate and interpret the meaning of the standard divisor and the standard quota.

- ☐ explain and carry out Hamilton's method of apportionment.

- ☐ explain and carry out Lowndes' method of apportionment.

- ☐ discuss the quota rule.

**By the time you finish studying Section 5.2, you should be able to:**

- ☐ explain how a divisor method differs from a quota method of apportionment.

- ☐ explain why a modified divisor is used in Jefferson's method of apportionment.

- ☐ explain and carry out Jefferson's method of apportionment.

- ☐ explain how Webster's method differs from Jefferson's method of apportionment.

- ☐ carry out Webster's method of apportionment.

**By the time you finish studying Section 5.3, you should be able to:**

- ☐ explain what it means for an apportionment method to obey the quota rule.

- ☐ describe the Alabama paradox and be able to recognize when the Alabama paradox occurs.

- ☐ describe the population paradox and be able to recognize when the population paradox occurs.

- ☐ describe the new-states paradox and be able to recognize when the new-states paradox occurs.

- ☐ discuss Balinski and Young's impossibility theorem and its importance in the debate over apportionment methods.

# Hints for Odd-Numbered Problems

## Section 5.1

1.  (a) The standard divisor is found by dividing the total population by the number of seats to be apportioned.
    (b) The standard quota for a state is found by dividing the state's population by the standard divisor.

3.  It might be helpful to write the formula for the standard quota in terms of $p$ by multiplying both sides of the equation by $D$.

5.  (a) Determine what is being apportioned and the total amount on which the apportionment is based.

7.  (b) There are 10 seats to be apportioned. Each class is initially allocated the number of seats equal to the integer part of its standard quota.
    (c) The remaining seats will be apportioned according to the size of the fractional parts in order from largest to smallest.

9.  (b) Divide the total apportionment population by the number of seats to be apportioned.
    (c) Divide each state's apportionment population by the standard divisor from part (b).
    (d) Determine the number of seats allocated based on the integer parts of the standard quotas.

11. (b) Recalculate the standard divisor based on the new total, and find the new standard quotas for the three friends.

13. To apply Hamilton's method, first find the standard divisor and use it to find the standard quota for each college. Add up the integer parts of the standard quotas to determine how many seats are accounted for and how many more must be apportioned.

15. The total population for the 15 states is given at the bottom of the table. Round each standard quota to the thousandths place.

17. (b) Use the formula for the standard divisor. The total population is known and the desired standard divisor is given. Solve the equation for the unknown number of legislative seats. Keep in mind that 30,000 is the *minimum* standard divisor.

19. To apply Hamilton's method, first find the standard divisor and use it to find the standard quota for each state. Add up the integer parts of the standard quotas to determine how many seats are accounted for and how many more must be apportioned.

21. (a) To find the relative fractional part divide the fractional part by the integer part.
    (b) Left over seats will be distributed according to the size of the relative fractional parts.
    (c) Left over seats will be distributed according to the size of the fractional parts.

23. (a) To find the relative fractional part divide the fractional part by the integer part.
    (b) Left over seats will be distributed according to the size of the relative fractional parts.
    (c) Left over seats will be distributed according to the size of the fractional parts.

25. (a) Divide the total population in 1990 by 25,000 to determine the number of representatives that were needed.
    (b) Calculate the standard divisor by dividing the total population by the number of representatives in part (a).
    (c) Divide the fractional part by the integer part as long as the integer part is not zero.
    (d) Lowndes' method guarantees every county at least 1 seat.

27. Calculate the standard divisor and standard quota for each county. Remember that Hamilton's method allocates left over seats according to the size of the fractional part of the standard quota.

29. Calculate the standard divisor and standard quota for each county. Remember that Hamilton's method allocates left over seats according to the size of the fractional part of the standard quota.

31. Calculate the standard divisor and standard quota for each state. Remember that Lowndes' method allocates left over seats according to the size of the relative fractional part of the standard quota.

33. (a) The standard divisor is calculated by dividing total population in 2001 by the number of officers to be apportioned. Left over officers are apportioned according to the size of the fractional part.
    (b) The standard divisor is calculated by dividing the total area by the number of officers to be apportioned. Left over officers are apportioned according to the size of the fractional part.

35. (a) The standard divisor is calculated by dividing total population in 2001 by the number of officers to be apportioned. Left over officers arc apportioned according to the size of the relative fractional part.
    (b) The standard divisor is calculated by dividing the total area by the number of officers to be apportioned. Left over officers are apportioned according to the size of the relative fractional part.

37. Calculate the standard divisor and standard quota for each province. Remember that Hamilton's method allocates left over seats according to the size of the fractional part of the standard quota.

## Section 5.2

1. Compare two fractions in which the numerators are the same, but the denominators differ. Which one is larger?

3. The numerators of each fraction pair are the same. Compare the denominators.

5. The difference between a standard quota and a modified quota is the denominator. In order to create a modified quota that is larger than the standard quota, the denominator must be decreased.

7. Jefferson's method is completed when the integer parts of the modified quotas add to the number of seats to be apportioned. If they differ, then modify the divisor up or down.

9. (b) Was the sum of the integer parts of the standard quotas from part (a) larger or smaller than the number of seats to be apportioned? Will the modified quotas need to be increased or decreased?

11. (a) If you modify the divisor, and the sum of the integer parts of the modified quotas is too small, then you must modify the divisor again. Then if the sum of the integer parts of the modified quotas is too large, the next modified divisor must be in between the two previous modified divisors. There is no standard rule for generating the next guess. You could average the previous two modified divisors or just guess a number between them.

13. It may take several guesses at a modified divisor before you find one that will produce modified quotas that have integer parts that add to the number of seats to be apportioned.

15. Always round modified divisors down to the nearest integer.

17. Follow the suggested method in the text to find modified divisors. Once the sum of the integer parts of the modified quotas is too large, it will be wise to think carefully about your choice of the next modified divisor. Study the first modified quotas. What small change in one of the state's modified quota would produce integer parts that add to 105? What modified divisor would change the state's modified quota in this way?

19. Follow the suggested method in the text to find the first modified divisor. Then determine the range of values available for the second modified divisor, try several different values, and pay attention to the integer parts of the resulting modified quotas. Will the integer parts of the modified quotas ever add to 6?

21. Remember that Jefferson's method allocates the number of classes equal to the integer part of the modified quotas. Webster's method rounds modified quotas to the nearest integer.

23. Follow the suggested method in the text to find modified divisors. Once the sum of the modified quotas rounded to the nearest integer is too large, there is no rule to generate the next guess at a new modified divisor. Try a modified divisor between the two previous divisors and pay attention to how the modified quotas round to help you make your next guess.

25. Follow the suggested method in the text to find modified divisors. Round modified divisors down.

27. Follow the suggested method in the text to find modified divisors. Round modified divisors down.

29. Follow the suggested method in the text to find modified divisors. Round modified divisors down. Each county must be apportioned at least one seat.

31. Round modified divisors down. Remember that Webster's method rounds standard quotas and modified quotas to the nearest integer. Each county must be apportioned at least one seat.

## Section 5.3
1. Which of the three paradoxes is possible under the circumstances?

3. Which of the three paradoxes is possible under the circumstances?

5. Consider the various ways modified quotas were rounded for different apportionment methods.

7. Remember that Hamilton's method initially allocates to each state the integer part of the standard quota, and left over seats are given out depending on the size of the fractional parts of the standard quotas.

9. Remember that Jefferson's method apportions to each state the integer part of the modified quotas.

11. (b) Noticing a change in apportionment after a population increase is not enough to conclude that the population paradox occurred. You must consider how the populations of each state changed compared to the original size. Find the percent increase for each state. See Example 5.11.

13. (c) The new state was added to the country and brought 7 seats to the legislature. How many seats were apportioned to the new state by Hamilton's method? Did the apportionments for the other states change?

15. (c) The new state was added to the country and brought 7 seats to the legislature. How many seats were apportioned to the new state by Jefferson's method? Did the apportionments for the other states change?

17. Consider how Hamilton's method assigns seats to a state. How are any left over seats assigned?

19. Remember that Jefferson's method often requires a modified divisor that increases modified quotas.

21. Consider which provinces lose seats, if any, as the size of the legislature changes.

23. Consider which provinces lose seats, if any, as the size of the legislature changes.

25. (b) Divide the population of City E by the standard divisor to determine how many new officers should be hired. Round to the nearest integer.

(c) Find a new standard divisor for the reapportionment. Do not forget to include the population of the new city and the additional officers that will be apportioned.

27. See Example 5.11.

29. Noticing a change in apportionment after a population increase is not enough to conclude that the population paradox occurred. Study the percent increases from problem 27.

31. Noticing a change in apportionment after a population increase is not enough to conclude that the population paradox occurred. Study the percent increases from problem 27.

33. Is there a change in the seats apportioned for the original provinces?

35. Is there a change in the seats apportioned for the original provinces?

37. (b) Find a new standard divisor.
    (c) One more patrol was added. Did any areas lose a patrol?

39. Answers will vary. Left over seats are apportioned according to the size of the fractional parts of the standard quotas. When the number of seats increases, the standard quotas will change. To cause the Alabama paradox to occur, the fractional parts of the standard quotas must change so that different states get the left over seats in the reapportionment.

41. Jefferson's method apportions seats according to the quotas. In any reapportionment due to an increase in the number of seats, the standard divisor will decrease. New standard quotas will increase.

## Solutions to Odd-Numbered Problems in Section 5.1

1. Let $P = 85,000$ and $M = 25$.

   (a) The standard divisor is $D = \dfrac{P}{M} = \dfrac{85,000}{25} = 3400$. The standard divisor is the number of people per legislative seat. In other words, each legislator would represent 3400 people.

   (b) To find the standard quota, divide the population of each state, $p$, by the standard divisor from part (a).

   For the state with population $p = 30,600$, the standard quota is $\dfrac{p}{D} = \dfrac{30,600}{3400} = 9$.

   For the state with population $p = 54,400$, the standard quota is $\dfrac{p}{D} = \dfrac{54,400}{3400} = 16$.

   In each case, the standard quota is the number of legislators that should be assigned to represent each state.

3. The total population of a city is 57,200, and the number of seats to be apportioned is 20, so the standard divisor can be calculated as follows: $D = \dfrac{P}{M} = \dfrac{57,200}{20} = 2860$.

   The standard quota is defined as $Q = \dfrac{p}{D}$. Solve the equation for $p$, the population of a district, by multiplying both sides of the equation by $D$ as shown next.

$$Q = \frac{p}{D}$$

$$Q \times D = \frac{p}{D} \times D$$

$$Q \times D = \frac{p}{\not{D}} \times \not{D}$$

$$Q \times D = p$$

Use the formula $p = Q \times D$ to find the approximate population for each of the four districts by substituting the given value of the standard quota and the standard divisor into the formula.

| District | Standard Quota | District Population $p = Q \times D$ |
|----------|----------------|--------------------------------------|
| (I) | 3.67 | $p = 3.67 \times 2860 \approx 10{,}496$ |
| (II) | 4.53 | $p = 4.53 \times 2860 \approx 12{,}956$ |
| (III) | 6.05 | $p = 6.05 \times 2860 = 17{,}303$ |
| (IV) | 5.75 | $p = 5.75 \times 2860 = 16{,}445$ |

5. **(a)** In order to calculate the standard divisor, the total number of hours spent helping the grandmother will be divided by the number of shares of stock to be apportioned as follows: $D = \frac{P}{M} = \frac{72}{36} = 2$.

Therefore, every two hours that was spent helping the grandmother is worth a share of stock.

**(b)** To find each teenager's standard quota, the time each spent helping the grandmother will be divided by the standard divisor.

| Teenager | Time Spent Helping Grandmother Each Month | Standard Quota $Q = \dfrac{p}{D}$ |
|----------|-------------------------------------------|-----------------------------------|
| Daphne | 44 | $\dfrac{44}{2} = 22$ |
| Mike | 8 | $\dfrac{8}{2} = 4$ |
| Melinda | 20 | $\dfrac{20}{2} = 10$ |

The standard quota represents the number of shares each will inherit. Daphne will inherit 22 shares of stock. Mike will inherit 4 shares of stock. Melinda will inherit 10 shares of stock.

7. **(a)** Ten representative seats will be apportioned using Hamilton's method.

| Class | Standard Quota | Integer Part | Fractional Part |
|-------|----------------|--------------|-----------------|
| Freshman | 1.39 | 1 | 0.39 |
| Sophomore | 3.50 | 3 | 0.50 |
| Junior | 3.23 | 3 | 0.23 |
| Senior | 1.88 | 1 | 0.88 |

**(b)** Add the column of integer parts from part (a) to obtain $1 + 3 + 3 + 1 = 8$. Thus, 8 representative seats are apportioned according to the integer parts. Two more seats remain out of the 10 seats and will be apportioned based on the size of the fractional parts.

**(c)** Hamilton's method allocates the two remaining seats according to the size of the fractional part beginning with the class with the largest fractional part. The senior class has the largest fractional part, 0.88, and the sophomore class has the second largest fractional part, 0.50. Therefore, the senior and sophomore classes will each receive an additional seat. The final apportionment gives the freshman class 1 seat, the sophomore class 4 seats, the junior class 3 seats, and the senior class 2 seats.

9. **(a)** The total apportionment population of the four states is 2,852,927 + 4,461,130 + 8,206,975 + 16,028,890 = 31,549,922.

   **(b)** The total apportionment population is 31,549,922, and the number of seats to be apportioned is 435, so the standard divisor can be calculated as follows: $D = \dfrac{P}{M} = \dfrac{31,549,922}{435} \approx 72,528.556$.

   **(c)** To find the standard quota, divide the apportionment population of each state, $p$, by the standard divisor, $D$, from part (b).

   For Mississippi, $p = 2,852,927$, and the standard quota is $Q = \dfrac{p}{D} = \dfrac{2,852,927}{72,528.556} \approx 39.335$.

   For Alabama, $p = 4,461,130$, and the standard quota is $Q = \dfrac{p}{D} = \dfrac{4,461,130}{72,528.556} \approx 61.509$.

   For Georgia, $p = 8,206,975$, and the standard quota is $Q = \dfrac{p}{D} = \dfrac{8,206,975}{72,528.556} \approx 113.155$.

   For Florida, $p = 16,028,890$, and the standard quota is $Q = \dfrac{p}{D} = \dfrac{16,028,890}{72,528.556} \approx 221.001$.

   **(d)** Each state is initially allocated the number of seats equal to the integer part of its standard quota, so 39 + 61 + 113 + 221 = 434 seats are accounted for. There is one seat left, and it will be given to the state with the largest fractional part. Alabama has the largest fractional part, so Alabama is given the one remaining seat. According to Hamilton's method, the seats are apportioned as follows: Mississippi, 39; Alabama, 62; Georgia, 113; and Florida, 221.

11. **(a)** To find the standard divisor, divide the total amount of money contributed by the number of bottles of wine to be apportioned: $D = \dfrac{P}{M} = \dfrac{295+205+390}{20} = \dfrac{890}{20} = 44.5$.

   To find the standard quota for each friend, divide the contribution, $p$, by the standard divisor, $D$.

   For Jaron, $p = \$295$, so the standard quota is $Q = \dfrac{p}{D} = \dfrac{295}{44.5} \approx 6.629$.

   For Mikkel, $p = \$205$, so the standard quota is $Q = \dfrac{p}{D} = \dfrac{205}{44.5} \approx 4.607$.

   For Robert, $p = \$390$, so the standard quota is $Q = \dfrac{p}{D} = \dfrac{390}{44.5} \approx 8.764$.

   Each friend is initially allocated the number of bottles equal to the integer part of his standard quota, so 6 + 4 + 8 = 18 bottles are accounted for. There are 2 bottles left, and they will be given to the men with the largest fractional parts of the standard quota. Robert has the largest fractional part, so Robert is given one of the bottles. Jaron has the second largest fractional part, so Jaron is given the other bottle. According to Hamilton's method, the bottles are apportioned as follows: Jaron, 7; Mikkel, 4; and Robert, 9.

   **(b)** With the additional money from Mikkel, the standard divisor would change: $D = \dfrac{P}{M} = \dfrac{295+225+390}{20} = \dfrac{910}{20} = 45.5$. Recalculate the standard quotas for the friends.

   For Jaron, $p = \$295$, so the standard quota is $Q = \dfrac{p}{D} = \dfrac{295}{45.5} \approx 6.484$.

   For Mikkel, $p = \$225$, so the standard quota is $Q = \dfrac{p}{D} = \dfrac{225}{45.5} \approx 4.945$.

For Robert, $p = \$390$, so the standard quota is $Q = \dfrac{p}{D} = \dfrac{390}{45.5} \approx 8.571$.

Each friend is initially allocated the number of bottles equal to the integer part of his standard quota, so $6 + 4 + 8 = 18$ bottles are accounted for. There are 2 bottles left, and they will be given to the men with the largest fractional parts of the standard quota. This time, Mikkel has the largest fractional part, and Robert has the second largest fractional part, so they will receive the 2 remaining bottles. According to Hamilton's method, the bottles are apportioned as follows: Jaron, 6; Mikkel, 5; and Robert, 9.

13. To apply Hamilton's method, find the standard divisor and the standard quotas. To find the standard divisor, divide the total enrollment of the University by the number of seats to be apportioned:
$D = \dfrac{P}{M} = \dfrac{2540+3580+1410+1830+750}{30} = \dfrac{10,110}{30} = 337.$

To find the standard quota for each college, divide the college's enrollment, $p$, by the standard divisor, $D$.

For the college of Fine and Performing Arts, $p = 2540$, so $Q = \dfrac{p}{D} = \dfrac{2540}{337} \approx 7.537.$

For the college of Math and Physical Science, $p = 3580$, so $Q = \dfrac{p}{D} = \dfrac{3580}{337} \approx 10.623.$

For the college of Engineering, $p = 1410$, so $Q = \dfrac{p}{D} = \dfrac{1410}{337} \approx 4.184.$

For the college of Social Science, $p = 1830$, so $Q = \dfrac{p}{D} = \dfrac{1830}{337} \approx 5.430.$

For the college of Agriculture, $p = 750$, so $Q = \dfrac{p}{D} = \dfrac{750}{337} \approx 2.226.$

Each college is initially allocated the number of seats equal to the integer part of its standard quota, so $7 + 10 + 4 + 5 + 2 = 28$ seats are accounted for. There are 2 seats left, and they will be given to the colleges with the largest fractional parts of the standard quota. The college of Fine and Performing Arts has the largest fractional part, and the college of Math and Physical Science has the second largest fractional part, so those two colleges will each receive one of the left over seats. According to Hamilton's method, the seats for the 2001 faculty senate are apportioned as follows: Fine and Performing Arts, 8; Math and Physical Science, 11; Engineering, 4; Social Science, 5; and Agriculture, 2.

15. (a) The standard divisor is $D = \dfrac{P}{M} = \dfrac{3,615,920}{105} \approx 34,437.333$. The standard divisor is the number of people per legislative seat. In other words, each legislator would represent approximately 34,437 people.

(b) The standard quota for each state is calculated and presented in the following table:

| State | Population in 1790 | Standard Quota $Q = \dfrac{p}{D}$ | Integer Part | Fractional Part | Final Apportionment |
|---|---|---|---|---|---|
| Connecticut | 236,841 | $\dfrac{236,841}{34,437.333} \approx 6.877$ | 6 | 0.877* | 7 |
| Delaware | 55,540 | $\dfrac{55,540}{34,437.333} \approx 1.613$ | 1 | 0.613* | 2 |
| Georgia | 70,835 | $\dfrac{70,835}{34,437.333} \approx 2.057$ | 2 | 0.057 | 2 |

| Kentucky | 68,705 | $\dfrac{68,705}{34,437.333} \approx 1.995$ | 1 | 0.995* | 2 |
|---|---|---|---|---|---|
| Maryland | 278,514 | $\dfrac{278,514}{34,437.333} \approx 8.088$ | 8 | 0.088 | 8 |
| Massachusetts | 475,327 | $\dfrac{475,327}{34,437.333} \approx 13.803$ | 13 | 0.803* | 14 |
| New Hampshire | 141,822 | $\dfrac{141,822}{34,437.333} \approx 4.118$ | 4 | 0.118 | 4 |
| New Jersey | 179,570 | $\dfrac{179,570}{34,437.333} \approx 5.214$ | 5 | 0.214 | 5 |
| New York | 331,589 | $\dfrac{331,589}{34,437.333} \approx 9.629$ | 9 | 0.629* | 10 |
| North Carolina | 353,523 | $\dfrac{353,523}{34,437.333} \approx 10.266$ | 10 | 0.266 | 10 |
| Pennsylvania | 432,879 | $\dfrac{432,879}{34,437.333} \approx 12.570$ | 12 | 0.570* | 13 |
| Rhode Island | 68,446 | $\dfrac{68,446}{34,437.333} \approx 1.988$ | 1 | 0.988* | 2 |
| South Carolina | 206,236 | $\dfrac{206,236}{34,437.333} \approx 5.989$ | 5 | 0.989* | 6 |
| Vermont | 85,533 | $\dfrac{85,533}{34,437.333} \approx 2.484$ | 2 | 0.484 | 2 |
| Virginia | 630,560 | $\dfrac{630,560}{34,437.333} \approx 18.310$ | 18 | 0.310 | 18 |

The sum of the integer parts of the standard quotas is 97. Each state is initially allocated the number of seats equal to the integer part of the state's standard quota. There are $105 - 97 = 8$ additional seats that must be apportioned. Compare the fractional parts of the standard quotas. The eight states with the largest fractional parts have their fractional parts marked with an asterisk in the table. Those states will each receive one additional seat. The final apportionment is listed in the final column.

**17. (a)** The standard divisor is $D = \dfrac{P}{M} = \dfrac{15,908,376}{223} \approx 71,338.009$. The standard divisor is the number of people per legislative seat. In other words, each legislator would represent approximately 71,338 people.

**(b)** The number of legislative seats is unknown. The apportionment population is 15,908,376 and we want the standard divisor to be 30,000. Use the formula for the standard divisor. Substitute in the known values and solve for $M$, the number of legislative seats.

$$D = \frac{P}{M}$$

$$30,000 = \frac{15,908,376}{M}$$

$$30,000 \times M = \frac{15,908,376}{M} \times M$$

$$30,000 \times M = 15,908,376$$

$$M = \frac{15,908,376}{30,000}$$

$$M = 530.2792$$

Therefore, 530 seats would have been required in the year 1840 to maintain a standard divisor of 30,000. Remember that 30,000 is the minimum standard divisor. If 531 seats were used, then the standard divisor would have been less than 30,000 which would violate the Constitution.

19. (a) Add the populations of the states to find the standard divisor. The standard divisor is $D = \frac{P}{M} = \frac{77,500+77,500+45,000}{6} = \frac{200,000}{6} \approx 33,333.333$. The standard divisor is the number of people per legislative seat. In other words, each legislator would represent approximately 33,333 people.

(b) The formula for the standard quota is $Q = \frac{p}{D}$. The population of each state is given, and the standard divisor was calculated in part (a). The standard quotas are calculated in the following table:

| State Population | Standard Quota $Q = \frac{p}{D}$ | Integer Part | Fractional Part |
|---|---|---|---|
| 77,500 | $\frac{77,500}{33,333.33} \approx 2.325$ | 2 | 0.325 |
| 77,500 | $\frac{77,500}{33,333.33} \approx 2.325$ | 2 | 0.325 |
| 45,000 | $\frac{45,000}{33,333.33} \approx 1.350$ | 1 | 0.350 |

(c) The sum of the integer parts of the standard quotas is 5. Each state is initially allocated the number of seats equal to the integer part of the state's standard quota. There is one additional seat that must be apportioned. Compare the fractional parts of the standard quotas. The state with the largest fractional part is the one with the smallest population. Hamilton's method will give each state two seats.

21. (a)

| Class Section | Standard Quota | Integer Part | Fractional Part | Relative Fractional Part: $\frac{\text{Fractional Part}}{\text{Integer Part}}$ |
|---|---|---|---|---|
| A | 3.23 | 3 | 0.23 | $\frac{0.23}{3} \approx 0.077$ |
| B | 2.25 | 2 | 0.25 | $\frac{0.25}{2} = 0.125$ |

| | | | | |
|---|---|---|---|---|
| C | 1.49 | 1 | 0.49 | $\dfrac{0.49}{1} = 0.49$ |
| D | 2.21 | 2 | 0.21 | $\dfrac{0.21}{2} = 0.105$ |
| E | 1.82 | 1 | 0.82 | $\dfrac{0.82}{1} = 0.82$ |

From largest to smallest, the class sections ordered according to the relative fractional parts are E, C, B, D, and A.

(b) According to Lowndes' method, each class is initially allocated the number of assistants equal to the integer part of the class' standard quota. The sum of the integer parts is 9, so 9 assistants have been accounted for. To allocate the remaining 2 teaching assistants, compare the relative fractional parts. Class sections E and C have the largest relative fractional parts, so they will each receive an additional assistant. The apportionment according to Lowndes' method gives 3 assistants to class section A and 2 assistants to each of the other class sections.

(c) According to Hamilton's method, each class is initially allocated the number of assistants equal to the integer part of the class' standard quota. The sum of the integer parts is 9, so 9 assistants have been accounted for. There are 2 additional assistants that must be apportioned. Compare the fractional parts of the standard quotas. The two classes with the largest fractional parts are class sections E and C. Hamilton's method will give 3 assistants to class section A and 2 assistants to each of the other classes.

23. (a)

| State | Standard Quota | Integer Part | Fractional Part | Relative Fractional Part: $\dfrac{\text{Fractional Part}}{\text{Integer Part}}$ |
|---|---|---|---|---|
| Florida | 24.78 | 24 | 0.78 | $\dfrac{0.78}{24} \approx 0.033$ |
| Mississippi | 4.41 | 4 | 0.41 | $\dfrac{0.41}{4} \approx 0.103$ |
| Alabama | 6.90 | 6 | 0.90 | $\dfrac{0.90}{6} = 0.15$ |
| Georgia | 12.69 | 12 | 0.69 | $\dfrac{0.69}{12} \approx 0.058$ |
| Tennessee | 8.81 | 8 | 0.81 | $\dfrac{0.81}{8} \approx 0.101$ |

(b) According to Lowndes' method, each state is initially given the number of seats equal to the integer part of the state's standard quota. A total of $24 + 4 + 6 + 12 + 8 = 54$ seats have been allocated. The remaining 4 seats will be given according to the size of the relative fractional part. The 4 states with the largest relative fractional parts are Alabama, Mississippi, Tennessee, and Georgia so each of these states receives one additional seat. The apportionment according to Lowndes' method gives 24 seats to Florida, 5 seats to Mississippi, 7 seats to Alabama, 13 seats to Georgia, and 9 seats to Tennessee.

(c) According to Hamilton's method, each state is initially given the number of seats equal to the integer part of the state's standard quota. As we noted in part (b), 54 seats have been allocated, so the remaining 4 seats will be given according to the size of the fractional part. The four states with the largest fractional parts are Alabama, Tennessee, Florida, and Georgia. Each of these states will receive one additional seat. The apportionment according to Hamilton's method gives 25 seats to Florida, 4 seats to Mississippi, 7 seats to Alabama, 13 seats to Georgia, and 9 seats to Tennessee.

25. **(a)** The total population of Hawaii in 1990 was 1,108,229. It was decided that every 25,000 people would be represented by 1 seat on the commission. Divide the total population by 25,000 to determine the number of representatives needed.: $\dfrac{1,108,229}{25,000} \approx 44.329$. There were 44 representatives.

**(b)** From part (a), the number of representatives needed was determined to be 44. Now we use the total population of Hawaii in 1990 and the number of representatives to determine the standard divisor. The standard divisor is calculated as $D = \dfrac{P}{M} = \dfrac{1,108,229}{44} \approx 25,187.023$. The standard quotas for each county are calculated in the following table:

| County | Population 1990 | Standard Quota $Q = \dfrac{P}{D}$ |
|---|---|---|
| Hawaii | 120,317 | $\dfrac{120,317}{25,187.023} \approx 4.777$ |
| Honolulu | 836,231 | $\dfrac{836,231}{25,187.023} \approx 33.201$ |
| Kalawao | 130 | $\dfrac{130}{25,187.023} \approx 0.005$ |
| Kauai | 51,177 | $\dfrac{51,177}{25,187.023} \approx 2.032$ |
| Maui | 100,374 | $\dfrac{100,374}{25,187.023} \approx 3.985$ |

**(c)** The relative fractional parts are calculated in the following table. Notice that because the population of Kalawao is so small, the standard quota is less than 1. We cannot calculate the relative fractional part for Kalawao because we cannot divide by 0, the integer part. Kalawao will be granted 1 seat. Lowndes' method guarantees that every state will receive at least 1 seat.

| County | Standard Quota | Integer Part | Fractional Part | Relative Fractional Part: $\dfrac{\text{Fractional Part}}{\text{Integer Part}}$ |
|---|---|---|---|---|
| Hawaii | 4.777 | 4 | 0.777 | $\dfrac{0.777}{4} \approx 0.194$ |
| Honolulu | 33.201 | 33 | 0.201 | $\dfrac{0.201}{33} \approx 0.006$ |
| Kalawao | 0.005 | 0 | 0.005 | Undefined |
| Kauai | 2.032 | 2 | 0.032 | $\dfrac{0.032}{2} = 0.016$ |
| Maui | 3.985 | 3 | 0.985 | $\dfrac{0.985}{3} \approx 0.328$ |

**(d)** According to Lowndes' method, each county is initially given the number of seats equal to the integer part of the county's standard quota, and Kalawao is granted 1 seat. A total of $4 + 33 + 1 + 2 + 3 = 43$ seats have been accounted for. There is 1 seat left over which will be given to Maui because it has the largest relative fractional part. The apportionment according to Lowndes' method gives 4 seats to Hawaii, 33 seats to Honolulu, 1 seat to Kalawao, 2 seats to Kauai, and 4 seats to Maui.

**27.** If there were 50 representatives in the year 1990, then the standard divisor would have been calculated as follows: $D = \dfrac{P}{M} = \dfrac{1,108,229}{50} = 22,164.58$. Find the standard quota, the integer part, and the fractional part for each county.

| County | Population 1990 | Standard Quota $Q = \dfrac{p}{D}$ | Integer Part | Fractional Part |
|---|---|---|---|---|
| Hawaii | 120,317 | $\dfrac{120,317}{22,164.58} \approx 5.428$ | 5 | 0.428 |
| Honolulu | 836,231 | $\dfrac{836,231}{22,164.58} \approx 37.728$ | 37 | 0.728 |
| Kalawao | 130 | $\dfrac{130}{22,164.58} \approx 0.006$ | 0 | 0.006 |
| Kauai | 51,177 | $\dfrac{51,177}{22,164.58} \approx 2.309$ | 2 | 0.309 |
| Maui | 100,374 | $\dfrac{100,374}{22,164.58} \approx 4.529$ | 4 | 0.529 |

Each county is initially given the number of seats equal to the integer part of its standard quota, and Kalawao is given 1 seat. A total of $5 + 37 + 1 + 2 + 4 = 49$ seats are accounted for. There is 1 left over seat which will be given to the county with the largest fractional part. Honolulu will be given the left over seat. The apportionment by Hamilton's method gives 5 seats to Hawaii, 38 seats to Honolulu, 1 seat to Kalawao, 2 seats to Kauai, and 4 seats to Maui.

**29.** If there were 54 representatives in the year 2000, then the standard divisor would have been calculated as follows: $D = \dfrac{P}{M} = \dfrac{1,211,537}{54} \approx 22,435.870$. Find the standard quota, the integer part, and the fractional part for each county.

| County | Population 2000 | Standard Quota $Q = \dfrac{p}{D}$ | Integer Part | Fractional Part |
|---|---|---|---|---|
| Hawaii | 148,677 | $\dfrac{148,677}{22,435.870} \approx 6.627$ | 6 | 0.627 |
| Honolulu | 876,156 | $\dfrac{876,156}{22,435.870} \approx 39.052$ | 39 | 0.052 |
| Kalawao | 147 | $\dfrac{147}{22,435.870} \approx 0.007$ | 0 | 0.007 |
| Kauai | 58,463 | $\dfrac{58,463}{22,435.870} \approx 2.606$ | 2 | 0.606 |
| Maui | 128,094 | $\dfrac{128,094}{22,435.870} \approx 5.709$ | 5 | 0.709 |

Each county is initially given the number of seats equal to the integer part of its standard quota, and Kalawao is given 1 seat. A total of $6 + 39 + 1 + 2 + 5 = 53$ seats are accounted for. There is 1 left over seat which will be given to the county with the largest fractional part. Maui will be given the left over seat. The apportionment by Hamilton's method gives 6 seats to Hawaii, 39 seats to Honolulu, 1 seat to Kalawao, 2 seats to Kauai, and 6 seats to Maui.

**31.** To apportion the 105 seats in the House of Representatives in 1790, we must calculate the standard divisor, standard quotas for each state and determine the relative fractional parts. The standard divisor is $D = \dfrac{P}{M} = \dfrac{3,615,920}{105} \approx 34,437.333$.

| State | Standard Quota $Q = \dfrac{p}{D}$ | Integer Part | Fractional Part | Relative Fractional Part: $\dfrac{\text{Fractional Part}}{\text{Integer Part}}$ | Final Apportionment |
|---|---|---|---|---|---|
| Connecticut | $\dfrac{236,841}{34,437.333} \approx 6.877$ | 6 | 0.877 | $\dfrac{0.877}{6} \approx 0.146 *$ | 7 |
| Delaware | $\dfrac{55,540}{34,437.333} \approx 1.613$ | 1 | 0.613 | $\dfrac{0.613}{1} \approx 0.613 *$ | 2 |
| Georgia | $\dfrac{70,835}{34,437.333} \approx 2.057$ | 2 | 0.057 | $\dfrac{0.057}{2} \approx 0.029$ | 2 |
| Kentucky | $\dfrac{68,705}{34,437.333} \approx 1.995$ | 1 | 0.995 | $\dfrac{0.995}{1} \approx 0.995 *$ | 2 |
| Maryland | $\dfrac{278,514}{34,437.333} \approx 8.088$ | 8 | 0.088 | $\dfrac{0.088}{8} \approx 0.011$ | 8 |
| Massachusetts | $\dfrac{475,327}{34,437.333} \approx 13.803$ | 13 | 0.803 | $\dfrac{0.803}{13} \approx 0.062 *$ | 14 |
| New Hampshir | $\dfrac{141,822}{34,437.333} \approx 4.118$ | 4 | 0.118 | $\dfrac{0.118}{4} \approx 0.030$ | 4 |
| New Jersey | $\dfrac{179,570}{34,437.333} \approx 5.214$ | 5 | 0.214 | $\dfrac{0.214}{5} \approx 0.043$ | 5 |
| New York | $\dfrac{331,589}{34,437.333} \approx 9.629$ | 9 | 0.629 | $\dfrac{0.629}{9} \approx 0.070 *$ | 10 |
| North Carolina | $\dfrac{353,523}{34,437.333} \approx 10.266$ | 10 | 0.266 | $\dfrac{0.266}{10} \approx 0.027$ | 10 |
| Pennsylvania | $\dfrac{432,879}{34,437.333} \approx 12.570$ | 12 | 0.570 | $\dfrac{0.570}{12} \approx 0.048$ | 12 |
| Rhode Island | $\dfrac{68,446}{34,437.333} \approx 1.988$ | 1 | 0.988 | $\dfrac{0.988}{1} = 0.988 *$ | 2 |
| South Carolina | $\dfrac{206,236}{34,437.33} \approx 5.989$ | 5 | 0.989 | $\dfrac{0.989}{5} \approx 0.198 *$ | 6 |
| Vermont | $\dfrac{85,533}{34,437.33} \approx 2.484$ | 2 | 0.484 | $\dfrac{0.484}{2} = 0.242 *$ | 3 |
| Virginia | $\dfrac{630,560}{34,437.33} \approx 18.310$ | 18 | 0.310 | $\dfrac{0.310}{18} \approx 0.017$ | 18 |

The sum of the integer parts of the standard quotas is 97. Each state is initially allocated the number of seats equal to the integer part of the state's standard quota. There are $105 - 97 = 8$ additional seats that must be apportioned. Compare the relative fractional parts. The eight states with the largest relative fractional parts have their relative fractional parts marked with an asterisk in the table. Those states will each receive one additional seat. The final apportionment according to Lowndes' method is listed in the final column.

**33. (a)** The standard divisor is calculated using the total population and the 360 officers that will be apportioned. The standard divisor is $D = \dfrac{P}{M} = \dfrac{796,599}{360} = 2212.775$.

| County | 2001 Population | Standard Quota $Q = \dfrac{p}{D}$ | Integer Part | Fractional Part |
|---|---|---|---|---|
| Kent | 128,822 | $\dfrac{128,822}{2212.775} \approx 58.217$ | 58 | 0.217 |
| New Castle | 507,085 | $\dfrac{507,085}{2212.775} \approx 229.162$ | 229 | 0.162 |
| Sussex | 160,692 | $\dfrac{160,692}{2212.775} \approx 72.620$ | 72 | 0.620 |

Each county is initially given the number of officers equal to the integer part of the county's standard quota. A total of $58 + 229 + 72 = 359$ officers have been accounted for. The remaining officer will be given to the county with the largest fractional part. Sussex has the largest fractional part and will receive the 1 left over officer. The apportionment according to Hamilton's method gives 58 officers to Kent County, 229 officers to New Castle County, and 73 officers to Sussex County.

**(b)** The standard divisor is calculated using the total area and the 360 officers that will be apportioned. The standard divisor is $D = \dfrac{P}{M} = \dfrac{1954}{360} \approx 5.428$.

| County | Area in Square Miles | Standard Quota $Q = \dfrac{p}{D}$ | Integer Part | Fractional Part |
|---|---|---|---|---|
| Kent | 590 | $\dfrac{590}{5.428} \approx 108.696$ | 108 | 0.696 |
| New Castle | 426 | $\dfrac{426}{5.428} \approx 78.482$ | 78 | 0.482 |
| Sussex | 938 | $\dfrac{938}{5.428} \approx 172.808$ | 172 | 0.808 |

Each county is initially given the number of officers equal to the integer part of the county's standard quota. A total of $108 + 78 + 172 = 358$ officers have been accounted for. The remaining 2 officers will be given to the counties with the largest fractional parts. Sussex and Kent have the largest fractional parts and will each receive 1 additional officer. The apportionment according to Hamilton's method gives 109 officers to Kent County, 78 officers to New Castle County, and 173 officers to Sussex County.

**35. (a)** The standard divisor is calculated using the total population and the 360 officers that will be apportioned. The standard divisor is $D = \dfrac{P}{M} = \dfrac{796,599}{360} = 2212.775$.

| County | 2001 Population | Standard Quota $Q = \dfrac{p}{D}$ | Integer Part | Fractional Part | Relative Fractional Part: $\dfrac{\text{Fractional Part}}{\text{Integer Part}}$ |
|---|---|---|---|---|---|
| Kent | 128,822 | $\dfrac{128,822}{2212.775} \approx 58.217$ | 58 | 0.217 | $\dfrac{0.217}{58} \approx 0.004$ |

| New Castle | 507,085 | $\dfrac{507,085}{2212.775} \approx 229.162$ | 229 | 0.162 | $\dfrac{0.162}{229} \approx 0.001$ |
|---|---|---|---|---|---|
| Sussex | 160,692 | $\dfrac{160,692}{2212.775} \approx 72.620$ | 72 | 0.620 | $\dfrac{0.620}{72} \approx 0.009$ |

Each county is initially given the number of officers equal to the integer part of the county's standard quota. A total of $58 + 229 + 72 = 359$ officers have been accounted for. The remaining officer will be given to the county with the largest relative fractional part. Sussex has the largest relative fractional part and will receive the 1 left over officer. The apportionment according to Lowndes' method gives 58 officers to Kent County, 229 officers to New Castle County, and 73 officers to Sussex County.

**(b)** The standard divisor is calculated using the total area and the 360 officers that will be apportioned.

The standard divisor is $D = \dfrac{P}{M} = \dfrac{1954}{360} \approx 5.428.$

| County | Area in Square Miles | Standard Quota $Q = \dfrac{P}{D}$ | Integer Part | Fractional Part | Relative Fractional Part: $\dfrac{\text{Fractional Part}}{\text{Integer Part}}$ |
|---|---|---|---|---|---|
| Kent | 590 | $\dfrac{590}{5.428} \approx 108.696$ | 108 | 0.696 | $\dfrac{0.696}{108} \approx 0.006$ |
| New Castle | 426 | $\dfrac{426}{5.428} \approx 78.482$ | 78 | 0.482 | $\dfrac{0.482}{78} \approx 0.006$ |
| Sussex | 938 | $\dfrac{938}{5.428} \approx 172.808$ | 172 | 0.808 | $\dfrac{0.808}{172} \approx 0.005$ |

Each county is initially given the number of officers equal to the integer part of the county's standard quota. A total of $108 + 78 + 172 = 358$ officers have been accounted for. The remaining 2 officers will be given to the counties with the largest relative fractional parts. Kent and New Castle have the largest relative fractional parts and will each receive 1 additional officer. The apportionment according to Lowndes' method gives 109 officers to Kent County, 79 officers to New Castle County, and 172 officers to Sussex County.

**37.** The standard divisor is calculated using the total population and the 57 seats that will be apportioned.
The standard divisor is $D = \dfrac{P}{M} = \dfrac{3,810,179}{57} \approx 66,845.246.$

| Province | Approximate 2002 Population | Standard Quota $Q = \dfrac{P}{D}$ | Integer Part | Fractional Part | Final Apportionment |
|---|---|---|---|---|---|
| Alajuela | 716,286 | $\dfrac{716,286}{66,845.246} \approx 10.716$ | 10 | 0.716 | 11 |
| Cartago | 432,395 | $\dfrac{432,395}{66,845.246} \approx 6.469$ | 6 | 0.469 | 7 |
| Guanacaste | 264,238 | $\dfrac{264,238}{66,845.246} \approx 3.953$ | 3 | 0.953 | 4 |
| Heredia | 354,732 | $\dfrac{354,732}{66,845.246} \approx 5.307$ | 5 | 0.307 | 5 |
| Limon | 339,295 | $\dfrac{339,295}{66,845.246} \approx 5.076$ | 5 | 0.076 | 5 |

| Puntarenas | 357,483 | $\dfrac{357,483}{66,845.246} \approx 5.348$ | 5 | 0.348 | 5 |
|---|---|---|---|---|---|
| San Jose | 1,345,750 | $\dfrac{1,345,750}{66,845.246} \approx 20.132$ | 20 | 0.132 | 20 |

Each province is initially given the number of seats equal to the integer part of the province's standard quota. A total of 54 seats have been accounted for. The remaining 3 seats will be given to the provinces with the largest fractional parts. Guanacaste, Alajuela, and Cartago have the largest fractional parts and will each receive 1 additional seat. The apportionment according to Hamilton's method is listed in the final column of the table.

## Solutions to Odd-Numbered Problems in Section 5.2

1.  In this case, the denominator must decrease in order for the value of the fraction to increase. Check for yourself by comparing several fractions for which you leave the numerator unchanged but you decrease the denominator. For example, for a numerator of 12, compare the values of the fractions that result when the denominators change from 6 to 4 to 3 and finally to 2. Notice in the following fractions, the value of the fraction increases as the denominator decreases: $\dfrac{12}{6}=2$, $\dfrac{12}{4}=3$, $\dfrac{12}{3}=4$, and $\dfrac{12}{2}=6$.

3.  Without using a calculator, determine which fraction is larger. Notice the numerators are the same in each fraction pair. Compare the denominators. When the numerators are the same, the larger fraction will be the one with the smaller denominator.
    (a) $\dfrac{500}{6} > \dfrac{500}{7}$
    (b) $\dfrac{7500}{37} > \dfrac{7500}{37+1}$
    (c) $\dfrac{2,345,674}{357+2} < \dfrac{2,345,674}{357+1}$

5.  In order to increase the modified quotas, the modified divisor must decrease in comparison to the standard divisor. The standard quota of a state, for example, divides the population of the state by the standard divisor. For a modified quota to be larger than the standard quota, the population of the state must be divided by a smaller number. Thus the modified divisor must be smaller than the standard divisor.

7.  (a) There are 27 school board seats available. Jefferson's method assigns to each zone the number of seats equal to the integer part of its modified quota. The total of the integer parts of the given modified quotas is $6 + 8 + 4 + 3 + 3 = 24$. The sum of the integer parts is less than the 27 seats available. The next step will be to increase the modified quotas. This will be accomplished by decreasing the modified divisor. Making the denominators of the modified quotas smaller will increase them so that, hopefully, the integer parts of the modified quotas will add to 27.
    (b) There are 27 school board seats available. Jefferson's method assigns to each zone the number of seats equal to the integer part of its modified quota. The total of the integer parts of the given modified quotas is $7 + 9 + 5 + 3 + 3 = 27$. The sum of the integer parts is exactly the 27 seats available. The next step is to apportion the seats according to the integer parts given.

9.  (a) The total population is $101 + 109 + 150 = 360$. The standard divisor is $D = \dfrac{P}{M} = \dfrac{360}{6} = 60$.

    The standard quotas are calculated in the following table:

| State | Population | Standard Quota $Q = \dfrac{p}{D}$ | Integer Part |
|-------|-----------|-----------------------------------|--------------|
| A | 101 | $\dfrac{101}{60} \approx 1.683$ | 1 |
| B | 109 | $\dfrac{109}{60} \approx 1.817$ | 1 |
| C | 150 | $\dfrac{150}{60} = 2.5$ | 2 |

The integer parts of the standard quotas add to 4. There are 6 seats available.

(b) Because the integer parts of the standard quotas add to a number less than the number of seats available, modified quotas must be created that are larger than the standard quotas. Larger modified quotas can be created by decreasing the denominator in the standard quotas. We must modify the divisor so that it is smaller than the standard divisor. Use the guess and test strategy as was explained in the text. The largest state is C with a population of 150. Let $I = 2$, the integer part of the largest state's standard quota. Our first guess at a modified divisor will be

$$d = \frac{\text{largest state's population}}{I+1} = \frac{150}{2+1} = \frac{150}{3} = 50.$$

(c)

| Modified Divisor | Modified Quota State A $mQ = \dfrac{p}{d}$ | Modified Quota State B $mQ = \dfrac{p}{d}$ | Modified Quota State C $mQ = \dfrac{p}{d}$ | Sum of Integer Parts |
|------------------|-------------------------------------------|-------------------------------------------|-------------------------------------------|----------------------|
| $d = 50$ | $\dfrac{101}{50} = 2.02$ | $\dfrac{109}{50} = 2.18$ | $\dfrac{150}{50} = 3$ | 7 too large |
| $d = 51$ | $\dfrac{101}{51} \approx 1.980$ | $\dfrac{109}{51} \approx 2.137$ | $\dfrac{150}{51} \approx 2.941$ | 5 too small |
| $d = 50.5$ | $\dfrac{101}{50.5} = 2$ | $\dfrac{109}{50.5} \approx 2.158$ | $\dfrac{150}{50.5} \approx 2.970$ | 6 |

(d) When the modified divisor in part (c) was 50.5, the integer parts of the modified quotas added to 6 which is exactly the number of seats to be apportioned. According to Jefferson's method, each state will receive 2 seats.

11. (a) First we must calculate the standard divisor: $D = \dfrac{P}{M} = \dfrac{295 + 205 + 390}{20} = \dfrac{890}{20} = 44.5$.

| Contribution | Standard Quota $Q = \dfrac{p}{D}$ | Modified Quota $mQ = \dfrac{p}{43}$ | Modified Quota $mQ = \dfrac{p}{39}$ | Modified Quota $mQ = \dfrac{p}{42}$ |
|--------------|-----------------------------------|-------------------------------------|-------------------------------------|-------------------------------------|
| Jaron $295 | $\dfrac{295}{44.5} \approx 6.629$ | $\dfrac{295}{43} \approx 6.860$ | $\dfrac{295}{39} \approx 7.564$ | $\dfrac{295}{42} \approx 7.024$ |
| Mikkel $205 | $\dfrac{205}{44.5} \approx 4.607$ | $\dfrac{205}{43} \approx 4.767$ | $\dfrac{205}{39} \approx 5.256$ | $\dfrac{205}{42} \approx 4.881$ |
| Robert $390 | $\dfrac{390}{44.5} \approx 8.764$ | $\dfrac{390}{43} \approx 9.070$ | $\dfrac{390}{39} = 10$ | $\dfrac{390}{42} \approx 9.286$ |
| Sum of Integer Parts | 18 too small | 19 too small | 22 too large | 20 |

The sum of the integer parts of the standard quotas is too small, so the standard quotas must be modified to be larger. To accomplish this, a modified divisor must be created that is smaller than the standard divisor. Let $I = 8$, the integer part of the largest contributor's standard quota.

The first guess at a modified divisor will be $d = \dfrac{\text{largest contribution}}{I+1} = \dfrac{390}{8+1} = \dfrac{390}{9} \approx 43$.

Remember that we round the modified divisor down. The first modified quotas are calculated with a modified divisor of $d = 43$.

The second guess at a modified divisor will be $d = \dfrac{\text{largest contribution}}{I+2} = \dfrac{390}{8+2} = \dfrac{390}{10} = 39$.

The second modified quotas are calculated with a modified divisor of $d = 39$.

The second guess at a modified divisor made the sum of the integer part of the modified quotas too large. The next modified divisor must be between 39 and 43. We will try letting $d = 42$. With a modified divisor of 42, the integer parts of the modified quotas added to 20, which is the number of bottles to be apportioned.

(b) The apportionment according to Jefferson's method gives 7 bottles to Jaron, 4 bottles to Mikkel, and 9 bottles to Robert.

13. First calculate the standard divisor by dividing the total enrollment by the number of seats in the senate: $D = \dfrac{P}{M} = \dfrac{10{,}110}{30} = 337$. The standard quotas for each college are calculated in the following table. Modified quotas are also given in the table. Reasoning behind the choice of each modified divisor is given after the table. Round each modified divisor down.

| College ($p$ = enrollment) | Standard Quota $Q = \dfrac{p}{337}$ | First Modified Quota $mQ = \dfrac{p}{325}$ | Second Modified Quota $mQ = \dfrac{p}{298}$ | Third Modified Quota $mQ = \dfrac{p}{310}$ |
|---|---|---|---|---|
| Fine and Performing Arts $p = 2540$ | $\dfrac{2540}{337} \approx 7.537$ | $\dfrac{2540}{325} \approx 7.815$ | $\dfrac{2540}{298} \approx 8.523$ | $\dfrac{2540}{310} \approx 8.194$ |
| Math and Physical Science $p = 3580$ | $\dfrac{3580}{337} \approx 10.623$ | $\dfrac{3580}{325} \approx 11.015$ | $\dfrac{3580}{298} \approx 12.013$ | $\dfrac{3580}{310} \approx 11.548$ |
| Engineering $p = 1410$ | $\dfrac{1410}{337} \approx 4.184$ | $\dfrac{1410}{325} \approx 4.338$ | $\dfrac{1410}{298} \approx 4.732$ | $\dfrac{1410}{310} \approx 4.548$ |
| Social Science $p = 1830$ | $\dfrac{1830}{337} \approx 5.430$ | $\dfrac{1830}{325} \approx 5.631$ | $\dfrac{1830}{298} \approx 6.141$ | $\dfrac{1830}{310} \approx 5.903$ |
| Agriculture $p = 750$ | $\dfrac{750}{337} \approx 2.226$ | $\dfrac{750}{325} \approx 2.308$ | $\dfrac{750}{298} \approx 2.517$ | $\dfrac{750}{310} \approx 2.419$ |
| Sum of Integer Parts | 28 too small | 29 too small | 32 too large | 30 |

The sum of the integer parts of the standard quotas is too small, so the standard quotas must be modified to be larger. To accomplish this, a modified divisor must be created that is smaller than the standard divisor. Let $I = 10$, the integer part of the largest college's standard quota.

The first guess at a modified divisor will be $d = \dfrac{\text{largest enrollment}}{I+1} = \dfrac{3580}{10+1} = \dfrac{3580}{11} \approx 325$. The first modified quotas are calculated with a modified divisor of $d = 325$. The sum of the integer parts of the first modified quotas is too small, so they must be modified to be even larger.

The second guess at a modified divisor will be $d = \dfrac{\text{largest enrollment}}{I+2} = \dfrac{3580}{10+2} = \dfrac{3580}{12} \approx 298$. The second modified quotas are calculated with a modified divisor of $d = 298$. This time, the sum of the integer parts of the second modified quotas is too large, so the third modified quotas must be smaller. We must select a modified divisor between 298 and 325. We will try letting $d = 310$, however, it would be perfectly reasonable to select any other number between 298 and 325 to use as the next modified divisor. With a modified divisor of 310, the integer parts of the third modified quotas add to 30, which is the number of seats to be apportioned.

According to Jefferson's method, each college will receive the number of seats equal to the integer part of the college's modified quota. The College of Fine and Performing Arts will receive 8 seats; Math and Physical Science, 11; Engineering, 4; Social Science, 5; and Agriculture, 2.

15. First calculate the standard divisor by dividing the total population by the number of officers: $D = \dfrac{P}{M} = \dfrac{807,385}{360} \approx 2242.736$. The standard quotas for each county are calculated in the following table. Modified quotas are also given in the table. Reasoning behind the choice of the modified divisor is given after the table. Modified divisors are always rounded down.

| County ($p$ = population) | Standard Quota $Q = \dfrac{p}{2242.736}$ | First Modified Quota $mQ = \dfrac{p}{2237}$ |
|---|---|---|
| Kent $p = 131,069$ | $\dfrac{131,069}{2242.736} \approx 58.442$ | $\dfrac{131,069}{2237} \approx 58.591$ |
| New Castle $p = 512,370$ | $\dfrac{512,370}{2242.736} \approx 228.458$ | $\dfrac{512,370}{2237} \approx 229.043$ |
| Sussex $p = 163,946$ | $\dfrac{163,946}{2242.736} \approx 73.101$ | $\dfrac{163,946}{2237} \approx 73.288$ |
| Sum of Integer Parts | 359  too small | 360 |

The sum of the integer parts of the standard quotas is too small, so the standard quotas must be modified to be larger. To accomplish this, a modified divisor must be created that is smaller than the standard divisor. Let $I = 228$, the integer part of the largest county's standard quota.

The first guess at a modified divisor will be $d = \dfrac{\text{largest population}}{I+1} = \dfrac{512,370}{228+1} = \dfrac{512,370}{229} \approx 2237$.

The first modified quotas are calculated with a modified divisor of $d = 2237$. With a modified divisor of 2237, the integer parts of the second modified quotas add to 360, which is the number of officers to be apportioned.

According to Jefferson's method, each county will receive the number of officers equal to the integer part of the college's modified quota. Kent County will receive 58 officers; New Castle County, 229; and Sussex County, 73.

17. First calculate the standard divisor by dividing the total population in 1790 by the number of seats to be apportioned: $D = \dfrac{P}{M} = \dfrac{3,615,920}{105} \approx 34,437.333$. The standard quotas for each state are calculated in the following table. Modified quotas are also given in the table. Reasoning behind the choices of the modified divisors is given after the table. Modified divisors are always rounded down.

| State | Standard Quota $Q = \dfrac{p}{D}$ | First Modified Quota $mQ = \dfrac{p}{33,187}$ | Second Modified Quota $mQ = \dfrac{p}{31,528}$ | Third Modified Quota $mQ = \dfrac{p}{33,150}$ | Final Apportionment |
|---|---|---|---|---|---|
| Connecticut | $\dfrac{236,841}{34,437.333}$ $\approx 6.877$ | $\dfrac{236,841}{33,187}$ $\approx 7.137$ | $\dfrac{236,841}{31,528}$ $\approx 7.512$ | $\dfrac{236,841}{33,150}$ $\approx 7.145$ | 7 |
| Delaware | $\dfrac{55,540}{34,437.333}$ $\approx 1.613$ | $\dfrac{55,540}{33,187}$ $\approx 1.674$ | $\dfrac{55,540}{31,528}$ $\approx 1.762$ | $\dfrac{55,540}{33,150}$ $\approx 1.675$ | 1 |
| Georgia | $\dfrac{70,835}{34,437.333}$ $\approx 2.057$ | $\dfrac{70,835}{33,187}$ $\approx 2.134$ | $\dfrac{70,835}{31,528}$ $\approx 2.247$ | $\dfrac{70,835}{33,150}$ $\approx 2.137$ | 2 |
| Kentucky | $\dfrac{68,705}{34,437.333}$ $\approx 1.995$ | $\dfrac{68,705}{33,187}$ $\approx 2.070$ | $\dfrac{68,705}{31,528}$ $\approx 2.179$ | $\dfrac{68,705}{33,150}$ $\approx 2.073$ | 2 |
| Maryland | $\dfrac{278,514}{34,437.333}$ $\approx 8.088$ | $\dfrac{278,514}{33,187}$ $\approx 8.392$ | $\dfrac{278,514}{31,528}$ $\approx 8.834$ | $\dfrac{278,514}{33,150}$ $\approx 8.402$ | 8 |
| Massachusetts | $\dfrac{475,327}{34,437.333}$ $\approx 13.803$ | $\dfrac{475,327}{33,187}$ $\approx 14.323$ | $\dfrac{475,327}{31,528}$ $\approx 15.076$ | $\dfrac{475,327}{33,150}$ $\approx 14.339$ | 14 |
| New Hampshire | $\dfrac{141,822}{34,437.333}$ $\approx 4.118$ | $\dfrac{141,822}{33,187}$ $\approx 4.273$ | $\dfrac{141,822}{31,528}$ $\approx 4.498$ | $\dfrac{141,822}{33,150}$ $\approx 4.278$ | 4 |
| New Jersey | $\dfrac{179,570}{34,437.333}$ $\approx 5.214$ | $\dfrac{179,570}{33,187}$ $\approx 5.411$ | $\dfrac{179,570}{31,528}$ $\approx 5.696$ | $\dfrac{179,570}{33,150}$ $\approx 5.417$ | 5 |
| New York | $\dfrac{331,589}{34,437.333}$ $\approx 9.629$ | $\dfrac{331,589}{33,187}$ $\approx 9.992$ | $\dfrac{331,589}{31,528}$ $\approx 10.517$ | $\dfrac{331,589}{33,150}$ $\approx 10.003$ | 10 |
| North Carolina | $\dfrac{353,523}{34,437.333}$ $\approx 10.266$ | $\dfrac{353,523}{33,187}$ $\approx 10.652$ | $\dfrac{353,523}{31,528}$ $\approx 11.213$ | $\dfrac{353,523}{33,150}$ $\approx 10.664$ | 10 |
| Pennsylvania | $\dfrac{432,879}{34,437.333}$ $\approx 12.570$ | $\dfrac{432,879}{33,187}$ $\approx 13.044$ | $\dfrac{432,879}{31,528}$ $\approx 13.730$ | $\dfrac{432,879}{33,150}$ $\approx 13.058$ | 13 |
| Rhode Island | $\dfrac{68,446}{34,437.333}$ $\approx 1.988$ | $\dfrac{68,446}{33,187}$ $\approx 2.062$ | $\dfrac{68,446}{31,528}$ $\approx 2.171$ | $\dfrac{68,446}{33,150}$ $\approx 2.065$ | 2 |
| South Carolina | $\dfrac{206,236}{34,437.333}$ $\approx 5.989$ | $\dfrac{206,236}{33,187}$ $\approx 6.214$ | $\dfrac{206,236}{31,528}$ $\approx 6.541$ | $\dfrac{206,236}{33,150}$ $\approx 6.221$ | 6 |

| | | | | | |
|---|---|---|---|---|---|
| Vermont | $\dfrac{85{,}533}{34{,}437.333}$ $\approx 2.484$ | $\dfrac{85{,}533}{33{,}187}$ $\approx 2.577$ | $\dfrac{85{,}533}{31{,}528}$ $\approx 2.713$ | $\dfrac{85{,}533}{33{,}150}$ $\approx 2.580$ | 2 |
| Virginia | $\dfrac{630{,}560}{34{,}437.333}$ $\approx 18.310$ | $\dfrac{630{,}560}{33{,}187}$ $\approx 19.000$ | $\dfrac{630{,}560}{31{,}528}$ $= 20$ | $\dfrac{630{,}560}{33{,}150}$ $= 19.021$ | 19 |
| Sum of Integer Parts | 97  too small | 104  too small | 108  too large | 105 | |

The sum of the integer parts of the standard quotas is 97 which is too small, so the standard quotas must be modified to be larger. To accomplish this, a modified divisor must be created that is smaller than the standard divisor. Let $I = 18$, the integer part of Virginia's, the largest state's standard quota.

The first guess at a modified divisor is $d = \dfrac{\text{largest population}}{I+1} = \dfrac{630{,}560}{18+1} = \dfrac{630{,}560}{19} \approx 33{,}187.$  The first modified quotas are calculated with a modified divisor of $d = 33{,}187$. The sum of the integer parts of the first modified quotas is 104 which is too small, so they must be modified to be even larger.

The second guess at a modified divisor is $d = \dfrac{\text{largest population}}{I+2} = \dfrac{630{,}560}{18+2} = \dfrac{630{,}560}{20} = 31{,}528.$

The second modified quotas are calculated with a modified divisor of $d = 31{,}528$. This time, the sum of the integer parts of the second modified quotas is 108 which is too large, so the third modified quotas must be smaller. We must select a modified divisor between 31,528 and 33,187. Rather than randomly guess, notice that the sum of the integer parts of the first modified quotas was very close to the 105 seats we want to apportion. In particular, notice that New York's first modified quota is 9.992. If we could find a third modified divisor that increases New York's third modified quota to just a little more than 10 without increasing the other states' third modified quotas past the next largest integer, then we would have the 105 seats apportioned. For the third modified divisor, try $d = 33{,}150$ because it will increase New York's third modified quota to 10.003. With a third modified divisor of 33,150, the integer parts of the third modified quotas add to 105, which is the number of seats to be apportioned. The final apportionment according to Jefferson's method is given in the last column of the table.

19. First calculate the standard divisor by dividing the total population by the number of seats to be apportioned:  $D = \dfrac{P}{M} = \dfrac{100+110+150}{6} = \dfrac{360}{6} = 60.$  The standard quotas for each zone are calculated in the following table. Modified quotas are also given in the table. Reasoning behind the choice of the modified divisor is given after the table.

| Zone Population | Standard Quota $Q = \dfrac{P}{D}$ | First Modified Quota $mQ = \dfrac{P}{50}$ |
|---|---|---|
| 100 | $\dfrac{100}{60} \approx 1.667$ | $\dfrac{100}{50} = 2$ |
| 110 | $\dfrac{110}{60} \approx 1.833$ | $\dfrac{110}{50} = 2.2$ |
| 150 | $\dfrac{150}{60} = 2.5$ | $\dfrac{150}{50} = 3$ |
| Sum of Integer Parts | 4  too small | 7  too large |

The sum of the integer parts of the standard quotas is 4 which is too small, so the standard quotas must be modified to be larger. To accomplish this, a modified divisor must be created that is smaller than the standard divisor. Let $I = 2$, the integer part of the largest state's standard quota.

The first guess at a modified divisor will be $d = \dfrac{\text{largest population}}{I+1} = \dfrac{150}{2+1} = \dfrac{150}{3} = 50$. The first modified quotas are calculated with a modified divisor of $d = 50$. The sum of the integer parts of the first modified quotas is 7 which is too large, so the second modified quotas must be smaller. We must select a modified divisor between 50 and 60. Here is where we have a problem. No matter what modified divisor between 50 and 60 is selected next, the sum of the integer parts of the modified quotas will be 4 or 5. For any modified divisor between 50 and 60, the integer part of the modified quota of the zone with population 100 will be 1; the integer part of the modified quota of the zone with population 110 will be 1 or 2; and the integer part of the modified quota of the zone with population 150 will be 2. There is no way to obtain a sum of 6 for the integer parts of the modified quotas. Jefferson's method fails to obtain an apportionment of the 6 seats in this case.

21. Apportion the 5 classes using Jefferson's method and then using Webster's method.

**Jefferson's Method:**

First calculate the standard divisor by dividing the total enrollment by the number of classes to be apportioned: $D = \dfrac{P}{M} = \dfrac{102}{5} = 20.4$.

| Course and Enrollment | Standard Quota $Q = \dfrac{p}{20.4}$ | First Modified Quota $mQ = \dfrac{p}{17}$ | Second Modified Quota $mQ = \dfrac{p}{17.5}$ |
|---|---|---|---|
| Beginning Spanish, 53 | $\dfrac{53}{20.4} \approx 2.598$ | $\dfrac{53}{17} \approx 3.118$ | $\dfrac{53}{17.5} \approx 3.029$ |
| Intermediate Spanish, 34 | $\dfrac{34}{20.4} \approx 1.667$ | $\dfrac{34}{17} = 2$ | $\dfrac{34}{17.5} \approx 1.943$ |
| Conversational Spanish, 15 | $\dfrac{15}{20.4} \approx 0.735$ | $\dfrac{15}{17} \approx 0.882$ | $\dfrac{15}{17.5} \approx 0.857$ |

Remember that each course must be given at least one class even if the standard quota for the course is less than one. The sum of the integer parts of the standard quotas (with Conversational Spanish receiving 1 class) is $2 + 1 + 1 = 4$ which is too small so the standard quotas must be modified to be larger. Let $I = 2$, the integer part of the largest course's standard quota.

The first guess at a modified divisor (rounded down) will be $d = \dfrac{\text{largest enrollment}}{I+1} = \dfrac{53}{2+1} = \dfrac{53}{3} \approx 17$. The first modified quotas are calculated with a modified divisor of $d = 17$. The sum of the integer parts of the first modified quotas is $3 + 2 + 1 = 6$ which is too large so they must be modified to be smaller. We must select a second modified divisor between 17 and 20.4. You may try several values before you find one that works. If we let the second modified divisor be $d = 17.5$, then the sum of the integer parts of the second modified quotas is $3 + 1 + 1 = 5$ which is the number of classes to be apportioned. Jefferson's method would allocate 3 classes to Beginning Spanish, 1 class to Intermediate Spanish, and 1 class to Conversational Spanish.

**Webster's Method:**

The standard divisor was calculated to be $D = 20.4$ when we worked through Jefferson's method. If the standard quotas, in the following table, are rounded to the nearest integer, then $3 + 2 + 1 = 6$ classes will be allocated which it too large. We need to modify the divisor to be larger than the standard divisor of 20.4 so that the modified quotas will be smaller than the standard quotas. There is no nice rule for this. Try a modified divisor of $d = 22$. When the first modified quotas are rounded to the nearest integer, they add to 5 which is the number of classes we have to allocate. Webster's method would allocate 2 classes to Beginning Spanish, 2 classes to Intermediate Spanish, and 1 class to Conversational Spanish.

| Course and Enrollment | Standard Quota $Q = \dfrac{p}{20.4}$ | First Modified Quota $mQ = \dfrac{p}{22}$ |
|---|---|---|
| Beginning Spanish, 53 | $\dfrac{53}{20.4} \approx 2.598$ | $\dfrac{53}{22} \approx 2.409$ |
| Intermediate Spanish, 34 | $\dfrac{34}{20.4} \approx 1.667$ | $\dfrac{34}{22} \approx 1.545$ |
| Conversational Spanish, 15 | $\dfrac{15}{20.4} \approx 0.735$ | $\dfrac{15}{22} \approx 0.682$ |

| Course | Number of Students | Standard Quota | Jefferson Apportionment | Webster Apportionment |
|---|---|---|---|---|
| Beginning Spanish | 53 | 2.598 | 3 | 2 |
| Intermediate Spanish | 34 | 1.667 | 1 | 2 |
| Conversational Spanish | 15 | 0.735 | 1 | 1 |

Webster's method keeps class sizes smaller so it may be the best apportionment. Jefferson's method gives the Intermediate Spanish group one class which has 34 students.

23. First calculate the standard divisor by dividing the total population by the number of officers: $D = \dfrac{P}{M} = \dfrac{807,385}{360} \approx 2242.736$. The standard quotas for each county are calculated in the following table. Modified quotas are also given in the table. Reasoning behind the choices of the modified divisors is given after the table. Modified divisors are always rounded down.

| County ($p$ = population) | Standard Quota $Q = \dfrac{p}{2242.736}$ | First Modified Quota $mQ = \dfrac{p}{2237}$ | Second Modified Quota $mQ = \dfrac{p}{2241}$ |
|---|---|---|---|
| Kent $p = 131,069$ | $\dfrac{131,069}{2242.736} \approx 58.442$ | $\dfrac{131,069}{2237} \approx 58.591$ | $\dfrac{131,069}{2241} \approx 58.487$ |
| New Castle $p = 512,370$ | $\dfrac{512,370}{2242.736} \approx 228.458$ | $\dfrac{512,370}{2237} \approx 229.043$ | $\dfrac{512,370}{2241} \approx 228.635$ |
| Sussex $p = 163,946$ | $\dfrac{163,946}{2242.736} \approx 73.101$ | $\dfrac{163,946}{2237} \approx 73.288$ | $\dfrac{163,946}{2241} \approx 73.158$ |
| Sum of Quotas Rounded to the Nearest Integer | 359 too small | 361 too large | 360 |

If we round the standard quotas to the nearest integer, then the sum is 359 which is too small, so the standard quotas must be modified to be larger. To accomplish this, a modified divisor must be created that is smaller than the standard divisor. Let $I = 228$, the integer part of the largest county's standard quota.

The first guess at a modified divisor will be $d = \dfrac{\text{largest population}}{I+1} = \dfrac{512,370}{228+1} = \dfrac{512,370}{229} \approx 2237$.

The first modified quotas are calculated with a modified divisor of $d = 2237$. The sum of the first modified quotas rounded to the nearest integer is 361 which is too large, so the second modified quotas must be smaller. We must select a modified divisor between 2237 and 2242.736. We will try letting

$d = 2241$. With a modified divisor of 2241, the second modified quotas rounded to the nearest integer add to 360, which is the number of officers to be apportioned.

According to Webster's method, Kent County will receive 58 officers; New Castle County, 229; and Sussex County, 73.

**25.** First calculate the standard divisor by dividing the total population by the number of seats: $D = \dfrac{P}{M} = \dfrac{11,060,000}{200} = 55,300$. The standard quotas for each state are calculated in the following table. Modified quotas are also given in the table. Reasoning behind the choice of each modified divisor is given after the table. Round each modified divisor down.

| State (p=population) | Standard Quota $Q = \dfrac{p}{55,300}$ | First Modified Quota $mQ = \dfrac{p}{55,171}$ | Second Modified Quota $mQ = \dfrac{p}{54,620}$ | Second Modified Quota $mQ = \dfrac{p}{54,625}$ |
|---|---|---|---|---|
| A $p = 1,592,000$ | $\dfrac{1,592,000}{55,300}$ $\approx 28.788$ | $\dfrac{1,592,000}{55,171}$ $\approx 28.856$ | $\dfrac{1,592,000}{54,620}$ $\approx 29.147$ | $\dfrac{1,592,000}{54,625}$ $\approx 29.144$ |
| B $p = 1,596,000$ | $\dfrac{1,596,000}{55,300}$ $\approx 28.861$ | $\dfrac{1,596,000}{55,171}$ $\approx 28.928$ | $\dfrac{1,596,000}{54,620}$ $\approx 29.220$ | $\dfrac{1,596,000}{54,625}$ $\approx 29.217$ |
| C $p = 5,462,000$ | $\dfrac{5,462,000}{55,300}$ $\approx 98.770$ | $\dfrac{5,462,000}{55,171}$ $\approx 99.001$ | $\dfrac{5,462,000}{54,620}$ $= 100$ | $\dfrac{5,462,000}{54,625}$ $\approx 99.991$ |
| D $p = 1,323,000$ | $\dfrac{1,323,000}{55,300}$ $\approx 23.924$ | $\dfrac{1,323,000}{55,171}$ $\approx 23.980$ | $\dfrac{1,323,000}{54,620}$ $\approx 24.222$ | $\dfrac{1,323,000}{54,625}$ $\approx 24.220$ |
| E $p = 1,087,000$ | $\dfrac{1,087,000}{55,300}$ $\approx 19.656$ | $\dfrac{1,087,000}{55,171}$ $\approx 19.702$ | $\dfrac{1,087,000}{54,620}$ $\approx 19.901$ | $\dfrac{1,087,000}{54,625}$ $\approx 19.899$ |
| Sum of Integer Parts | 196 too small | 197 too small | 201 too large | 200 |

The sum of the integer parts of the standard quotas is 196 which is too small, so the standard quotas must be modified to be larger. To accomplish this, a modified divisor must be created that is smaller than the standard divisor. Let $I = 98$, the integer part of the largest state's standard quota. The first guess at a modified divisor (rounded down) will be $d = \dfrac{\text{largest population}}{I + 1} = \dfrac{5,462,000}{98 + 1} = \dfrac{5,462,000}{99} \approx 55,171$. The first modified quotas are calculated with a modified divisor of $d = 55,171$. The sum of the integer parts of the first modified quotas is 197 which is too small, so they must be modified to be even larger.

The second guess at a modified divisor will be $d = \dfrac{\text{largest population}}{I + 2} = \dfrac{5,462,000}{98 + 2} = \dfrac{5,462,000}{100} = 54,620$. The second modified quotas are calculated with a modified divisor of $d = 54,620$. The sum of the integer parts of the second modified quotas is 201 which is too large, so they must be modified to be a little bit smaller. We must select a third modified divisor between 54,620 and 55,171. Notice that state C's second modified quota is 100.000. If we could find a third modified divisor that would force C's third modified quota to be a little less than 100 without changing the integer parts of the other state's second modified quota, then we would have apportioned the 200 seats. Try a third modified divisor of $d = 54,625$. The sum of the integer parts of the third modified quotas is 200 which is the number of seats we are trying to apportion. Jefferson's method gives 29 seats to state A, 29 seats to state B, 99 seats to state C, 24 seats to state D, and 19 seats to state E.

27. Add the populations of each state in the given table. The total population of Freedonia is 11,060,000. To maintain the ratio of one vote for every 50,000 citizens, divide the total population by 50,000. The number of seats needed is $\dfrac{11,060,000}{50,000} = 221.2$. There should be 221 seats. In order to apportion the 221 seats using Jefferson's method, we need to find the standard divisor: $D = \dfrac{P}{M} = \dfrac{11,060,000}{221} \approx 50,045.249$.

| State (p = population) | Standard Quota $Q = \dfrac{p}{50,045.249}$ | Modified Quota $mQ = \dfrac{p}{49,654}$ |
|---|---|---|
| A $p = 1,592,000$ | $\dfrac{1,592,000}{50,045.249} \approx 31.811$ | $\dfrac{1,592,000}{49,654} \approx 32.062$ |
| B $p = 1,596,000$ | $\dfrac{1,596,000}{50,045.249} \approx 31.891$ | $\dfrac{1,596,000}{49,654} \approx 32.142$ |
| C $p = 5,462,000$ | $\dfrac{5,462,000}{50,045.249} \approx 109.141$ | $\dfrac{5,462,000}{49,654} \approx 110.001$ |
| D $p = 1,323,000$ | $\dfrac{1,323,000}{50,045.249} \approx 26.436$ | $\dfrac{1,323,000}{49.654} \approx 26.644$ |
| E $p = 1,087,000$ | $\dfrac{1,087,000}{50,045.249} \approx 21.720$ | $\dfrac{1,087,000}{49,654} \approx 21.891$ |
| Sum of Integer Parts | 218 too small | 221 |

The sum of the integer parts of the standard quotas is 218 which is too small, so the standard quotas must be modified to be larger. To accomplish this, a modified divisor must be created that is smaller than the standard divisor. Let $I = 109$, the integer part of the largest state's standard quota. The first guess at a modified divisor (rounded down) will be $d = \dfrac{\text{largest population}}{I+1} = \dfrac{5,462,000}{109+1} = \dfrac{5,462,000}{110} \approx 49,654$. The first modified quotas are calculated with a modified divisor of $d = 49,654$. The sum of the integer parts of the first modified quotas is 221 which is the number of seats we are trying to apportion. Jefferson's method gives 32 seats to state A, 32 seats to state B, 110 seats to state C, 26 seats to state D, and 21 seats to state E.

29. We will apply Jefferson's method to apportion the 50 seats according to the 1990 population. The standard divisor is $D = \dfrac{P}{M} = \dfrac{1,108,229}{50} = 22,164.58$. The standard quotas for each county are calculated in the following table. Modified quotas are calculated in the table also, and reasoning behind the modified divisor is given after the table.

| County (p = population) | Standard Quota $Q = \dfrac{p}{22,164.58}$ | Modified Quota $mQ = \dfrac{p}{22,006}$ |
|---|---|---|
| Hawaii $p = 120,317$ | $\dfrac{120,317}{22,164.58} \approx 5.428$ | $\dfrac{120,317}{22,006} \approx 5.467$ |
| Honolulu $p = 836,231$ | $\dfrac{836,231}{22,164.58} \approx 37.728$ | $\dfrac{836,231}{22,006} \approx 38.000$ |

| Kalawao $p = 130$ | $\dfrac{130}{22,164.58} \approx 0.006$ | $\dfrac{130}{22,006} \approx 0.006$ |
|---|---|---|
| Kauai $p = 51,177$ | $\dfrac{51,177}{22,164.58} \approx 2.309$ | $\dfrac{51,177}{22,006} \approx 2.326$ |
| Maui $p = 100,374$ | $\dfrac{100,374}{22,164.58} \approx 4.529$ | $\dfrac{100,374}{22,006} \approx 4.561$ |
| Sum of Integer Parts | 49 too small | 50 |

Remember that every county must have at least one seat even if the integer part of the standard quota is zero. Therefore, Kalawao must be given one seat. The sum of the integer parts of the standard quotas, with Kalawao apportioned 1 seat, is 49 which is too small, so the standard quotas must be modified to be larger. To accomplish this, a modified divisor must be created that is smaller than the standard divisor. Let $I = 37$, the integer part of the largest county's standard quota. The first guess at a modified divisor (rounded down) will be $d = \dfrac{\text{largest population}}{I+1} = \dfrac{836,231}{37+1} = \dfrac{836,231}{38} \approx 22,006.$

The modified quotas are calculated with a modified divisor of $d = 22,006$. This time the sum of the integer parts of the modified quotas, with Kalawao apportioned 1 seat, is 50 which is the number of seats available. Jefferson's method apportions 5 seats to Hawaii, 38 seats to Honolulu, 1 seat to Kalawao, 2 seats to Kauai, and 4 seats to Maui.

**31.** We will apply Webster's method to apportion the 50 seats according to the 1990 population. The standard divisor is $D = \dfrac{P}{M} = \dfrac{1,108,229}{50} = 22,164.58.$ The standard quotas for each county are calculated in the following table. Modified quotas are calculated in the table also.

| County ($p$ = population) | Standard Quota $Q = \dfrac{p}{22,164.58}$ | Modified Quota $mQ = \dfrac{p}{22,300}$ |
|---|---|---|
| Hawaii $p = 120,317$ | $\dfrac{120,317}{22,164.58} \approx 5.428$ | $\dfrac{120,317}{22,300} \approx 5.395$ |
| Honolulu $p = 836,231$ | $\dfrac{836,231}{22,164.58} \approx 37.728$ | $\dfrac{836,231}{22,300} \approx 37.499$ |
| Kalawao $p = 130$ | $\dfrac{130}{22,164.58} \approx 0.006$ | $\dfrac{130}{22,300} \approx 0.006$ |
| Kauai $p = 51,177$ | $\dfrac{51,177}{22,164.58} \approx 2.309$ | $\dfrac{51,177}{22,300} \approx 2.295$ |
| Maui $p = 100,374$ | $\dfrac{100,374}{22,164.58} \approx 4.529$ | $\dfrac{100,374}{22,300} \approx 4.501$ |
| Sum of Quotas Rounded to the Nearest Integer | 51 too large | 50 |

The sum of the standard quotas rounded to the nearest integer, with Kalawao apportioned 1 seat, is 51 which is too large. Modified quotas must be created that are slightly smaller than the standard quotas. To accomplish this, we must find a modified divisor that is larger than the standard divisor. Guess and test modified divisors larger than 22,164.58. If we let $d = 24,300$, then the sum of the modified quotas rounded to the nearest integer, with Kalawao apportioned 1 seat, is 50. Webster's method apportions 5 seats to Hawaii, 37 seats to Honolulu, 1 seat to Kalawao, 2 seats to Kauai, and 5 seats to Maui. Webster's method and Jefferson's method from problem 29 yield different apportionments. By using Jefferson's method, Honolulu gets one additional seat while Maui loses a seat.

## Solutions to Odd-Numbered Problems in Section 5.3

**1.** This is an example of the Alabama paradox. It occurs when adding an additional seat or seats results in a loss for a state. In this case, two new assistants became available yet Seven Oak lost an assistant while three other schools gained an assistant.

**3.** This is an example of the new-states paradox. It occurs when a recalculation of the apportionment as a result of a new state joining the union results in a change in the apportionment of some of the other states. In this case Maryland and Virginia experienced a change in their apportionment. Maryland lost a seat to Virginia.

**5.** **(a)** Notice that modified quotas appear in the table. Hamilton's method is based on standard quotas, so Hamilton's method could not have been used. Also notice that state B's modified quota was rounded up, and 19 seats were apportioned to state B. Jefferson's method rounds modified quotas down, so Jefferson's method could not have been used. Webster's method rounds modified quotas to the nearest integer. Each modified quota was rounded to the nearest integer, so Webster's method was used in this case.

   **(b)** The standard quotas are not given, so there is no way to tell if a state was apportioned a number of seats equal to the whole number just below or just above the state's standard quota.

**7.** Apportion the seats using Hamilton's method.

   **(a)** Calculate the standard divisor: $D = \dfrac{3,760,000}{24} \approx 156,666.667$.

| State | Standard Quota | Apportionment |
|---|---|---|
| Medina | 3.383 | 4 |
| Alvare | 6.319 | 6 |
| Loranne | 14.298 | 14 |

   **(b)** Calculate the standard divisor: $D = \dfrac{3,760,000}{25} = 150,400$.

| State | Standard Quota | Apportionment |
|---|---|---|
| Medina | 3.524 | 3 |
| Alvare | 6.582 | 7 |
| Loranne | 14.894 | 15 |

   **(c)** A seat was added to the national assembly and the seats were reapportioned. As a result of the reapportionment, Medina lost a seat. This is an example of the Alabama paradox.

**9.** **(a)** Apportion the 24 seats using Jefferson's method.

   Calculate the standard divisor: $D = \dfrac{3,760,000}{24} \approx 156,666.667$.

   The sum of the integer parts of the standard quotas is 23 which is too small, so a modified divisor is needed that is smaller than the standard divisor.

   Calculate the modified divisor: $d = \dfrac{2,240,000}{15} \approx 149,333$.

| State | Standard Quota | Modified Quota | Apportionment |
|---|---|---|---|
| Medina | 3.383 | 3.549 | 3 |
| Alvare | 6.319 | 6.629 | 6 |
| Loranne | 14.298 | 15.000 | 15 |

   **(b)** Apportion the 25 seats using Jefferson's method.

Calculate the standard divisor: $D = \dfrac{3,760,000}{25} = 150,400$.

The sum of the integer parts of the standard quotas is 23 which is too small, so a modified divisor is needed that is smaller than the standard divisor.

Calculate the modified divisor: $d = \dfrac{2,240,000}{15} \approx 149,333$.

The sum of the integer parts of the first modified quotas is 24 which is still too small, so another modified divisor is needed that is smaller than the first modified divisor.

Calculate the second modified divisor: $d = \dfrac{2,240,000}{16} = 140,000$.

The sum of the integer parts of the second modified quotas is 26 which is too large. Select another modified divisor between 140,000 and 149,333. Try $d = 141,000$. The sum of the integer parts of the third modified quotas is 25.

| State | Standard Quota | First Modified Quota | Second Modified Quota | Third Modified Quota | Apportionment |
|---|---|---|---|---|---|
| Medina | 3.524 | 3.549 | 3.786 | 3.759 | 3 |
| Alvare | 6.582 | 6.629 | 7.071 | 7.021 | 7 |
| Loranne | 14.894 | 15.000 | 16.000 | 15.887 | 15 |

The Alabama paradox does not occur. No state lost a seat in the reapportionment.

(c) Apportion the 26 seats using Jefferson's method.

Calculate the standard divisor: $D = \dfrac{3,760,000}{26} \approx 144,615.385$.

The sum of the integer parts of the standard quotas is 24 which is too small, so a modified divisor is needed that is smaller than the standard divisor.

Calculate the modified divisor: $d = \dfrac{2,240,000}{16} = 140,000$.

The sum of the integer parts of the modified quotas is 26.

| State | Standard Quota | Modified Quota | Apportionment |
|---|---|---|---|
| Medina | 3.665 | 3.786 | 3 |
| Alvare | 6.846 | 7.071 | 7 |
| Loranne | 15.489 | 16.000 | 16 |

The Alabama paradox does not occur. No state lost a seat in the reapportionment.

**11.** Apportion the 24 seats using Hamilton's method.

(a) Calculate the standard divisor: $D = \dfrac{4,500,000}{24} = 187,500$.

| State | Standard Quota | Apportionment |
|---|---|---|
| Medina | 3.627 | 3 |
| Alvare | 6.667 | 7 |
| Loranne | 13.707 | 14 |

(b) When the populations were smaller, Medina had 4 seats, Alvare had 6 seats, and Loranne had 14 seats. As a result of the reapportionment, Medina lost a seat to Alvare. Take a look at the population growth in more detail.

**Percent increase for Medina:**

$$\frac{\text{new population} - \text{old population}}{\text{old population}} = \frac{680,000 - 530,000}{530,000}$$

$$\approx 0.283$$

$$= 28.3\%$$

**Percent increase for Alvare:**

$$\frac{\text{new population} - \text{old population}}{\text{old population}} = \frac{1,250,000 - 990,000}{990,000}$$

$$\approx 0.263$$

$$= 26.3\%$$

Medina experienced a faster growth rate, yet it lost a seat to Alvare. This is an example of the population paradox.

**13. (a)** The standard divisor is calculated as $D = \dfrac{P}{M} = \dfrac{800,000}{50} = 16,000$. Each seat in the legislature represents 16,000 people. Apportion the seats using Hamilton's method.

| State | Standard Quota | Apportionment |
|-------|----------------|---------------|
| A | 6.188 | 6 |
| B | 30.438 | 31 |
| C | 13.375 | 13 |

**(b)** Let the state with population 116,000 be called state D. The standard divisor is calculated as $D = \dfrac{P}{M} = \dfrac{916,000}{57} \approx 16,070.175$. Apportion the seats using Hamilton's method.

| State | Standard Quota | Apportionment |
|-------|----------------|---------------|
| A | 6.160 | 6 |
| B | 30.305 | 30 |
| C | 13.317 | 14 |
| D | 7.218 | 7 |

**(c)** A new state was added and a reapportionment was completed. As a result of the new apportionment, state B lost a seat to state C. This is an example of the new-states paradox.

**15. (a)** The standard divisor is calculated as $D = \dfrac{P}{M} = \dfrac{800,000}{50} = 16,000$. Each seat in the legislature represents 16,000 people. Apportion the seats using Jefferson's method. The sum of the integer parts of the standard quotas is 49 which is too small. A modified divisor is needed.

Calculate the first modified divisor: $d = \dfrac{487,000}{31} \approx 15,709$.

The sum of the integer parts of the modified quotients is 50.

| State | Standard Quota | Modified Quotient | Apportionment |
|-------|----------------|-------------------|---------------|
| A | 6.188 | 6.302 | 6 |
| B | 30.438 | 31.001 | 31 |
| C | 13.375 | 13.623 | 13 |

**(b)** Let the state with population 116,000 be called state D. The standard divisor is calculated as $D = \dfrac{P}{M} = \dfrac{916,000}{57} \approx 16,070.175$. Apportion the seats using Jefferson's method. The sum of the integer parts of the standard quotas is 56 which is too small, so a modified divisor is needed that is smaller than the standard divisor.

Calculate the first modified divisor: $d = \dfrac{487,000}{31} \approx 15,709$.

The sum of the integer parts of the modified quotas is 57.

| State | Standard Quota | Modified Quota | Apportionment |
|-------|----------------|----------------|---------------|
| A | 6.160 | 6.302 | 6 |
| B | 30.305 | 31.001 | 31 |
| C | 13.317 | 13.623 | 13 |
| D | 7.218 | 7.384 | 7 |

(c) No state lost a seat after the reapportionment using Jefferson's method. The new-states paradox did not occur.

17. Hamilton's method initially assigns the number of seats to a state equal to the integer part of the state's standard quota. Any left over seats get assigned according to the size of the fractional part of the standard quota. At most, a state will receive one extra seat. Thus every state is apportioned a number of seats equal to the whole number just above or just below the state's standard quota so the quota rule will never be violated.

19. Jefferson's method uses a modified divisor when the sum of the integer parts of the standard quotas is too small. The modified divisor is smaller than the standard divisor forcing the modified quotas to be larger than the standard quotas. Sometimes, a modified quota can be increased beyond the whole number just above the upper quota. Then, when each state's modified quota is rounded down, the state is apportioned a number of seats equal to a whole number greater than the upper quota. Thus, if Jefferson's method violates the quota rule, it will be an upper quota violation.

21. Compare the apportionment with 315 seats to the apportionment with 314 seats. No province loses a seat, so the Alabama paradox does not occur. Compare the apportionment with 316 seats to the apportionment with 315 seats. Notice the South province loses a seat to the North province when an extra seat is added to the legislature. The Alabama paradox occurred.

23. Compare the apportionment with 315 seats to the apportionment with 314 seats. No province loses a seat, so the Alabama paradox does not occur. Compare the apportionment with 316 seats to the apportionment with 315 seats. No province loses a seat, so the Alabama paradox does not occur.

25. (a) The standard divisor is $D = \dfrac{244,000}{100} = 2440$. The following table contains the standard quotas and apportionment based on Hamilton's method. The sum of the integer parts of the standard quotas is 99. City B receives the left over seat because it has the largest fractional part.

| City | Standard Quota | Apportionment |
|------|----------------|---------------|
| A | 10.348 | 10 |
| B | 58.402 | 59 |
| C | 25.205 | 25 |
| D | 6.045 | 6 |

(b) The standard divisor is 2440. The new city has a population of 49,440. City E will require $\dfrac{49,440}{2440} \approx 20$ officers.

(c) The new standard divisor is $D = \dfrac{293,440}{120} \approx 2445.333$. The following table contains the standard quotas and apportionment based on Hamilton's method. The sum of the integer parts of the standard quotas is 119. City A receives the left over seat because it has the largest fractional part.

| City | Standard Quota | Apportionment |
|------|----------------|---------------|
| A | 10.326 | 11 |
| B | 58.274 | 58 |
| C | 25.150 | 25 |
| D | 6.032 | 6 |
| E | 20.218 | 20 |

(d) In the reapportionment, City B lost a seat to City A. This is an example of the new-states paradox.

27. Percent increase for state A: $\dfrac{1,370,000 - 1,320,000}{1,320,000} = \dfrac{50,000}{1,320,000} \approx 3.79\%$

Percent increase for state B: $\dfrac{1,565,000 - 1,515,000}{1,515,000} = \dfrac{50,000}{1,515,000} \approx 3.30\%$

Percent increase for state C: $\dfrac{5,035,000 - 4,935,000}{4,935,000} = \dfrac{100,000}{4,935,000} \approx 2.03\%$

Percent increase for state D: $\dfrac{1,218,000 - 1,118,000}{1,118,000} = \dfrac{100,000}{1,118,000} \approx 8.94\%$

Percent increase for state E: $\dfrac{1,212,000 - 1,112,000}{1,112,000} = \dfrac{100,000}{1,112,000} \approx 8.99\%$

29. Apportion the 200 seats based on the old population using Lowndes' method.

The standard divisor based on the old population is $D = \dfrac{10,000,000}{200} = 50,000$. The sum of the integer parts of the standard quotas is 198. The two left over seats go to the states with the largest relative fractional parts, states A and D. Results are given in the following table.

| State | Standard Quota | Relative Fractional Part | Apportionment |
|-------|----------------|--------------------------|---------------|
| A | 26.400 | 0.015 | 27 |
| B | 30.300 | 0.01 | 30 |
| C | 98.700 | 0.007 | 98 |
| D | 22.360 | 0.016 | 23 |
| E | 22.240 | 0.011 | 22 |

Apportion the 200 seats based on the new population using Lowndes' method.

The standard divisor based on the new population is $D = \dfrac{10,400,000}{200} = 52,000$. The sum of the integer parts of the standard quotas is 198, so the two left over seats will be given to the states with the largest relative fractional parts, states D and E. Results are given in the following table.

| State | Standard Quota | Relative Fractional Part | Apportionment |
|-------|----------------|--------------------------|---------------|
| A | 26.346 | 0.0133 | 26 |
| B | 30.096 | 0.003 | 30 |
| C | 96.827 | 0.009 | 96 |
| D | 23.423 | 0.018 | 24 |
| E | 23.308 | 0.0134 | 24 |

In the reapportionment, state A lost a seat and state C lost two seats while both state D and state E gained seats. From problem 27, we see that state D and E were the fastest growing states so it makes sense that they gained seats. Notice, however, that state A grew faster than state B, yet state A lost a seat while state B retained all of its seats. This is an example of the population paradox.

31. Apportion the 200 seats based on the old population using Jefferson's method.

The standard divisor based on the old population is $D = \dfrac{10,000,000}{200} = 50,000$. The sum of the integer parts of the standard quotas is 198. Modify the divisor to be larger than the standard divisor. The first modified divisor is $d = \dfrac{4,935,000}{99} \approx 49,848$. The sum of the integer parts of the first modified quotas is 199. Modify the divisor again. The second modified divisor is $d = \dfrac{4,935,000}{100} = 49,350$. The sum of the integer parts of the second modified quotas is 200. Results are given in the following table.

| State | Standard Quota | First Modified Quota | Second Modified Quota | Apportionment |
|-------|----------------|----------------------|-----------------------|---------------|
| A | 26.400 | 26.481 | 26.748 | 26 |
| B | 30.300 | 30.392 | 30.699 | 30 |
| C | 98.700 | 99.001 | 100.000 | 100 |
| D | 22.360 | 22.428 | 22.655 | 22 |
| E | 22.240 | 22.308 | 22.533 | 22 |

Apportion the 200 seats based on the new population using Jefferson's method.

The standard divisor based on the new population is $D = \dfrac{10,400,000}{200} = 52,000$. The sum of the integer parts of the standard quotas is 198. Modify the divisor to be larger than the standard divisor. The first modified divisor is $d = \dfrac{5,035,000}{97} \approx 51,907$. The sum of the integer parts of the first modified quotas is 199. Modify the divisor again. The second modified divisor is $d = \dfrac{5,035,000}{98} \approx 51,377$. The sum of the integer parts of the second modified quotas is 200. Results are given in the following table.

| State | Standard Quota | First Modified Quota | Second Modified Quota | Apportionment |
|-------|----------------|----------------------|-----------------------|---------------|
| A | 26.346 | 26.393 | 26.666 | 26 |
| B | 30.096 | 30.150 | 30.461 | 30 |
| C | 96.827 | 97.000 | 98.001 | 98 |
| D | 23.423 | 23.465 | 23.707 | 23 |
| E | 23.308 | 23.349 | 23.590 | 23 |

In the reapportionment, state C lost two seats while both state D and state E gained a seat. From problem 27, we see that state D and E were the fastest growing states, and state C grew the slowest, so the seat transfer makes sense. States A and B both retained their seats. This is not an example of the population paradox.

33. Apportion the 338 seats of the new legislature using Hamilton's method.

The standard divisor is $D = \dfrac{3,385,000}{338} \approx 10,014.793$. The sum of the integer parts of the standard quotas is 337. The left over seat goes to the South province because it has the largest fractional part of the standard quota. Results are given in the following table.

| Province | Standard Quota | Apportionment |
|----------|---------------|---------------|
| North | 89.068 | 89 |
| South | 42.337 | 43 |
| West | 66.302 | 66 |
| East | 116.028 | 116 |
| Northwest | 24.264 | 24 |

Compare the results of the apportionment with 338 seats to the apportionment from problem 21 with 314 seats. Notice that none of the original provinces lost a seat in the reapportionment. The new-states paradox does not occur in this case.

**35.** Apportion the 338 seats of the new legislature using Webster's method.

The standard divisor is $D = \dfrac{3,385,000}{338} \approx 10,014.793$. The sum of the standard quotas after they have

been rounded to the nearest integer is 337. A modified divisor is needed. Try $d = 9980$. The sum of the modified quotas after they have been rounded to the nearest integer is 338. Results are given in the following table.

| Province | Standard Quota | Modified Quota | Apportionment |
|----------|---------------|----------------|---------------|
| North | 89.068 | 89.379 | 89 |
| South | 42.337 | 42.485 | 42 |
| West | 66.302 | 66.533 | 67 |
| East | 116.028 | 116.433 | 116 |
| Northwest | 24.264 | 24.349 | 24 |

Compare the results of the apportionment with 338 seats to the apportionment from problem 24 with 314 seats. Notice that none of the original provinces lost a seat in the reapportionment. The new-states paradox does not occur in this case.

**37. (a)** Apportion the 96 police patrols using Hamilton's method.

The standard divisor is $D = \dfrac{40+82+285}{96} = \dfrac{407}{96} \approx 4.240$. The sum of the integer parts of the

standard quotas is 95. The left over patrol goes to Area I because it has the largest fractional part of the standard quota. Results are given in the following table.

| Area | Standard Quota | Apportionment |
|------|---------------|---------------|
| I | 9.434 | 10 |
| II | 19.340 | 19 |
| III | 67.217 | 67 |

**(b)** Apportion the 97 police patrols using Hamilton's method.

The standard divisor is $D = \dfrac{40+82+285}{97} = \dfrac{407}{97} \approx 4.196$. The sum of the integer parts of the

standard quotas is 95. The two left over patrols go to Area III and Area II because they have the largest fractional parts of the standard quotas. Results are given in the following table.

| Area | Standard Quota | Apportionment |
|------|---------------|---------------|
| I | 9.533 | 9 |
| II | 19.542 | 20 |
| III | 67.922 | 68 |

**(c)** When the number of patrols was increased, Area I lost a patrol to Area III. This is an example of the Alabama paradox.

**39.** Answers will vary.

**41.** Jefferson's method will never produce the Alabama paradox. With Jefferson's method, each state is assigned the integer part of its modified quota, or we can think of Jefferson's method rounding modified quotas down to the nearest integer. When the total number of seats to be apportioned increases, the standard divisor decreases. A smaller standard divisor, in turn, produces larger standard quotas. No state's standard quota can decrease, so none of the original states can lose a seat to another one of the original states even when modified quotas are rounded down to the nearest integer.

**43.** Webster's method will never produce the new-states paradox. States involved in an apportionment have standard quotas based on a standard divisor. Any new state that joins will have a certain number of seats added to the apportionmend total based on that same standard divisor. In the reapportionment, the standard quotas are calculated from the standard divisor, so none of the original states will lose a seat to another one of the original states.

## Solutions to Chapter 5 Review Problems

**1.** **(a)** The standard divisor is $D = \dfrac{2327 + 3412 + 1980 + 1162}{10} = \dfrac{8881}{10} = 888.1$. Each school board seat represents 888.1 people.

**(b)** The standard quotas, given in the following table, represent the number of seats each zone should be apportioned if fractional seats would be allowed.

| Zone | Standard Quota $Q = \dfrac{P}{D}$ |
|------|------------------------------------|
| I | $\dfrac{2327}{888.1} \approx 2.620$ |
| II | $\dfrac{3412}{888.1} \approx 3.842$ |
| III | $\dfrac{1980}{888.1} \approx 2.229$ |
| IV | $\dfrac{1162}{888.1} \approx 1.308$ |

**2.** Each zone is initially given the number of seats equal to the integer part of the zone's standard quota. The sum of the integer parts of the standard quotas from problem 1(b) is $2 + 3 + 2 + 1 = 8$. Two seats are left over and will be given to the zones with the greatest fractional parts of the standard quotas. Zones II and I have the greatest fractional parts, so they will each receive an additional seat. According to Hamilton's method, Zone I receives 3 seats, Zone II receives 4 seats, Zone III receives 2 seats, and Zone IV receives 1 seat.

**3.** Each zone is initially given the number of seats equal to the integer part of the zone's standard quota. The sum of the integer parts of the standard quotas from problem 1(b) is $2 + 3 + 2 + 1 = 8$. Two seats are left over and will be given to the zones with the greatest relative fractional parts. Zones I and IV have the greatest relative fractional parts so they will each receive an additional seat. The final apportionment according to Lowndes' method is given in the table.

| Zone | Standard Quota $Q = \dfrac{P}{D}$ | Relative Fractional Part | Apportionment |
|---|---|---|---|
| I | 2.620 | 0.310 | 3 |
| II | 3.842 | 0.281 | 3 |
| III | 2.229 | 0.115 | 2 |
| IV | 1.308 | 0.308 | 2 |

4. Recalculate the standard divisor to include the new population total:

$$D = \frac{2327 + 3412 + 1980 + 1483}{10} = \frac{9202}{10} = 920.2.$$

(a) Apportion the 10 seats using Hamilton's method. The sum of the integer parts of the standard quotas is $2 + 3 + 2 + 1 = 8$. The 2 left over seats will go to the zones with the largest fractional parts of the standard quotas. Zones IV and II have the largest fractional parts, so they will each receive an additional seat. The apportionment according to Hamilton's method is given in the following table.

| Zone $p$ = population | Standard Quota $Q = \dfrac{P}{D}$ | Apportionment |
|---|---|---|
| I $p = 2327$ | $\dfrac{2327}{920.2} \approx 2.529$ | 2 |
| II $p = 3412$ | $\dfrac{3412}{920.2} \approx 3.708$ | 4 |
| III $p = 1980$ | $\dfrac{1980}{920.2} \approx 2.152$ | 2 |
| IV $p = 1483$ | $\dfrac{1843}{920.2} \approx 1.612$ | 2 |

(b) Apportion the 10 seats using Lowndes' method. The sum of the integer parts of the standard quotas is $2 + 3 + 2 + 1 = 8$. The 2 left over seats will go to the zones with the largest relative fractional parts of the standard quotas. Zones IV and I have the largest relative fractional parts, so they will each receive an additional seat. The apportionment according to Lowndes' method is given in the following table.

| Zone $p$ = population | Standard Quota $Q = \dfrac{P}{D}$ | Relative Fractional Part | Apportionment |
|---|---|---|---|
| I $p = 2327$ | $\dfrac{2327}{920.2} \approx 2.529$ | 0.265 | 3 |
| II $p = 3412$ | $\dfrac{3412}{920.2} \approx 3.708$ | 0.236 | 3 |
| III $p = 1980$ | $\dfrac{1980}{920.2} \approx 2.152$ | 0.076 | 2 |
| IV $p = 1483$ | $\dfrac{1843}{920.2} \approx 1.612$ | 0.612 | 2 |

5. (a) The standard divisor is the number of crimes per patrol.

$$D = \frac{11 + 47 + 6 + 23 + 31 + 20}{35} = \frac{138}{35} \approx 3.943.$$

(b) Lowndes' method initially allocates to each area the number of patrols equal to the integer part of the standard quota. The sum of the integer parts of the standard quotas is 31. The 4 left over patrols will be given to areas with the largest relative fractional parts. In order from largest to smallest, areas C, A, D, and E will each receive one additional patrol. The final apportionment is given in the following table.

| Area $p$ = Crimes | Standard Quota $Q = \dfrac{p}{D}$ | Relative Fractional Part | Apportionment |
|---|---|---|---|
| A $p = 11$ | $\dfrac{11}{3.943} \approx 2.790$ | 0.395 | 3 |
| B $p = 47$ | $\dfrac{47}{3.943} \approx 11.920$ | 0.084 | 11 |
| C $p = 6$ | $\dfrac{6}{3.943} \approx 1.522$ | 0.522 | 2 |
| D $p = 23$ | $\dfrac{23}{3.943} \approx 5.833$ | 0.167 | 6 |
| E $p = 31$ | $\dfrac{31}{3.943} \approx 7.862$ | 0.123 | 8 |
| F $p = 20$ | $\dfrac{20}{3.943} \approx 5.072$ | 0.014 | 5 |

6.  (a) Jefferson's method allocates the number of seats equal to the integer part of the quota (standard or modified). The sum of the integer parts of the standard quotas is $2 + 4 + 1 = 7$ which is too small. There are 9 seats to apportion. A modified divisor must be found that is smaller than the standard divisor so that the modified quotas will be larger than the standard quotas.

   (b) Let $I = 4$, the integer part of the most populated zone's standard quota. The first guess at a modified divisor, rounded down, will be $d = \dfrac{\text{largest population}}{I+1} = \dfrac{946}{4+1} = \dfrac{946}{5} \approx 189$. The sum of the integer parts of the modified divisors is 9.

| Zone Populations | Standard Quotas | Modified Quotas $mQ = \dfrac{p}{d}$ |
|---|---|---|
| Zone 1:  567 | 2.93 | $\dfrac{567}{189} = 3$ |
| Zone 2:  946 | 4.88 | $\dfrac{946}{189} \approx 5.005$ |
| Zone 3:  231 | 1.19 | $\dfrac{231}{189} \approx 1.222$ |

   (c) Jefferson's method apportions 3 seats to Zone 1, 5 seats to Zone 2, and 1 seats to Zone 3.

7.  (a) Webster's method allocates to each zone the number of seats equal to the integer nearest its modified (or standard) quota. The sum of the rounded standard quotas is $2 + 4 + 1 + 1 = 8$ which is too small. There are 9 seats to apportion. A modified divisor is needed.

   (b) Let $I = 4$, the integer part of the most populated zone's standard quota. The first guess at a modified divisor, rounded down, will be $d = \dfrac{\text{largest population}}{I+1} = \dfrac{811}{4+1} = \dfrac{811}{5} \approx 162$. The sum of the rounded modified divisors is 11 which is too large. We must modify the divisor again and pick a value between 162 and 189. Try $d = 183$. This time the sum of the rounded modified quotas is 9.

| Zone Populations | Standard Quotas | First Modified Quotas $mQ = \dfrac{p}{162}$ | Second Modified Quotas $mQ = \dfrac{p}{183}$ |
|---|---|---|---|
| Zone 1: 428 | 2.21 | $\dfrac{428}{162} \approx 2.642$ | $\dfrac{428}{183} \approx 2.339$ |
| Zone 2: 811 | 4.19 | $\dfrac{811}{162} \approx 5.006$ | $\dfrac{811}{183} \approx 4.432$ |
| Zone 3: 230 | 1.19 | $\dfrac{230}{162} \approx 1.420$ | $\dfrac{230}{183} \approx 1.257$ |
| Zone 4: 275 | 1.42 | $\dfrac{275}{162} \approx 1.698$ | $\dfrac{275}{183} \approx 1.503$ |

(c) Webster's method apportions 2 seats to Zone 1, 4 seats to Zone 2, 1 seat to Zone 3, and 2 seats to Zone 4.

8. Apportion the 100 representative seats using Hamilton's method. The standard divisor is $D = \dfrac{1,058,920}{100} = 10,589.2$. The sum of the integer parts of the standard quotas is 97. The 3 left over seats will be given to the counties with the largest fractional parts of the standard quotas: Kent, Washington, and Bristol.

| County $p$ = population | Standard Quota $Q = \dfrac{p}{D}$ | Apportionment |
|---|---|---|
| Bristol $p = 51,173$ | $\dfrac{51,173}{10,589.2} \approx 4.833$ | 5 |
| Kent $p = 169,224$ | $\dfrac{169,224}{10,589.2} \approx 15.981$ | 16 |
| Newport $p = 85,218$ | $\dfrac{85,218}{10,589.2} \approx 8.048$ | 8 |
| Providence $p = 627,314$ | $\dfrac{627,314}{10,589.2} \approx 59.241$ | 59 |
| Washington $p = 125,991$ | $\dfrac{125,991}{10,589.2} \approx 11.898$ | 12 |

(a) Divide the population of each county by its apportioned number of seats and compare. Newport County ends up with the largest number of people per representative. In Newport County one seat represents 10,652.25 people.

(b) Bristol County ends up with the smallest number of people per representative. In Bristol County one representative represents 10,234.6 people. Notice Bristol is the least populated county.

9. Apportion the 100 representative seats using Jefferson's method. The standard divisor is $D = \dfrac{1,058,920}{100} = 10,589.2$. The sum of the integer parts of the standard quotas is 97 which is too small. Modify the divisor to be larger than the standard divisor. Let $I = 59$, the integer part of the most populated county's standard quota. The modified divisor, rounded down, is $d = \dfrac{\text{largest population}}{I+1} = \dfrac{627,314}{59+1} \approx 10,455$. The sum of the integer parts of the modified quotas is 100.

| County $p$ = population | Standard Quota $Q = \dfrac{p}{D}$ | Modified Quotas $mQ = \dfrac{p}{10,455}$ | Apportionment |
|---|---|---|---|
| Bristol $p = 51,173$ | $\dfrac{51,173}{10,589.2} \approx 4.833$ | $\dfrac{51,173}{10,455} \approx 4.895$ | 4 |
| Kent $p = 169,224$ | $\dfrac{169,224}{10,589.2} \approx 15.981$ | $\dfrac{169,224}{10,455} \approx 16.186$ | 16 |
| Newport $p = 85,218$ | $\dfrac{85,218}{10,589.2} \approx 8.048$ | $\dfrac{85,218}{10,455} \approx 8.151$ | 8 |
| Providence $p = 627,314$ | $\dfrac{627,314}{10,589.2} \approx 59.241$ | $\dfrac{627,314}{10,455} \approx 60.001$ | 60 |
| Washington $p = 125,991$ | $\dfrac{125,991}{10,589.2} \approx 11.898$ | $\dfrac{125,991}{10,455} \approx 12.051$ | 12 |

(a) Divide the population of each county by its apportioned number of seats and compare. Bristol County ends up with the largest number of people per representative. In Bristol County one seat represents 12,793.25 people. Notice that Bristol is the least populated county.

(b) Providence County ends up with the smallest number of people per representative. In Providence County one representative represents approximately 10,455.23 people. Notice that Providence is the most populated county.

10. Apportion the 100 representatives using Webster's method. The standard quotas are given in the solution to the previous problem. The sum of the rounded standard quotas is 100. Webster's method apportions 5 seats to Bristol, 16 seats Kent, 8 seats to Newport, 59 seats to Providence, and 12 seats to Washington. The apportionment based on Webster's method is the same as for Jefferson's method.

11. If the method used was a quota method, then the apportionment for each county would be the whole number above or below the standard quota. The standard quota for Kent County is $Q = \dfrac{169,224}{10,589.2} \approx 15.981$ yet the method apportioned 17 seats to Kent County. This violates the quota rule. Therefore, the method could not have been a quota method.

12. (a) The standard divisor is $D = \dfrac{\text{Total Enrollment}}{\text{Number of Volunteers}} = \dfrac{143}{15} \approx 9.533$. The standard divisor represents the number of children per volunteer. The standard quotas for each class are given in the table in part (b).

(b) Apportion the volunteers to the classes based on enrollment. The sum of the integer parts of the standard quotas is 12 which is too small. Modified quotas are needed that are larger than the standard quotas. To accomplish this, a modified divisor is needed that is smaller than the standard divisor. Let $I = 3$, the integer part of the standard quota for the class with the largest enrollment. The first guess at a modified divisor is $d = \dfrac{\text{largest enrollment}}{I+1} = \dfrac{30}{3+1} = 7.5$. The sum of the integer parts of the first modified quotas is 17 which is too large. Another modified divisor is needed between 7.5 and 9.533. Try $d = 8.3$. The sum of the integer parts of the second modified quotas is 15. The apportionment based on Jefferson's method is given in the following table.

| Class $p$ = enrollment | Standard Quota $Q = \dfrac{p}{D}$ | First Modified Quota $mQ = \dfrac{p}{7.5}$ | Second Modified Quota $mQ = \dfrac{p}{8.3}$ | Apportionment |
|---|---|---|---|---|

| Kinder $p = 26$ | $\frac{26}{9.533} \approx 2.727$ | $\frac{26}{7.5} \approx 3.467$ | $\frac{26}{8.3} \approx 3.133$ | 3 |
|---|---|---|---|---|
| Grade 1 $p = 18$ | $\frac{18}{9.533} \approx 1.888$ | $\frac{18}{7.5} = 2.4$ | $\frac{18}{8.3} = 2.169$ | 2 |
| Grade 2 $p = 23$ | $\frac{23}{9.533} \approx 2.413$ | $\frac{23}{7.5} \approx 3.067$ | $\frac{23}{8.3} \approx 2.771$ | 2 |
| Grade 3 $p = 30$ | $\frac{30}{9.533} \approx 3.147$ | $\frac{30}{7.5} = 4$ | $\frac{30}{8.3} \approx 3.614$ | 3 |
| Grade 4 $p = 21$ | $\frac{21}{9.533} \approx 2.203$ | $\frac{21}{7.5} = 2.8$ | $\frac{21}{8.3} \approx 2.530$ | 2 |
| Grade 5 $p = 25$ | $\frac{25}{9.533} \approx 2.622$ | $\frac{25}{7.5} \approx 3.333$ | $\frac{25}{8.3} \approx 3.012$ | 3 |
| *Sum of the Integer Parts* | *12 too small* | *17 too large* | *15* | |

(c) The standard quotas are given in the table in part (b). For kindergarten through grade 5, the standard quotas rounded to the nearest integer are 3, 2, 2, 3, 2, and 3 respectively. The sum of the rounded standard quotas is 15 which is the number of volunteers. The apportionment according to Webster's method is the same as the one obtained by using Jefferson's method which is given in the table in part (b).

13. Hamilton's and Lowndes' methods are both quota methods. Both would apportion a number of seats equal to the whole number just above or just below the standard quota. Notice that for state B, the apportioned value was 8 while the standard quota was 6.81. The quota rule has been violated. Neither Hamilton's nor Lowndes' method could have been used. Jefferson's method must have been used.

14. Notice the table contains modified quotas. Hamilton's method is based on standard quotas, so Hamilton's method was not used. The apportioned values are all the values you would get by rounding each of the modified quotas to the nearest integer. Jefferson's method would have rounded each modified quota down, so Webster's method must have been used.

15. (a) Apportion the 30 packages using Hamilton's method. The standard divisor is $D = \frac{10{,}110}{30} = 337$.

The sum of the integer parts of the standard quotas is 28. The 2 left over packages will be given to the divisions with the largest fractional parts of the standard quotas. The Mathematics Division and the Science and Engineering Division have the largest fractional parts, so they each receive one additional package. The apportionment according to Hamilton's method is given in the following table.

| Division $p$ = enrollment | Standard Quota $Q = \frac{p}{D}$ | Apportionment |
|---|---|---|
| Health Sciences $p = 1830$ | $\frac{1830}{337} \approx 5.430$ | 5 |
| Science and Engineering $p = 3520$ | $\frac{3520}{337} \approx 10.445$ | 11 |
| Mathematics $p = 1510$ | $\frac{1510}{337} \approx 4.481$ | 5 |

| | | |
|---|---|---|
| Social Sciences $p = 2500$ | $\dfrac{2500}{337} \approx 7.418$ | 7 |
| Foreign Language $p = 750$ | $\dfrac{750}{337} \approx 2.226$ | 2 |

**(b)** Apportion the 31 packages using Hamilton's method. The standard divisor is $D = \dfrac{10{,}110}{31} \approx 326.129$. The sum of the integer parts of the standard quotas is 28. The 3 left over packages will be given to the divisions with the largest fractional parts of the standard quotas. The Science and Engineering Division, the Social Sciences Division, and the Mathematics Division have the largest fractional parts, so they each receive one additional package. The apportionment according to Hamilton's method is given in the following table.

| Division $p$ = enrollment | Standard Quota $Q = \dfrac{p}{D}$ | Apportionment |
|---|---|---|
| Health Sciences $p = 1830$ | $\dfrac{1830}{326.129} \approx 5.611$ | 5 |
| Science and Engineering $p = 3520$ | $\dfrac{3520}{326.129} \approx 10.793$ | 11 |
| Mathematics $p = 1510$ | $\dfrac{1510}{326.129} \approx 4.630$ | 5 |
| Social Sciences $p = 2500$ | $\dfrac{2500}{326.129} \approx 7.666$ | 8 |
| Foreign Language $p = 750$ | $\dfrac{750}{326.129} \approx 2.300$ | 2 |

**(c)** In the reapportionment, the Social Science Division received the new package. No division lost a package in the reapportionment. The Alabama paradox did not occur.

**16. (a)** Apportion the 25 silver dollars using Hamilton's method. The standard divisor is $D = \dfrac{388}{25} = 15.52$. The sum of the integer parts of the standard quotas is 23. The 2 left over silver dollars will be given to the children with the largest fractional parts of the standard quotas. Josh and Jack have the largest fractional parts, so they each receive one additional silver dollar. The apportionment according to Hamilton's method is given in the following table.

| Child $p$ = hours | Standard Quota $Q = \dfrac{p}{D}$ | Apportionment |
|---|---|---|
| Josh $p = 73$ | $\dfrac{73}{15.52} \approx 4.704$ | 5 |

| Paula $p = 41$ | $\dfrac{41}{15.52} \approx 2.642$ | 2 |
|---|---|---|
| Jack $p = 274$ | $\dfrac{274}{15.52} \approx 17.655$ | 18 |

**(b)** Apportion the 25 silver dollars using Hamilton's method. The standard divisor is $D = \dfrac{391}{25} = 15.64$. The sum of the integer parts of the standard quotas is 23. The 2 left over silver dollars will be given to the children with the largest fractional parts of the standard quotas. Josh and Paula have the largest fractional parts, so they each receive one additional silver dollar. The apportionment according to Hamilton's method is given in the following table.

| Child $p = $ hours | Standard Quota $Q = \dfrac{p}{D}$ | Apportionment |
|---|---|---|
| Josh $p = 75$ | $\dfrac{75}{15.64} \approx 4.795$ | 5 |
| Paula $p = 41$ | $\dfrac{41}{15.64} \approx 2.621$ | 3 |
| Jack $p = 275$ | $\dfrac{275}{15.64} \approx 17.583$ | 17 |

**(c)** In the reapportionment, both Josh and Jack gained hours, yet Jack lost a silver dollar to Paula. This is an example of the population paradox.

**17. (a)** Apportion the 24 task force members using Hamilton's method. The standard divisor is $D = \dfrac{450,000}{24} = 18,750$. The sum of the integer parts of the standard quotas is 23. The left over task force member will be given to the county with the largest fractional part of the standard quota. Lane County has the largest fractional part, so it receives one additional task force member. The apportionment according to Hamilton's method is given in the following table.

| County $p = $ population | Standard Quota $Q = \dfrac{p}{D}$ | Apportionment |
|---|---|---|
| Lane $p = 46,000$ | $\dfrac{46,000}{18,750} \approx 2.453$ | 3 |
| Marion $p = 321,000$ | $\dfrac{321,000}{18,750} = 17.12$ | 17 |
| Linn $p = 83,000$ | $\dfrac{83,000}{18,750} \approx 4.427$ | 4 |

**(b)** Apportion the 29 task force members using Hamilton's method. The standard divisor is $D = \dfrac{536,000}{29} \approx 18,482.759$. The sum of the integer parts of the standard quotas is 27. The 2 left over task force members will be given to the counties with the largest fractional parts of the standard quotas. Benton County and Linn County have the largest fractional parts, so they each receive one additional task force member. The apportionment according to Hamilton's method is given in the following table.

| County $p = $ population | Standard Quota $Q = \dfrac{p}{D}$ | Apportionment |
|---|---|---|

| | | |
|---|---|---|
| Lane<br>$p = 46,000$ | $\dfrac{46,000}{18,482.759} \approx 2.489$ | 2 |
| Marion<br>$p = 321,000$ | $\dfrac{321,000}{18,482.759} \approx 17.368$ | 17 |
| Linn<br>$p = 83,000$ | $\dfrac{83,000}{18,482.759} \approx 4.491$ | 5 |
| Benton<br>$p = 86,000$ | $\dfrac{86,000}{18,482.759} \approx 4.653$ | 5 |

(c) In the reapportionment, Lane County lost a member to Linn County. This is an example of the new-states paradox.

# *Chapter 6:* Routes and Networks

## *Key Concepts*

**By the time you finish studying Section 6.1, you should be able to:**

☐ determine the number of edges, the degree of each vertex, and list the adjacent vertices in a graph.

☐ draw a graph given information about adjacent vertices.

☐ state the relationship between the number of edges and the sum of the degrees in a graph.

☐ determine whether a path is an Euler path, a circuit, or an Euler circuit.

☐ determine whether a graph is connected.

☐ find the components of a disconnected graph.

☐ use Euler's theorem to decide whether a graph has an Euler circuit or an Euler path.

☐ use Fleury's algorithm to find an Euler circuit in a graph that contains at least one Euler circuit.

**By the time you finish studying Section 6.2, you should be able to:**

☐ create a weighted graph to represent a situation.

☐ explain how to tell whether a graph has redundant connections.

☐ determine whether a graph is a tree.

☐ describe the difference between a spanning tree and a minimal spanning tree.

☐ construct a minimal spanning tree using Kruskal's algorithm.

**By the time you finish studying Section 6.3, you should be able to:**

☐ determine whether a path is a Hamiltonian path or a Hamiltonian circuit.

☐ determine whether a graph is complete.

☐ calculate the number of edges in a complete graph with $n$ vertices.

☐ list all Hamiltonian paths and circuits for a graph.

☐ calculate the number of Hamiltonian paths in a complete graph.

☐ describe the Traveling-Salesperson Problem.

☐ construct a complete weighted graph.

☐ apply the brute-force algorithm to determine the Hamiltonian circuit of least cost.

☐ apply the nearest-neighbor algorithm to find an approximate least-cost Hamiltonian circuit.

☐ apply the cheapest-link algorithm to find an approximate least-cost Hamiltonian circuit.

☐ apply the repetitive nearest-neighbor algorithm to find an approximate least-cost Hamiltonian circuit.

## *Hints for Odd-Numbered Problems*

### Section 6.1

1. Let vertices in your graph represent cities. If a highway connects two cities on the map, then an edge should connect the two vertices in your graph.

3. (i) The degree of a vertex in a graph is the total number of edges at that vertex.
   (ii) Vertices are adjacent if there is an edge connecting them.

5. (i) The degree of a vertex in a graph is the total number of edges at that vertex. Remember that a loop contributes 2 to the degree of a vertex.
   (ii) Vertices are adjacent if there is an edge connecting them.

7. Draw all vertices first, and then connect adjacent vertices. There are many graphs possible.

9. In a connected graph, there is a path from each vertex to every other vertex. However, there may be vertices that are not adjacent.

11. A path does not need to pass through each vertex.

13. An edge is a bridge if its removal from the graph would leave behind a graph that is not connected.

15. An edge is a bridge if its removal from the graph would leave behind a graph that is not connected.

17. See Example 6.6.

19. See Example 6.6.

21. Consider Euler's theorem.

23. To be an Euler circuit, the path must begin and end at the same vertex and each edge must be traveled only once.

25. Many Euler circuits are possible.

27. Many Euler circuits are possible.

29. See Example 6.10.

31. See Example 6.10.

33. What is required in order for a graph to have an Euler circuit? See Euler's theorem.

35. At which vertices can an Euler path begin? At which vertices can an Euler path end?

**37.** What is required in order for a graph to have an Euler circuit? See Euler's theorem.

**39. (a)** Begin in Pendleton where the main offices are located.

**41.** Try to form an Euler path.

**43. (a)** How many vertices with an odd degree are there?

## Section 6.2

**1.** The lengths of the edges do not have to be proportional to the weights.

**3.** There will be no edge between cities with no bus service between them.

**5. (b)** A subgraph is a set of vertices and edges chosen from the original graph.

**7. (b)** A subgraph is a set of vertices and edges chosen from the original graph.

**9.** The graph must be connected and it must contain no circuits.

**11.** A spanning tree is a subgraph that contains all of the original vertices, is connected, and contains no circuits.

**13.** A spanning tree is a subgraph that contains all of the original vertices, is connected, and contains no circuits.

**15. (i)** Identify any circuits. How many edges must be removed to eliminate any circuits?
**(ii)** Do not remove an edge that is a bridge.

**17. (i)** Identify any circuits. How many edges must be removed to eliminate any circuits?
**(ii)** Do not remove an edge that is a bridge.

**19.** Determine how many edges must be removed. Avoid removing any bridges.

**21.** Determine how many edges must be removed. Avoid removing any bridges.

**23.** Determine how many edges must be removed. Avoid removing any bridges.

**25.** Determine how many edges must be removed. Avoid removing any bridges.

**27.** See example 6.14.

**29.** See example 6.14.

**31.** Consider the acceptable edges and choose the edge with the largest weight to add to the subgraph.

**33.** Consider the acceptable edges and choose the edge with the largest weight to add to the subgraph.

**35.** See the solution to the initial problem.

**37.** See the solution to the initial problem.

## Section 6.3

**1.** A Hamiltonian path visits each vertex exactly once.

**3.** A Hamiltonian path visits each vertex exactly once.

**5. (b)** What is the degree of vertex $A$? Is it possible to begin at vertex $A$ and return to vertex $A$ without visiting any vertex more than once?

**7. (b)** Is it possible to begin at vertex $A$ and return to vertex $A$ without visiting any vertex more than once?

**9. (b)** Is it possible to begin at vertex $A$ and return to vertex $A$ without visiting any vertex more than once?

**11. (b)** Is it possible to begin at vertex $A$ and return to vertex $A$ without visiting any vertex more than once?

**13.** Refer to the theorem that gives the number of edges in a complete graph.

**15.** A complete graph is one in which every pair of vertices is connected by exactly one edge.

17. A graph is not complete if you can find a pair of vertices with no edge connecting them.

19. Refer to the theorem that gives the number of Hamiltonian circuits in a complete graph.

21. A Hamiltonian path visits each vertex exactly once.

23. For a complete graph with $n$ vertices, there are $n!$ Hamiltonian paths. Try different values for $n$ until $n!$ matches the number of paths given.

25. You should be able to find 24 Hamiltonian paths.

27. (a) Use the Hamiltonian paths found in problem 25, and add one vertex to each to complete the circuit.
    (c) Mirror image circuits traverse the same edges in opposite directions.

29. Remember that an Euler path traverses each edge exactly once.

31. Many graphs are possible.

33. (a) Recall Euler's theorem to determine whether the graph has an Euler circuit.

35. (a) List all Hamiltonian circuits that begin and end at $A$. Determine the circuit with the least cost.
    (b) Vertex $A$ is the starting vertex.

37. Apply the nearest-neighbor algorithm five times.

39. Vertex $X$ is the starting vertex.

41. Apply the cheapest-link algorithm and list the resulting Hamiltonian circuit so that it begins and ends in Gaborone.

43. Apply the nearest-neighbor algorithm eight times.

## Solutions to Odd-Numbered Problems in Section 6.1

1. (a) The vertices of the graph will represent the cities, and the edges will represent the highways. The edges may be drawn curved or straight. The following graph is one representation. Your graph may differ.

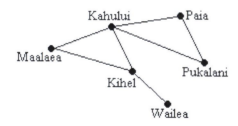

   (b) The vertex labeled Kihel has three edges connecting it to three other vertices. The three vertices adjacent to Kihel are Maalaea, Kahului, and Wailea.
   (c) The vertex labeled Kahului has four edges connecting it to four other vertices. The four vertices adjacent to Kahului are Paia, Pukalani, Kihel, and Maalaea.
   (d) The degree of a vertex is the total number of edges at that vertex. For each vertex in the graph from part (a), count the number of edges at the vertex. The degree of Maalaea is 2; Kahului, 4; Kihel, 3; Wailea, 1; Paia, 2; and Pukalani, 2.

**3.** **(i)** For each vertex, count the number of edges at that vertex. Adjacent pairs of vertices are connected by an edge.

| Vertex | Degree |
|--------|--------|
| T | 3 |
| U | 3 |
| K | 3 |
| R | 3 |
| J | 2 |

**(ii)** The adjacent pairs are $R$ and $T$, $R$ and $K$, $R$ and $J$, $J$ and $K$, $U$ and $K$, and $T$ and $U$. The adjacent pair $T$ and $U$ are joined by two edges.

**(iii)** The sum of the degrees is $3 + 3 + 3 + 3 + 2 = 14$. There are 7 edges in the graph. The sum of the degrees of the vertices is twice the sum of the edges.

**5.** **(i)** For each vertex, count the number of edges at that vertex. Adjacent pairs of vertices are connected by an edge.

| Vertex | Degree |
|--------|--------|
| A | 2 |
| B | 4 |
| C | 3 |
| D | 3 |
| E | 2 |

**(ii)** The adjacent pairs are $A$ and $E$, $A$ and $C$, $C$ and $D$, $B$ and $D$, $B$ and $B$, and $B$ and $E$. The adjacent pair $C$ and $D$ are joined by two edges.

**(iii)** The sum of the degrees is $2 + 4 + 3 + 3 + 2 = 14$. There are 7 edges in the graph. The sum of the degrees of the vertices is twice the sum of the edges.

**7** **(a)** Two possible graphs:

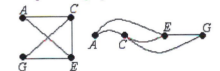

**(b)** Two possible graphs:

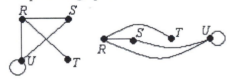

**9.** A graph is connected if, for every pair of vertices, there is a path that contains them. There may be pairs of vertices that are not adjacent, and there may be pairs of vertices with more than one edge connecting them.

**(a)** Graphical representations may differ.

**(b)** Graphical representations may differ.

**11. (a)** Your answers may differ. Three possible paths include $FGH$, $FKH$, and $FKGH$.
   **(b)** Your answers may differ. Three possible paths include $RT$, $RUT$, and $RSUT$.

**13.** The only edge that is a bridge is $BC$. Removing any other edge leaves a connected graph. Removing edge $BC$ leaves the following two components.

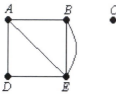

**15.** The edges $CD$, $DE$, and $DF$ are bridges. Removing any of those three edges will cause the graph to not be connected.
   The following two components result from the removal of bridge $CD$:

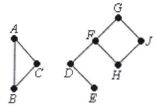

   The following two components result from the removal of bridge $DE$:

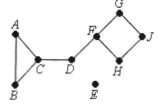

   The following two components result from the removal of bridge $DF$:

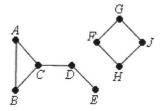

**17. (a)** It is a path. It is a route passing from vertex to adjacent vertex such that each edge connecting adjacent vertices is used at most once.
   **(b)** It is a circuit. It is a path that begins and ends at the same vertex such that each edge is used only once. Notice that there are two edges connecting vertices $R$ and $U$. The list of vertices $RUR$ forms a circuit only if both edges are used.

(c) This is not a path. There is only one edge connecting vertex $Y$ to vertex $R$. Edge $YR$ is used twice, so a path is not formed.

(d) This is a circuit, however, it is not an Euler circuit. The path does begin and end at the same vertex, but one of the edges between vertex $U$ and vertex $R$ does not get used.

19. (a) It is a path. It is a route passing from vertex to adjacent vertex such that each edge connecting adjacent vertices is used at most once.

(b) This is not a path. There is no edge connecting vertex $M$ to vertex $K$.

(c) This is an Euler circuit. The path begins and ends at vertex $R$, and each edge is used only once.

(d) This is an Euler circuit. The path begins and ends at vertex $K$, and each edge is used only once.

21. According to Euler's theorem, if the graph has exactly two vertices of odd degree, then it has at least one Euler path, but it does not have an Euler circuit. The graph in problem 17 has exactly two vertices of odd degree: vertex $U$ and vertex $R$.

23. There are many graphs possible. Yours may differ.

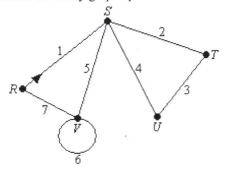

25. There are many graphs possible. Yours may differ.

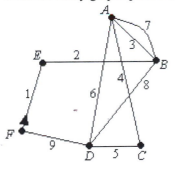

27. (a) The list of vertices in the order traveled is $EDCEFGCBGAADE$.

(b) Your circuit may differ. A possible Euler circuit is $GBCGFECDEDAAG$.

29. There are exactly two vertices that have an odd degree: vertex $U$ and vertex $R$. By Euler's theorem, we know the graph has at least one Euler path, but it does not have an Euler circuit. The Euler path must start at one of the two vertices with an odd degree and end at the other. Use Fleury's algorithm to find an Euler path. Your Euler path may differ.

Draw a copy of the original graph and label it "Unnumbered Edges". Draw a second copy of the vertices without the edges and label it "Numbered Edges". Suppose vertex $U$ is the selected vertex.

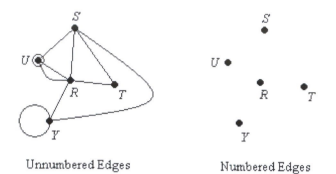

Unnumbered Edges                    Numbered Edges

Notice none of the three edges connected to vertex $U$ is a bridge, so select any of the three edges to shift to the graph with numbered edges. Suppose edge $US$ is selected and moved to the new graph. Label $US$ with a 1. Vertex $S$ becomes the selected vertex.

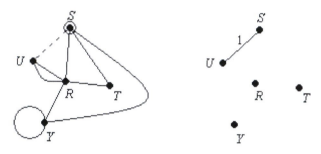

None of the three edges connected to vertex $S$ is a bridge, so select any of the three edges. Suppose edge $SY$ is selected and moved to the new graph. Label $SY$ with a 2. Vertex $Y$ becomes the new selected vertex. Notice the edge $YR$ is a bridge.

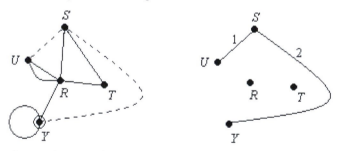

We cannot select edge $YR$ because its removal would leave behind a disconnected graph. We must select the loop at vertex $Y$ and move it to the new graph. Label loop $YY$ with a 3. Vertex $Y$ remains the selected vertex.

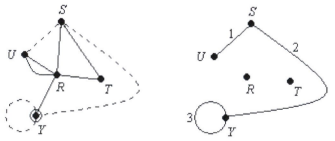

The only edge connected to vertex $Y$ is a bridge. We move edge $YR$ to the new graph. Label $YR$ with a 4. Vertex $R$ becomes the selected vertex.

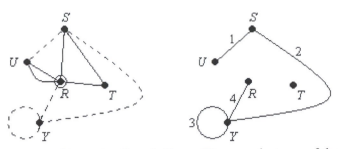

None of the edges connected to vertex $R$ are bridges. We can select any of them to move to the new graph. Suppose edge $RU$ is moved to the new graph. Label $RU$ with a 5. Vertex $U$ becomes the selected vertex. Notice edge $UR$ is a bridge.

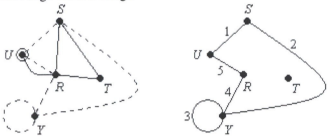

The only edge connected to vertex $U$ is a bridge. We move $UR$ to the new graph. Label $UR$ with a 6. Vertex $R$ becomes the selected vertex.

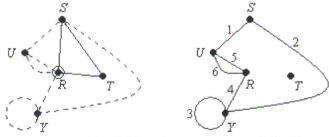

Neither edge connected to vertex $R$ is a bridge. Suppose we move $RS$ to the new graph. Label $RS$ with a 7. Vertex $S$ becomes the selected vertex.

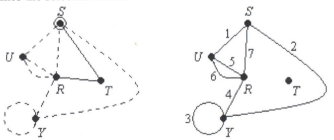

Once edges $ST$ and $TR$ are added to the graph and labeled with an 8 and 9 respectively, Fleury's algorithm is complete. The vertices are traversed in the following order: $USYYRURSTR$.

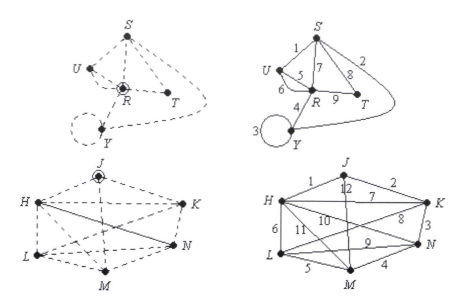

**31.** There are exactly two vertices that have an odd degree: vertex $K$ and vertex $M$. By Euler's theorem, we know the graph has at least one Euler path, but it does not have an Euler circuit. The Euler path must start at one of the two vertices with an odd degree and end at the other. Use Fleury's algorithm to find an Euler path. Your Euler path may differ.

Draw a copy of the original graph and label it "Unnumbered Edges". Draw a second copy of the vertices without the edges and label it "Numbered Edges". Suppose vertex $M$ is the selected vertex.

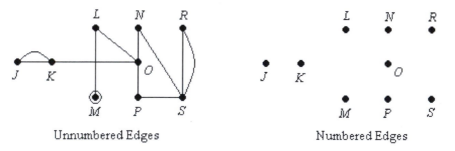

Unnumbered Edges                    Numbered Edges

The only edge connected to $M$ is a bridge. Move edge $ML$ to the new graph and label it with a 1. Vertex $L$ becomes the selected vertex.

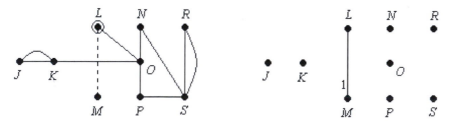

Edge $LO$ is a bridge, but it is the only edge connected to vertex $L$, so we must add $LO$ to the new graph next and label it with a 2. Vertex $O$ becomes the selected vertex.

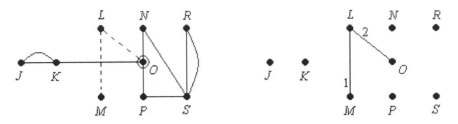

We must avoid the bridge *KO* and select either edge *ON* or edge *OP*. Suppose we select *ON* to move to the new graph next. Label *ON* with a 3. Vertex *N* becomes the selected vertex and we will be forced to add *NS* in the next step because it will be the only edge connected to *N*. Once *NS* is added to the new graph and labeled with a 4, vertex *S* becomes the selected vertex.

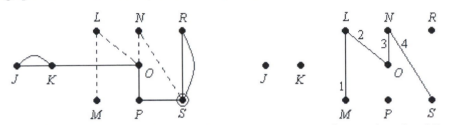

Edge *PS* is a bridge, so we must select one of the two edges connecting vertices *S* and *R* to add to the new graph. Once one of them is added to the new graph and labeled with a 5, the next edges to be added are *RS*, *SP*, *PO*, and *OK* in that order (labeled 6 through 9) because they will be the only choices of edges to add at each step. Once *OK* is added to the new graph, vertex *K* will become the selected vertex.

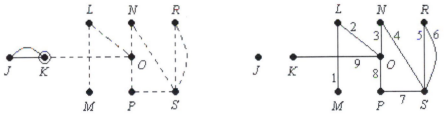

The two remaining edges can be added in either order. Fleury's algorithm is complete. The vertices are traversed in the following order: *MLONSRSPOKJK*.

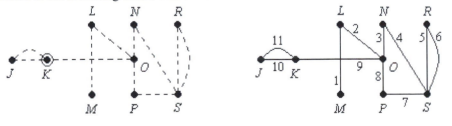

33. **(a)** No matter which edge is removed, the graph remains connected. No edge is a bridge.
   **(b)** In order to be able to create an Euler circuit, the graph must have no vertices with odd degrees. The graph that is equivalent to the system of bridges has four vertices. Each vertex has an odd degree. If one new bridge is built between two vertices, that would change the degree of only two vertices. There would still be two vertices with an odd degree. Therefore, there is no location for a new bridge that would allow an Euler circuit to be created.

**35.** Many Euler paths are possible. Your answer may differ. One possible Euler path passes through each of the following cities in order:
Wichita, Dodge City, Garden City, Oakley, Wakeeney, Hill City, Stockton, Hays, Wakeeney, Dodge City, Kinsley, Hays, Salina, Manhattan, Topeka, Salina, Newton, Witchita, Emporia, Kansas City, and Topeka.

**37.** This is not possible because there are two vertices with odd degree (Wichita and Emporia). No Euler circuit can be found according to Euler's theorem.

**39. (a)** The courier may cover each road exactly once, however, the courier cannot begin and end in the same city. Many routes are possible.

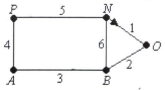

**(b)** The graph has exactly two vertices with an odd degree. By Euler's theorem, there is at least one Euler path.

**41.** The vertex at Portland has an even degree. If the courier begins trips in Portland, then the courier will have to travel a road more than once to get back to Portland.

**43. (a)** Because there are 10 vertices of odd degree, it is not possible for the paperboy to find an Euler circuit.
**(b)** Many graphs are possible. Your answer may differ.

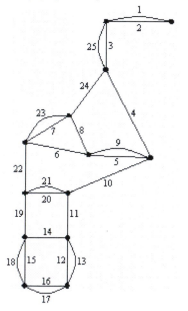

## Solutions to Odd-Numbered Problems in Section 6.2

1.  Many different representations are possible.
    (a) Create four vertices in the graph to represent the college ($C$), the clothing store ($S$), the apartment ($A$), and the rest home ($R$). The weights in the weighted graph represent the time it takes in minutes to get from one place to another.

    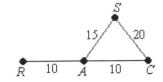

    (b) The vertex $S$ will no longer be in the graph.

    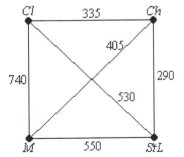

3.  Many different representations are possible.
    (a)

    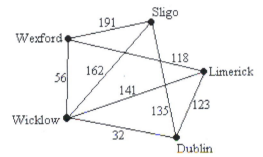

    (b) Many routes are possible.

    | Route | Distance in Miles |
    |---|---|
    | Wexford, Limerick | 118 |
    | Wexford, Wicklow, Limerick | 56 + 141 = 197 |
    | Wexford, Wicklow, Dublin, Limerick | 56 + 32 + 123 = 211 |

5.  Many representations are possible.
    (a) There are four vertices, and weights represent miles between cities.

    (b) The graph currently contains a circuit, so there are redundant edges. There are many ways to remove redundant edges. The subgraph you end up with must contain all four vertices, and there must be a path from any vertex to any other vertex in the graph. One possible solution is given.

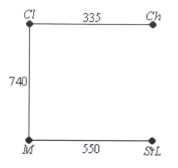

7. Many representations are possible.
   (a) There are five vertices in the graph.

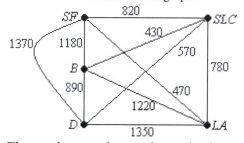

   (b) The graph currently contains a circuit, so there are redundant edges. There are many ways to remove redundant edges. The subgraph you end up with must contain all five vertices, and there must be a path from any vertex to any other vertex in the graph. One possible solution is given.

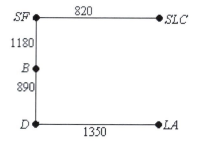

9. A graph is a tree if there is a path from any vertex to any other vertex, but there are no circuits.
   (a) There are no circuits, and the graph is connected. The graph is a tree.
   (b) The path *LMNL* is a circuit. The graph is not a tree.
   (c) There are no circuits in the graph, but there is no way to get to vertex *B* and vertex *E* from any of the other vertices. The graph is not connected. The graph is not a tree.

11. A subgraph that contains all the original vertices of a graph, is connected, and contains no circuits is called a spanning tree. Many spanning trees are possible. You must create three. Possible answers include the following.

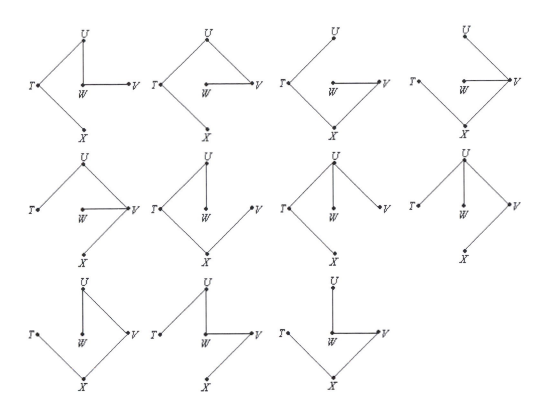

13. A subgraph that contains all the original vertices of a graph, is connected, and contains no circuits is called a spanning tree. Many spanning trees are possible. You must create three. Possible answers include the following.

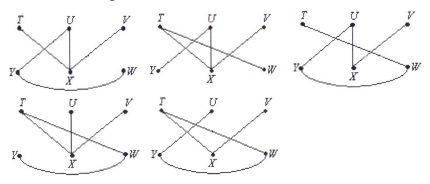

15. (i) An edge must be removed from each of the circuits *ABFA* and *BEDB*. Therefore, two edges must be removed.
   (ii) An edge cannot be removed if it will leave a disconnected graph. In other words, do not remove a bridge. Edge *CB* is a bridge, so do not remove that edge.
   (iii) If edge *AB* is removed, then one of the three edges *BD*, *DE*, or *BE* must also be removed.
      If edge *AF* is removed, then one of the three edges *BD*, *DE*, or *BE* must also be removed.
      If edge *FB* is removed, then one of the three edges *BD*, *DE*, or *BE* must also be removed.
      Therefore, there are 9 different spanning trees that can be produced from the graph.

17. (i) One edge must be removed from the circuit *BCEFB*.
   (ii) An edge cannot be removed if it will leave a disconnected graph. In other words, do not remove a bridge. Edges *AE* and *BD* are both bridges, so do not remove either edge.

**33.** The edges added in order are *DH*, *JE*, *AK*, *BC*, *HF*, *CJ*, *EF*, *AB*, and *FG*. The total weight of the maximal spanning tree is $17 + 15 + 12 + 11 + 10 + 9 + 8 + 7 + 2 = 91$.

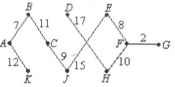

**35.** The total distance in the following minimal spanning tree is 2386 kilometers. Edges were added in the following order:
Serowe-Palapye,
Ramatlabama-Lobaste,
Palapye-Mahalapye,
Lobaste-Gaborone,
Jwaneng-Lobaste,
Selibe-Phikwe-Palapye,
Francistown-Selibe-Phikwe,
Nata-Francistown,
Gaborone-Mahalapye,
Orapa-Serowe,
Kang-Jwaneng,
Kang-Ghanzi,
Maun-Nata, and
Tsabong-Kang.

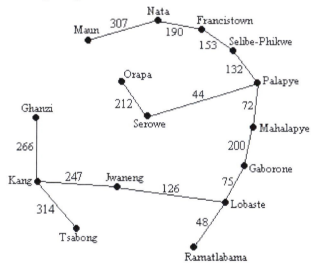

**37.** The total length of the following minimal spanning tree is 6640 feet. It will cost $40,000 per mile to install the power lines. It will cost $\dfrac{6640 \text{ ft}}{1} \times \dfrac{1 \text{ mi}}{5280 \text{ ft}} \times \dfrac{\$40,000}{1 \text{ mi}} \approx \$50,303$. Edges were added in the following order: *AB*, *BC*, *GF*, *HI*, *HB*, *IE*, *AD*, and *EF*.

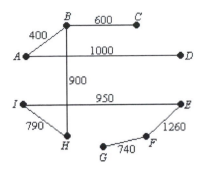

## Solutions to Odd-Numbered Problems in Section 6.3

1. **(a)** Each edge is traversed once. This is an Euler path.
   **(b)** Edge $AD$ is never traversed, and edge $CD$ is traversed twice. This is neither an Euler path nor a Hamiltonian path.
   **(c)** Each vertex is visited once. This is a Hamiltonian path.
   **(d)** Vertex $A$ is visited twice. Edge $DC$ is never traversed. This is neither an Euler path nor a Hamiltonian path.

3. Answers may vary.
   **(a)** One path is $AIEFHDCBG$.
   **(b)** One such path is $DHFEIGABC$.
   **(c)** One such path is $CBAGIEDFH$.

5. **(a)** It is possible to find a Hamiltonian path beginning at $A$. One such path is $AEBCD$.
   **(b)** It is impossible to find a Hamiltonian circuit beginning at $A$. The edge $AE$ is a bridge. Once $AE$ is traversed, it is impossible to get back to vertex $A$ without visiting vertex $E$ again.

7. **(a)** Many Hamiltonian paths exist. One such path is $AFECBD$.
   **(b)** A Hamiltonian circuit can be formed using the Hamiltonian path from part (a). A Hamiltonian circuit is $AFECBDA$.

9. **(a)** Many Hamiltonian paths exist. One such path is $AEDCBF$.
   **(b)** A Hamiltonian circuit can be formed using the Hamiltonian path from part (a). A Hamiltonian circuit is $AEDCBFA$.

11. **(a)** Many Hamiltonian paths exist. One such path is $ABCDEJGIFH$.
    **(b)** There is no way to form a Hamiltonian circuit. Once all of the vertices have been visited, there is no way to get back to vertex $A$ without revisiting a vertex.

13. A complete graph with $n$ vertices has $\dfrac{n(n-1)}{2}$ edges.
    **(a)** If a graph has 7 vertices, then $n = 7$. The number of edges in a complete graph is
    $$\frac{n(n-1)}{2} = \frac{7(7-1)}{2} = \frac{7(6)}{2} = \frac{42}{2} = 21.$$
    **(b)** If a graph has 10 vertices, then $n = 10$. The number of edges in a complete graph is
    $$\frac{n(n-1)}{2} = \frac{10(10-1)}{2} = \frac{10(9)}{2} = \frac{90}{2} = 45.$$

(c) If a graph has 11 vertices, then $n = 11$. The number of edges in a complete graph is
$$\frac{n(n-1)}{2} = \frac{11(11-1)}{2} = \frac{11(10)}{2} = \frac{110}{2} = 55.$$

15. A complete graph with 6 vertices has $\dfrac{n(n-1)}{2} = \dfrac{6(6-1)}{2} = \dfrac{6(5)}{2} = \dfrac{30}{2} = 15$ edges.

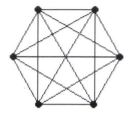

17. **(a)** The graph is complete. Every vertex is adjacent to every other vertex.
    **(b)** The graph is not complete. Every vertex is not adjacent to every other vertex. For example, there is no edge connecting vertices $E$ and $G$.

19. The number of Hamiltonian circuits in a complete graph with $n$ vertices is $n!$.
    **(a)** If a graph has 8 vertices, then $n = 8$. The number of Hamiltonian circuits is $8! = 8 \times 7 \times 6 \times 5 \times 4 \times 3 \times 2 \times 1 = 40,320$.
    **(b)** If a graph has 9 vertices, then $n = 9$. The number of Hamiltonian circuits is $9! = 9 \times 8 \times 7 \times 6 \times 5 \times 4 \times 3 \times 2 \times 1 = 362,880$.

21. A Hamiltonian path must visit each vertex once. Many Hamiltonian paths exist. Four examples of Hamiltonian paths include $CEDABF$, $CFABDE$, $CEDBAF$, and $CBFADE$.

23. The number of Hamiltonian paths in a complete graph with $n$ vertices is $n!$. In this case, we do not know how many vertices the graph has, but we do know the number of Hamiltonian paths is $39,916,800$. Guess and test values for $n$ until you find one that satisfies the equation $n! = 39,916,800$. In this case $n = 11$ because $11! = 11 \times 10 \times 9 \times 8 \times 7 \times 6 \times 5 \times 4 \times 3 \times 2 \times 1 = 39,916,800$. Therefore there are 11 vertices.

25. Systematically list all Hamiltonian paths. There are 4 vertices, so there are $4! = 4 \times 3 \times 2 \times 1 = 24$ Hamiltonian paths. Add the weights along each path to determine the cost of each path. Notice that there are two paths with the lowest cost of 65: $BCAD$ and $DACB$.

| Hamiltonian Path | Cost | Hamiltonian Path | Cost |
|---|---|---|---|
| ABCD | 81 | CADB | 73 |
| ABDC | 89 | CABD | 78 |
| ACBD | 67 | CBAD | 74 |
| ACDB | 80 | CBDA | 71 |
| ADBC | 71 | CDAB | 87 |
| ADCB | 76 | CDBA | 89 |
| BACD | 83 | DABC | 74 |
| BADC | 87 | DACB | 65 |
| BCAD | 65 | DBAC | 78 |
| BCDA | 76 | DBCA | 67 |
| BDCA | 80 | DCAB | 83 |
| BDAC | 73 | DCBA | 81 |

**27. (a) through (c)** Each of the 24 Hamiltonian paths from problem 25 can be turned into Hamiltonian circuits by adding one more vertex so that the ending and beginning vertices are the same. The following table contains all possible Hamiltonian circuits together with the cost of the circuit. The circuits have been arranged so that mirror image circuits are next to each other. Notice that mirror image circuits have the same cost because they travel the same path, just in a different order. The least-cost circuits have a cost of 92 and are shaded in the following table.

| All Possible Hamiltonian Circuits | |
|---|---|
| Circuit and Cost | Mirror Image Circuit and Cost |
| ABCDA 106 | ADCBA *106* |
| ABDCA 110 | ACDBA *110* |
| ACBDA 92 | ADBCA 92 |
| BACDB 110 | BDCAB *110* |
| BADCB 106 | BCDAB *106* |
| BCADB 92 | BDACB 92 |
| CABDC 110 | CDBAC *110* |
| CBADC 106 | CDABC *106* |
| CBDAC 92 | CADBC 92 |
| DCBAD 106 | DABCD *106* |
| DBCAD 92 | DACBD 92 |
| DCABD *110* | DBACD *110* |

**29.** It is possible to have an Euler path that is not a Hamiltonian path. Consider the following graph:

The path $ADCABC$ is an Euler path because each edge has been traversed once. Notice, however, vertex $A$ has been visited twice. Therefore, the path is not a Hamiltonian path.

**31.** Answers may vary.
   **(a)** A path that is both an Euler path and a Hamiltonian path is $AB$.

   **(b)** A path that is both an Euler path and a Hamiltonian path is $ABC$.

   **(c)** A path that is both an Euler path and a Hamiltonian path is $ABCD$.

**33. (a)** There are exactly two vertices with odd degree. By Euler's theorem, there are no Euler circuits in the graph. If an edge is added to connect the two vertices with odd degree, vertices $E$ and $F$, then each vertex will have an even degree, and there will be at least one Euler circuit. One Euler circuit is $AEFBCFEDA$.

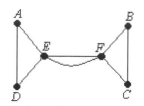

**(b)** Notice edge *EF* is a bridge. Once the bridge is traversed, it is impossible to get back to the beginning vertex without revisiting a vertex. No Hamiltonian circuit exists. If an edge is added that connects vertices *A* and *B*, then a Hamiltonian circuit can be formed. One Hamiltonian circuit is *ADEFCBA*.

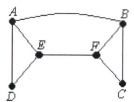

**35. (a)** Use the brute-force algorithm. Systematically list all Hamiltonian circuits that begin at vertex *A*.

| Hamiltonian Circuit | Cost |
|---|---|
| *ABDCEA* | 12 + 13 + 16 + 15 + 10 = 66 |
| *ACDBEA* | 20 + 16 + 13 + 14 + 10 = 73 |
| *AEBDCA* | 10 + 14 + 13 + 16 + 20 = 73 |
| *AECDBA* | 10 + 15 + 16 + 13 + 12 = 66 |

There are two lowest-cost Hamiltonian circuits that begin at *A* and have a cost of 66: *ABDCEA* and *AECDBA*. These circuits are mirror images of each other.

**(b)** Use the nearest-neighbor algorithm. Begin at vertex *A*.

Select the least-cost edge from the following edges: *AB*, 12; *AC*, 20; and *AE*, 10.

Add edge *AE* to the subgraph, and move to vertex *E*.

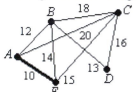

Select the least-cost edge from the following edges: *EB*, 14 and *EC*, 15.

Add edge *EB* to the subgraph, and move to vertex *B*.

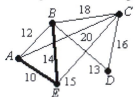

Notice edge *BA* cannot be considered because it would form a circuit if added to the subgraph and edge *BE* cannot be considered because it has already been added to the subgraph.

Select the least-cost edge from the following edges: *BC*, 18 and *BD*, 13.

Add edge *BD* to the subgraph, and move to vertex *D*.

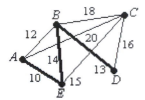

One vertex remains. Add vertex *C* to the graph and then return to vertex *A*. The Hamiltonian circuit created is *AEBDCA* and has a cost of 10 + 14 + 13 + 16 + 20 = 73. The mirror-image circuit, *ACDBEA*, has the same cost.

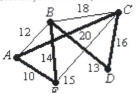

**(c)** Use the cheapest-link algorithm.

All edges are acceptable. No edges have been selected. Choose edge *AE* which is the edge with the least cost of 10.

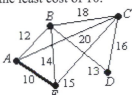

All edges other than *AE* are acceptable. Of the remaining acceptable edges, the edge with the least weight is *AB*. Choose edge *AB* with a cost of 12.

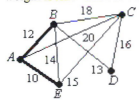

Edge *BE* is not acceptable because it would complete a circuit. Edge *AC* is not acceptable because it would add a third edge at vertex *A*. Edges *AB* and *AE* have already been selected. Of the remaining acceptable edges, the edge with the least weight is *BD*. Select edge *BD* with a cost of 13.

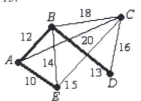

There are only two acceptable edges: *EC* and *CD*. All of the other remaining edges have been selected already, would complete a circuit, or share a vertex with two selected edges. Choose *EC* with a cost of 15. Once *EC* is selected, the only acceptable edge left is the one that completes the Hamiltonian circuit, edge *DC*. The approximate least-cost Hamiltonian circuit starting at vertex *A* is *AECDBA* or its mirror image circuit *ABDCEA*. It has a cost of 66.

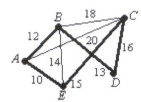

**37.** Use the repetitive nearest-neighbor algorithm. Apply the nearest-neighbor algorithm five times beginning at a different vertex each time the algorithm is applied.

**Begin at vertex _A_.** The edge in the following table with the least cost is _AE_. Select _AE_.

| Current Vertex | Edges from Current Vertex to Unused Vertices | Cost |
|---|---|---|
| A | AB | 13 |
| | AC | 27 |
| | AD | 10 |
| | AE | 9 |

Vertex _E_ becomes the current vertex. The edge in the following table with the least cost is _EB_. Select _EB_.

| Current Vertex | Edges from Current Vertex to Unused Vertices | Cost |
|---|---|---|
| E | EB | 11 |
| | EC | 21 |
| | ED | 20 |

Vertex _B_ becomes the current vertex. The edge in the following table with the least cost is _BC_. Select _BC_.

| Current Vertex | Edges from Current Vertex to Unused Vertices | Cost |
|---|---|---|
| B | BD | 35 |
| | BC | 12 |

Vertex _C_ becomes the current vertex. The last unvisited vertex is _D_. Select _CD_ and return to vertex _A_. The low cost Hamiltonian circuit found by starting at vertex _A_ is _AEBCDA_ and has a total cost of $9 + 11 + 12 + 11 + 10 = 53$. Its mirror image circuit is _ADCBEA_ and has the same total cost.

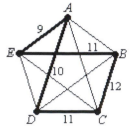

**Begin at vertex B.** The edge in the following table with the least cost is *BE*. Select *BE*.

| Current Vertex | Edges from Current Vertex to Unused Vertices | Cost |
|---|---|---|
| B | BA | 13 |
| | BD | 35 |
| | BC | 12 |
| | BE | 11 |

Vertex *E* becomes the current vertex. The edge in the following table with the least cost is *EA*. Select *EA*.

| Current Vertex | Edges from Current Vertex to Unused Vertices | Cost |
|---|---|---|
| E | EA | 9 |
| | EC | 21 |
| | ED | 20 |

Vertex *A* becomes the current vertex. The edge in the following table with the least cost is *AD*. Select *AD*.

| Current Vertex | Edges from Current Vertex to Unused Vertices | Cost |
|---|---|---|
| A | AD | 10 |
| | AC | 27 |

Vertex *D* becomes the current vertex. The last unvisited vertex is *C*. Select *DC* and return to vertex *B*. The low cost Hamiltonian circuit found by starting at vertex *B* is *BEADCB* and has a total cost of $11 + 9 + 10 + 11 + 12 = 53$. Its mirror image circuit is *BCDAEB* and has the same total cost.

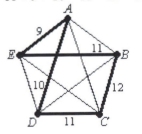

**Begin at vertex C.** The edge in the following table with the least cost is *CD*. Select *CD*.

| Current Vertex | Edges from Current Vertex to Unused Vertices | Cost |
|---|---|---|
| C | CA | 27 |
| | CB | 12 |
| | CD | 11 |
| | CE | 21 |

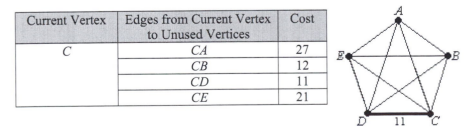

Vertex $D$ becomes the current vertex. The edge in the following table with the least cost is $DA$. Select $DA$.

| Current Vertex | Edges from Current Vertex to Unused Vertices | Cost |
|---|---|---|
| $D$ | $DA$ | 10 |
| | $DB$ | 35 |
| | $DE$ | 20 |

Vertex $A$ becomes the current vertex. The edge in the following table with the least cost is $AE$. Select $AE$.

| Current Vertex | Edges from Current Vertex to Unused Vertices | Cost |
|---|---|---|
| $A$ | $AE$ | 9 |
| | $AB$ | 13 |

Vertex $E$ becomes the current vertex. The last unvisited v    $D$    11    $C$    d return to vertex $C$. The low cost Hamiltonian circuit found by starting at vertex $C$ is $CDAEBC$ and has a total cost of $11 + 10 + 9 + 11 + 12 = 53$. Its mirror image circuit is $CBEADC$ and has the same total cost.

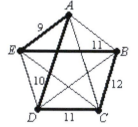

**Begin at vertex $D$.** The edge in the following table with the least cost is $DA$. Select $DA$.

| Current Vertex | Edges from Current Vertex to Unused Vertices | Cost |
|---|---|---|
| $D$ | $DA$ | 10 |
| | $DB$ | 35 |
| | $DC$ | 11 |
| | $DE$ | 20 |

Vertex $A$ becomes the current vertex. The edge in the following table with the least cost is $AE$. Select $AE$.

| Current Vertex | Edges from Current Vertex to Unused Vertices | Cost |
|---|---|---|
| $A$ | $AB$ | 13 |
| | $AC$ | 27 |
| | $AE$ | 9 |

Vertex $E$ becomes the current vertex. The edge in the following table with the least cost is $EB$. Select $EB$.

| Current Vertex | Edges from Current Vertex to Unused Vertices | Cost |
|---|---|---|
| $E$ | $EB$ | 11 |
| | $EC$ | 21 |

Vertex $B$ becomes the current vertex. The last unvisited vertex is $C$. Select $BC$ and return to vertex $D$. The low cost Hamiltonian circuit found by starting at vertex $D$ is $DAEBCD$ and has a total cost of $10 + 9 + 11 + 12 + 11 = 53$. Its mirror image circuit is $DCBEAD$ and has the same total cost.

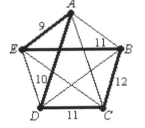

**Begin at vertex $E$.** The edge in the following table with the least cost is $EA$. Select $EA$.

| Current Vertex | Edges from Current Vertex to Unused Vertices | Cost |
|---|---|---|
| $E$ | $EA$ | 9 |
| | $EB$ | 11 |
| | $EC$ | 21 |
| | $ED$ | 20 |

Vertex $A$ becomes the current vertex. The edge in the following table with the least cost is $AD$. Select $AD$.

| Current Vertex | Edges from Current Vertex to Unused Vertices | Cost |
|---|---|---|
| $A$ | $AB$ | 13 |
| | $AC$ | 27 |
| | $AD$ | 10 |

Vertex $D$ becomes the current vertex. The edge in the following table with the least cost is $DC$. Select $DC$.

| Current Vertex | Edges from Current Vertex to Unused Vertices | Cost |
|---|---|---|
| $D$ | $DB$ | 35 |
| | $DC$ | 11 |

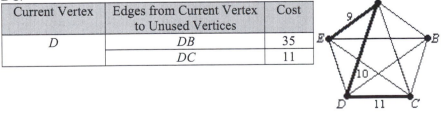

Vertex $C$ becomes the current vertex. The last unvisited vertex is $B$. Select $CB$ and return to vertex $E$. The low cost Hamiltonian circuit found by starting at vertex $E$ is $EADCBE$ and has a total cost of $9 + 10 + 11 + 12 + 11 = 53$. Its mirror image circuit is $EBCDAE$ and has the same total cost. Notice each of the Hamiltonian circuits obtained by the repetitive nearest-neighbor algorithm has the same total cost of 53 and traverses the same edges.

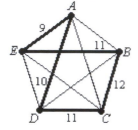

**39.** Use the nearest-neighbor algorithm and begin at vertex $X$. The edge in the following table with the smallest distance is $XD$. Select $XD$.

| Current Vertex | Edges from Current Vertex to Unused Vertices | Distance |
|---|---|---|
| $X$ | $XA$ | 33 |
| | $XB$ | 47 |
| | $XC$ | 30 |
| | $XD$ | 29 |
| | $XE$ | 50 |

Vertex $D$ becomes the current vertex. The edge in the following table with the smallest distance is $DE$. Select $DE$.

| Current Vertex | Edges from Current Vertex to Unused Vertices | Distance |
|---|---|---|
| $D$ | $DA$ | 31 |
| | $DB$ | 41 |
| | $DC$ | 40 |
| | $DE$ | 12 |

Vertex $E$ becomes the current vertex. The edge in the following table with the smallest distance is $EA$. Select $EA$.

| Current Vertex | Edges from Current Vertex to Unused Vertices | Distance |
|---|---|---|
| $E$ | $EA$ | 22 |
| | $EB$ | 46 |
| | $EC$ | 60 |

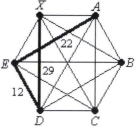

Vertex $A$ becomes the current vertex. The edge in the following table with the smallest distance is $AB$. Select $AB$.

| Current Vertex | Edges from Current Vertex to Unused Vertices | Distance |
|----------------|-----------------------------------|----------|
| $A$ | $AB$ | 32 |
|  | $AC$ | 55 |

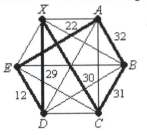

The last unvisited vertex is $C$. Select $BC$ and return to vertex $X$. The low cost Hamiltonian circuit found by starting at vertex $X$ is $XDEABCX$ and has a total distance of $29 + 12 + 22 + 32 + 31 + 30 = 156$ miles. Its mirror image circuit is $XCBAEDX$ and has the same total distance.

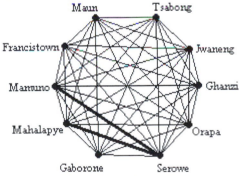

41. Use the cheapest-link algorithm. Initially, all edges are acceptable. The edge with the smallest weight of 44 kilometers is Mamuno - Serowe, so choose that edge. With only one edge in the subgraph, all of the other edges are acceptable, so choose edge Mahalapye - Serowe with weight 110 kilometers. These two edges are darkened in the following figure.

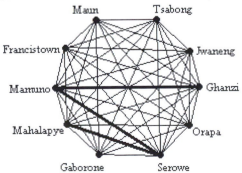

The edge Mamuno – Mahalapye is not acceptable because it would complete a circuit. No edge that includes the vertex Serowe is acceptable because adding one more edge at that vertex would create a vertex with three edges. All other edges are acceptable. The acceptable edge with the smallest weight of 163 kilometers is Mamuno - Ghanzi, so add that edge next.

More edges are now unacceptable. In addition to the unacceptable edges previously mentioned, no edge that includes the vertex Mamuno is acceptable because adding one more edge at that vertex would create a vertex with three edges. The edge Mahalapye - Ghanzi is unacceptable because adding it would create a circuit. All other edges are acceptable. The acceptable edge with the smallest weight of 198 kilometers is Mahalapye - Gaborone, so add that edge next.

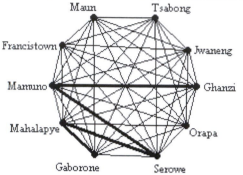

Currently the unacceptable edges include any edge with the following vertices as endpoints: Serowe, Mamuno, and Mahalapye. Additionally, the edge Gaborone - Ghanzi is unacceptable because adding it would create a circuit. All other edges are acceptable. The acceptable edge with the smallest weight of 202 kilometers is Gaborone - Jwaneng, so add that edge next.

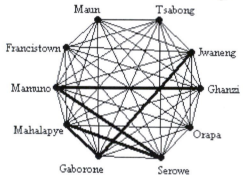

Currently the unacceptable edges include any edge with the following vertices as endpoints: Serowe, Mamuno, Mahalapye, and Gaborone. Additionally, the edge Jwaneng - Ghanzi is unacceptable because adding it would create a circuit. All other edges are acceptable. The acceptable edge with the smallest weight of 240 kilometers is Francistown - Orapa, so add that edge next.

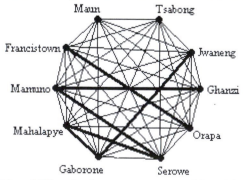

No additional edges are unacceptable. The next acceptable edge with the smallest weight of 255 kilometers is Maun - Orapa, so add that edge next.

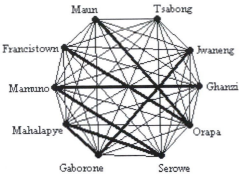

Currently the unacceptable edges include any edge with the following vertices as endpoints: Serowe, Mamuno, Mahalapye, Gaborone, and Orapa. Additionally, the edges Jwaneng - Ghanzi and Francistown - Maun are unacceptable because adding either of them would create a circuit. All other edges are acceptable. The acceptable edge with the smallest weight of 275 kilometers is Maun - Ghanzi, so add that edge next.

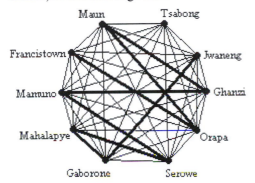

There is one remaining vertex that must be added to the subgraph. Add edges Jwaneng - Tsabong with a weight of 280 kilometers and Francistown - Tsabong with a weight of 790 kilometers.

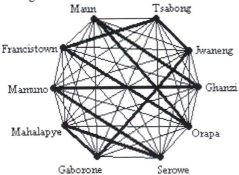

The Hamiltonian circuit must begin and end at Gaborone. Using the cheapest-link algorithm, a low cost circuit passes through the following vertices in order: Gaborone, Mahalapye, Serowe, Mamuno, Ghanzi, Maun, Orapa, Francistown, Tsabong, Jwaneng, and Gaborone. The total distance traveled is 2557 kilometers. The mirror image circuit has the same weight.

**43.** Use the repetitive nearest-neighbor algorithm. Apply the nearest-neighbor algorithm using each vertex as the start vertex. The algorithm will be applied 8 times.
**Use FedEx as the start vertex:**
The vertices will be added in the following order: FedEx, $G$, $B$, $F$, $E$, $C$, $D$, $A$, and FedEx. The total weight of the circuit is $14 + 6 + 7.5 + 7 + 6.5 + 7 + 23 + 27 = 98$ blocks.

**Use *A* as the start vertex:**
The vertices will be added in the following order: *A, B, G, F, E, C, D*, FedEx, and *A*. The total weight of the circuit is $8 + 6 + 10 + 7 + 6.5 + 7 + 31 + 27 = 102.5$ blocks.

**Use *B* as the start vertex:**
The vertices will be added in the following order: *B, G, F, E, C, D, A*, FedEx, and *B*. The total weight of the circuit is $6 + 10 + 7 + 6.5 + 7 + 23 + 27 + 19 = 105.5$ blocks.

**Use *C* as the start vertex:**
The vertices will be added in the following order: *C, E, F, B, G*, FedEx, *A, D*, and *C*. The total weight of the circuit is $6.5 + 7 + 7.5 + 6 + 14 + 27 + 23 + 7 = 98$ blocks.

**Use *D* as the start vertex:**
The vertices will be added in the following order: *D, C, E, F, B, G*, FedEx, *A*, and *D*. The total weight of the circuit is $7 + 6.5 + 7 + 7.5 + 6 + 14 + 27 + 23 = 98$ blocks.

**Use *E* as the start vertex:**
The vertices will be added in the following order: *E, C, D, G, B, F, A*, FedEx, and *E*. The total weight of the circuit is $6.5 + 7 + 18 + 6 + 7.5 + 10 + 27 + 24 = 106$ blocks.

**Use *F* as the start vertex:**
The vertices will be added in the following order: *F, E, C, D, G, B, A*, FedEx, and *F*. The total weight of the circuit is $7 + 6.5 + 7 + 18 + 6 + 8 + 27 + 17 = 96.5$ blocks.

**Use *G* as the start vertex:**
The vertices will be added in the following order: *G, B, F, E, C, D, A*, FedEx, and *G*. The total weight of the circuit is $6 + 7.5 + 7 + 6.5 + 7 + 23 + 27 + 14 = 98$ blocks.

There is one circuit with a total distance of 96.5 blocks. The approximate lowest-distance circuit will pass through the vertices in the following order: FedEx, *F, E, C, D, G, B, A*, FedEx. The mirror image circuit has the same weight.

# *Solutions to Chapter 6 Review Problems*

1. **(a)** Many graphs are possible.

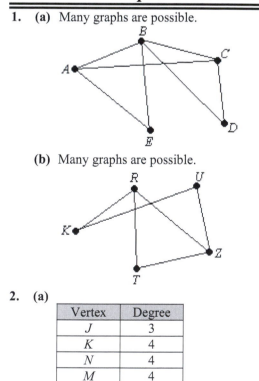

   **(b)** Many graphs are possible.

2. **(a)**

| Vertex | Degree |
|--------|--------|
| *J* | 3 |
| *K* | 4 |
| *N* | 4 |
| *M* | 4 |
| *L* | 4 |
| *H* | 5 |

**(b)** The sum of the degrees is 24.  There are 12 edges.  The number of edges is half of the sum of the degrees.

**3.** **(a)** The vertices adjacent to Newton are Kinsley, Salina, Emporia, and Wichita.

**(b)**

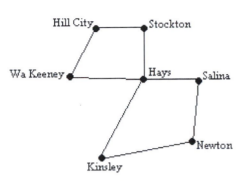

**(c)** Many paths are possible.
   1. Emporia, Newton, Salina, Hays, Wakeeney, Oakley
   2. Emporia, Wichita, Dodge City, Garden City, Oakley
   3. Emporia, Topeka, Salina, Hays, Wakeeney, Oakley
**(d)** Many circuits are possible.  Begin and end at Topeka.
   1. Topeka, Salina, Hays, Wakeeney, Dodge City, Wichita, Emporia, Topeka
   2. Topeka, Emporia, Wichita, Dodge City, Garden City, Oakley, Wakeeney, Hays, Salina, Topeka.

**4.** **(a)** The graph is connected.  For any pair of vertices, there is a path that contains them.
**(b)** There are four bridges:  *CD*, *ID*, *GI*, and *GH*.
Remove bridge *CD*:

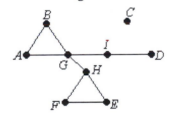

Remove bridge *ID*:

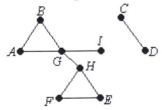

Remove bridge *GI*:

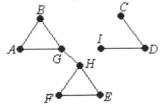

Remove bridge *GH*:

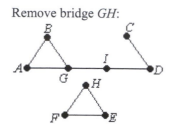

5. (a) There are several paths from *A* to *E*. Three paths include *AGHE*, *ABGHE*, and *AGHFE*.
   (b) At least one Euler path exists because there are exactly two vertices with an odd degree. The Euler path must begin at one of the vertices with odd degree and end at the other. One Euler path is *CDIGABGHEFH*.

6. (a) The Euler circuit begins at vertex *C*. The Euler circuit is *CDAEACEFBFDBC*.
   (b) Beginning at vertex *F*, an Euler circuit which traverses edge *AC* as the third edge is *FEACDFBDAECBF*.

7. The graph has exactly two vertices with an odd degree: vertex *H* and vertex *J*. Euler's theorem guarantees at least one Euler path and no Euler circuits. The Euler path must begin at one of the vertices with an odd degree and end at the other. There are many possible Euler paths. Two Euler paths are *HMJHLNMLKNHKJ* and *JKNMLHJMHNLKH*.

8. Each vertex has an even degree. Euler's theorem guarantees at least one Euler circuit exists which is also an Euler path. Use Fleury's algorithm to find an Euler circuit. Many Euler circuits are possible. Vertex *S* is the selected vertex. Copy the vertices for the new graph.

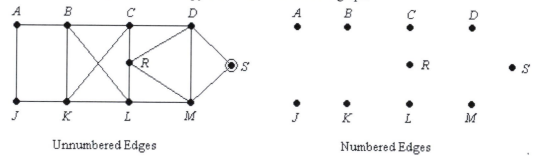

Unnumbered Edges                    Numbered Edges

Neither of the two edges connected to vertex *S* is a bridge, so select either of the two edges to shift to the new graph. Suppose we shift edge *SD* to the new graph and label it with a 1. Vertex *D* becomes the selected vertex, and edge *SM* becomes a bridge.

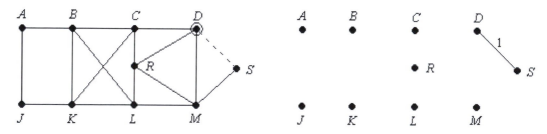

None of the three edges connected to vertex *D* is a bridge, so select one of them to shift to the new graph. Suppose we shift *DC* to the new graph followed by *CB*, *BA*, *AJ*, *JK*, *KL*, and *LM*. Label these edges with a 2, 3, 4, 5, 6, 7, and 8 respectively. Once edge *LM* is added to the new graph, vertex *M* becomes the selected vertex.

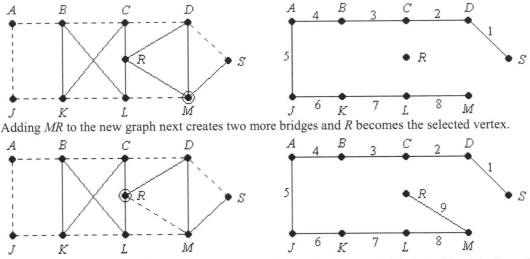

Adding *MR* to the new graph next creates two more bridges and *R* becomes the selected vertex.

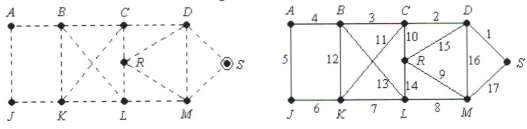

We can add either edge *RC* next or *RL*. Suppose *RC* is added next and labeled with a 10. Once *RC* is added, the remaining edges must be added in the following order: *CK, KB, BL, LR, RD, DM,* and *MS*. Label the edges 11 through 17 respectively. With the addition of *MS*, Fleury's algorithm is complete. The vertices are traversed in the following order: *SDCBAJKLMRCKBLRDMS*.

9. **(a)** Graphical representations may vary.

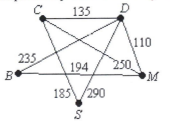

**(b)** The total distance in the trip is $290 + 185 + 250 + 194 = 919$ miles.

**(c)** A Hamiltonian circuit is needed if the desire is to visit each city (each vertex) and then return home. It is possible to find such a circuit. The traveler can pass through the circuit *CSDBMC* or *CMBDSC*.

**(d)** An Euler circuit is needed if the desire is to travel along each edge in the graph and then return home. It is impossible to find an Euler circuit because there are two vertices with an odd degree. By Euler's theorem, it would be possible to find an Euler path, but not an Euler circuit.

10. One edge must be removed from the circuit. This can be done in any of six ways. Do not remove bridge *DE*. There are six spanning trees.

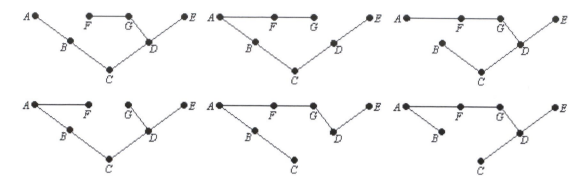

11. Let both river banks and each island be represented by vertices. Each bridge is represented by an edge in the graph.
    (a) The following is one representation of the graph that represents the system of bridges and land.

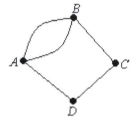

    (b) Many such paths are possible. One path that traverses each bridge exactly once is $BABCDA$.
    (c) It is impossible to traverse each bridge and begin and end on an island. Euler's theorem states that if a graph has exactly two vertices with an odd degree, then there is at least one Euler path, and the Euler path must begin at one vertex with odd degree and end at the other. Only one of the island vertices has an odd degree. Therefore it is impossible to both begin and end on an island.

12. It is possible to begin and end at the same point if each bridge is traversed twice. There are many paths that can be created. One such path is $BCDABADCBAB$.

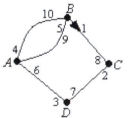

13. Crossing an edge twice is like doubling the number of edges in the graph. If every edge is doubled, then the degree at each vertex will be even. By Euler's theorem, there would always be at least one Euler circuit.

14. Using Kruskal's algorithm, the edges added in order are $DC$, $BD$, $ED$, and $AB$. The following graph represents the minimal spanning tree.

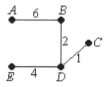

**15.** Each vertex in the graph has an even degree.  By Euler's theorem, we know the graph has at least one Euler circuit. Use Fleury's algorithm to find an Euler circuit.  Your Euler circuit may differ.  Suppose $A$ is the selected vertex in the original graph.

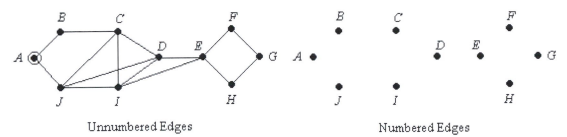

Unnumbered Edges                    Numbered Edges

We can select either edge $AJ$ or edge $AB$ to add to the new graph.  Suppose edge $AB$ is added and labeled with a 1.  Vertex $B$ becomes the selected vertex.  Notice $AJ$ and $BC$ are bridges.  Edge $BC$ will have to be the next edge added.  Once edge $BC$ is added to the new graph and labeled with a 2, vertex $C$ becomes the selected vertex.

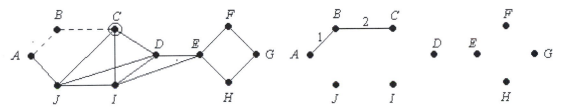

Any of the edges connected to $C$ can be added next.  Suppose we add edge $CD$ to the new graph labeled with a 3 followed by edge $DE$ labeled with a 4.  Vertex $E$ will become the selected vertex and edge $IE$ will become a bridge.

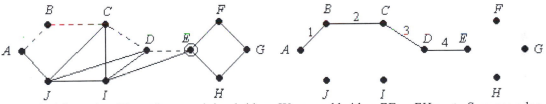

Avoid adding edge $IE$ next because it is a bridge.  We can add either $EF$ or $EH$ next.  Suppose edge $EF$ is added next and labeled with a 5.  Once $EF$ is added to the new graph, the next four edges must be added in the following order:  $FG$, $GH$, $HE$, and $EI$.  Label the edges with a 6, 7, 8, and 9 respectively.  Once $EI$ is added to the new graph, vertex $I$ becomes the selected vertex.

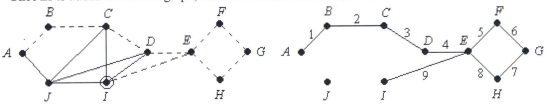

Any of the three edges connected to vertex $I$ can be added next.  Suppose edge $ID$ is added followed by $DJ$, $JC$, $CI$, $IJ$, and finally $JA$ numbered 10 through 15 respectively.  Fleury's algorithm is complete.  The vertices added in order were $ABCDEFGHEIDJCIJA$.

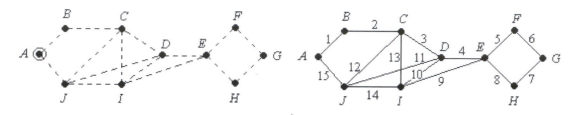

**16.** Use Kruskal's algorithm to find the minimal spanning tree. The edges should be added in the following order: *GF, BC, CH, FH, HE, CD,* and *AG.* The length of cable needed is 92.5 feet.

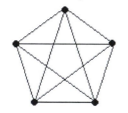

**17. (a)** The following is a complete graph with five vertices.

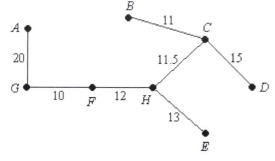

**(b)** A complete graph with 5 vertices has $\dfrac{n(n-1)}{2} = \dfrac{5(5-1)}{2} = \dfrac{5(4)}{2} = \dfrac{20}{2} = 10$ edges.

**(c)** The number of Hamiltonian paths in a complete graph with five vertices is $5! = 5 \times 4 \times 3 \times 2 \times 1 = 120$.

**(d)** For each of the Hamiltonian paths from part (c), if one more vertex is added to the path to end at the same vertex from which the path started, then a Hamiltonian circuit is formed. There are 120 Hamiltonian paths. Therefore, there are 120 Hamiltonian circuits.

**(e)** In general, a complete graph with *n* vertices has *n*! Hamiltonian paths. If we know there are 18! Hamiltonian paths, then there must be 18 vertices in the graph. There are $\dfrac{n(n-1)}{2} = \dfrac{18(18-1)}{2} = \dfrac{18(17)}{2} = \dfrac{306}{2} = 153$ edges.

**(f)** Guess and check the value of *n*! for different values of *n*. We see $11! = 39{,}916{,}800$. Therefore, there must be 11 vertices in the graph. There are $\dfrac{n(n-1)}{2} = \dfrac{11(11-1)}{2} = \dfrac{11(10)}{2} = \dfrac{110}{2} = 55$ edges.

18. For the given graph with five vertices, there are many Hamiltonian circuits. Systematically list all possible circuits from each vertex and determine the cost of each.

| Hamiltonian Circuit | Cost |
|---|---|
| ABCDEA | 8 + 12 + 15 + 6 + 10 = 51 |
| ABDCEA | 8 + 9 + 15 + 13 + 10 = 55 |
| AEDCBA | 10 + 6 + 15 + 12 + 8 = 51 |
| AECDBA | 10 + 13 + 15 + 9 + 8 = 55 |
| BAEDCB | 8 + 10 + 6 + 15 + 12 = 51 |
| BAECDB | 8 + 10 + 13 + 15 + 9 = 55 |
| BDCEAB | 9 + 15 + 13 + 10 + 8 = 55 |
| BCDEAB | 12 + 15 + 6 + 10 + 8 = 51 |
| CBAEDC | 12 + 8 + 10 + 6 + 15 = 51 |
| CEABDC | 13 + 10 + 8 + 9 + 15 = 55 |
| CDEABC | 15 + 6 + 10 + 8 + 12 = 51 |
| CDBAEC | 15 + 9 + 8 + 10 + 13 = 55 |
| DEABCD | 6 + 10 + 8 + 12 + 15 = 51 |
| DBAECD | 9 + 8 + 10 + 13 + 15 = 55 |
| DCBAED | 15 + 12 + 8 + 10 + 6 = 51 |
| DCEABD | 15 + 13 + 10 + 8 + 9 = 55 |
| EABCDE | 10 + 8 + 12 + 15 + 6 = 51 |
| EABDCE | 10 + 8 + 9 + 15 + 13 = 55 |
| ECDBAE | 13 + 15 + 9 + 8 + 10 = 55 |
| EDCBAE | 6 + 15 + 12 + 8 + 10 = 51 |

There are ten Hamiltonian circuits with a cost of 51. Five of the circuits are mirror image circuits of the other five. The circuits traverse the same edges, but they each start at a different vertex. The least-cost Hamiltonian circuit is illustrated with thicker edges in the following graph.

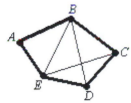

19. Use the nearest-neighbor algorithm and begin at vertex A. The edge in the following table with the smallest cost is AB. Select AB.

| Current Vertex | Edges from Current Vertex to Unused Vertices | Cost |
|---|---|---|
| A | AB | 8 |
|  | AC | 13 |
|  | AD | 17 |
|  | AE | 9 |
|  | AF | 16 |

Vertex B becomes the current vertex. The edge in the following table with the smallest cost is BC. Select BC.

| Current Vertex | Edges from Current Vertex to Unused Vertices | Cost |
|---|---|---|
| B | BC | 15 |
|  | BD | 22 |
|  | BE | 21 |
|  | BF | 19 |

Vertex *C* becomes the current vertex. The edge in the following table with the smallest cost is *CD*. Select *CD*.

| Current Vertex | Edges from Current Vertex to Unused Vertices | Cost |
|---|---|---|
| *C* | *CD* | 10 |
| | *CE* | 23 |
| | *CF* | 24 |

Vertex *D* becomes the current vertex. The edge in the following table with the smallest cost is *DF*. Select *DF*.

| Current Vertex | Edges from Current Vertex to Unused Vertices | Cost |
|---|---|---|
| *D* | *DE* | 26 |
| | *DF* | 18 |

The last unvisited vertex is *E*. Select *FE* and return to vertex *A*. The low cost Hamiltonian circuit found by starting at vertex *A* is *ABCDFEA* and has a total cost of 8 + 15 + 10 + 18 + 12 + 9 = 72. Its mirror image circuit is *AEFDCBA* and has the same total cost.

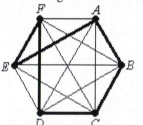

**20.** Use the cheapest-link algorithm. Initially all edges are acceptable. The least-cost edge is *AB* with a cost of 8. Add *AB* to the subgraph.

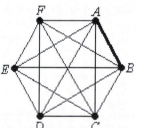

All edges other than *AB* are acceptable. The least-cost edge is *AE* with a cost of 9. Add *AE* to the subgraph.

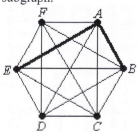

Currently no edge is acceptable with vertex *A* as an endpoint. There are already two edges at vertex *A*. Edge *EB* is also not acceptable because including it would create a circuit. The least-cost acceptable edge is *CD* with a cost of 10. Add *CD* to the subgraph. Adding edge *CD* does not create any new unacceptable edges. The next least-cost acceptable edge is *EF* with a cost of 12. Add *EF* to the subgraph.

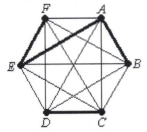

Currently no edge is acceptable with vertices *A* or *E* as an endpoint. There are already two edges at both of those vertices. Edge *FB* is unacceptable because including it would create a circuit. The least-cost acceptable edge is *BC* with a cost of 15. Add *BC* to the subgraph. Once edge *BC* is added, the only acceptable edge left to add is *FD*. The approximate least-cost Hamiltonian circuit is given in the following graph. The total cost of the circuit is 9 + 8 + 10 + 12 + 15 + 18 = 72.

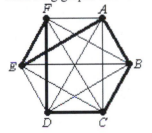

**21.** Use the repetitive nearest-neighbor algorithm. Apply the nearest-neighbor algorithm six times beginning at a different vertex each time the algorithm is applied.

**Use vertex *A* as the start vertex:**
The vertices will be added in the following order: *ABCDFEA*. The total cost is 8 + 15 + 10 + 18 + 12 + 9 = 72.

**Use vertex *B* as the start vertex:**
The vertices will be added in the following order: *BAEFDCB*. The total cost is 8 + 9 + 12 + 18 + 10 + 15 = 72.

**Use vertex *C* as the start vertex:**
The vertices will be added in the following order: *CDABFEC*. The total cost is 10 + 17 + 8 + 19 + 12 + 23 = 89.

**Use vertex *D* as the start vertex:**
The vertices will be added in the following order: *DCABFED*. The total cost is 10 + 13 + 8 + 19 + 12 + 26 = 88.

**Use vertex *E* as the start vertex:**
The vertices will be added in the following order: *EABCDFE*. The total cost is 9 + 8 + 15 + 10 + 18 + 12 = 72.

**Use vertex *F* as the start vertex:**
The vertices will be added in the following order: *FEABCDF*. The total cost is 12 + 9 + 8 + 15 + 10 + 18 = 72.

Four of the Hamiltonian circuits have a cost of 72. Each of the circuits and their mirror image circuits traverse the same edges as indicated in the following graph.

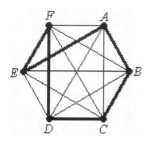

**22.** Use the nearest-neighbor algorithm and begin at vertex $A$. There are two edges with a low cost of 11. Either one can be selected. Suppose we select $AB$.

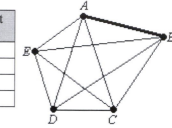

| Current Vertex | Edges from Current Vertex to Unused Vertices | Cost |
|---|---|---|
| $A$ | $AB$ | 11 |
|  | $AC$ | 12 |
|  | $AD$ | 11 |
|  | $AE$ | 12 |

Vertex $B$ becomes the current vertex. The edge in the following table with the smallest cost is $BC$. Select $BC$.

| Current Vertex | Edges from Current Vertex to Unused Vertices | Cost |
|---|---|---|
| $B$ | $BC$ | 13 |
|  | $BD$ | 18 |
|  | $BE$ | 14 |

Vertex $C$ becomes the current vertex. The edge in the following table with the smallest cost is $CD$. Select $CD$.

| Current Vertex | Edges from Current Vertex to Unused Vertices | Cost |
|---|---|---|
| $C$ | $CD$ | 7 |
|  | $CE$ | 16 |

The last unvisited vertex is $E$. Select $DE$ and return to vertex $A$. The low cost Hamiltonian circuit found by starting at vertex $A$ is $ABCDEA$ and has a total cost of $11 + 13 + 7 + 6 + 12 = 49$. Its mirror image circuit is $AEDCBA$ and has the same total cost.

**23.** Use the cheapest-link algorithm. Initially all edges are acceptable. The least-cost edge is *DE* with a cost of 6. Add *DE* to the subgraph.

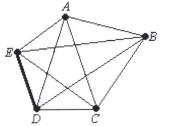

All edges other than *DE* are acceptable. The acceptable least-cost edge is *CD* with a cost of 7. Add *CD* to the subgraph.

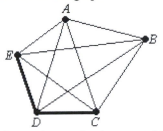

Currently no edge is acceptable with vertex *D* as an endpoint. There are already two edges at vertex *D*. Edge *EC* is also not acceptable because including it would create a circuit. The least-cost acceptable edge is *AB* with a cost of 11. Add *AB* to the subgraph. Adding edge *AB* does not create any new unacceptable edges. There are two acceptable edges with a cost of 12. Either one can be added next. Suppose edge *AC* is added.

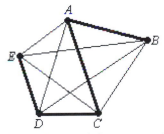

The only remaining acceptable edge to add is *BE* with a cost of 14. The approximate least-cost Hamiltonian circuit is given in the following graph. The total cost of the circuit is $11 + 14 + 6 + 7 + 12 = 50$.

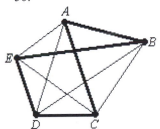

Notice that after edge *AB* was added, we chose to add *AC* with a cost of 12. We could have added *AE* with a cost of 12 instead. If *AE* had been added followed by *BC*, then the following lower cost Hamiltonian circuit would have resulted. It has a total cost of $11 + 13 + 7 + 6 + 12 = 49$.

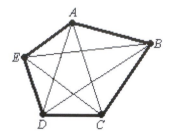

**24.** Use the repetitive nearest-neighbor algorithm. Apply the nearest-neighbor algorithm five times beginning at a different vertex each time the algorithm is applied.

**Use vertex *A* as the start vertex:**
The vertices will be added in the following order: *ABCDEA*. The total cost is 11 + 13 + 7 + 6 + 12 = 49.

**Use vertex *B* as the start vertex:**
The vertices will be added in the following order: *BAEDCB*. The total cost is 11 + 12 + 6 + 7 + 13 = 49.

**Use vertex *C* as the start vertex:**
The vertices will be added in the following order: *CDEABC*. The total cost is 7 + 6 + 12 + 11 + 13 = 49.

**Use vertex *D* as the start vertex:**
The vertices will be added in the following order: *DEABCD*. The total cost is 6 + 12 + 11 + 13 + 7 = 49.

**Use vertex *E* as the start vertex:**
The vertices will be added in the following order: *EDCABE*. The total cost is 6 + 7 + 12 + 11 + 14 = 50.

Four of the Hamiltonian circuits have a cost of 49. Each of the circuits and their mirror image circuits traverse the same edges as indicated in the following graph.

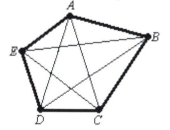

**25.** Using the **nearest-neighbor algorithm**, and starting at Montgomery, the cities are added in the following order: Selma, Birmingham, Anniston, Gadsden, and back to Montgomery. The total distance in the Hamiltonian circuit is 49 + 160 + 62 + 27 + 127 = 425 miles.

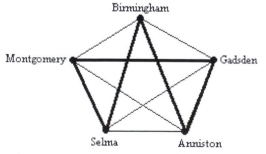

Using the **cheapest-link algorithm**, the edges would be added in the following order:
Anniston - Gadsden: 27 miles
Montgomery - Selma: 49 miles

Anniston - Birmingham:  62 miles
Birmingham - Montgomery:  91 miles
Gadsden - Selma:  224 miles
The total distance in the Hamiltonian circuit is 453 miles.

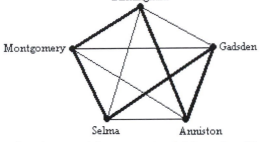

Using the **repetitive nearest-neighbor algorithm**, the following Hamilton circuits would result:
Montgomery, Selma, Birmingham, Anniston, Gadsden, and Montgomery; Total distance = 425 miles.
Birmingham, Anniston, Gadsden, Montgomery, Selma, and Birmingham; Total distance = 425 miles.
Gadsden, Anniston, Birmingham, Montgomery, Selma, and Gadsden; Total distance = 453 miles.
Anniston, Gadsden, Birmingham, Montgomery, Selma, and Anniston; Total distance = 453 miles.
Selma, Montgomery, Birmingham, Anniston, Gadsden, and Selma; Total distance = 453 miles.

# *Chapter 7:* Scheduling

## *Key Concepts*

**By the time you finish studying Section 7.1, you should be able to:**

☐ divide a project into tasks.

☐ describe what it means for two tasks to have an order requirement or precedence relation and what it means for tasks to be independent.

☐ construct an order-requirement digraph for a set of tasks.

☐ identify isolated vertices and vertices that are sinks or sources.

☐ list all possible paths for an order-requirement digraph and identify all maximal paths.

☐ construct a weighted order-requirement digraph for a project.

☐ determine the finishing time for a project.

☐ determine the weight of a path, identify the critical path and the critical time for a project.

☐ explain why a critical path must also be one of the maximal paths.

**By the time you finish studying Section 7.2, you should be able to:**

☐ create an increasing-time or a decreasing-time priority list for a project.

☐ interpret a Gantt chart for a project.

☐ apply the list-processing algorithm to assign tasks to processors.

☐ create a Gantt chart for a project.

☐ determine whether a schedule is optimal.

**By the time you finish studying Section 7.3, you should be able to:**

☐ apply the algorithm for forming the critical-path priority list.

☐ apply the list-processing scheduling algorithm using a critical-path priority list to schedule tasks to two or three processors.

# Hints for Odd-Numbered Problems

## Section 7.1

1. Divide the project into tasks that cannot easily be divided into smaller tasks. Several small tasks may be combined if they are to be completed together. There are many ways to divide the project into tasks.

3. (c) An isolated vertex has no precedence relations, and therefore no arcs attached to it.

5. The two arrows between the two tasks have meaning. What do the arrows mean?

7. (a) Create a vertex for each task. Identify which tasks must be completed before others.

9. Create a vertex for each task. Identify which tasks must be completed before others. Order-requirement digraphs may vary.

11. It may be helpful to identify the last step in the process and then work backward to determine which steps have an order requirement.

13. (a) You may find it helpful to begin with the vertices with the most prerequisite tasks.

15. Notice that two courses have Engr 201 as a prerequisite, so the vertex that represents Engr 201 must have two arrows leaving it. Two courses have Engr 211 as a prerequisite.

17. A path is any list of vertices such the list goes in the same direction as the arcs or arrows. Notice the longest path is made up of four vertices. Systematically list all two-vertex paths, three-vertex paths, and four-vertex paths.

19. A maximal path is a path that cannot be extended at either end.

21. (a) A sink must be at the end of a path, so no arcs will leave a sink.

(b) A source must be at the beginning of a path, so no arcs will enter a source.

(c) Systematically list all two-vertex paths and all three-vertex paths.

(d) There can be many maximal paths for a digraph. A maximal path must begin at a source and end at a sink. Consider the paths from part (c).

23. (a) Sinks have no arcs leaving them. All isolated vertices are sinks, but list only those sinks that are not isolated.

(b) Sources have no arcs entering them. All isolated vertices are sources, but list only those sources that are not isolated.

(c) Any maximal path must begin at a source and end at a sink.

25. (a) Sinks have no arcs leaving them.

(b) Sources have no arcs entering them.

(c) Any maximal path must begin at a source and end at a sink.

27. (a) Each vertex in a weighted order-requirement digraph must have a weight given. See Example 7.8.

(b) The finishing time is the time from the beginning of a project until the end of the project.

29. (d) The critical path is the maximal path with the largest weight.

(e) The critical time is the weight of the critical path.

31. Consider the weighted order-requirement digraph you constructed from problem 27. List the maximal paths and determine which path is the critical path.

33. (a) Any maximal path must begin at a source and end at a sink.

(b) The finishing time can be found by adding the weights of every vertex in the weighted order-requirement digraph.

(c) The critical path is the maximal path with the largest weight. The critical time is the weight of the critical path.

**35.** There are many correct solutions. Create a weighted order-requirement digraph for each part. Make sure the weights of the vertices in the critical path have a sum of 78 minutes. The finishing time for the project may be much longer.

**37.** There are twelve maximal paths. Without listing all of the maximal paths, can you reason out which one will be the critical path? Notice that tasks 7 through 12 will be a part of every maximal path.

## Section 7.2

**1.** **(a)** Arrange the tasks so that the completion times are in increasing order.

**(b)** Arrange the tasks so that the completion times are in decreasing order.

**3.** **(d)** Look for gaps in the schedule when processors are not working on a task.

**5.** **(b)** The ready tasks are tasks that are sources.

**7.** Using the list-processing algorithm with the decreasing-time priority list forces $T_2$ to be selected first from among the ready tasks because it is ahead of $T_1$ in the priority list.

**9.** **(a)** Create the increasing-time priority list. Were the highest-priority tasks scheduled to the lowest-numbered idle processor?

**(b)** Create the decreasing-time priority list.

**11.** Notice that there are times when $P_2$ is idle. $T_6$ could not be started until $T_3$ was finished. What does that tell you about the order requirement for those tasks?

**13.** **(b)** At the beginning of the project, ready tasks have no arrows entering them.

**(c)** From among the ready tasks, assign the lowest-numbered processor the highest-priority ready task first.

**(e)** Apply the list-processing algorithm. As you are learning to create Gantt charts, it will be helpful to write out

the steps of the list-processing algorithm as is done in Example 7.14.

**15.** **(a)** Apply the list-processing algorithm. Keep track of each step as is done in Example 7.14.

**17.** Remember from Section 7.1, the critical time is the weight of the critical path. The critical path is the path having the largest possible weight. List all maximal paths and determine which has the largest weight.

**19.** **(b)** Determine which tasks are ready and then assign the highest-priority ready task to the lowest-numbered idle processor.

**(c)** Study Example 7.14.

**21.** **(b)** Determine which tasks are ready and then assign the highest-priority ready task to the lowest-numbered idle processor.

**(c)** Study Example 7.14.

**23.** See Example 7.15.

**25.** **(b)** The schedule is optimal if the finishing time is as short as possible. If the finishing time matches the critical time, then the schedule is optimal. For some projects, it may not be possible to create a schedule to match the critical time.

**27.** **(a)** See Example 7.15.

**(b)** The schedule is optimal if the finishing time is as short as possible. If the finishing time matches the critical time, then the schedule is optimal. For some projects, it may not be possible to create a schedule to match the critical time.

**29.** **(b)** The schedule is optimal if the finishing time is as short as possible. If the finishing time matches the critical time, then the schedule is optimal. For some projects, it may not be possible to create a schedule to match the critical time.

**31.** **(a)** See Example 7.15.

**(b)** The schedule is optimal if the finishing time is as short as possible. If the finishing time matches the critical time, then the schedule is optimal. For some projects, it may not be possible to create a schedule to match the critical time.

**33. (b)** The schedule is optimal if the finishing time is as short as possible. If the finishing time matches the critical time, then the schedule is optimal. For some projects, it may not be possible to create a schedule to match the critical time.

**35. (a)** See Example 7.15.
**(b)** The schedule is optimal if the finishing time is as short as possible. If the finishing time matches the critical time, then the schedule is optimal. For some projects, it may not be possible to create a schedule to match the critical time.

**37. (b)** The schedule may not be optimal if the finishing time does not match the critical time from part (a).

## Section 7.3

**1. (a)** See Section 7.1.
**(c)** Remember that the critical time is the weight of the maximal path with the largest weight.
**(d)** There is no need to form each priority list to answer this question. Consider the ready tasks. An increasing-time priority list would force you to select the ready-task with the smallest weight.

**3. (d)** Refer to Section 7.1 for a discussion of critical time versus finishing time.

**5. (b)** You are told which tasks are assigned to each processor, but be sure you pay attention to the order requirements as you construct the Gantt chart.

**7. (a)** Study the algorithm for forming the critical-path priority list. See Step 2.
**(b)** Notice that you are performing Step 3 when $T_2$ is removed and then you are going back to Step 2.

**(d)** When one path is left, the remaining tasks must be placed in the critical-path priority list in that order.

**9. (b)** See Example 7.19.
**(c)** Apply the list-processing algorithm using the critical-path priority list. See Example 7.20.
**(d)** Compare the finishing time to the critical time from part (a).

**11. (b)** See Example 7.19.
**(c)** Apply the list-processing algorithm using the critical-path priority list. See Example 7.20.

**13.** Consider whether the weight change for a task changes the critical-path priority list. Recreate the critical-path priority list and Gantt chart in each case. Does a reduction in weight for a task cause the same reduction in the finishing time?

**15.** Reschedule the project. Will a 10-minute increase in one task cause a 10-minute increase in the finishing time?

**17.** The critical-path priority list will be used for problems 19 and 20 also.

**19.** The critical time was determined in problem 17.

**21.** The critical time was determined in problem 17. Is there more idle time with three processors than there was with two processors? Compare your results to the schedule for two processors in problem 19.

**23. (a)** We must assume the processor working on $T_1$ cannot do anything else at the same time.

*Solutions to Odd-Numbered Problems in Section 7.1*

1. Tasks may vary.
   $T_1$: Peel potatoes
   $T_2$: Dice potatoes
   $T_3$: Grate cheese
   $T_4$: Dice onions
   $T_5$: Add cheese, onion, soup, salt, pepper, and milk
   $T_6$: Stir
   $T_7$: Bake

3. **(a)** There are four vertices.
   **(b)** Yes, it is an order-requirement digraph. Raking the clippings can only be done after mowing and edging.
   **(c)** There is one isolated vertex. Planting flowers can be done at any point and does not depend on the other tasks. Notice there are no arrows attached to the vertex.

5. One arc implies that $T_2$ must be done after $T_1$ while the other arc implies that $T_1$ must be done before $T_2$. We do not allow cycles like this in digraphs.

7. **(a)** Digraphs may vary.

   Wash Face •
   Brush Teeth •
   Floss Teeth • ——→ Climb into Bed
   Put on Pajamas •
   Set Alarm •

   **(b)** There are six vertices. No vertex is isolated. Five tasks must be done before climbing into bed.

9. Answers may vary.

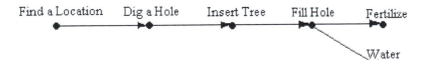

11. Answers may vary.

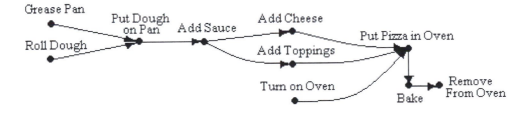

13. **(a)** Answers may vary.

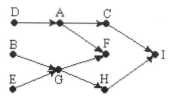

(b) Vertices D, B, and E appear only at the beginning of an arc.

(c) Vertices F and I appear only at the end of an arc.

(d) No vertices are isolated.

**15.**

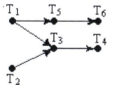

**17. Two-vertex paths:**

$T_1 \to T_2$

$T_1 \to T_3$

$T_2 \to T_5$

$T_3 \to T_4$

$T_4 \to T_5$

**Three-vertex paths:**

$T_1 \to T_2 \to T_5$

$T_1 \to T_3 \to T_4$

$T_3 \to T_4 \to T_5$

**Four-vertex paths:**

$T_1 \to T_3 \to T_4 \to T_5$

**19.** Order-requirement digraph:

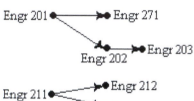

The following are maximal paths:

$T_1 \to T_5 \to T_6$

$T_1 \to T_3 \to T_4$

$T_2 \to T_3 \to T_4$

**21. (a)** Sinks are vertices that are not the start of any arc. There are two sinks: $T_4$ and $T_5$.

**(b)** Sources are vertices that are not the end of any arc. There are two sources: $T_1$ and $T_6$.

**(c) Two-vertex paths:**

$T_1 \to T_2$

$T_2 \to T_5$

$T_2 \to T_4$

$T_1 \to T_3$

$T_3 \to T_4$

$T_6 \to T_3$

**Three-vertex paths:**

$T_1 \to T_2 \to T_5$

$T_1 \to T_2 \to T_4$

$T_1 \to T_3 \to T_4$

$T_6 \to T_3 \to T_4$

(d) Maximal paths begin at vertices that are sources and end at vertices that are sinks. Sources were identified in part (b), and sinks were identified in part (a). Consider all paths that begin at $T_1$ or $T_6$ and end at $T_4$ or $T_5$.

$T_1 \rightarrow T_2 \rightarrow T_5$
$T_1 \rightarrow T_2 \rightarrow T_4$
$T_1 \rightarrow T_3 \rightarrow T_4$
$T_6 \rightarrow T_3 \rightarrow T_4$

23. (a) Sinks are vertices that are not the start of any arc. There is one isolated vertex that is a sink, but we are listing all sinks that are not isolated. There are three sinks that are not isolated vertices: $T_4$, $T_6$, and $T_7$.

(b) Sources are vertices that are not the end of any arc. There is one isolated vertex that is a source, but we are listing all sources that are not isolated. There are two sources that are not isolated: $T_1$ and $T_2$.

(c) Maximal paths begin at vertices that are sources and end at vertices that are sinks. Consider all paths that begin at $T_1$ or $T_2$ and end at $T_4$, $T_6$, or $T_7$.

$T_1 \rightarrow T_3 \rightarrow T_4$
$T_2 \rightarrow T_3 \rightarrow T_4$
$T_1 \rightarrow T_3 \rightarrow T_5 \rightarrow T_6$
$T_1 \rightarrow T_3 \rightarrow T_5 \rightarrow T_7$
$T_2 \rightarrow T_3 \rightarrow T_5 \rightarrow T_6$
$T_2 \rightarrow T_3 \rightarrow T_5 \rightarrow T_7$

25. (a) Sinks are vertices that are not the start of any arc. There are two sinks: $T_6$ and $T_7$.

(b) Sources are vertices that are not the end of any arc. There are two sources: $T_1$ and $T_4$.

(c) Maximal paths begin at vertices that are sources and end at vertices that are sinks. Consider all paths that begin at $T_1$ or $T_4$ and end at $T_6$ or $T_7$.

$T_1 \rightarrow T_2 \rightarrow T_7$
$T_4 \rightarrow T_5 \rightarrow T_6$
$T_1 \rightarrow T_2 \rightarrow T_5 \rightarrow T_6$
$T_1 \rightarrow T_3 \rightarrow T_5 \rightarrow T_6$

27. (a) Representations may vary.

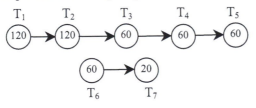

(b) With only one processor, the finishing time is the sum of all of the weights. It will take $120 + 120 + 60 + 60 + 60 + 60 + 20 = 500$ seconds or 8 minutes and 20 seconds.

29. (a) Add the weight of each vertex in the path. The weight of the path is $3 + 10 + 6 + 5 = 24$ hours.

(b) The finishing time is the sum of the weights: $3 + 1 + 10 + 6 + 12 + 5 + 9 + 5 = 51$ hours.

(c) There is only one source, namely $T_1$. There are three sinks, $T_6$, $T_7$, and $T_8$. A maximal path must begin at a source and end at a sink. The weight of each maximal path is the sum of the weights of each vertex in the path.

| Maximal Path: | Weight: |
| --- | --- |
| $T_1 \rightarrow T_2 \rightarrow T_7$ | 13 hours |
| $T_1 \rightarrow T_2 \rightarrow T_4 \rightarrow T_5 \rightarrow T_6$ | 27 hours |
| $T_1 \rightarrow T_2 \rightarrow T_4 \rightarrow T_8$ | 15 hours |
| $T_1 \rightarrow T_3 \rightarrow T_4 \rightarrow T_8$ | 24 hours |
| $T_1 \rightarrow T_3 \rightarrow T_4 \rightarrow T_5 \rightarrow T_6$ | 36 hours |

**(d)** The critical path is the maximal path with the largest weight. The path $T_1 \rightarrow T_3 \rightarrow T_4 \rightarrow T_5 \rightarrow T_6$ has a weight of 36 hours which is more than the weight of any other maximal path. Path $T_1 \rightarrow T_3 \rightarrow T_4 \rightarrow T_5 \rightarrow T_6$ is the critical path.

**(e)** The critical time is the weight of the critical path. From part (d), we know the weight of the critical path is 36 hours. Therefore, the critical time is 36 hours.

**31.** Refer to problem 27. The critical path is $T_1 \rightarrow T_2 \rightarrow T_3 \rightarrow T_4 \rightarrow T_5$. This path has a weight of $120 + 120 + 60 + 60 + 60 = 420$ seconds or 7 minutes.

**33. (a)** **Maximal Path:**                              **Weight:**

$T_1 \rightarrow T_2 \rightarrow T_3 \rightarrow T_6 \rightarrow T_8 \rightarrow T_{10}$     60 days

$T_1 \rightarrow T_2 \rightarrow T_3 \rightarrow T_6 \rightarrow T_7 \rightarrow T_{11}$     70 days

$T_9 \rightarrow T_{11}$                                   72 days

$T_5 \rightarrow T_3 \rightarrow T_6 \rightarrow T_8 \rightarrow T_{10}$          52 days

$T_5 \rightarrow T_3 \rightarrow T_6 \rightarrow T_7 \rightarrow T_{11}$          62 days

**(b)** The finishing time is the sum of all the weights, and in this case, the finishing time is 170 days.

**(c)** The critical path is the path with the largest weight. The critical path is $T_9 \rightarrow T_{11}$. The critical time is 72 days.

**35. (a)** Answers will vary.

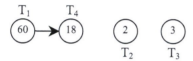

**(b)** Answers will vary.

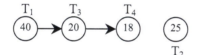

**(c)** Answers will vary.

**37.** The finishing time is the sum of all of the weights. For this project the finishing time is 995 minutes. The critical time is the weight of the critical path. There are four source vertices: $T_1$, $T_2$, $T_4$, and $T_6$. There are three sink vertices: $T_{14}$, $T_{15}$, and $T_{16}$. All maximal paths must begin at a source and end at a sink. Consider all maximal paths and determine which one has the largest weight. The critical path is $T_2 \rightarrow T_3 \rightarrow T_5 \rightarrow T_7 \rightarrow T_8 \rightarrow T_9 \rightarrow T_{10} \rightarrow T_{11} \rightarrow T_{12} \rightarrow T_{13} \rightarrow T_{15}$. The critical time is the weight of the critical path. The critical time is 865 minutes.

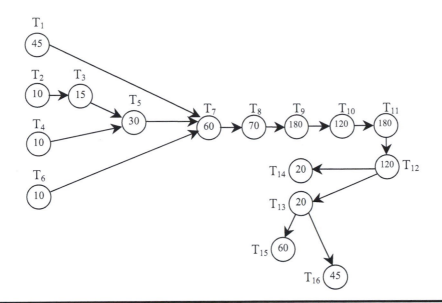

## Solutions to Odd-Numbered Problems in Section 7.2

1. **(a)** Order the tasks from shortest time to longest time. The increasing-time priority list is set alarm, floss teeth, wash face, change into pajamas, and brush teeth.
   **(b)** Order the tasks from longest time to shortest time. The decreasing-time priority list is brush teeth, change into pajamas, wash face, floss teeth, and set alarm.

3. **(a)** There are two processors.
   **(b)** There are six tasks. $T_1$ has a completion time of 10 minutes; $T_2$, 12 minutes; $T_3$, 15 minutes; $T_4$, 8 minutes; $T_5$, 5 minutes; and $T_6$, 3 minutes.
   **(c)** The finishing time is 30 minutes.
   **(d)** There are three instances in which $P_2$ is not actively working on a task. There are $3 + 2 + 2 = 7$ minutes of idle time.

5. **(a)** The increasing-time priority list is $T_6$, $T_1$, $T_5$, $T_2$, $T_4$, $T_3$.
   **(b)** The ready tasks are both represented by source vertices: $T_1$ and $T_2$.
   **(c)** One processor must complete each task. $T_1$ is selected first from among the ready tasks because it is of higher priority according to the increasing-time priority list.

| $P_1$ | $T_1$ | $T_2$ | $T_5$ | $T_4$ | $T_3$ | $T_6$ |
|---|---|---|---|---|---|---|

| Time (min) | 0 | 5 | 16 | 25 | 41 | 58 | 62 |
|---|---|---|---|---|---|---|---|

7. **(a)** The decreasing-time priority list is $T_3$, $T_4$, $T_2$, $T_5$, $T_1$, $T_6$.
   **(b)** The ready tasks are both represented by source vertices: $T_1$ and $T_2$.
   **(c)** One processor must complete each task. $T_2$ is selected first from among the ready tasks because it is of higher priority according to the decreasing-time priority list.

| $P_1$ | $T_2$ | $T_5$ | $T_1$ | $T_3$ | $T_4$ | $T_6$ |
|---|---|---|---|---|---|---|

| Time (min) | 0 | 11 | 20 | 25 | 42 | 58 | 62 |
|---|---|---|---|---|---|---|---|

9. (a) For this project, the increasing-time priority list is $T_1$, $T_5$, $T_2$, $T_3$, $T_4$. Notice that $T_1$ comes before $T_3$ in the increasing-time priority list. In this case, an increasing-time priority list was not used since $T_3$, with a weight of 10, was assigned to the first processor while, at the same time $T_1$, with a weight of 7, was assigned to the second processor. The highest-priority ready task must be assigned to the first processor.

   (b) For this project, the decreasing-time priority list is $T_4$, $T_2$, $T_3$, $T_5$, $T_1$. A decreasing-time priority list could have been used since $T_3$, with a weight of 10, was scheduled to processor 1, while $T_1$, with a weight of 7, was scheduled to processor 2. $T_4$ was the third task to be assigned and has a larger weight than $T_2$ or $T_5$. The highest-priority ready tasks were assigned to $P_1$ when both processors were assigned tasks at the same time.

   (c) Many digraphs are possible. The following is a possible weighted order-requirement digraph. The digraph together with a decreasing-time priority list would produce the Gantt chart given in the problem.

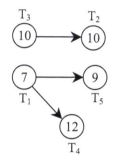

11. Answers will vary. A possible priority list is $T_3$, $T_2$, $T_5$, $T_6$, $T_1$, $T_4$. A possible order-requirement digraph follows. You should create a solution of your own.

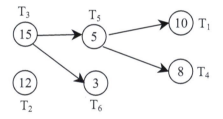

13. (a) The increasing-time priority list is $T_4$, $T_6$, $T_1$, $T_5$, $T_2$, $T_3$.

   (b) The ready tasks are the isolated tasks and the source vertices. There are four ready tasks: $T_1$, $T_5$, $T_6$, and $T_3$.

   (c) The highest-priority ready task must be assigned to the lowest-numbered idle processor. Assign $T_6$ to $P_1$ and $T_1$ to $P_2$.

   (d) $P_1$ is idle at 25 seconds at which time $T_3$ and $T_5$ are ready tasks, so $T_5$ will be assigned to $P_1$ next.

   (e) Use the list-processing algorithm.
   **At 0 seconds, Step 1 (Assignment of Processors):**
   $P_1$ and $P_2$ are idle at the start.
   $T_1$, $T_5$, $T_6$, and $T_3$ are ready.
   $P_1$ is assigned to $T_6$ (the highest-priority ready task).
   $P_2$ is assigned to $T_1$.
   **At 25 seconds, Step 2 (Status Check) and Step 1 (New Assignments):**
   $P_1$ is idle and $P_2$ is working on $T_1$.
   $T_3$ and $T_5$ are ready.
   $P_1$ is assigned to $T_5$ (the highest-priority ready task).

**At 30 seconds, Step 2 (Status Check) and Step 1 (New Assignments):**
$P_2$ is idle and $P_1$ is working on $T_5$.
$T_2$ and $T_3$ are ready.
$P_2$ is assigned to $T_2$.
**At 60 seconds, Step 2 (Status Check) and Step 1 (New Assignments):**
$P_1$ is idle and $P_2$ is working on $T_2$.
$T_3$ is ready.
$P_1$ is assigned to $T_3$.
**At 70 seconds, Step 2 (Status Check) and Step 1 (New Assignments):**
$P_2$ is idle and $P_1$ is working on $T_3$.
No tasks are ready until $T_3$ is completed.
$P_2$ remains idle.
**At 135 seconds, Step 2 (Status Check) and Step 1 (New Assignments):**
$P_2$ is idle and $P_1$ is idle.
$T_4$ is ready.
$P_1$ is assigned to $T_4$. $P_2$ remains idle.
**At 140 seconds, Step 2 (Status Check) and Step 1 (New Assignments):**
$P_1$ and $P_2$ are idle.
No tasks are ready.
All tasks have been completed.
The finishing time is 140 seconds.

15. (a) Use the list-processing algorithm.
**At 0 minutes, Step 1 (Assignment of Processors):**
$P_1$ and $P_2$ are idle at the start.
$T_2$, $T_3$, and $T_7$ are ready.
$P_1$ is assigned to $T_7$ (the highest-priority ready task).
$P_2$ is assigned to $T_3$.
**At 15 minutes, Step 2 (Status Check) and Step 1 (New Assignments):**
$P_2$ is idle and $P_1$ is working on $T_7$.
$T_2$ is ready.
$P_2$ is assigned to $T_2$.
**At 20 minutes, Step 2 (Status Check) and Step 1 (New Assignments):**
$P_1$ is idle and $P_2$ is working on $T_2$.
No tasks are ready.
$P_1$ remains idle.
**At 25 minutes, Step 2 (Status Check) and Step 1 (New Assignments):**
$P_1$ is idle and $P_2$ is idle.
$T_1$ and $T_4$ are ready.
$P_1$ is assigned to $T_1$ and $P_2$ is assigned to $T_4$.
**At 30 minutes, Step 2 (Status Check) and Step 1 (New Assignments):**
$P_1$ is idle and $P_2$ is working on $T_4$.
$T_6$ is ready.
$P_1$ is assigned to $T_6$.

**At 32 minutes, Step 2 (Status Check) and Step 1 (New Assignments):**
$P_1$ is idle and $P_2$ is working on $T_4$.
No tasks are ready.
$P_1$ remains idle.
**At 33 minutes, Step 2 (Status Check) and Step 1 (New Assignments):**
$P_1$ and $P_2$ are idle.
$T_5$ is ready.
$P_1$ is assigned to $T_5$ and $P_2$ remains idle.
**At 44 minutes, Step 2 (Status Check) and Step 1 (New Assignments):**
$P_1$ and $P_2$ are idle.
No tasks are ready.
All tasks have been completed.

**(b)** The finishing time is 44 minutes and there are 17 minutes of idle time.

17. The critical time for a project is the weight of the path having the largest weight. List all maximal paths and determine which has the largest weight.

| Maximal Path: | Weight: |
| --- | --- |
| $T_4 \to T_7$ | 14 minutes |
| $T_1 \to T_7$ | 17 minutes |
| $T_2 \to T_3 \to T_7$ | 22 minutes |
| $T_5 \to T_7$ | 11 minutes |
| $T_6 \to T_7$ | 13 minutes |

The critical time is 22 minutes. It would take one processor 45 minutes working alone.

19. **(a)** The increasing-time priority list is $T_5$, $T_3$, $T_6$, $T_4$, $T_7$, $T_1$, $T_2$.
 **(b)** Processor 1 should be assigned the highest-priority ready task, $T_5$. Processor 2 should be assigned the next ready task in the priority list, $T_6$.
 **(c)** Use the list-processing algorithm.
 **At 0 minutes, Step 1 (Assignment of Processors):**
 $P_1$ and $P_2$ are idle at the start.
 $T_1$, $T_2$, $T_4$, $T_5$, and $T_6$ are ready.
 $P_1$ is assigned to $T_5$ (the highest-priority ready task).
 $P_2$ is assigned to $T_6$.
 **At 3 minutes, Step 2 (Status Check) and Step 1 (New Assignments):**
 $P_1$ is idle and $P_2$ is working on $T_6$.
 $T_1$, $T_2$, and $T_4$ are ready.
 $P_1$ is assigned to $T_4$.
 **At 5 minutes, Step 2 (Status Check) and Step 1 (New Assignments):**
 $P_2$ is idle and $P_1$ is working on $T_4$.
 $T_1$ and $T_2$ are ready.
 $P_2$ is assigned to $T_1$.

**At 9 minutes, Step 2 (Status Check) and Step 1 (New Assignments):**
$P_1$ is idle and $P_2$ is working on $T_1$.
$T_2$ is ready.
$P_1$ is assigned to $T_2$.
**At 14 minutes, Step 2 (Status Check) and Step 1 (New Assignments):**
$P_2$ is idle and $P_1$ is working on $T_2$.
No tasks are ready.
$P_2$ remains idle.
**At 19 minutes, Step 2 (Status Check) and Step 1 (New Assignments):**
$P_1$ and $P_2$ are idle.
$T_3$ is ready.
$P_1$ is assigned to $T_3$ and $P_2$ remains idle.
**At 23 minutes, Step 2 (Status Check) and Step 1 (New Assignments):**
$P_1$ and $P_2$ are idle.
$T_7$ is ready.
$P_1$ is assigned to $T_7$ and $P_2$ remains idle.
**At 31 minutes, Step 2 (Status Check) and Step 1 (New Assignments):**
$P_1$ and $P_2$ are idle.
No tasks are ready.
All tasks have been completed.
It takes two processors 31 minutes to finish the project.

**21. (a)** The decreasing-time priority list is $T_2, T_1, T_7, T_4, T_6, T_3, T_5$.
**(b)** Processor 1 should be assigned the highest-priority ready task, $T_2$. Processor 2 should be assigned the next ready task in the priority list, $T_1$.
**(c)** Use the list-processing algorithm.
**At 0 minutes, Step 1 (Assignment of Processors):**
$P_1$ and $P_2$ are idle at the start.
$T_1, T_2, T_4, T_5,$ and $T_6$ are ready.
$P_1$ is assigned to $T_2$ (the highest-priority ready task).
$P_2$ is assigned to $T_1$.
**At 9 minutes, Step 2 (Status Check) and Step 1 (New Assignments):**
$P_2$ is idle and $P_1$ is working on $T_2$.
$T_4, T_5,$ and $T_6$ are ready.
$P_2$ is assigned to $T_4$.
**At 10 minutes, Step 2 (Status Check) and Step 1 (New Assignments):**
$P_1$ is idle and $P_2$ is working on $T_4$.
$T_3, T_5,$ and $T_6$ are ready.
$P_1$ is assigned to $T_6$.
**At 15 minutes, Step 2 (Status Check) and Step 1 (New Assignments):**
$P_1$ and $P_2$ are idle.
$T_3$ and $T_5$ are ready.
$P_1$ is assigned to $T_3$ and $P_2$ is assigned to $T_5$.

**At 18 minutes, Step 2 (Status Check) and Step 1 (New Assignments):**
$P_2$ is idle and $P_1$ is working on $T_3$.
No tasks are ready.
$P_2$ remains idle.
**At 19 minutes, Step 2 (Status Check) and Step 1 (New Assignments):**
$P_1$ and $P_2$ are idle.
$T_7$ is ready.
$P_1$ is assigned to $T_7$ and $P_2$ remains idle.
**At 27 minutes, Step 2 (Status Check) and Step 1 (New Assignments):**
$P_1$ and $P_2$ are idle.
No tasks are ready.
All tasks have been completed.
It takes two processors 27 minutes to finish the project.

**23.** It will take three processors 65 minutes to finish the project.

**25. (a)** The increasing-time priority list is $T_2$, $T_1$, $T_6$, $T_4$, $T_7$, $T_3$, $T_5$. It will take two processors 36 minutes to finish the project.

**(b)** The critical path is $T_1 \to T_3 \to T_4 \to T_5 \to T_6$. The weight of the critical path is $3 + 10 + 6 + 12 + 5 = 36$ minutes. The critical time for this project is 36 minutes. From the Gantt chart in part (a), we see the finishing time is 36 minutes. The schedule is optimal.

**27. (a)** The finishing time and Gantt chart are exactly the same as the finishing time and the Gantt chart for two processors in problem 25. The third processor never will be assigned a task.

**(b)** The schedule was optimal with two processors. Adding another processor did not decrease the finishing time at all. In fact, it is a waste of the third processor's time to be included. That processor is idle the entire time.

**29. (a)** The increasing-time priority list is $T_2$, $T_3$, $T_8$, $T_{10}$, $T_4$, $T_7$, $T_1$, $T_5$, $T_6$, $T_9$. The finishing time for the project is 45 minutes when two processors are used.

| $P_1$ | $T_4$ | $T_5$ | $T_7$ | $T_9$ | $T_{10}$ |
|---|---|---|---|---|---|
| $P_2$ | $T_1$ | $T_2$ | $T_3$ | $T_6$ | $T_8$ |

Time (min)  0   9  10   15   19 20   28   32   37 40   45

**(b)** The critical path for this project is $T_1 \rightarrow T_2 \rightarrow T_3 \rightarrow T_6 \rightarrow T_8 \rightarrow T_{10}$. The critical time for this project is 42 minutes. The schedule in part (a) may not be optimal.

**(c)** There are 8 minutes of idle time in this schedule.

**31. (a)** The increasing-time priority list is $T_2$, $T_3$, $T_8$, $T_{10}$, $T_4$, $T_7$, $T_1$, $T_5$, $T_6$, $T_9$. The finishing time is 42 minutes.

| $P_1$ | $T_4$ | $T_5$ | $T_6$ | $T_8$ | $T_{10}$ |
|---|---|---|---|---|---|
| $P_2$ | $T_1$ | $T_2$ | $T_3$ | | |
| $P_3$ | $T_9$ | $T_7$ | | | |

Time (min)  0   9 10 12  15  19 20  24   32   37   42

**(b)** The finishing time is the same as the critical time, 42 minutes, so the schedule is optimal.

**(c)** There is a total of 44 minutes of idle time.

**33. (a)** The increasing-time priority list is $T_3$, $T_4$, $T_8$, $T_{10}$, $T_7$, $T_6$, $T_{11}$, $T_1$, $T_2$, $T_5$, $T_9$. The finishing time is 102 minutes.

| $P_1$ | $T_4$ | $T_5$ | $T_9$ | $T_{11}$ |
|---|---|---|---|---|
| $P_2$ | $T_1$ | $T_2$ | $T_3$ $T_6$ $T_8$ $T_{10}$ $T_7$ | |

Time (min)  0  5   15   30 33 38   50 55 60  68   90   102

**(b)** The critical path for this project is $T_9 \rightarrow T_{11}$. The critical time for this project is 72 minutes. This schedule may not be optimal.

**(c)** There are 34 minutes of idle time in the schedule.

35. **(a)** The increasing-time priority list is $T_3$, $T_4$, $T_8$, $T_{10}$, $T_7$, $T_6$, $T_{11}$, $T_1$, $T_2$, $T_5$, $T_9$. The finishing time is 77 minutes.

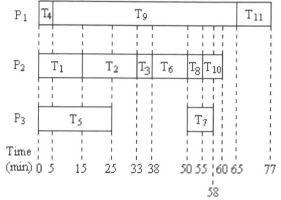

**(b)** The critical path for this project is $T_9 \rightarrow T_{11}$. The critical time for this project is 72 minutes. This schedule may not be optimal.

**(c)** There are 61 minutes of idle time in this schedule.

37. **(a)** The finishing time for one processor is 182 minutes. The critical time is 90 minutes.

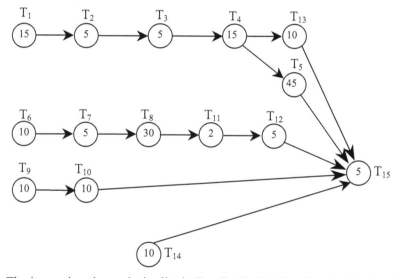

**(b)** The increasing-time priority list is $T_{11}$, $T_2$, $T_3$, $T_7$, $T_{12}$, $T_{15}$, $T_6$, $T_9$, $T_{10}$, $T_{13}$, $T_{14}$, $T_1$, $T_4$, $T_8$, $T_5$. It will take two processors 112 minutes to finish the project. Begin the project by 4:08 p.m. to be finished by 6 p.m.. This schedule may not be optimal.

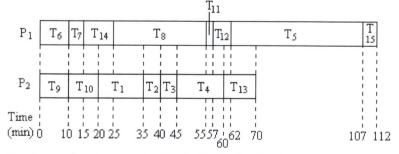

**(c)** The decreasing-time priority list is $T_5$, $T_8$, $T_4$, $T_1$, $T_{14}$, $T_{13}$, $T_{10}$, $T_9$, $T_6$, $T_{15}$, $T_{12}$, $T_7$, $T_3$, $T_2$, $T_{11}$. It will take two processors 105 minutes to finish the project. Begin the project at 4:15 p.m. to be finished by 6 p.m.. This schedule may not be optimal.

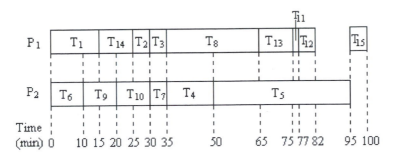

# Solutions to Odd-Numbered Problems in Section 7.3

1. (a) Of the nonisolated vertices, the sources are $T_1$ and $T_4$ and the sink is $T_5$. There are two maximal paths.

   **Maximal Path:**  **Weight:**
   $T_1 \to T_3 \to T_5$    $8 + 4 + 21 = 33$
   $T_4 \to T_3 \to T_5$    $16 + 4 + 21 = 41$

   (b) There are two isolated vertices. $T_2$ has a weight of 17. $T_6$ has a weight of 2.

   (c) From part (a), the largest weight of a maximal path is 41, so the critical time is 41 minutes.

   (d) There are four ready tasks: $T_1$, $T_2$, $T_4$, and $T_6$. $T_6$ would be scheduled first according to an increasing-time priority list because it is the ready task with the smallest weight. $T_2$ would be scheduled first according to a decreasing-time priority list because it has the largest weight. $T_4$ would be scheduled first according to a critical-path priority list because it is the first task in the critical path and is a ready task at the start of the project.

3. (a) Add the weight of each task. The finishing time for one processor is 104 minutes.

   (b) There are no isolated vertices. The sources are $T_1$, $T_2$, and $T_7$. The sink is $T_6$.

   **Maximal Path:**          **Weight:**
   $T_1 \to T_3 \to T_4 \to T_6$     $20 + 5 + 11 + 10 = 46$
   $T_1 \to T_3 \to T_5 \to T_6$     $20 + 5 + 6 + 10 = 41$
   $T_2 \to T_3 \to T_4 \to T_6$     $17 + 5 + 11 + 10 = 43$
   $T_2 \to T_3 \to T_5 \to T_6$     $17 + 5 + 6 + 10 = 38$
   $T_7 \to T_5 \to T_6$          $35 + 6 + 10 = 51$

   (c) The critical path is $T_7 \to T_5 \to T_6$, and the critical time is 51 minutes.

   (d) The critical time is the minimum time in which the project can be completed.

5. (a) The critical path is $T_2 \to T_5 \to T_4 \to T_6$, and the critical time is 66 minutes.

   (b) The finishing time is 66 minutes.

| $P_1$ | $T_2$ | | $T_5$ | $T_4$ | $T_6$ |
|---|---|---|---|---|---|
| $P_2$ | $T_1$ | $T_3$ | | | |
| Time (min) 0 | 15 | 22 25 | 40 | 61 | 66 |

   (c) The finishing time is 69 minutes.

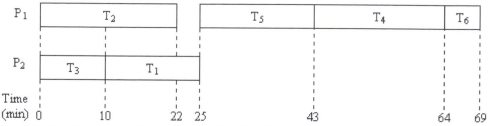

(d) The schedule from part (b) is optimal. The finishing time is the same as the critical time. Notice that the schedule from part (c) has idle time since $T_5$ must wait until $T_1$ is complete. It is not an optimal schedule.

7. (a) The critical path is the one with the weight of 66 minutes. $T_2$ is the first task in the critical path, so it will be placed first in the critical-path list.

(b) In the following table, $T_1$ heads the path with the greatest weight. $T_1$ will be placed next in the critical-path list.

| Maximal Path | Weight (Time in Minutes) |
|---|---|
| $T_1 \rightarrow T_5 \rightarrow T_4 \rightarrow T_6$ | $15 + 18 + 21 + 5 = 59$ |
| $T_3 \rightarrow T_4 \rightarrow T_6$ | $10 + 21 + 5 = 36$ |

(c) In the following table, $T_5$ heads the path with the greatest weight. $T_5$ will be placed next in the critical-path list.

| Maximal Path | Weight (Time in Minutes) |
|---|---|
| $T_5 \rightarrow T_4 \rightarrow T_6$ | $18 + 21 + 5 = 44$ |
| $T_3 \rightarrow T_4 \rightarrow T_6$ | $10 + 21 + 5 = 36$ |

(d) Once $T_5$ is removed, we see $T_3$ heads the only remaining path as shown in the following table, so $T_3$, $T_4$, and $T_6$ will be placed next in the critical-path list, in that order.

| Maximal Path | Weight (Time in Minutes) |
|---|---|
| $T_3 \rightarrow T_4 \rightarrow T_6$ | $10 + 21 + 5 = 36$ |

9. (a) The following table contains all of the maximal paths and their weights. There are no isolated vertices. The critical path is $T_2 \rightarrow T_3 \rightarrow T_4 \rightarrow T_6$. The critical time is 43 hours.

| Maximal Path | Weight (Time in Hours) |
|---|---|
| $T_2 \rightarrow T_1 \rightarrow T_4 \rightarrow T_6$ | $8 + 10 + 13 + 7 = 38$ |
| $T_2 \rightarrow T_3 \rightarrow T_4 \rightarrow T_6$ | $8 + 15 + 13 + 7 = 43$ |
| $T_5 \rightarrow T_3 \rightarrow T_4 \rightarrow T_6$ | $5 + 15 + 13 + 7 = 40$ |

(b) Apply the algorithm for forming the critical-path priority list.

**STEP 1:**

| Maximal Path | Weight (Time in Hours) |
|---|---|
| $T_2 \rightarrow T_1 \rightarrow T_4 \rightarrow T_6$ | $8 + 10 + 13 + 7 = 38$ |
| $T_2 \rightarrow T_3 \rightarrow T_4 \rightarrow T_6$ | $8 + 15 + 13 + 7 = 43$ |
| $T_5 \rightarrow T_3 \rightarrow T_4 \rightarrow T_6$ | $5 + 15 + 13 + 7 = 40$ |

**STEP 2 Case A:**

$T_2$ will be placed first in the critical-path priority list because it is at the head of the critical path.

**STEP 3:** Remove $T_2$ and all attached edges. Notice when $T_2$ is removed from $T_2 \to T_3 \to T_4 \to T_6$ the resulting path is part of a longer path, so it must be deleted from the table.

| Maximal Path | Weight (Time in Hours) |
|---|---|
| $T_1 \to T_4 \to T_6$ | $10 + 13 + 7 = 30$ |
| $T_5 \to T_3 \to T_4 \to T_6$ | $5 + 15 + 13 + 7 = 40$ |

**STEP 2 Case A:**
$T_5$ will be placed next in the critical-path priority list because it is at the head of the path with the greatest weight.

**STEP 3:** Remove $T_5$ and all attached edges.

| Maximal Path | Weight (Time in Hours) |
|---|---|
| $T_1 \to T_4 \to T_6$ | $10 + 13 + 7 = 30$ |
| $T_3 \to T_4 \to T_6$ | $15 + 13 + 7 = 35$ |

**STEP 2 Case A:**
$T_3$ will be placed next in the critical-path priority list because it is at the head of the path with the greatest weight.

**STEP 3:** Remove $T_3$ and all attached edges. Notice that when $T_3$ is removed from $T_3 \to T_4 \to T_6$ the resulting path is part of a longer path, so it must be deleted from the table.

| Maximal Path | Weight (Time in Hours) |
|---|---|
| $T_1 \to T_4 \to T_6$ | $10 + 13 + 7 = 30$ |

**STEP 2 Case A:**
$T_1$ will be placed next in the critical-path priority list because it is at the head of the path with the greatest weight followed by $T_4$ and $T_6$.

The critical-path priority list is $T_2, T_5, T_3, T_1, T_4, T_6$.

**(c)**

**(d)** The finishing time is the same as the critical time so this schedule is optimal.

**11. (a)** The following table contains all of the maximal paths and their weights. There are no isolated vertices. The critical path is $T_8 \to T_5 \to T_6$. The critical time is 51 minutes.

| Maximal Path | Weight (Time in Minutes) |
|---|---|
| $T_1 \to T_3 \to T_4 \to T_6$ | $22 + 5 + 13 + 10 = 50$ |
| $T_2 \to T_3 \to T_4 \to T_6$ | $18 + 5 + 13 + 10 = 46$ |
| $T_7 \to T_5 \to T_6$ | $9 + 14 + 10 = 33$ |
| $T_8 \to T_5 \to T_6$ | $27 + 14 + 10 = 51$ |

**(b)** Apply the algorithm for forming the critical-path priority list.
**STEP 1:** See part (a) for the list of all maximal paths and weights.
**STEP 2 Case A:** $T_8$ heads the critical path and should be placed first in the critical-path priority list.

**STEP 3:** Remove $T_8$ and all attached edges. Notice when $T_8$ is removed, the resulting path is part of a longer path, so it must be deleted from the table.

| Maximal Path | Weight (Time in Minutes) |
|---|---|
| $T_1 \rightarrow T_3 \rightarrow T_4 \rightarrow T_6$ | $22 + 5 + 13 + 10 = 50$ |
| $T_2 \rightarrow T_3 \rightarrow T_4 \rightarrow T_6$ | $18 + 5 + 13 + 10 = 46$ |
| $T_7 \rightarrow T_5 \rightarrow T_6$ | $9 + 14 + 10 = 33$ |

**STEP 2 Case A:**

$T_1$ will be placed next in the critical-path priority list because it is at the head of the path with the greatest weight.

**STEP 3:** Remove $T_1$ and all attached edges. Notice when $T_1$ is removed, the resulting path is part of a longer path, so it must be deleted from the table.

| Maximal Path | Weight (Time in Minutes) |
|---|---|
| $T_2 \rightarrow T_3 \rightarrow T_4 \rightarrow T_6$ | $18 + 5 + 13 + 10 = 46$ |
| $T_7 \rightarrow T_5 \rightarrow T_6$ | $9 + 14 + 10 = 33$ |

**STEP 2 Case A:**

$T_2$ will be placed next in the critical-path priority list because it is at the head of the path with the greatest weight.

**STEP 3:** Remove $T_2$ and all attached edges.

| Maximal Path | Weight (Time in Minutes) |
|---|---|
| $T_3 \rightarrow T_4 \rightarrow T_6$ | $5 + 13 + 10 = 28$ |
| $T_7 \rightarrow T_5 \rightarrow T_6$ | $9 + 14 + 10 = 33$ |

**STEP 2 Case A:**

$T_7$ will be placed next in the critical-path priority list because it is at the head of the path with the greatest weight.

**STEP 3:** Remove $T_7$ and all attached edges.

| Maximal Path | Weight (Time in Minutes) |
|---|---|
| $T_3 \rightarrow T_4 \rightarrow T_6$ | $5 + 13 + 10 = 28$ |
| $T_5 \rightarrow T_6$ | $14 + 10 = 24$ |

**STEP 2 Case A:**

$T_3$ will be placed next in the critical-path priority list because it is at the head of the path with the greatest weight.

**STEP 3:** Remove $T_3$ and all attached edges.

| Maximal Path | Weight (Time in Minutes) |
|---|---|
| $T_4 \rightarrow T_6$ | $13 + 10 = 23$ |
| $T_5 \rightarrow T_6$ | $14 + 10 = 24$ |

**STEP 2 Case A:**

$T_5$ will be placed next in the critical-path priority list because it is at the head of the path with the greatest weight.

**STEP 3:** Remove $T_5$ and all attached edges. Notice when $T_5$ is removed, the resulting vertex is part of a longer path, so it must be deleted from the table.

| Maximal Path | Weight (Time in Minutes) |
|---|---|
| $T_4 \rightarrow T_6$ | $13 + 10 = 23$ |

**STEP 2 Case A:**

$T_4$ will be placed next in the critical-path priority list because it is at the head of the only remaining path followed by $T_6$.

The critical-path priority list is $T_8$, $T_1$, $T_2$, $T_7$, $T_3$, $T_5$, $T_4$, $T_6$.

**(c)** It will take two processors 68 minutes to finish the project.

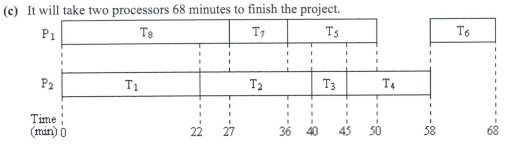

13. **(a)** Work through the steps of the algorithm for forming the critical-path priority list from problem 11 again, but this time, replace the weight of $T_4$ with 8. The critical-path priority list is $T_8$, $T_1$, $T_2$, $T_7$, $T_5$, $T_3$, $T_4$, $T_6$. However, the way in which the tasks are assigned to the processors will not change. $T_4$ is completed during idle time for $P_1$, so decreasing the weight of $T_4$ by 5 minutes only decreases the finishing time by 5 minutes.

   **(b)** Work through the steps of the algorithm for forming the critical-path priority list from problem 11 again, but this time, change the weight of $T_8$ from 27 to 20. The critical-path priority list changes in this case. The critical-path priority list is $T_1$, $T_2$, $T_8$, $T_7$, $T_3$, $T_5$, $T_4$, $T_6$. This time $P_1$ works on $T_1$, $T_7$, $T_3$, $T_4$, and $T_6$. The finishing time is 62 minutes. Notice that there was a 7-minute decrease in the weight of $T_8$ but the finishing time decreased only 6 minutes.

15. The schedules are the same in that the same tasks are assigned to the same processors in the same order. The delay in the first task in the critical path eliminates the idle time that used to be in the schedule from problem 11. The finishing time for the project is 70 minutes which is two minutes longer than the schedule from problem 11.

17. **(a)** The critical-path priority list is $T_1$, $T_3$, $T_2$, $T_6$, $T_5$, $T_4$, $T_7$, $T_8$.
    **(b)** The critical path is $T_1 \rightarrow T_3 \rightarrow T_6 \rightarrow T_7 \rightarrow T_8$. The critical time is 53 minutes.

19. $P_1$ will be assigned tasks 1, 3, 6, 7, and 8. $P_2$ will be assigned tasks 2, 4, and 5. There are 33 minutes of idle time in this schedule. The schedule may not be optimal. The critical time was 53 minutes, and two processors would finish the job in 62 minutes.

| $P_1$ | $T_1$ | $T_3$ | $T_6$ | | $T_7$ | $T_8$ |
|---|---|---|---|---|---|---|
| $P_2$ | | $T_2$ | $T_4$ | $T_5$ | | |

Time (min)  0   10   16   25   36  39   48   57   62

21. There are 68 minutes of idle time in the schedule. The schedule is optimal. The critical time for this project is 53 minutes.

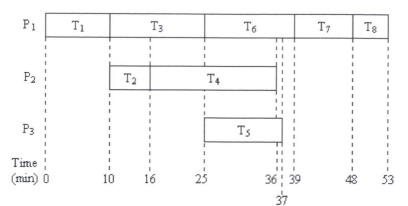

**23. (a)** Time is in minutes.

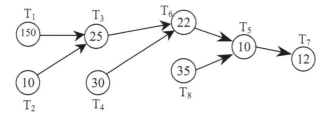

**(b)** The critical path is $T_1 \rightarrow T_3 \rightarrow T_6 \rightarrow T_5 \rightarrow T_7$. The critical time is 219 minutes.

**(c)** The critical-path priority list is $T_1, T_2, T_4, T_3, T_8, T_6, T_5, T_7$.

**(d)** The finishing time is 219 minutes.

| P₁ | | | | | | T₁ | | | T₃ | T₆ | T₅ | T₇ |

| P₂ | T₂ | T₄ | T₈ | | | | | | | | | |

Time (min)  0  10  40  75  150  175  197 207  219

**(e)** The schedule is optimal. The finishing time is the same as the critical time.

## Solutions to Chapter 7 Review Problems

**1. (a)** Answers may vary.
  T₁: Preheat oven to 350°
  T₂: Butter a loaf pan
  T₃: Mix flour, brown sugar, baking powder, and salt
  T₄: Add the egg, milk, and butter
  T₅: Stir
  T₆: Add nuts
  T₇: Spoon into pan
  T₈: Bake
  T₉: Remove from pan

**(b)**

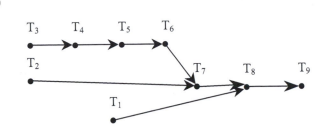

2. **(a)** Task $E$ may be completed at any time because it is an isolated vertex. Task $C$ and Task $G$ are both source vertices and must be completed before Task $M$, Task $R$, and Task $B$. Task $M$ must be completed after Task $C$. Task $R$ must be completed after Task $C$, Task $M$, and Task $G$ are completed. Task $B$ cannot be completed until Task $C$, Task $M$, Task $G$, and Task $R$ are completed.

   **(b)** The vertex for Task $E$ is isolated. There are no arrows entering it or leaving it.

   **(c)** There are 8 paths:

   | | |
   |---|---|
   | $T_C \rightarrow T_M$ | $T_C \rightarrow T_M \rightarrow T_R$ |
   | $T_M \rightarrow T_R$ | $T_G \rightarrow T_R \rightarrow T_B$ |
   | $T_G \rightarrow T_R$ | $T_M \rightarrow T_R \rightarrow T_B$ |
   | $T_R \rightarrow T_B$ | $T_C \rightarrow T_M \rightarrow T_R \rightarrow T_B$ |

3. **(a)** $T_1$: Remove blankets and sheets
   $T_2$: Put on clean fitted sheet
   $T_3$: Put on clean flat sheet
   $T_4$: Put on blanket
   $T_5$: Remove old pillow case
   $T_6$: Put on clean pillow case
   $T_7$: Put pillow on bed
   $T_8$: Put on bedspread

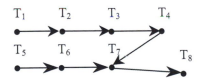

   **(b)** No vertices are isolated. Every task is part of a precedence relation.

4. **(a)** There are two sinks: $T_3$ and $T_8$.
   **(b)** There are two sources: $T_1$ and $T_4$.
   **(c)** There are four maximal paths:
   $T_1 \rightarrow T_2 \rightarrow T_5 \rightarrow T_7 \rightarrow T_8$
   $T_1 \rightarrow T_3$
   $T_4 \rightarrow T_5 \rightarrow T_7 \rightarrow T_8$
   $T_4 \rightarrow T_6 \rightarrow T_8$

5. There are six maximal paths:
   $T_1 \rightarrow T_3 \rightarrow T_5 \rightarrow T_7 \rightarrow T_8$
   $T_1 \rightarrow T_3 \rightarrow T_5 \rightarrow T_6$
   $T_2 \rightarrow T_3 \rightarrow T_5 \rightarrow T_7 \rightarrow T_8$
   $T_2 \rightarrow T_3 \rightarrow T_5 \rightarrow T_6$
   $T_4 \rightarrow T_5 \rightarrow T_7 \rightarrow T_8$
   $T_4 \rightarrow T_5 \rightarrow T_6$

**6.**

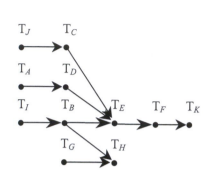

**7. (a)** The critical path is $T_5 \to T_6 \to T_7$. The earliest completion time, the critical time, is 30 minutes
   **(b)** The decreasing-time priority list is $T_6$, $T_5$, $T_7$, $T_3$, $T_4$, $T_2$, $T_1$. The finishing time is 30 minutes.

| $P_1$ | $T_5$ | | $T_6$ | | $T_7$ | |

```
P1 |  T5   |  T6   |    T7    |

P2 |T1| T2 |  T3   |  T4  |

Time
(min) 0   3    8 10   16   21 23    30
```

   **(c)** Task 3 is not part of the critical path so shortening its completion time will not affect the shortest completion time for the project. The finishing time is not affected either. The only difference is the amount of idle time in the schedule. Processor 2 ends up being idle for three more minutes compared to the schedule from part (b).

**8. (a)** The increasing-time priority list is $T_7$, $T_1$, $T_6$, $T_5$, $T_3$, $T_2$, $T_4$.
   **(b)** $T_1$ and $T_7$ are ready at the start of the project.
   **(c)**

```
P1 |T7|  T2  |  T3  |  T4  | T6 |

P2 | T1 |                 | T5 |

Time
(min) 0  2 3   12     20      30  34 35
```

**9. (a)** There are five maximal paths.

| Maximal Path: | Weight: |
|---|---|
| $T_1 \to T_2 \to T_7$ | 32 minutes |
| $T_9 \to T_{11}$ | 20 minutes |
| $T_8 \to T_{10} \to T_{11}$ | 62 minutes |
| $T_4 \to T_3 \to T_5 \to T_6 \to T_7$ | 126 minutes |
| $T_4 \to T_3 \to T_5 \to T_{11}$ | 76 minutes |

   **(b)** The finishing time for one processor is 201 minutes.
   **(c)** The critical path is $T_4 \to T_3 \to T_5 \to T_6 \to T_7$, and the critical time is 126 minutes.

**10. (a)** The increasing-time priority list is $T_8$, $T_9$, $T_2$, $T_1$, $T_4$, $T_{11}$, $T_7$, $T_3$, $T_5$, $T_{10}$, $T_6$.
   **(b)** One processor would complete the tasks in the following order: $T_8$, $T_9$, $T_1$, $T_2$, $T_4$, $T_3$, $T_5$, $T_{10}$, $T_{11}$, $T_6$, and $T_7$.

**11. (a)** The decreasing-time priority list is $T_6$, $T_{10}$, $T_5$, $T_3$, $T_7$, $T_{11}$, $T_4$, $T_1$, $T_2$, $T_9$, $T_8$.

**(b)**

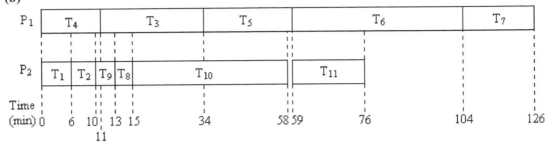

**(c)** The schedule from part (b) is optimal. From problem 9 (c) we know the critical time is 126 minutes. The finishing time is the same as the critical time.

**12. (a)** The critical-path priority list is $T_4$, $T_3$, $T_5$, $T_6$, $T_8$, $T_{10}$, $T_1$, $T_2$, $T_7$, $T_9$, $T_{11}$.

**(b)**

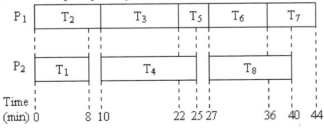

**(c)** The schedule from part (b) is optimal. From problem 9 (c) we know the critical time is 126 minutes. The finishing time is the same as the critical time.

**13. (a)** The decreasing-time priority list is $T_4$, $T_8$, $T_3$, $T_2$, $T_6$, $T_1$, $T_7$, $T_5$.

**(b)** The finishing time is 44 minutes. The critical path is $T_2 \rightarrow T_3 \rightarrow T_5 \rightarrow T_6 \rightarrow T_7$ and the critical time is 44 minutes. The schedule is optimal.

**14.** The critical-path priority list is $T_2$, $T_1$, $T_3$, $T_4$, $T_5$, $T_6$, $T_8$, $T_7$.

**15.** The critical path is $T_1 \rightarrow T_3 \rightarrow T_5$ and the critical time is 20 weeks. The critical-path priority list is $T_1$, $T_2$, $T_4$, $T_3$, $T_6$, $T_5$. Using the critical-path priority list, $P_1$ is assigned $T_1$, $T_3$, and $T_5$ while $P_2$ is assigned $T_2$, $T_4$, and $T_6$. The finishing time is 20 weeks. This schedule is optimal since the finishing time is the same as the critical time.

**16.** The critical path does not change even though the weight of $T_3$ is reduced from 9 weeks to 6 weeks. The critical path is $T_1 \rightarrow T_3 \rightarrow T_5$ and the critical time is 17 weeks. The critical-path priority list is $T_1$, $T_4$, $T_2$, $T_6$, $T_3$, $T_5$. Using the critical-path priority list, $P_1$ is assigned $T_1$, $T_6$, and $T_5$ while $P_2$ is assigned $T_4$, $T_2$, and $T_3$. The finishing time is 18 weeks. This schedule may not be optimal since the finishing time is longer than the critical time. Notice a 3-week reduction in the weight of a task did not reduce the finishing time by 3 weeks.

**17. (a)** The increasing-time priority list is $T_3$, $T_4$, $T_2$, $T_7$, $T_9$, $T_5$, $T_1$, $T_6$, $T_8$. The finishing time is 103 minutes.

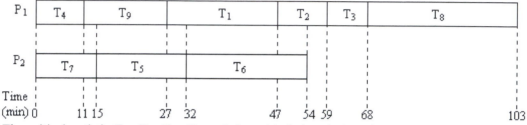

**(b)** The critical path is $T_4 \rightarrow T_5 \rightarrow T_6 \rightarrow T_8$ and the critical time is 85 minutes. The schedule is not optimal.

**(c)** There are 49 minutes of idle time in the schedule.

**18. (a)** The decreasing-time priority list is $T_8$, $T_6$, $T_1$, $T_5$, $T_9$, $T_7$, $T_2$, $T_4$, $T_3$. The finishing time is 116 minutes.

| $P_1$ | $T_1$ | $T_2$ | $T_3$ | | $T_5$ | $T_6$ | $T_8$ |
| $P_2$ | $T_9$ | $T_7$ | $T_4$ | | | | |

Time (min) 0    16 20    31 32    41 42    59    81    116

**(b)** The critical path is $T_4 \rightarrow T_5 \rightarrow T_6 \rightarrow T_8$ and the critical time is 85 minutes. The schedule may not be optimal.

**(c)** There are 75 minutes of idle time in the schedule.

**19. (a)** The critical-path priority list is $T_4$, $T_1$, $T_5$, $T_7$, $T_6$, $T_2$, $T_9$, $T_3$, $T_8$. The finishing time is 100 minutes.

| $P_1$ | $T_4$ | $T_5$ | $T_2$ | $T_9$ | $T_3$ | $T_8$ |
| $P_2$ | $T_1$ | | $T_7$ | $T_6$ | | |

Time (min) 0    11    20    28    35    40    56 57    65    100

**(b)** The critical path is $T_4 \rightarrow T_5 \rightarrow T_6 \rightarrow T_8$ and the critical time is 85 minutes. The schedule is not optimal.

**(c)** There are 43 minutes of idle time in the schedule.

**20. (a)** The completion time is 85 minutes.

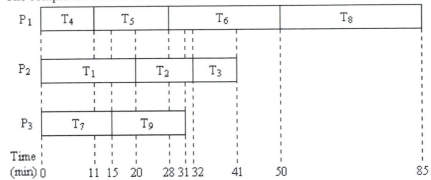

**(b)** The schedule is optimal. The finishing time is the same as the critical time.

**(c)** There are 98 minutes of idle time in the schedule.

# *Chapter 8:* Descriptive Statistics— Data and Patterns

## *Key Concepts*

**By the time you finish studying Section 8.1, you should be able to:**

☐ create a dot plot for a set of data and describe for what types of data dot plots are particularly useful.

☐ describe patterns or interesting features of data that is displayed in a dot plot.

☐ create a stem-and-leaf plot for a set of data and describe for what types of data stem-and-leaf plots are particularly useful.

☐ interpret a stem-and-leaf plot by identifying any clusters, gaps, and outliers present.

☐ create a frequency table and a histogram for a set of data and describe for what types of data histograms are particularly useful.

☐ describe the features of a histogram, in particular, explain what a measurement class is, what the height of the bars represents, and what the width of the bars represents.

☐ describe how to determine the relative frequency for data in a measurement class and create a relative frequency histogram.

☐ describe one drawback in using a histogram compared to a dot plot or a stem-and-leaf plot.

☐ create a bar graph for a set of data.

☐ summarize information presented in a bar graph.

☐ create a line graph for a set of data and describe for what types of data line graphs are particularly useful.

☐ summarize and interpret information presented in a line graph.

☐ create a pie chart for a set of data and describe for what types of data pie charts are particularly useful.

☐ summarize and interpret information presented in a pie chart.

**By the time you finish studying Section 8.2, you should be able to:**

☐ compare two data sets using a double stem-and-leaf plot.

☐ create a comparison histogram for two data sets.

☐ compare data sets using multiple-bar graphs.

☐ compare data sets using multiple-line graphs.

☐ compare data sets using multiple pie charts.

☐ create a proportional bar graph for a data set.

☐ determine which type of graph would be the best display of two data sets.

**By the time you finish studying Section 8.3, you should be able to:**

☐ create a bar graph to emphasize differences or to mislead by manipulating the vertical axis or reversing the orientation of an axis.

☐ identify misleading features of a bar graph.

☐ create a line graph to emphasize differences or to mislead by cropping.

☐ identify misleading features of a line graph.

☐ identify misleading features of three-dimensional graphs.

☐ create and identify misleading features of exploding pie charts.

☐ create and identify misleading features of pictographs.

☐ interpret important features and identify misleading aspects of graphical maps.

## *Skills Brush-Ups*

Problems in Section 8.3 involving misleading graphs will require that you be comfortable working with ratios and proportions.

### *RATIO:*

A ratio is an ordered pair of numbers. We will express the ratio of the number $a$ to the number $b$ as $a$ to $b$ or as $a:b$. A ratio may also be written as a fraction $\frac{a}{b}$ in which case $b \neq 0$. Ratios can give a part-to-part comparison or a part-to-whole comparison.

*Example:* Suppose there are 14 women in a class and 13 men. The ratio of women to men is 14 to 13, 14:13, or $\frac{14}{13}$. This ratio gives a part-to-part comparison because women make up part of the class and men make up part of the class. By knowing the ratio of women to men, we can express the fraction of women in the class as $\frac{14}{14+13} = \frac{14}{27}$, and the fraction of men in the class as $\frac{13}{14+13} = \frac{13}{27}$. Both of these ratios give a part-to-whole comparison because the number of women **or** men is compared to the total number of women **and** men.

## PROPORTION:

If two ratios $\frac{a}{b}$ and $\frac{c}{d}$ are equal, then the equation $\frac{a}{b} = \frac{c}{d}$ is a proportion. For proportion $\frac{a}{b} = \frac{c}{d}$, $a$ and $d$ are called the **extremes**. The numbers $b$ and $c$ are called the **means**. For any proportion, the product of the extremes is equal to the product of the means; that is, for proportion $\frac{a}{b} = \frac{c}{d}$, $ad = bc$. This is also called the cross product property. Sometimes a number in a proportion is unknown. In Section 8.3, you may be asked to solve proportions for an unknown value. Using the cross product property will allow you to solve a proportion for an unknown value.

*Example:* To solve the proportion $\frac{3.5}{27} = \frac{4.1}{y}$ for the unknown value $y$, use the cross product property.

$$\frac{3.5}{27} = \frac{4.1}{y}$$
$$3.5 \times y = 27 \times 4.1$$
$$y = \frac{27 \times 4.1}{3.5}$$
$$y = \frac{110.7}{3.5}$$
$$y \approx 31.629$$

*Example:* Suppose you sketch a dog on a piece of paper. The real dog is 2 feet tall, but the dog you sketched is 1.5 inches tall. You would like to sketch a child on the same piece of paper. If the child is 3.4 feet tall, then how tall must the sketch of the child be if you want actual heights and sketched heights to be proportional? To answer this question, let $y$ represent the height of the child in the sketch. Set up a proportion and solve for the unknown value.

$$\frac{\text{height of dog}}{\text{height of child}} = \frac{\text{height of dog sketch}}{\text{height of child sketch}}$$
$$\frac{2}{3.4} = \frac{1.5}{y}$$
$$2 \times y = 3.4 \times 1.5$$
$$y = \frac{3.4 \times 1.5}{2}$$
$$y = \frac{5.1}{2}$$
$$y = 2.55$$

Therefore, the sketch of the child should be 2.55 inches tall.

## Skills Practice

1. A bag contains 27 red marbles and 15 blue marbles.
   (a) What is the red to blue marble ratio?
   (b) What is the blue to red marble ratio?
   (c) What fraction of the bag of marbles is red?
   (d) What fraction of the bag of marbles is blue?

2. Solve the proportion for the unknown value. Round answers to the nearest thousandth.

   (a) $\dfrac{y}{11} = \dfrac{2.5}{15}$  (b) $\dfrac{2.31}{75} = \dfrac{4.22}{x}$  (c) $\dfrac{12.3}{a} = \dfrac{25}{5.1}$  (d) $\dfrac{1.78}{85} = \dfrac{b}{153}$

3. Set up a proportion and solve the proportion for the unknown value.
   (a) A woman who is 5.5 feet tall has a foot that is 9 inches long. How long should the foot of a child be if the child is 3.2 feet tall and foot length is proportional to height?
   (b) A bar used in a graph is 0.58 centimeters long and represents the 63% of people who use a cellular phone daily. How long should the bar be that represents the 31% of people who use a cellular phone weekly if bar lengths are proportional to the percentages they represent?

*Answers to Skills Practice:*

1. (a) 27 to 15 or 27:15 (b) 15 to 27 or 15:27 (c) $\dfrac{27}{42}$  (d) $\dfrac{15}{42}$

2. (a) 1.833  (b) 137.013  (c) 2.509  (d) 3.204

3. (a) $\dfrac{5.5}{3.2} = \dfrac{9}{y}$, y ≈ 5.2 inches  (b) $\dfrac{0.58}{y} = \dfrac{63}{31}$, y ≈ 0.29 centimeters

## Hints for Odd-Numbered Problems

### Section 8.1

1. (c) Count the number of dots in the dot plot.

3. (a) Count the leaves in the stem-and-leaf plot.
   (b) The stems represent the ten's digit of the number of medals earned by a country. The leaves represent the digit in the one's place.

5. (a) Use stems that show how many thousands of dollars have been earned. A leaf of 2 would stand for $200.

7. Every one of the batting champions had an average of at least 0.3, so using a stem of 0.3 would not reveal any useful information. Try using stems in tenths and hundredths and leaves in thousandths.

9. (a) The intervals in the frequency table should not overlap.

(d) How many students earned at least $9.00 per hour? Divide the number of students who earned at least $9.00 per hour by the total number of students who worked during the summer and then multiply by 100%.

11. See Example 8.2.

13. The maximum score on the exam is 80 points. The bin size is 10, so the first bin should include scores from 0 to 9.

15. Determine the total number of men who ran the New York Marathon in 2003.
   (c) Divide the number of men in each interval by the total number of men who ran the race.

17. (c) A score greater than zero is a score above the national average.

**19. (c)** Consider Figures 8.3, 8.4, and 8.5 from the section and the discussion about bin size.

**21. (d)** Compare consecutive bars (decades) in the graph. For which pair of bars is the change in height the greatest?

**23. (c)** Divide the average number of children per minivan by the average number of children per SUV and by the average number of children per subcompact sports car.

**25. (b)** The total world population is not given.

**27.** The vertical scale should represent the cost per person in dollars.

**29. (b)** Look for increases compared to the previous year.
   **(c)** Look for decreases compared to the previous year.

**31. (b)** Use your estimates from part (a). For the percentage of children vaccinated each year beginning in 1998, subtract the percentage of children vaccinated in the previous year.

**33. (a)** Label the vertical axis in thousands.

**35. (a)** Percentages are given for each source. Convert each percentage to a decimal and multiply by the total of all state and local taxes in 2002.
   **(c)** Percentages are given for each source. Convert each percentage to a decimal and multiply by 360°.

**37. (a)** For each category, find $\dfrac{\text{amount spent}}{\$100}$ and convert to a percent.
   **(b)** Use percentages found in part (a). Convert each percentage to a decimal and multiply by 360°.

**39.** Use the percentages given to calculate the degree measure of each corresponding sector of the pie chart.

**41.** Consider each type of graph and the uses in Table 8.7.

**43.** Consider each type of graph and the uses in Table 8.7.

**45.** Choose a graph that would show a trend over time.

## Section 8.2

**1. (a)** Count the leaves for each class.
   **(c)** Consider such features as the high and low scores, the concentration of the data sets, and gaps.

**3. (a)** Use years as stems and months as leaves.

**5.** Use the digits in the ten's places as the stems and digits in the one's places as the leaves.

**7. (b)** Add up the frequencies represented by the height of each bar.
   **(c)** Find the total number of Dallas offense players who weigh at least 250 pounds and divide by the total number of Dallas offense players.

**9.** Use bins of length 10 years.

**11. (b)** Subtract bar heights, percentages, for each category.

**13. (a)** Label the vertical axis with percentages.

**15. (b)** For each year from 1990 to 2000 calculate (salary for a graduate with a bachelor's degree) − (salary for a graduate with a master's degree).

**17. (a)** Label the vertical axis with the CPI values.

**19. (a)** Consider the overall trend and any significant increases or decreases.

**21. (c)** Look for two consecutive years during which the greatest increase occurred.
   **(d)** Look for two consecutive years during which the greatest decrease occurred.

23. (b) Identify time periods of increasing or decreasing trends and comment on how the line representing males compares to that of females.

25. (d) When did SUVs become popular?

27. (a) Label the horizontal axis with years and the vertical axis with income.

29. (a) Could you create a pie chart to represent the percentages of children living with a single mother? Consider the percentages at the top of each bar and what they represent.

31. (a) Multiply each percent by the total number of operating school districts.
    (b) Multiply each percent by the total number of students enrolled.

33. Notice how similar the percentages are in each category of each pie chart.

35. For which region(s) were there increases? For which were there decreases?

37. (d) Consider consecutive years only.

39. (d) Consider the important features you noticed in part (c). Which graph demonstrates these best?

41. (a) Notice that Tennessee's increase also includes a voter-approved increase in taxes.

43. Compare income levels as well as the trend over time.

## Section 8.3

1. What do the heights of the bars represent?

3. (b) Which flavor stands out the most?

5. (c) Set up at least two proportions.

7. (a) Do not stretch or shrink the dollar.
   (c) Calculate the area of the original dollar bill. If the length is twice the width then $l = 2w$.

9. (b) For each missing height, set up and solve a proportion.

11. (a) For each bar, divide the percent by the number of computer monitors in the bar.
    (b) Determine whether bar heights are proportional to the percentages they represent. Also consider other features of the graph.

13. Notice that the number of people 65 years old and over are represented as well as the percentage of the population that are 65 years old and over.

15. (b) Set up a proportion and solve for the unknown height.

17. (a) Multiply each percentage by 360°.

19. (b) Study the vertical scale. Notice 0 is at the top of the graph.

21. (a) Expand the vertical scale beyond what is needed to display the data.
    (b) Crop the graph so the vertical scale extends just beyond the largest rate in the table.

23. What feature of the egg is being used to represent the dollars distributed?

25. (b) Notice the legend.
    (c) Is there a scale for the thickness?
    (f) Let each bar in the proportional bar graph represent one of the four categories of drug costs.

27. (b) Compare the ratio of land area support for each candidate and the ratio of electoral vote support for each candidate. Are they the same?

## Solutions to Odd-Numbered Problems in Section 8.1

**1.** (a) The horizontal axis represents the midterm exam scores for students in the biology class.
   (b) The vertical axis represents the number of students who received each exam score.
   (c) Count the frequencies for each score. There were 37 scores.
   (d) The tallest column of dots is centered over the score of 80.
   (e) The high score was 100, and the low score was 20.

**3.** (a) There are 25 leaves in the stem-and-leaf plot. Therefore, 25 countries earned medals.
   (b) There were a total of $1 + 1 + 2 + 2 + 2 + 2 + 2 + 3 + 3 + 3 + 4 + 4 + 6 + 7 + 8 + 8 + 11 + 11 + 12 + 15 + 17 + 17 + 24 + 34 + 35 = 234$ medals awarded.
   (c) The largest stem is 3, and the largest leaf for that stem is 5, so the largest number of medals is 35.

**5.** (a) The stems are thousands of dollars, and the leaves are hundreds.

| | |
|---|---|
| 35 | 4 |
| 34 | |
| 33 | |
| 32 | |
| 31 | 0 |
| 30 | 2 3 |
| 29 | 4 5 5 |
| 28 | 2 2 8 |
| 27 | 1 8 8 |
| 26 | 8 8 |
| 25 | 6 |

   (b) Most of the data values cluster between $25,600 and $31,000.
   (c) There is a large gap between $31,000 and $35,400. The $35,400 salary appears to be an outlier.

**7.** The batting averages are concentrated between 0.326 and 0.372. The only outlier seems to be the 0.390. Therefore, you can expect that the American League batting average champion will have a batting average in the 0.326 to 0.372 range.

| | |
|---|---|
| 0.39 | 0 |
| 0.38 | |
| 0.37 | 2 |
| 0.36 | 1 3 3 6 8 |
| 0.35 | 0 6 7 7 8 9 |
| 0.34 | 1 3 3 7 9 |
| 0.33 | 2 3 3 6 9 9 |
| 0.32 | 6 8 |

**9.** (a)

| Interval (Dollars) | Frequency |
|---|---|
| 7.00 - 7.49 | 2 |
| 7.50 - 7.99 | 5 |
| 8.00 - 8.49 | 6 |
| 8.50 - 8.99 | 4 |
| 9.00 - 9.49 | 4 |
| 9.50 - 9.99 | 5 |
| 10.00 - 10.49 | 3 |
| 10.50 - 10.99 | 1 |

   (b) Add the frequencies. A total of 30 students worked during the summer.

**(c)** The most frequent hourly pay range was $8.00 to $8.49. The least frequent hourly pay range was $10.50 to $10.99.

**(d)** A total of $4 + 5 + 3 + 1 = 13$ students earned at least $9.00 per hour. Therefore, $\dfrac{13}{30} \approx 0.4333$ or approximately 43.33% earned at least $9.00 per hour.

**(e)** A total of $6 + 4 + 4 + 5 = 19$ students earned between $8.00 and $10.00 per hour. Therefore, $\dfrac{19}{30} \approx 0.6333$ or approximately 63.33% earned between $8.00 and $10.00 per hour.

**11. (a)** 10 bins will be needed.

| Interval (Hours) | Frequency |
|---|---|
| 0 - 2.9 | 2 |
| 3 - 5.9 | 0 |
| 6 - 8.9 | 0 |
| 9 -11.9 | 1 |
| 12 - 14.9 | 1 |
| 15 - 17.9 | 6 |
| 18 - 20.9 | 0 |
| 21 - 23.9 | 0 |
| 24 - 26.9 | 8 |
| 27 - 29.9 | 3 |

**(b)** 6 bins will be needed.

| Interval (Hours) | Frequency |
|---|---|
| 0 - 4.9 | 2 |
| 5 - 9.9 | 1 |
| 10 - 14.9 | 1 |
| 15 - 19.9 | 6 |
| 20 - 24.9 | 3 |
| 25 - 29.9 | 8 |

**(c)** 3 bins will be needed.

| Interval (Hours) | Frequency |
|---|---|
| 0 - 9.9 | 3 |
| 10 - 19.9 | 7 |
| 20 - 29.9 | 11 |

**(d)** Answers will vary. The best display of the data will be with 10 bins. This histogram shows that there are three groups of students: those who watch less than 3 hours of television per week, those who watch at least 9 but less than 18 hours per week, and those who watch more than 24 hours a week. The 6-hour gaps are not visible in the other two histograms.

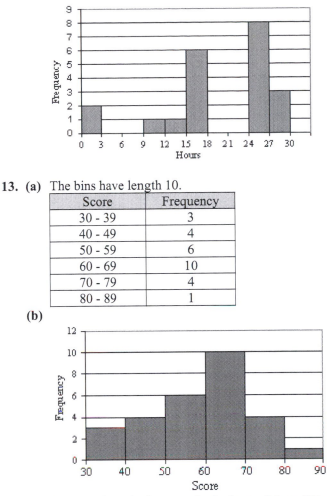

13. (a) The bins have length 10.

| Score | Frequency |
|-------|-----------|
| 30 - 39 | 3 |
| 40 - 49 | 4 |
| 50 - 59 | 6 |
| 60 - 69 | 10 |
| 70 - 79 | 4 |
| 80 - 89 | 1 |

(b)

(c) More students had a score in the interval from 60 to 69 than in any other interval.

(d) Answers will vary. The bin length may not be appropriate. There are many scores in the interval from 60 to 69. By looking at the histogram, it is impossible to tell whether they are closer to 60 or closer to 69 in general. A smaller bin size might be more appropriate.

15. (a) Divide the number of male runners who were between the ages of 30 and 39 by the total number of male runners. There were 8483 men who were between the ages of 30 and 39, and 22,948 men ran the race in 2003, so $\frac{8483}{22,948} \approx 0.3697$ or approximately 37% of male runners were between the ages of 30 and 39.

(b) Divide the number of male runners who were age 60 or older by the total number of male runners. There were $856 + 117 + 3 + 1 = 977$ men who were age 60 or older, and 22,948 men ran the race in 2003, so $\frac{977}{22,948} \approx 0.0426$ or approximately 4.3% of male runners were age 60 or older.

(d) Minivans had $\dfrac{365}{609} \approx 0.60$ the child fatalities compared to SUVs yet carried twice as many children. Minivans had $\dfrac{365}{494} \approx 0.74$, or about three-fourths, the child fatalities compared to subcompact sports cars. Minivans appear to be a safer option than SUVs.

**25. (a)**

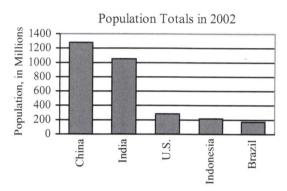

(b) Pie charts are used to show relative quantities. These data are given as population totals rather than percentages and the total world's population is not given, so percentages cannot be determined. A pie chart would not be used in this case.

**27.**

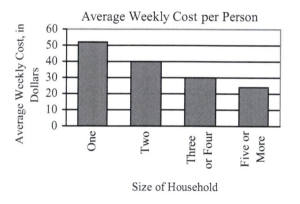

**29. (a)** Estimates will vary.
Estimates for the years 1994 through 2003 are 58%, 60%, 61%, 62%, 68%, 67%, 71%, 74%, 76%, and 79% respectively.
**(b)** During 1997 the greatest increase occurred.
**(c)** In the years 1998 and 1999 there were decreases in seat belt use.
**(d)** In 1994, approximately 58% of the population used seat belts while approximately 79% used seat belts in 2003. The change in seat belt use from 1994 to 2003 is 79% − 58% = 21%.
**(e)** In the year 2001, approximately 75% of people used their seat belts.

**31. (a)** Estimates will vary. The percentage of children vaccinated from the year 1997 to the year 2002 are 25%, 44%, 58%, 69%, 76%, and 80% respectively.
**(b)** Use the estimates found in part (a) and subtract estimates for consecutive years. From 1997 to 1998, the percentage of children vaccinated increased by 44% − 25% = 19%. From 1998 to 1999, there was an increase of 58% − 44% = 14%. From 1999 to 2000, there was an increase of 69% −

58% = 11%. From 2000 to 2001, there was an increase of 76% − 69% = 7%. From 2001 to 2002, there was an increase of 80% − 76% = 4%. Each year, there is an increase, but the increases are decreasing as time passes.

(c) Currently, the percentage of children vaccinated each year is increasing. The percentage cannot increase indefinitely, and a 100% vaccination rate is not likely. Notice from part (b) that the percent increase is decreasing, so the percentage of children vaccinated each year appears to be leveling off.

**33. (a)**

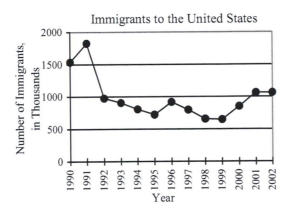

(b) The greatest increase appears to have taken place in 1991. The greatest decrease occurred in 1992.
(c) Significant increases occurred in 1991, 1996, 2000, and 2001. A significant decrease occurred in 1992.
(d) There seems to be a general pattern of several years of decreasing immigrant totals followed by a surge in immigrants coming to the United States.

**35. (a)** For each source, convert the given percentage to a decimal and multiply by the total tax.
Corporate income tax: 0.018 × $4,277,900,000 = $77,002,200
Sales tax: 0.195 × $4,277,900,000 = $834,190,500
Other state taxes: 0.159 × $4,277,900,000 =$680,186,100
Property taxes: 0.337 × $4,277,900,000 = $1,441,652,300
Other local taxes: 0.039 × $4,277,900,000 = $166,838,100
Individual income taxes: 0.251 × $4,277,900,000 =$1,073,752,900

**(b)** Property taxes and individual income taxes are the two sources that contributed the most. Corporate income taxes contributed the least.

**(c)** For each source convert the given percentage to a decimal and multiply by 360°. Round each answer to the nearest tenth of a degree.
Corporate income tax: 0.018 × 360° ≈ 6.5°
Sales tax: 0.195 × 360° = 70.2°
Other state taxes: 0.159 × 360° ≈ 57.2°
Property taxes: 0.337 × 360° ≈ 121.3°
Other local taxes: 0.039 × 360° ≈ 14°
Individual income taxes: 0.251 × 360° ≈ 90.4°

**37. (a)** For each source, divide the amount spent by $100 and multiply by 100%.

Perishables: $\dfrac{\$50.42}{\$100} \times 100\% = 50.42\%$

Beverages: $\dfrac{\$10.71}{\$100} \times 100\% = 10.71\%$

Misc. grocery: $\dfrac{\$5.34}{\$100} \times 100\% = 5.34\%$

Non-food grocery: $\dfrac{\$9.03}{\$100} \times 100\% = 9.03\%$

Snack foods: $\dfrac{\$6.25}{\$100} \times 100\% = 6.25\%$

Main meal items: $\dfrac{\$8.25}{\$100} \times 100\% = 8.25\%$

Health & beauty care: $\dfrac{\$3.96}{\$100} \times 100\% = 3.96\%$

General merchandise: $\dfrac{\$3.39}{\$100} \times 100\% = 3.39\%$

Pharmacy/unclassified: $\dfrac{\$2.65}{\$100} \times 100\% = 2.65\%$

**(b)** For each source convert the percentage from part (a) to a decimal and multiply by 360°. Round each answer to the nearest tenth of a degree.
Perishables: $0.5042 \times 360° \approx 181.5°$
Beverages: $0.1071 \times 360° \approx 38.6°$
Misc. grocery: $0.0534 \times 360° \approx 19.2°$
Non-food grocery: $0.0903 \times 360° \approx 32.5°$
Snack foods: $0.0625 \times 360° = 22.5°$
Main meal items: $0.0825 \times 360° = 29.7°$
Health & beauty care: $0.0396 \times 360° \approx 14.3°$
General merchandise: $0.0339 \times 360° \approx 12.2°$
Pharmacy/unclassified: $0.0265 \times 360° \approx 9.5°$

**(c)** Percents have been rounded to the nearest 1%.

**How $100 is Spent**

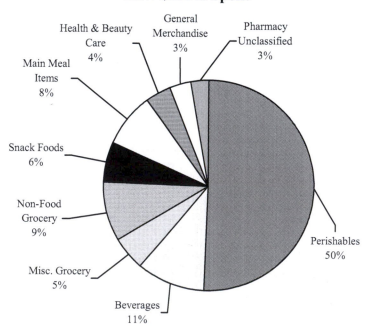

**39.**

**Reasons for Being Fired**

Incompetence 39%

Other 8%

Fail to Follow Instructions 7%

Lack of Motivation 7%

Negative Attitude 10%

Dishonesty or Lying 12%

Inability to Get Along with Others 17%

**41. (a)** No, a histogram would not be an appropriate display for this data because there are no measurement classes.

**(b)** A bar graph would be an appropriate display for this data because the heights of the bars will represent the frequency that each category was selected.

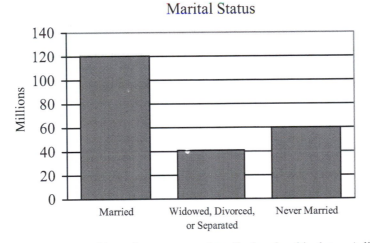

**Marital Status**

**(c)** A line chart would not be an appropriate display for this data. A line chart is useful for showing trends over time.

**(d)** A pie chart would be an appropriate display for the data. Each person age 15 and older falls into one of the categories so relative amounts can be determined.

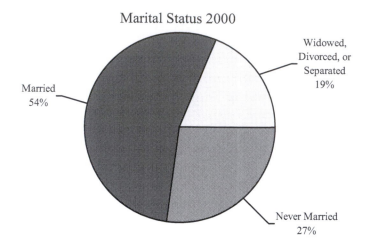

Marital Status 2000

43. **(a)** No, a histogram would not be an appropriate display for this data because there are no measurement classes.
   **(b)** A bar graph would be an appropriate display for this data. Bar graphs can be used to display trends over time.

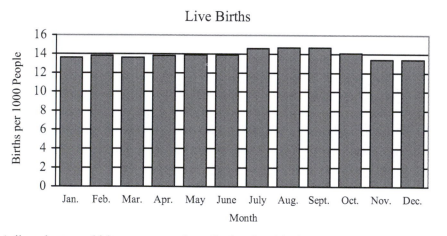

Live Births

   **(c)** A line chart would be an appropriate display for this data because we may want to see how the number of live births per 1000 people changes over time.

Live Births

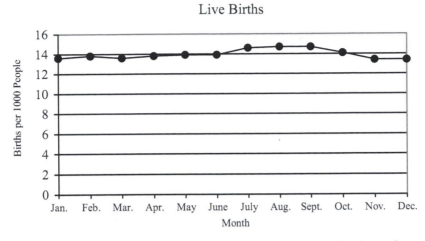

(d) A pie chart would not be an appropriate display for this data. Pie charts do not show time trends.

45. Answers may vary. A line chart would convey this information in a visual way most effectively.

Victims of Crime per 1000 People
Age 12 or Over

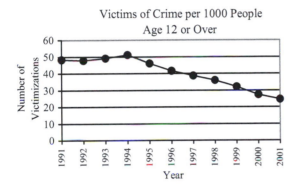

The number of crime victims per 1000 age 12 or over stayed relatively steady from 1991 through 1994 and peaked at 51.2 per thousand in 1994. It then began a steady decline through 2001 where it reached a low of 24.7 per thousand.

## Solutions to Odd-Numbered Problems in Section 8.2

1. (a) Count the leaves for each class. There are 26 students in the morning class and 21 students in the afternoon class.

   (b) The stems represent tens while the leaves represent ones. The morning class' high score was 96 and the low score was 19. The afternoon class' high score was 100 and the low score was 22.

   (c) Answers will vary. The afternoon class did much better on the exam. There were two scores of 100 in the afternoon class and no scores of 100 in the morning class. Half of the morning class scored below 70 on the exam. Only 3 out of 21 students from the afternoon class, or approximately 14%, scored below 70 on the exam.

**3. (a)**

| First Class | | Second Class |
|---:|:---:|:---|
| | 8 | 1 |
| | 7 | 2 5 |
| 9 2 1 | 6 | 2 3 8 |
| 9 8 8 7 6 6 4 4 3 3 3 2 2 1 | 5 | 0 0 4 5 6 6 |
| 9 9 3 | 4 | 1 4 7 7 7 9 |
| | 3 | 6 7 |

**(b)** Answers will vary. The classes are similar in that most of the scores are between 4.1 and 6.9, but the second class had more high scores and more low scores. The second class has more variability in the scores.

**(c)** There are no large gaps in the scores for either class. There are no outliers. It is striking that 14 out of 20 or 70% of the first class is reading at the fifth-grade level.

**5.**

| Babe Ruth | | Mickey Mantle |
|---:|:---:|:---|
| 0 | 6 | |
| 9 4 4 | 5 | 2 4 |
| 9 7 6 6 6 1 1 | 4 | 0 2 |
| 5 4 | 3 | 0 1 4 5 7 |
| 5 2 | 2 | 1 2 3 3 7 |
| | 1 | 3 5 8 9 |

Babe Ruth was a better home-run hitter than Mickey Mantle. In 11 of 15 years, Ruth hit more than 40 home runs per year. In 14 of 18 years, Mantle hit fewer than 40 home runs per year.

**7. (a)** The measurement classes have length 25 pounds.

**(b)** Dallas has 29 offense players and Minnesota has 26 offense players.

**(c)** There are $4 + 1 + 7 + 3 + 1 = 16$ Dallas offensive players who weigh at least 250 pounds. Therefore, $\frac{16}{29}$ or approximately 55% of the Dallas offensive players weigh at least 250 pounds.

**(d)** There are $5 + 1 + 4 + 3 = 13$ Minnesota offensive players who weigh at least 250 pounds. Therefore, $\frac{13}{26}$ or 50% of the Minnesota offensive players weigh at least 250 pounds.

**(e)** The average weight of the offensive players for Dallas is greater than the average weight of the offensive players for Minnesota.

9. (a)

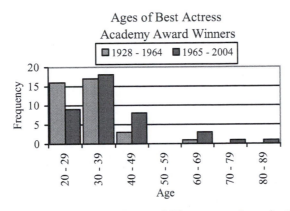

Ages of Best Actress
Academy Award Winners

- 1928 - 1964   ■ 1965 - 2004

(b) From 1928 to 1964, 4 out of 37 or approximately 11% of actresses who won an Oscar for best actress in a leading role were at least 40 at the time of the award. From 1965 to 2004, 13 out of 40 or 32.5% of actresses who won an Oscar for best actress in a leading role were at least 40 at the time of the award.

(c) In the past 40 years, the winners of the best actress Oscar tended to be older than they were more than 40 years ago.

11. (a) The percentage of households that cook at least once a day dropped in every category from 1993 to 2001. In one-person households it dropped about 4%, in two-people households it dropped 8%, in three-people households it dropped 9%, in four-people households it dropped 7%, in five-people households it dropped 8%, and in households with 6 or more people it dropped 7%.

(b) Households having 3 people experienced the greatest decrease in the percentage of households that cook at least once a day.

13. (a)

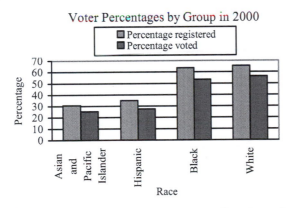

Voter Percentages by Group in 2000

- Percentage registered
- Percentage voted

(b) Whites had the highest percentage of registered voters who voted, while Asian and Pacific Islanders had the lowest percentage registered.

(c) When you divide the percent who voted by the percentage who were registered, you get the percentage of those who were registered that actually voted. If we do this, the percentage of Asian and Pacific Islanders is about 83%, the percentage of Hispanics is about 79%, the percentage of Blacks is about 84% and for Whites it is 86%. So, of those who could vote Whites had the highest turnout and Hispanics had the lowest.

15. **(a)** Estimates will vary.

The median salary for chemistry graduates with a bachelor's degree in 1990 was about $22,000. In 1997 it was about $30,000, and in 2000, it was about $35,000. From 1990 to 1996 the median salary for a chemistry graduate with a bachelor's degree did not increase very much.

**(b)** Estimates will vary.

The median salary for chemistry graduates with a master's degree in 1990 was about $31,000; in 1995, about $38,000; and in 2000, about $43,000. The approximate differences between the salaries for graduates with a bachelor's degree and the salaries for graduates with a master's degree from 1990 through 2000 were − $9000; −$10,000; −$8000; −$10,000; −$6000; −$15,000; −$10,000; −$10,000; −$10,000; −$11,000; and −$9000 respectively. The difference appears to stay steady except for 1994 through 1996 when the difference fluctuated considerably.

**(c)** Estimates will vary.

The median salary for chemistry graduates with a doctoral degree in 1990 was about $43,000; in 1997, about $55,000; and in 2000, about $64,000. Salaries for graduates with a doctoral degree increased from 1991 to 1993, 1995, and from 1997 to 2000.

17. **(a)**

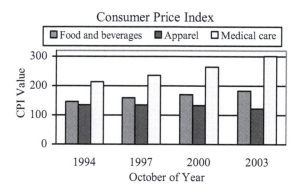

**(b)** The largest change in the CPI for Food and Beverages occurred from October 1994 to October 1997 with an increase of 13.1. The other items did not experience their largest changes during the same 3-year period.

**(c)** The CPI for medical care increased in each three year period. From 1994 to 1997 it increased 21.8; from 1997 to 2000 it increased 27.9; and from 2000 to 2003 it increased 36.8. So the value not only increased, but at an increasing rate.

**(d)** The CPI for apparel decreased by 0.3 from 1994 to 1997. It decreased by 2.1 from 1997 to 2000, and it decreased by 11.3 from 2000 to 2003. So it not only decreased, but decreased more rapidly each 3-year period.

19. **(a)** The percentage of men who smoke decreased during each of the 5-year periods. The general trend for women was similar, only during one of the 5-year periods did the percentage increase.

**(b)** Estimates will vary.

| Year | Percentage of Men Who Smoked | Percentage of Women Who Smoked |
|------|------------------------------|--------------------------------|
| 1965 | 52 | 34 |
| 1970 | 44 | 31 |
| 1975 | 43 | 32 |
| 1980 | 38 | 30 |
| 1985 | 32 | 29 |
| 1990 | 29 | 23 |
| 1995 | 28 | 23 |
| 2000 | 26 | 21 |

(c) The 5-year period from 1965 to 1970 showed the greatest decrease in the percentage of men who smoked. The 5-year period from 1985 to 1990 showed the greatest decrease in the percentage of women who smoked. The percentage of men who smoke has consistently been greater than the percentage of women who smoke, but the difference has become smaller.

**21. (a)** From 1990 through 2001 the number of CDs sold increased every year except two while the number of cassettes sold declined every year except two.

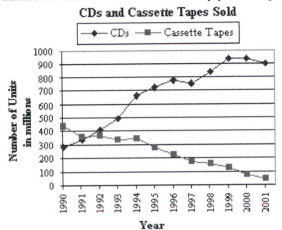

(b) Between 1991 and 1992 the number of CDs and cassettes sold were about the same.

(c) Between 1993 and 1994, CD total sales increased the most.

(d) Between 1990 to 1991 cassette tape sales decreased the most.

**23. (a)**

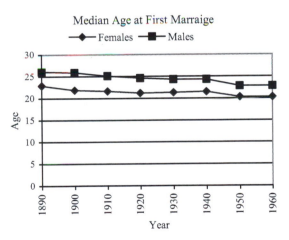

(b) The median age at first marriage for females declined gradually from 1890 through 1920. It rose slightly during the 1930s and 1940s and then dropped sharply in 1950 and remained steady through 1960. The median age at first marriage for males declined gradually from 1890 through 1930. It stayed steady from 1930 to 1940 and then dropped in 1950 and remained steady to 1960. The median ages when males and females marry has consistently stayed 2 to 4 years apart during the 70-year time period.

**25. (a)** Estimates will vary.

| Type | 1991 | 1994 | 1997 | 2000 |
|------|------|------|------|------|
| Pickup Trucks | 2550 | 2500 | 2550 | 2600 |
| SUVs | 850 | 1100 | 1475 | 2000 |
| Vans | 490 | 550 | 700 | 750 |

**(b)** Estimates will vary. For SUVs, 1200. For pickup trucks, 400. For vans, 250.

**(c)** For SUVs, the number of fatal rollover crashes increased steadily every year after 1992, and over the 10-year period, SUVs experienced a large increase in the number of these types of crashes. For vans, there were slight increases each year until 1999 when there was a decrease. For pickup trucks, there are periods of decrease followed by periods of increase, however, over the 10-year period, the numbers remain fairly stable.

**(d)** One other reason for the increase in fatal rollover crashes for SUVs would be their popularity. The number of SUVs sold each year increased rapidly during the 1990s. More of these types of vehicles on the road would lead to an increase in the number of these types of crashes.

**27. (a)** Notice that the line graphs for Alabama and Oklahoma are almost identical and therefore are hard to differentiate on the graph.

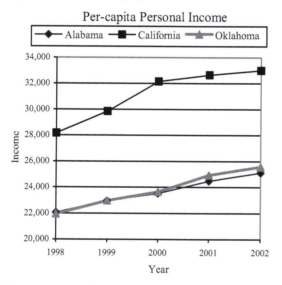

**(b)** The percentage increase in per-capita personal income from 1998 to 2002 for Alabama was approximately 14%, for California about 17%, and for Oklahoma about 16.5%. Per-capita income in California had the largest percentage increase and Alabama had the smallest.

**(c)** 2000, 2001, and 2002

**(d)** For Oklahoma, the percentage increase from 1999 to 2000 was 3%, from 2000 to 2001 was 5.5%, and from 2001 to 2002 it was 2.5%. For Alabama the percentage increase from 1999 to 2000 was 2.4%, from 2000 to 2001 was 4.1%, and from 2001 to 2002 it was 2.7%. For both states the largest percent increase was from 2000 to 2001. Oklahoma had the larger increases from 1999 to 2000 and from 2000 to 2001, but Alabama had a slightly larger increase from 2001 to 2002. In general, it appears that Alabama tends to have smaller increases, so per-capita income for Alabama residents may not exceed that of Oklahoma in the near future.

**29. (a)** No, you cannot redraw using pie charts. The percentages do not add to 100% because they are listed by ethnic group, not percentages from the same population. Notice, for example, in the bar graph representing children living with a single mother, 23% of all children live with a single mother. However, 48% of all black children live with a single mother. The percentages are of each of the four race categories.

(b) The number above each bar represents the percentage of children, in the race represented by the bar, that live with a single parent of the given sex.

(c) The group with the largest percentage of children who live with a single mother were blacks. The group with the smallest percentage of children who live with a single mother were Asian and Pacific islanders.

(d) The third sector will be labeled "other" because we do not know the living situation of the other children.

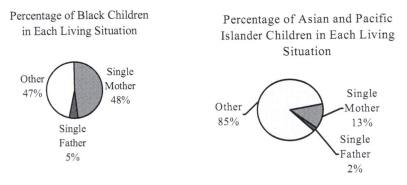

31. (a) For each region convert the percentage to a decimal and multiply by 15,215.

| Region | Number of Operating School Districts |
|---|---|
| New England | $0.08 \times 15{,}215 \approx 1217$ |
| Mid East | $0.12 \times 15{,}215 \approx 1826$ |
| Southeast | $0.11 \times 15{,}215 \approx 1674$ |
| Great Lakes | $0.22 \times 15{,}215 \approx 3347$ |
| Plains | $0.16 \times 15{,}215 \approx 2434$ |
| Southwest | $0.16 \times 15{,}215 \approx 2434$ |
| Rocky Mountains | $0.05 \times 15{,}215 \approx 761$ |
| Far West | $0.10 \times 15{,}215 \approx 1522$ |

(b) For each region convert the percentage to a decimal and multiply by 47,792,369.

| Region | Number of Students Enrolled |
|---|---|
| New England | $0.05 \times 47{,}792{,}369 \approx 2{,}389{,}618$ |
| Mid East | $0.15 \times 47{,}792{,}369 \approx 7{,}168{,}855$ |
| Southeast | $0.23 \times 47{,}792{,}369 \approx 10{,}992{,}245$ |
| Great Lakes | $0.16 \times 47{,}792{,}369 \approx 7{,}646{,}779$ |
| Plains | $0.07 \times 47{,}792{,}369 \approx 3{,}345{,}466$ |
| Southwest | $0.13 \times 47{,}792{,}369 \approx 6{,}213{,}008$ |
| Rocky Mountains | $0.04 \times 47{,}792{,}369 \approx 1{,}911{,}695$ |
| Far West | $0.17 \times 47{,}792{,}369 \approx 8{,}124{,}703$ |

(c) The regions with a higher percentage of school districts and lower percentages of students are New England, Great Lakes, Plains, Southwest, and the Rocky Mountains. The regions with a lower percentage of school districts and higher percentages of students are Mid East, Southeast, and Far West.

33. From these three pie charts it might be concluded that if someone approved of how Bill Clinton was doing his job, that person would not want him to be impeached, and even if he were impeached by the House of Representatives, that person would not want him convicted by the Senate.

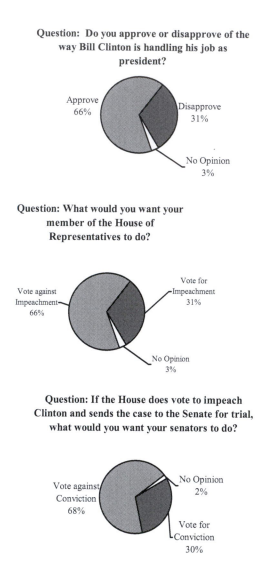

Question: Do you approve or disapprove of the way Bill Clinton is handling his job as president?

Approve 66%
Disapprove 31%
No Opinion 3%

Question: What would you want your member of the House of Representatives to do?

Vote against Impeachment 66%
Vote for Impeachment 31%
No Opinion 3%

Question: If the House does vote to impeach Clinton and sends the case to the Senate for trial, what would you want your senators to do?

Vote against Conviction 68%
No Opinion 2%
Vote for Conviction 30%

35. The percentage of the population for each region stayed relatively constant. Notice there was a 1% drop from 1990 to 2000 in the midwest and in the northeast. Both the west and the south gained in population. The small changes seem to indicate that people moved from the midwest and northeast to the west and south.

37. (a) The source that contributes the least is the Federal government. The source that contributes the most is the state government except in the years 1993, 1994, and 2002 when the local government contributed the most.
    (b) The source that is most stable is the Federal government.
    (c) In 1995 the percentage contributed by the local source was the smallest.
    (d) Between 1994 and 1995 the percentage contributed by the state changed the most, and it was an increase in contribution.

**39. (a)**

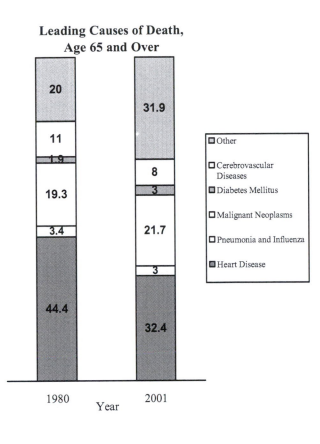

**Leading Causes of Death,
Age 65 and Over**

1980: 20, 11, 1.9, 19.3, 3.4, 44.4

2001: 31.9, 8, 3, 21.7, 3, 32.4

Legend:
- Other
- Cerebrovascular Diseases
- Diabetes Mellitus
- Malignant Neoplasms
- Pneumonia and Influenza
- Heart Disease

Year

**(b)**

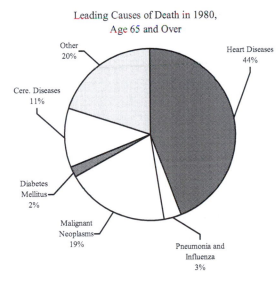

Leading Causes of Death in 1980,
Age 65 and Over

Other
20%

Heart Diseases
44%

Cere. Diseases
11%

Diabetes
Mellitus
2%

Malignant
Neoplasms
19%

Pneumonia and
Influenza
3%

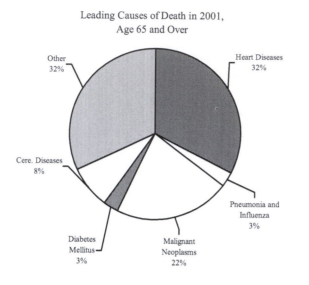

Leading Causes of Death in 2001,
Age 65 and Over

Other
32%

Heart Diseases
32%

Cere. Diseases
8%

Pneumonia and
Influenza
3%

Diabetes
Mellitus
3%

Malignant
Neoplasms
22%

(c) Deaths by heart disease decreased significantly in 2001 compared to 1980. Other categories remained about the same except for the "Other" category which gained what the heart disease category lost in 2001.

(d) Answers will vary. Both graphs do a nice job. Proportional bar graphs give the bars side-by-side which makes comparisons easier.

41. (a) Tennessee would experience an increase of $174 per capita, and Alabama would experience an increase of $270 per capita. All other states would remain the same. The increase for Tennessee includes a voter-passed tax increase, so it is impossible to tell if the increase for Tennessee is due to the voter-passed tax increase or due to the proposed tax increase. Notice Alabama had the lowest revenue per capita of the 11 states initially. After the proposed tax increase, Alabama would have had the 7th highest revenue per capita. Alabama would have been adversely affected by the proposed tax increase.

(b) Answers will vary. A double-bar graph will emphasize the fact that most states will have the same tax revenue per capita under the new plan as they do currently. This graph shows how costly the new plan is for people in Alabama.

Current versus Proposed Tax Plan
Based on Income

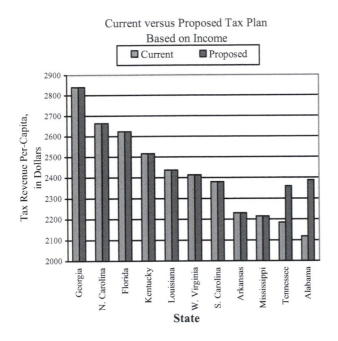

43. (a) Graph choices may vary. A double-bar graph makes comparison from year to year at each income level very easy because the bars are side by side. The double-line graph, however, seems to emphasize the trend that smaller percentages of wealthier people shop at dollar stores.

Percentage Who Shop at the Dollar Store,
Based on Income

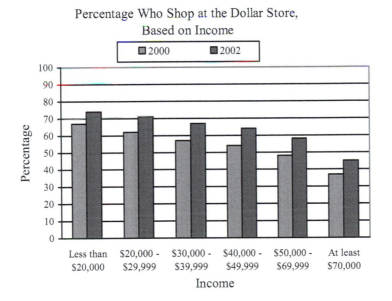

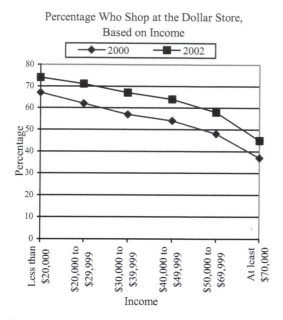

Percentage Who Shop at the Dollar Store, Based on Income

(b) The percentage of people who shop at a dollar store decreases as their income increases. At each income level, the percentage of people who shop at a dollar store increased several percentage points between the year 2000 and 2002.

## Solutions to Odd-Numbered Problems in Section 8.3

1. The graph indicates that Solei diamonds are rated higher for superior overall beauty and have more sparkle, but there is no vertical scale for the bar graph. There is no way to tell if the vertical axis begins at 0 or another value.

3. (a) The company probably produces cherry soda.
   (b) Grape soda and root beer are each preferred by 30% of consumers. However, the wedge for cherry soda looks larger because it is separated and placed in the foreground.

5. (a) Calculate $\dfrac{\text{height of bar 2}}{\text{height of bar 1}} = \dfrac{8.5}{1} = 8.5$ and compare it to $\dfrac{\text{value of bar 2}}{\text{value of bar 1}} = \dfrac{\$5950}{\$700} = 8.5$. Both

   ratios are the same, so we conclude the bar heights and the numeric values they represent are proportional.

   (b) Calculate $\dfrac{\text{height of bar 1}}{\text{height of bar 2}} = \dfrac{1.3}{0.4} = 3.25$ and compare it to $\dfrac{\text{value of bar 1}}{\text{value of bar 2}} = \dfrac{52\%}{15\%} \approx 3.47$. Both

   ratios are not the same, so we conclude the bar heights and the numeric values they represent are not proportional.

   (c) Calculate $\dfrac{\text{height of bar 2}}{\text{height of bar 1}} = \dfrac{27}{3} = 9$ and compare it to $\dfrac{\text{value of bar 2}}{\text{value of bar 1}} = \dfrac{11,250}{1250} = 9$. Both ratios

   are the same. Calculate $\dfrac{\text{height of bar 3}}{\text{height of bar 1}} = \dfrac{35}{3} \approx 11.67$ and compare it to

   $\dfrac{\text{value of bar 3}}{\text{value of bar 1}} = \dfrac{14,950}{1250} = 11.96$. Both of these ratios are not the same. Notice also

$\dfrac{\text{height of bar 3}}{\text{height of bar 2}} = \dfrac{35}{27} \approx 1.296$ and $\dfrac{\text{value of bar 3}}{\text{value of bar 2}} = \dfrac{14,950}{11,250} \approx 1.329$ are not the same. We conclude the bar heights and the numeric values they represent are not proportional.

7. **(a)** Use 3 bills of the same size.

Total Cost in 2002

**(b)** Originally the bill had a length of 40 millimeters and one bill represented the total cost of an item in 2001. In 2002, the item cost $\dfrac{\$4000}{\$5000} = 0.8$ or 80% of what it cost in 2001. The length of the bill should be 80% of its original length. The length of the bill should be changed to $0.80 \times 40 = 32$ millimeters.

20 mm — 32 mm — = Total Cost in 2002

**(c)** Let $l$ represent the length of the 2002 dollar bill and let $w$ represent the width of the 2002 dollar bill. We know $l = 2w$. In 2001, the area of the dollar bill was $l \times w = 40 \times 20 = 800$ square millimeters. In 2002, the area should be double or 1600 square millimeters.

$\text{Area} = l \times w$

$1600 = 2w \times w$

$1600 = 2w^2$

$800 = w^2$

$\sqrt{800} = w$

The width should change to $\sqrt{800}$ or approximately 28.3 millimeters. The length would change to $\dfrac{1600}{\sqrt{800}} \approx 56.6$ millimeters.

9. **(a)** To see if bar heights are proportional to the percentages they represent, check to see if the following equation is true for a pair of bars: $\dfrac{\text{value of bar } x}{\text{value of bar } y} = \dfrac{\text{height of bar } x}{\text{height of bar } y}$. If the equation is true for the first pair of bars, then check a second pair. In this case, the bar heights are not proportional to the percentages they represent.

**(b)** If the bar that represents the 40% of Americans who would spend their rebate money was drawn 5 centimeters tall, then let $y$ be the height of the bar that represents the 75% of Americans who would give their rebate away. Set up a proportion and solve for the unknown height.

$$\frac{\text{value of spend it bar}}{\text{value of give it away bar}} = \frac{\text{height of spend it bar}}{\text{height of give it away bar}}$$

$$\frac{40\%}{75\%} = \frac{5 \text{ cm}}{y}$$

$$40 \times y = 75 \times 5$$

$$y = \frac{75 \times 5}{40}$$

$$y = 9.375$$

Therefore, the bar representing those Americans who would give the rebate away should be drawn 9.4 cm tall. Similarly, the bar representing those Americans who would save or invest the rebate should be drawn 5.5 cm tall.

11. **(a)** The 1997 bar contains 4 monitors and represents 38% of farms. Therefore, each monitor represents $\frac{38\%}{4} = 9.5\%$. In 1999, each monitor represents $\frac{47\%}{5} = 9.4\%$. In 2001, each monitor represents $\frac{55\%}{6} \approx 9.2\%$. In 2003, each monitor represents $\frac{58\%}{6} \approx 9.7\%$. Each monitor represents a different number of percentage points in each stack.

   **(b)** Answers will vary. From part (a), we see that a computer monitor represents a different number of percentage points in each stack. No partial computer monitors appear. Bars are drawn as if each percent was rounded up, so bar heights are not proportional to the percentages they represent. Notice the stacks of computer monitors was drawn on a slant. The increases each year do not appear as large due to the slant. From 2001 to 2003, there is an increase of 3%, yet the bar for 2001 appears taller than the bar for 2003. At first glance, it appears that there is a decrease from 2001 to 2003.

13. **(a)** The height of each figure represents the number of people 65 years or older in the United States measured in millions.

   **(b)** In 1990, there were approximately 32 million Americans age 65 or older. It is predicted that in 2030 there will be approximately 72 million Americans age 65 or older. The predicted value for 2030 is more than twice as much as the 1990 value.

   **(c)** The area for the 2030 figure is more than four times the area of the 1990 figure. The area is not meaningful in this graph.

   **(d)** Answers will vary. In addition to the heights of the figures changing from year to year, the widths also change. Changing the widths make the increases in the population of people age 65 years or older appear much greater than the actual increases.

15. **(a)** The federal debt per person in 2005 was $26,600. If the stack of coins contained only dimes then there would have to be $\frac{\$26,600}{1} \times \frac{10 \text{ dimes}}{\$1} = 266,000$ dimes.

   **(b)** If stack height is proportional to the amount represented, then set up a proportion and solve for the unknown height.

$$\frac{\text{1840 debt per person}}{\text{2005 debt per person}} = \frac{\text{1840 stack height}}{\text{2005 stack height}}$$

$$\frac{\$0.21}{\$26{,}600} = \frac{0.5 \text{ cm}}{\text{2005 stack height}}$$

$$0.21 \times (\text{2005 stack height}) = 26{,}600 \times 0.5$$

$$\text{2005 stack height} = \frac{26{,}600 \times 0.5}{0.21}$$

$$\text{2005 stack height} \approx 63{,}333$$

The stack of coins that represents the federal debt per person in 2005 should be approximately 63,333 cm tall.

(c) Answers will vary. It is unclear from the graph whether the height of each stack or the dollar amount contained in each stack is what is being used to represent the federal debt per person. In either case, the stack representing the federal debt per person in 2005 is not drawn accurately. Notice also that the stack appears to disappear into the clouds at the top of the graph.

17. (a) Multiply each percentage by 360°. Round to the nearest degree.
Married: $0.534 \times 360° \approx 192°$
Never married: $0.29 \times 360° = 104°$
Widowed, divorced, separated: $0.177 \times 360° \approx 64°$

(b) For the category married, the angle in the graph appears to be too large. The angles in the graph appear to be too small for the other two categories.

(c) We are looking at the cakes at an angle which creates a misleading effect. The sectors are not accurately drawn. The sides of the pieces of cake are visible in the graph making the percentage who were married appear even greater.

(d)

U.S. Citizen Marital Status in 2004

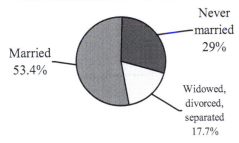

19. (a) Answers will vary. At first glance, it looks like there is an overall budget surplus with or without the Social Security surplus because usually the vertical axis is scaled so that 0 is at the bottom.

(b) The values along the vertical axis range from −$900 billion at the bottom to $0 billion at the top. Negative numbers represent budget deficits. The line graph is misleading because the lower the line extends, the larger the budget deficit.

**(c)**

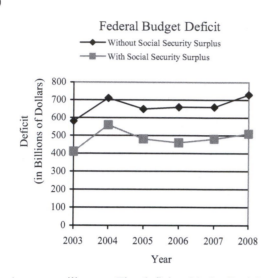

Federal Budget Deficit

**(d)** Answers will vary. The deficit with the Social Security surplus is generally $200 billion less than the deficit without the surplus. With or without the surplus, there was a sharp increase in the deficit from 2003 to 2004. Without the surplus, the deficit is projected to increase more sharply from 2007 to 2008 than it would with the surplus.

**21. (a)** To downplay the differences, extend the vertical scale.

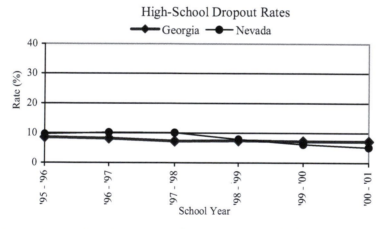

High-School Dropout Rates

**(b)** To emphasize the differences, select a vertical scale that does not start at 0. Select a range of values so that lines in the graph extend from the top to the bottom.

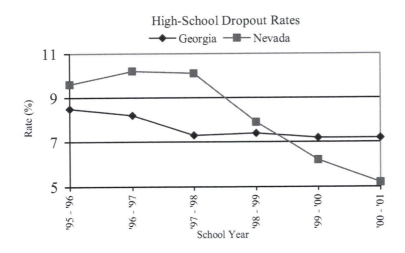

High-School Dropout Rates

**23.** Answers may vary.  The darkened sectors are placed over the narrow part of the oval egg making the shaded regions appear smaller than they really are.  It appears that the heights of the eggs do represent the relative sizes of the pension plans, however it is the area of the eggs that is most noticed.  From 1991 to 1993, the height doubled and the size of the pension plans doubled, yet the area of the egg representing 1993 is about four times as large as that of the egg representing 1991.

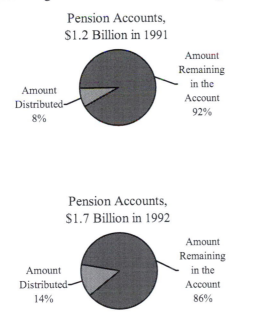

Pension Accounts,
$2.4 Billion in 1993

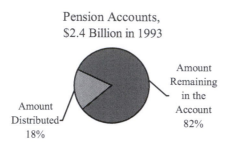

Amount
Remaining
in the
Account
82%

Amount
Distributed
18%

By making all of the circles the same size, the relative amounts distributed in each year are more accurately displayed. However, the amount in the pension accounts is not represented visually.

25. (a) The horizontal axis represents drug costs in dollars.
    (b) The darker shaded region represents the percentage the recipient pays. Notice that this region increases from a drug cost of $250 to a drug cost of $2000. The lighter shaded region represents the percentage Medicare pays. As costs increase, the shaded regions slant upward to reflect the increase, but between a cost of $250 and $2250, Medicare pays 75% while the recipient pays 25%. The shading should reflect these percentages. The shading is misleading. The text on the graph indicates that Medicare pays 75% up to a cost of $2250, yet the graph is labeled at $2000.
    (c) The thickness is meaningless.
    (d) That portion should be darkened because the recipient pays all costs up to $250.
    (e) The recipient must pay if there is no coverage so that portion should be darkened. By not darkening the graph, it makes it appear that the recipient pays less than they really have to pay.
    (f) The proportional bar graph makes clear the percentage paid by Medicare and the percentage paid by the recipient for each level of drug costs. Also, the thickness of the three-dimensional graph does not mean anything, so it is eliminated on the proportional bar graph. Eliminating the thickness makes the graph easier to interpret.

Medicare's Prescription Drug Coverage
Percentages Paid

□ Medicare □ Recipient

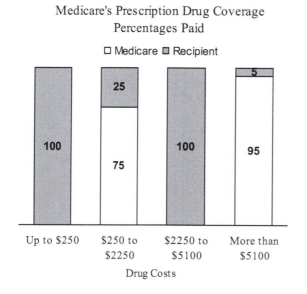

| Up to $250 | $250 to $2250 | $2250 to $5100 | More than $5100 |

Drug Costs

**27. (a)** Just looking at land area, it appears that each candidate was supported by half of the United States. Notice, however, that Alaska and Hawaii are not drawn to scale. In general, Dole was supported in Alaska, the Midwest and much of the South. Clinton was supported in the West, the Southwest, and much of the eastern half of the nation.

**(b)** Compare $\dfrac{\text{Area of Support for Clinton}}{\text{Area of Support for Dole}} \approx \dfrac{1}{1}$ to $\dfrac{\text{Electoral Votes for Clinton}}{\text{Electoral Votes for Dole}} = \dfrac{379}{159} \approx 2.4$ and note

that they are not the same. The area of support for each candidate is not proportional to the vote total.

## Solutions to Chapter 8 Review Problems

**1. (a)** There is a small cluster of states between 8.5% and 10.6%, and a large cluster of states between 11.0% and 15.6%. There are gaps between 5.7% and 8.5% and also between 15.6% and 17.6%. There are two outliers: 5.7% and 17.6%. The state with the highest percentage of residents who were at least 65 years old is probably Florida, because of their high concentration of retired people who reside there.

**(b)** In 90% of the states, more than 10% of the population is at least 65 years old. In 82% of the states, less than 14% of the population is at least 65 years old.

**2. (a)**

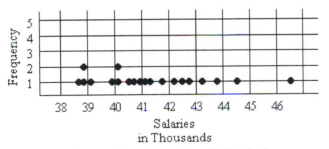

**(b)** Stems are in thousands and leaves are in hundreds.

```
46 | 5
45 |
44 | 5
43 | 2 8
42 | 2 5 7
41 | 1 2 7
40 | 1 1 6 8 9
39 | 1 9
38 | 8 9 9
```

**(c)** Most of the salaries cluster between $38,800 and $44,500. There is a $2000 gap between the top salary and the next-lowest salary. The salary $46,500 is an outlier.

3. **(a)** Stems are in thousands and leaves are in hundreds.

| Master's | | Bachelor's |
|---:|:---:|:---|
| 5 | 46 | |
| | 45 | |
| 5 | 44 | 1 |
| 8 2 | 43 | |
| 7 5 2 | 42 | |
| 7 2 1 | 41 | 4 |
| 9 8 6 1 1 | 40 | 3 |
| 9 1 | 39 | 3 5 |
| 9 9 8 | 38 | 1 1 3 |
| | 37 | 0 3 6 8 9 |
| | 36 | 0 5 5 7 |
| | 35 | 2 5 8 |

**(b)** Most of the salaries for graduates with bachelor's degrees are clustered between $35,200 and $38,300. There is a large gap between $41,400 and $44,100. $44,100 is an outlier.

**(c)** This double stem-and-leaf plot shows that in general, a graduate with a master's degree will have a starting salary that is about $3000 more than the starting salary of a graduate with a bachelor's degree.

4. **(a)**

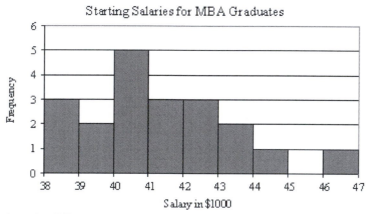

**(b)** The only difference between the histogram in part (a) and the relative frequency histogram is the vertical scale.

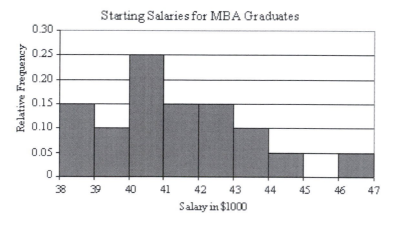

5. (a) The following table gives the change in turkey production for each 5-year period. The greatest change was an increase and occurred from 1985 to 1990.

| Years | Change in Turkey Production |
|---|---|
| 1975 to 1980 | 165,200,000 − 124,200,000 = 41,000,000 |
| 1980 to 1985 | 185,000,000 − 165,200,000 = 19,800,000 |
| 1985 to 1990 | 282,445,000 − 185,000,000 = 97,445,000 |
| 1990 to 1995 | 292,856,000 − 282,445,000 = 10,411,000 |
| 1995 to 2000 | 269,969,000 − 292,856,000 = −22,887,000 |

(b)

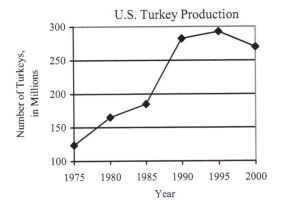

6. (a) The percentage represents the percentage of youth who use the Internet for each particular activity.
   (b) The majority of youths use the Internet to complete school work, send emails or instant messages, and play games. A third of youths use the Internet to obtain news, weather, or sports information and to search for products. One in five youths use the Internet to participate in chat rooms, watch online TV, movies or listen to the radio. Only 10.6% make purchases online probably due to the lack of credit cards. Making phone calls is a recent development in Internet use and only 3.2% of youths make phone calls over the Internet.
   (c) A single pie chart is not an appropriate display for this data. Each bar represents the percentage of youths out of all youths who participate in the activity and a single youth could be included in each category. Nine different pie charts could be used however.

7. (a) Gasoline prices were rising in October, January, February, March, and August. Gasoline prices were fairly steady in September, May, June, and July.
   (b) Gasoline prices were rising in March and April. Gasoline prices were fairly steady in September, January, February, May, June, July, and August.
   (c) From the period from September 2002 through August 2003 the greatest increase occurred in August. From the period from September 2001 through August 2002 the greatest increase occurred in March.

**(d)**

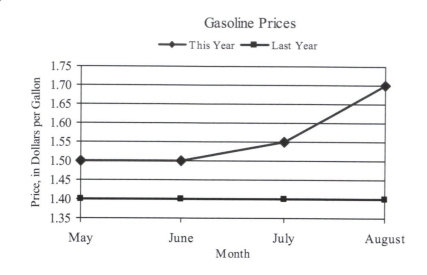

Gasoline Prices

8. **(a)** It appears that Pepsi and Dr Pepper hold about half of the market share. The Pepsi sector is exploded from the graph making it appear larger than it really is. The slant to the pie chart makes the Coca-Cola sector appear much smaller than it really is.

   **(b)**

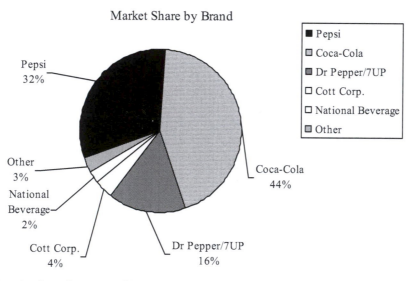

Market Share by Brand

   **(c)** The pie chart from part (b) more clearly demonstrates that Coca-Cola dominates Pepsi in market share compared to the three-dimensional pie chart.

9. **(a)** According to the table, one serving of Cheezy Crackers contributes 7% of the calories to a 2000-calorie diet. There are $0.07 \times 2000 = 140$ calories in one serving of Reduced-Fat Cheezy Crackers.

(b)

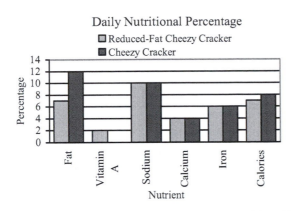

Daily Nutritional Percentage

The two types of crackers are the same nutritionally for sodium, calcium and iron. The regular Cheezy Crackers have more calories, less Vitamin A which is probably added to the Reduced-Fat Cheezy Crackers, and much more fat.

(c) The amount of fat in one serving of the regular Cheezy Crackers is 12% of the daily requirement. We know $0.25 \times 12\%$ is 3%, so if a serving of the new cracker contains at most $12\% - 3\% = 9\%$ fat based on a 2000-calorie diet, then it is a reduced-fat food. Reduced-Fat Cheezy Crackers do qualify as a reduced-fat food. They contain 7% of the daily requirement of fat based on a 2000-calorie diet. It has $\frac{7}{12} \approx 0.58$ or approximately 58% of the fat of the regular cracker. Therefore, it has approximately 42% less fat than the regular Cheezy Crackers.

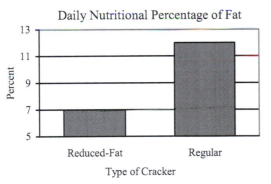

Daily Nutritional Percentage of Fat

10. Answers may vary. A proportional bar graph would most clearly communicate this information. A proportional bar graph shows relative amounts and trends at the same time.

11. Answers may vary. A double-line graph would most clearly communicate this information. Line graphs show trends over time.

**12. (a)**

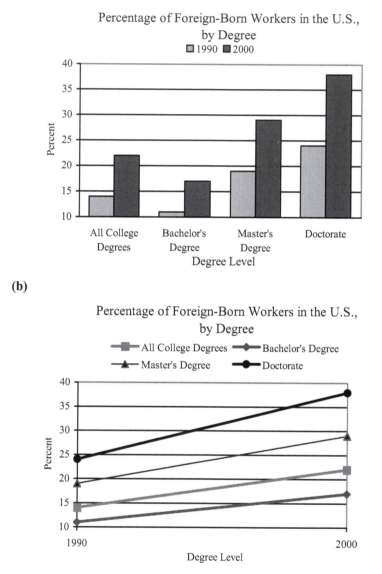

Percentage of Foreign-Born Workers in the U.S., by Degree

**(b)**

Percentage of Foreign-Born Workers in the U.S., by Degree

**(c)** Answers will vary. The bar graph from part (a) may be the best choice. It emphasizes the gap between percentages in 1990 and 2000 more effectively than does the line graph.

**13. (a)**

Male - Female Population Comparison
in the U.S. by Age

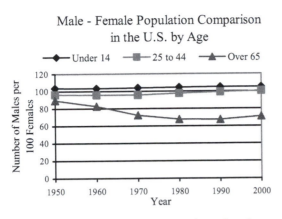

**(b)** In 1990 there were the fewest number of males per 100 females in the "Over 65" age group. In 2000 there were about the same number of males and females in the "25 to 44" age group.

**(c)** There are more males under 14 years of age than females under 14 years of age, and the number has stayed fairly constant for 50 years. The number of males per 100 females in the "25 to 44" category decreased slightly from 1950 to 1970 but has been slowly increasing for the past 30 years and in the year 2000 there were actually more males than females in this category. From 1950 to 1990 the number of males per 100 females in the "Over 65" category decreased substantially, but has increased slightly in the past 10 years.

**(d)** Females tend to live longer than males.

**14. (a)** Approximately 33% of male drivers aged 21 to 24 who were involved in fatal accidents were speeding. The greatest decrease in the percentage of fatal car crashes involving speeding male drivers occurred between the 21 to 24 year category and the 25 to 34 year category.

**(b)** For the 45 to 54 year age group, approximately 10% of female drivers involved in fatal accidents were speeding. The percentages of drivers in fatal crashes that were speeding for males and females are nearly equal in the 65 to 74 year age group.

**(c)** For both males and females, the percentage of fatal car crashes involving speeding drivers decreases with age. For females involved in fatal crashes, at most 25% involved speeding drivers in any age category. For males involved in fatal crashes who were younger than 25 years of age, between 33% and 35% involved speeding drivers which might explain why insurance costs are higher for males. In every age category, the percentage of males involved in fatal car crashes who were speeding is greater than for females.

**15. (a)** The time period represented by the graph is 1980 to 2001.

**(b)** From 1980 through 1986 the average annual consumption of beef exceeded 75 pounds per person.

**(c)** In 1992 the average annual per-capita consumption of chicken was the same as that of beef. In 1985 the average annual per-capita consumption of chicken was the same as that of pork.

**(d)**

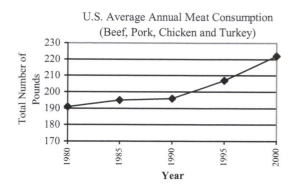

U.S. Average Annual Meat Consumption
(Beef, Pork, Chicken and Turkey)

There has been an increase in total meat consumption per capita for the past 20 years. Even though there has been a gradual decrease in beef consumption, the large increase in chicken consumption, especially in the last 10 years, has caused an increase in the total consumption.

**16. (a)** The impression is that the Prius takes much longer to accelerate from 0 to 60 mph than the other two cars. This impression is caused by two characteristics of the bar graph. First, the horizontal scale begins at 8 seconds instead of 0 seconds which exaggerates the time differences. Second, by putting the bar for the Prius closer to the bar for the faster accelerating car, the Corolla, and farther from the slower accelerating car, the Camry, the difference is again accentuated.

   **(b)** Changing the horizontal scale reduces the impression that the Prius accelerates at a much slower rate. In the modified graph, the Prius seems to accelerate comparatively quickly.

Acceleration (0-60 mph)

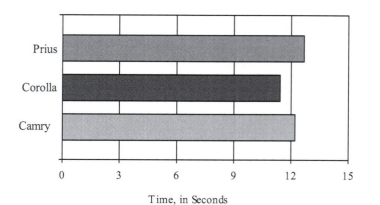

Time, in Seconds

**17.**

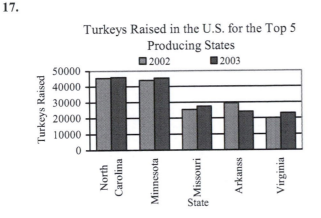

Turkeys Raised in the U.S. for the Top 5
Producing States

18. (a) Turkey exports increased steadily from 1990 until 1997. There was a sharp decrease in 1998 and exports have fluctuated but remained fairly stable since.
   (b) In 1993 approximately 5% of the total turkey production was exported. More than 8% of the turkey production was exported in the years 1997, 1998, 2000, and 2001.
   (c) The largest percentage of the turkey production was exported in 1997.
   (d) There was a decrease from 1997 to 1998 and another decrease from 1998 to 1999. The percentage increased from 1999 to 2000 and from 2000 to 2001 followed by another decrease from 2001 to 2002.

**19.**

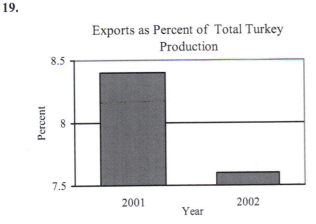

Exports as Percent of Total Turkey
Production

**20** The percentage of the annual turkey production not exported has decreased from 1990 to the low point of approximately 89% in 1997. There was an increase in the percentage of total turkey production not exported from 1997 to 1999. Since then the percentage not exported has remained fairly steady.

Percent of Total Turkey Production Not Exported

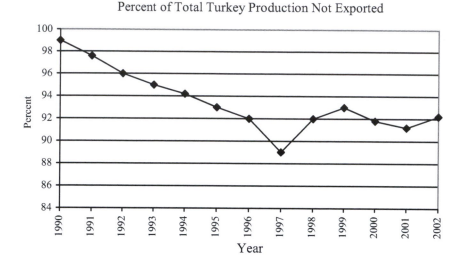

**21. (a)** The coffee grower receives $\dfrac{0.035}{3.75} \approx 0.0093$ or approximately 0.9%.

The coffee millers, importers, exporters and roasters receive $\dfrac{0.175}{3.75} \approx 0.0467$ or approximately 4.7%.

The milk is $\dfrac{0.40}{3.75} \approx 0.1067$ or approximately 10.7% of the total cost.

The cup is $\dfrac{0.07}{3.75} \approx 0.0187$ or approximately 1.9% of the total cost.

The labor at the coffee shop is $\dfrac{1.35}{3.75} = 0.36$ or 36% of the total cost.

The shop rent, marketing, and general administration is $\dfrac{1.29}{3.75} = 0.344$ or 34.4% of the total cost.

The initial investment is $\dfrac{0.18}{3.75} = 0.048$ or 4.8% of the total cost.

The profit for the shop owner is $\dfrac{0.25}{3.75} \approx 0.0667$ or approximately 6.7% of the total cost.

**(b)** Coffee grower: $0.009 \times 360° \approx 3.2°$
Coffee millers, importers, exporters and roasters: $0.047 \times 360° \approx 16.9°$
Milk: $0.107 \times 360° \approx 38.5°$
Cup: $0.019 \times 360° \approx 6.8°$
Labor costs: $0.36 \times 360° = 129.6°$
Rent, marketing and administration: $0.344 \times 360° \approx 123.8°$
Inital investment: $0.048 \times 360° \approx 17.3°$
Profit: $0.067 \times 360° \approx 24.1°$.

Where the Money goes for your Double Cappucino

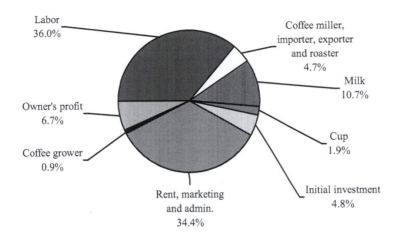

Labor
36.0%

Coffee miller,
importer, exporter
and roaster
4.7%

Milk
10.7%

Owner's profit
6.7%

Cup
1.9%

Coffee grower
0.9%

Initial investment
4.8%

Rent, marketing
and admin.
34.4%

(c) Answers will vary. The smallest amount of the cost of the cappucino goes to the grower. The two largest portions of the cost goes to pay for the labor, and renting the space for the shop, marketing and administration. The owner's profit is even less than the cost of the milk used in making the espresso.

22. (a) The percentage of one-person households headed by males under age 65 has steadily increased since 1970 with the largest increase between 1970 and 1980. The percentage of one-person households headed by males 65 or older decreased slightly between 1960 and 1970, dropped dramatically between 1970 and 1980 and has slowly increased since that time.

(b) In 1960 the percentage of one-person households headed by males was 37.2%, in 1970 it was 35.7%, in 1980 it was 38.8%, in 1990 it was 40.7%, and in 2000 it increased to 43.3%. The percentage of one-person households headed by males has increased steadily since 1980. In 1960 the percentage of one-person households headed by females was 62.8%, in 1970 it was 64.3%, in 1980 it was 61.2%, in 1990 it was 59.3%, and in 2000 it dropped to 56.7%. The percentage of one-person households headed by females has decreased steadily since 1980.

(c) The percentage of one-person households headed by females under age 65 steadily decreased from 1960 through 1990, but increased from 1990 to 2000. The percentage of one-person households headed by females 65 or older increased rapidly between 1960 and 1970, dropped almost as dramatically from 1970 to 1980, stayed steady from 1980 to 1990 and decreased dramatically between 1990 and 2000.

23. (a)

Highway Gas Mileages for Four-Wheel Drive Minivans

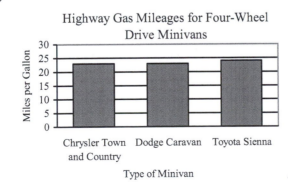

Chrysler Town and Country   Dodge Caravan   Toyota Sienna

Type of Minivan

**(b)**

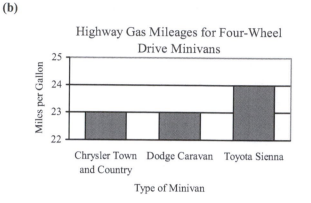

Highway Gas Mileages for Four-Wheel Drive Minivans

24. **(a)** The pictograph uses a rectangular credit card rather than a circle so the sector representing the 25-day grace period and the sector representing the 20-day grace period appear much larger than they should.

   **(b)** Multiply each percentage by 360°. Round to the nearest degree.
   25-day grace period: $0.51 \times 360° \approx 184°$
   20-day grace period: $0.38 \times 360° \approx 137°$
   20 to 25-day grace period: $0.10 \times 360 = 36°$
   No grace period: $0.01 \times 360° \approx 4°$

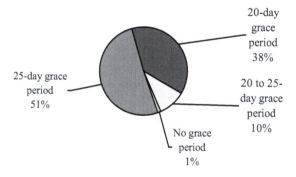

Credit Card Grace Periods

25. **(a)** Find the ratio of bar lengths and compare it to the ratio of the number of computers they represent.

   For example, $\dfrac{\text{length of KaZaA bar}}{\text{length of WinMX bar}} \approx 5.1$ but $\dfrac{\text{KaZaA number of computers}}{\text{WinMX numer of computers}} \approx 3.3$. Therefore, bar lengths are not proportional to the number of computers that they represent.

   **(b)** If bar lengths are proportional to the quantities they represent, then to find the length needed for KaZaA, set up and solve the following proportion:

   $$\frac{\text{length of KaZaA bar}}{\text{length of LimeWire bar}} = \frac{\text{number of KaZaA computers}}{\text{number of LimeWire computers}}$$

   $$\frac{\text{length of KaZaA bar}}{1 \text{ mm}} = \frac{7.60 \text{ million}}{0.05 \text{ million}}$$

   length of KaZaA bar $= 152$ mm

   The length of the bar that is used to represent KaZaA should be 152 millimeters long. The lengths of the other bars are found similarly. The bar for WinMX should be 46 millimeters; iMesh, 17.8 millimeters; Morpheus, 8.8millimeters; and Grokster, 6 millimeters.

**26. (a)** The largest percentages of people 5 years and over who spoke a language other than English were found in Southern California, Arizona, New Mexico, and Texas. The lowest percentages were found in the eastern half of the United States especially along the Mississippi River and in the surrounding states.

**(b)** It makes sense that larger percentages of people 5 years and over who spoke a language other than English would be concentrated along the border with Mexico.

# *Chapter 9:* Collecting and Interpreting Data

## *Key Concepts*

**By the time you finish studying Section 9.1, you should be able to:**

☐ describe the population and sample in a survey.

☐ identify the variable being measured in a sample as quantitative or qualitative.

☐ discuss possible sources of bias in a sample.

☐ define simple random sample.

☐ use a random-number table to generate a simple random sample.

**By the time you finish studying Section 9.2, you should be able to:**

☐ generate an independent sample.

☐ explain why the size of an independent sample cannot be fixed ahead of time.

☐ generate a 1-in-$k$ systematic sample.

☐ generate a quota sample.

☐ define a stratum and generate a stratified random sample.

☐ define a sampling unit and a frame and generate a cluster sample.

☐ explain the differences between stratified sampling and cluster sampling.

**By the time you finish studying Section 9.3, you should be able to:**

☐ calculate the mean for a given set of data values or for a given histogram.

☐ calculate the median for a given set of data values or for a given histogram.

☐ explain what it means for a distribution to be skewed left or skewed right.

☐ determine the mode for a given set of data values or for a given histogram.

☐ calculate a weighted mean.

☐ calculate the range for a set of data values.

☐ define and calculate the first quartile, third quartile, and the interquartile range.

☐ give the five-number summary for a data set.

☐ create a box-and-whisker plot.

☐ calculate the sample standard deviation for a sample.

☐ explain how the sample variance and sample standard deviation are related.

☐ calculate the population variance and population standard deviation.

## *Hints for Odd-Numbered Problems*

### Section 9.1

1. The population is the entire set being studied. The sample is a subset of the population. The variable of interest is the characteristic of the population that is being measured.

3. The population is the entire set being studied. The sample is a subset of the population. The variable of interest is the characteristic of the population that is being measured.

5. The population is the entire set being studied. The sample is a subset of the population. The variable of interest is the characteristic of the population that is being measured.

7. The population is the entire set being studied. The sample is a subset of the population. The variable of interest is the characteristic of the population that is being measured.

9. Numeric variables are quantitative. Data not measured on a numeric scale is qualitative.

11. Ordinal data can be ordered or ranked.

13. Consider the five common sources of bias in surveys. Which might apply in this case?

15. Consider the five common sources of bias in surveys. Which might apply in this case?

17. In this case, consider the sampling method and discuss in what ways the method could be a source of bias.

19. In this case, consider the sampling method and discuss in what ways the method could be a source of bias.

21. (a) (i) Who is most likely to be at the library on a Saturday morning.
(ii) How might residents in a single city block be similar?
(iii) Who is most likely to call?

23. See the table in Figure 9.4. Eliminate repeated numbers and numbers greater than 35.

25. See Example 9.6.

27. See Example 9.6.

29. (b) After exhausting the numbers in column 5, proceed to the top of column 6 and continue.

### Section 9.2

1. (a) Notice the random selection was done in each of two distinct groups.
(c) It is important to notice that all of the corn plants in a selected region are examined.
(d) Consider the percentage of women in the sample who had cesarean sections.

3. To find the desired percentage divide the number of people in the sample by the total

number of coin flips, and multiply the result by 100%.

5. Will a 20% independent sample always contain 20% of the population?

7. See Example 9.9.

9. There are 50 governors, so you will be looking through 10 rows (5 digits each) of random digits.

11. **(b)** To find the percentage of governors in the sample, divide the number of governors in the sample by 50 and multiply the result by 100%.

13. What numbers were assigned to the governors from Illinois and Missouri? Figure out what digit(s) are in the corresponding positions in the list of random digits.

15. **(c)** Begin with the 9th printer and select every 15th printer after that.

17. The digit you obtain from the random-number table will be the number of the governor with which you begin the sample.

19. Consider the random number in column 3 and row 102. Reading from left to right, select the first digit that is 1, 2, or 3. Use that to begin the systematic sample.

21. **(b)** Assume the sampling would continue until the theater is empty.

23. Divide the number of males by the total population and divide the number of females by the total population.

25. Would a sample from only Florida be representative of the south in general?

27. **(b)** If you run out of numbers in a column, then move to the top of the next column.

29. **(b)** It is reasonable to find the percentage of students in each stratum and use it to determine the sample size to be taken from each stratum.

31. **(a)** Students are grouped in a convenient way.

33. In the map, the public waters are numbered from 1 to 29. In order to use the random-number table, think of the public waters as being labeled 01 to 29 and read the second and third digits off of the table as directed. If you run out of random numbers in column 2, then move to the top of column 3.

35. When reading the numbers off of the random-number table, read across the row rather than down the column. When you run out of numbers in row 109, move to row 110.

37. See Table 9.3.

39. Answers will vary. Consider adults with children. Could there be differences in tax opinions based on political party affiliation?

41. Answers will vary. There is likely to be a big difference between opinions of smokers and nonsmokers.

## Section 9.3

1. **(c)** Where did Marcus Banks' height occur in the ordered list of heights? Think about how each statistic is calculated.

3. **(c)** You might want to choose values that will show the greatest decrease from one time period to the other.

5. See Figure 9.12 and refer to the definition of skewed distributions.

7. **(a)** Make a list of all the data values. The height of each bar indicates the number of times the data value is observed.

9. **(a)** Think about what could affect normal body temperature and how the mean, median, and mode are calculated.
   **(b)** Select a measure of central tendency that takes on one of the data values.

11. See Example 9.19.

13. **(a)** How do you find the median if there are five values? The median is 55, so you know one of the data values. You are

told the mode is 61, so how many values of 61 must you have?

(b) Pick a smallest value to be in your data set. Knowing the range allows you to determine the largest data value in the set. If there is no mode, then what do you know about the values?

15. There is an odd number of data values. Identify the middle value and then split the data values into a lower half and an upper half.

17. There is an even number of data values. Split the data values into a lower half and an upper half.

19. (a) See Figure 9.16.
(b) Compare the distance between the first quartile and the median to the distance between the median and the third quartile. To which side are the data values more concentrated?

21. As you compare the box-and-whisker plots, it may be helpful to calculate the range and the interquartile range for each data set.

23. (c) Compare the distance between the first quartile and the median to the distance between the median and the third quartile. To which side are the data values more concentrated?

25. The data are already ordered in the table. There is an even number of data values, so the median is the mean of the middle two values.

27. (e) Compare the widths of each plot. What information does that give you? Calculate the range and interquartile range for each data set.

29. To make sure that the mode is the same as the mean and median, which value must be the most frequently observed value?

31. Use a variable. Let $x$ represent the missing score. Set up the calculation for the mean and solve for $x$.

33. (b) The deviation from the mean for each data value is found by subtracting the mean from each data value.

35. Set up calculations in a table as is done in Example 9.25.

37. (c) Consider how the mean of a set of data values such as $x_1, x_2, \ldots, x_N$ is calculated. Add 5 to each of these data values and calculate the mean.
(d) Think about what standard deviation measures. If each data value is altered by adding 5, then has the spread of the data been changed?

39. (c) Consider how the mean of a set of data values such as $x_1, x_2, \ldots, x_N$ is calculated. Multiply each of these data values by 3 and calculate the mean.
(d) Use the definition of the sample standard deviation and the information from part (c).

41. See Example 9.25 and use a table to organize your work.

43. (a) Think about what the population variance measures. What would have to happen for the population variance to be zero?
(b) Think about how the population variance is calculated. What is done to each of the deviations from the mean?

## Solutions to Odd-Numbered Problems in Section 9.1

1. The population is the set of light bulbs manufactured by this company. The sample is the set of eight bulbs purchased. The variable measured is whether or not the bulb burned out before 2000 hours.

3. The population is the set of all full-time students at the university. The sample is the set of 100 full-time students selected. The variable measured is whether or not the student commutes on a regular basis.

5. The population is the set of all Hewlett-Packard customers in 2003. The sample is the set of 4100 Hewlett-Packard customers who were sent a survey. The variable measured is whether or not the customer felt they had a good relationship with Hewlett-Packard.

7. The population is the set of 7140 registered voters in the city. The sample is the set of $185 + 210 + 25 = 420$ registered voters selected. The variable measured is the political affiliation of the voter.

9. The quantitative variables are numerically measured and include age and identification number. The qualitative variables are not numeric and include country of origin and profession.

11. The quantitative variables are numerically measured and include age and height. The qualitative variables are not numeric and include location, variety, and health. There is a natural ordering to the health variable as one could rank the trees by level of health, so health is an ordinal variable. The other two qualitative variables, variety and location, are both nominal. [An argument could be made that the variable location is quantitative. A global positioning system could be used, for example, to identify the coordinates of a certain tree.]

13. Dentists in the Minneapolis - St. Paul area may be more likely to support a local company so questioning only local dentists will be a source of bias. The questionnaires that are sent may contain poorly worded questions to influence the answers.

15. People who want to be on television may be more likely to select the lemon-lime drink. Honest preferences may not be obtained.

17. The population is the set of all the fish in the lake. The sample is the set of 500 fish caught and examined. Bias can result from the tagging procedure. If the tagged fish do not mix in with the rest of the fish before the sampling is done, the tagged fish could be over or under represented. Tagged fish could die before the sampling is done. If it is possible for a tagged fish to have any memory of being tagged, then it might avoid being caught again.

19. The population is the set of all doctors. The sample is the set of 20 doctors selected. Bias results from the drug company sampling until they get the results that they want.

21. Answers may vary.
    (a) For option (i), it could be that more residents with small children go to the library on a Saturday. These residents may be more interested in having fluoridated water. Selecting residents at a single location may produce a very biased sample.
    For option (ii), residents in a single city block may be more alike because of their similar economic status. More families with children may live in a city block and might favor adding fluoride more than the population in general.
    For option (iii), requiring residents to call a telephone number means that only those residents who have a strong opinion will call. Residents who do not subscribe to the newspaper will not even participate.
    Of the three options, (ii) would likely yield the most representative sample in this case because the ten city blocks are randomly selected throughout the city.
    (b) The city already has a list of all residents who are using the city's water supply. The city could select a simple random sample of residents and call or visit those residents.

23. The random number in column 3 of row 115 is 50011. Proceed down the column noting the last two digits of each random number. Each student is represented by a number from 00 to 35. Keep the two-digit numbers from 00 to 35, and do not list repeated numbers. The students who are selected to be in the sample are represented by the numbers 11, 16, 28, 18, and 32.

**25.** The random number in column 2 of row 110 is 57601. Proceed down the column noting the first three digits of each random number. Each student is represented by a number from 000 to 249. Keep the three-digit numbers from 000 to 249, and do not list repeated numbers. The students that are selected to be in the sample are represented by the numbers 121, 066, 146, 060, 025, 236, 075, 139, 042, and 201.

**27.** Answers will vary. The 14 players in the list can be represented by the numbers 00 to 13. Let Scott Schoeneweis be represented by 00. Number the rest of the players consecutively so that Vernon Wells is represented by the number 13. Pick a row and column in the random-number table. Decide which two digits of each random number to read. Read down the column keeping numbers from 00 to 13 and ignoring repeats. List the five players who are selected.

**29. (a)** The random number in column 2 of row 117 is 73603. Proceed down the column noting the second and third digits of each random number. Each city is represented by a number from 00 to 29. Keep the two-digit numbers from 00 to 29, and do not list repeated numbers. The cities that are selected to be in the sample are represented by the numbers 21, 25, 09, 29, and 27. The cities are Nashville-Davidson, Tennessee; Charlotte, North Carolina; Detroit, Michigan; Tucson, Arizona; and Portland, Oregon.

**(b)** The random number in column 5 of row 140 is 43906. Proceed down the column noting the last two digits of each random number. Each city is represented by a number from 00 to 29. Keep the two-digit numbers from 00 to 29, and do not list repeated numbers. The cities that are selected to be in the sample are represented by the numbers 06, 22, 29, 03, and 05. The cities are San Diego, California; El Paso, Texas; Tucson, Arizona; Houston, Texas; and Phoenix, Arizona.

**(c)** The random number in column 1 of row 102 is 10041. Proceed down the column noting the third and fourth digits of each random number. Each city is represented by a number from 00 to 29. Keep the two-digit numbers from 00 to 29, and do not list repeated numbers. The cities that are selected to be in the sample are represented by the numbers 04, 01, 10, 28, and 11. The cities are Philadelphia, Pennsylvania; Los Angeles, California; San Jose, California; Oklahoma City, Oklahoma; and Indianapolis, Indiana.

**(d)** The fraction of cities that are west coast cities is $\frac{6}{30} = \frac{1}{5}$. The sample in part (a) contains 1 west coast city, so $\frac{1}{5}$ of the cities in that sample are west coast cities. The sample in part (b) contains 1 west coast city, so $\frac{1}{5}$ of the cities in that sample are west coast cities. The sample in part (c) contains 2 west coast cities, so $\frac{2}{5}$ of the cities in that sample are west coast cities.

**(e)** The cities were selected by conducting a simple random sample. There will be samples with more than one west coast city. A sample is not necessarily biased if it contains more than one west coast city.

## Solutions to Odd-Numbered Problems in Section 9.2

**1. (a)** The population has been divided into two nonoverlapping groups: urban and rural. Eighty residents from each group are selected. This is a stratified random sample.

**(b)** Every 7th person is selected so this is a 1-in-7 systematic sample.

**(c)** Every corn plant in a selected region is examined. The regions are the clusters or sampling units. This is a cluster sample.

**(d)** The women are separated into two nonoverlapping groups. Forty percent of women had cesarean sections and 40% of the women in the sample will be women who have had cesarean sections. This is a quota sample.

**3. (a)** Out of the first 200 people, 75 are selected to be included in the sample. Thus, $\frac{75}{200} \times 100\% = 0.375 \times 100\% = 37.5\%$ of the first 200 people are included in the sample.

**(b)** Yes, this is a 50% independent sample. Each person had the same probability of being included in the sample. A person was included or excluded as the result of a coin flip. The probability a person was included was 0.5, or we could say each person had a 50% chance of being included.

**5.** In a 20% independent sample, each registered voter would have a 20% chance of being selected for jury duty. There is no guarantee, however, that the total number of registered voters in the sample will be exactly 20% of all of the registered voters in the town. The sample might contain fewer or more voters. If a fixed number of voters are required in the sample, a different method should be used.

**7.** The random number in row 1 and column 3 is 47772. We will consider each digit and note the positions in which 0s occur. Reading down the column, notice the first occurrence of a 0 is in fourth row. This is the 19th digit. Continuing to read down the column, we see 0s occur in the following positions: 19, 23, 28, 29, 55, 57, 60, 65, 72, 73, and 76. Therefore, a total of eleven cars from the set of 100 cars are chosen.

**9.** The governors are numbered according to their position when listed alphabetically by state. The random number in row 101 and column 2 is 77195. Note the position of every 0 and 1 as you read down the column, and include the governors in those positions. The governors in positions 3, 7, 23, 25, 31, 32, 38, 40, 49, and 50 should be selected. The governors included are the governors from Arizona, Connecticut, Minnesota, Missouri, New Mexico, New York, Pennsylvania, South Carolina, Wisconsin, and Wyoming.

**11. (a)** The random number in row 130 and column 2 is 77175. Note the position of every 1, 2, and 3 as you read down the column, and include the governors in those positions. The governors in positions 3, 9, 11, 12, 14, 26, 27, 29, 30, 40, 41, 46, and 47 should be selected. The governors included are governors from Arizona, Florida, Hawaii, Idaho, Indiana, Montana, Nebraska, New Hampshire, New Jersey, South Carolina, South Dakota, Virginia, and Washington.

**(b)** A total of 13 out of the 50 governors are included in the sample. Thus, $\frac{13}{50} \times 100\% = 0.26 \times 100\% = 26\%$ of the governors are included.

**13.** Listed alphabetically, the Illinois governor is in position 13, and the Missouri governor is in position 25. In the list of random digits, the digit in the 13th position is a 5. The digit in the 25th position is also a 5. Notice that the digit 5 does not appear anywhere else in the list of random numbers. It appears that the digit 5 was the digit that meant a governor was to be selected. Thus one digit out of 10 determined whether a governor was included. This sample was a $\frac{1}{10} \times 100\% = 0.1 \times 100\% = 10\%$ independent sample.

**15. (a)** The value of $k$ in this case is 15. This is a 1-in-15 systematic sample. One printer will be selected out of every 15.

**(b)** The number chosen at random, in this case 9, is the value of $r$. The 9th printer out of every 15 printers will be selected.

**(c)** The sample will include the 9th printer, the $9 + 15 = 24$th printer, the $9 + 2(15) = 39$th printer, and so on until the 180th printer is reached. The printers included in the sample will be printers 9, 24, 39, 54, 69, 84, 99, 114, 129, 144, 159, and 174.

17. The fourth digit from column 1 in row 128 is a 3. Every 3rd governor out of 10 will be selected to be included in the sample. The governors in position 3, 13, 23, 33, and 43 will be selected. The governors selected are from Arizona, Illinois, Minnesota, North Carolina, and Texas.

19. The random number in column 3 and row 102 is 63857. Notice the second digit is a 3. For the 1-in-5 systematic sample, the third letter out of every 5 letters will be included in the sample. Letters in positions 3, 8, 13, 18, and 23 will be included. The letters selected are C, H, M, R, and W.

21. **(a)** Notice that the 9th adult is selected and then the 19th adult. There is a 19 − 9 = 10 adult difference between selected adults. Every 10th adult is selected to be in the sample. This is a 1-in-10 systematic sample beginning with the 9th adult.
   **(b)** Because this was a 1-in-10 systematic sample, every 10th adult was handed a questionnaire. If there had been a 69th adult in the theater, that person would have received a questionnaire also. The last adult included was the 59th adult. Therefore, there were at least 59 but fewer than 69 adults in the theater.

23. We must determine the percentage of males and the percentage of females in the United States in 2002. The total number of males and females in the United States in 2002 was 141,661,000 + 146,708,000 = 288,369,000.

$$\text{Percentage of Males in the Population} = \frac{141,661,000}{288,369,000} \times 100\% \approx 0.4912 \times 100\% = 49.12\%$$

$$\text{Percentage of Females in the Population} = \frac{146,708,000}{288,369,000} \times 100\% \approx 0.5088 \times 100\% = 50.88\%$$

Therefore, out of 800 people in the sample, 49.12% should be males and 50.88% should be females.
Percentage of Males in the Sample = 49.12% of 800

$$= 0.4912 \times 800$$
$$= 392.96$$
$$\approx 393$$

Percentage of Females in the Sample = 50.88% of 800

$$= 0.5088 \times 800$$
$$= 407.04$$
$$\approx 407$$

Therefore, 393 males and 407 females should be selected.

25. The opinions of people living in Florida cannot be assumed to be representative of all the people living in the south. The results of the survey would be biased. Florida has a large percentage of older residents.

27. **(a)** There are two strata, one consisting of men, and one consisting of women.
   **(b)** The random number in row 113 and column 2 is 91130. Use the second and third digits and eliminate repetitions. The men in the sample include those with the following numbers: 11, 64, 24, 33, 36, 55, 21, 54, 66, and 43.
   The random number in row 113 and column 3 is 74270. Use the second and third digits and eliminate repetitions. The women in the sample include those with the following numbers: 42, 21, 59, 33, 78, 65, 46, 75, 29, and 64.

29. **(a)** Stratum I is made up of freshmen and sophomores. Stratum II is made up of juniors and seniors. Stratum III is made up of graduate students. It may be reasonable to group the classes into three strata. Students in the freshman and sophomore classes who are just starting out in college may be less likely to support another fee, so their opinions may be more homogeneous than the general population of students. Likewise, students who have a year or two until graduation, those in the

stratum containing juniors and seniors may be more likely to support a new fee because they will not have to pay it for long and they will be able to use the upgraded computer facilities.

**(b)** Randomly select a sample from each stratum in proportion to the number of students in each stratum.

Percent of Students in Stratum I $= \dfrac{950}{2000} \times 100\% = 47.5\%$

Percent of Students in Stratum II $= \dfrac{800}{2000} \times 100\% = 40\%$

Percent of Students in Stratum III $= \dfrac{250}{2000} \times 100\% = 12.5\%$

There will be 40 students in the sample. Select $0.475 \times 40 = 19$ from Stratum I. Select $0.40 \times 40 = 16$ from Stratum II. Select $0.125 \times 40 = 5$ from Stratum III.

**(c)** **Sampling from Stratum I:** The random number in column 1 and row 105 is 43857. Use the first three digits and read down the column. The 19 students selected are students 438, 918, 340, 491, 724, 842, 615, 253, 401, 584, 180, 527, 052, 760, 290, 456, 076, 949, and 505.

**Sampling from Stratum II:** The random number in column 3 and row 105 is 49026. Use the first three digits and read down the column. The 16 students selected are students 490, 370, 486, 359, 784, 413, 742, 284, 500, 021, 159, 633, 178, 365, 492, and 046.

**Sampling from Stratum III:** The random number in column 5 of row 105 is 51382. Use the first three digits and read down the column. The 5 students selected are students 238, 067, 115, 152, and 093.

**31. (a)** The sampling units are the dormitory rooms. If a total of 60 residents will be selected, then 20 sampling units must be selected because there are 3 residents living in each room.

**(b)** The random number in column 2 and row 115 is 32457. Use the second and third digits and read down the column. The rooms in the sample are those with numbers 24, 33, 36, 55, 21, 54, 66, 43, 46, 60, 39, 25, 41, 71, 09, 77, 58, 29, 75, and 56.

**33. (a)** If regions are used as sampling units, then there are five sampling units. A random sample of the regions could be taken and all of the public waters in the region investigated. Travel time would be reduced because all of the regions are not included in the sample.

**(b)** If counties are the clusters and all public waters are investigated in each county selected, then there may be several counties in the list which contain no public waters. Sampling counties all over the state would increase travel time and cost.

**(c)** There are 29 public waters in central Ohio. The public waters are marked and numbered from 01 to 29. The random number in column 2 and row 125 is 14648. The 10 public waters included in the sample are 25, 09, 29, 27, 01, 22, 13, 21, 24, and 20.

**35. (a)** The 48 random digits, reading across the row, beginning in column 1 of row 109 are as follows: 72479 24246 35932 33358 34853 77573 84281 57601 78425 362. Select trees that have a 1, 2, 3, or 4 in the same position as the tree number. The trees that will be selected to be in the sample are trees number 02, 03, 06, 07, 08, 09, 11, 14, 15, 16, 17, 18, 21, 22, 25, 30, 32, 33, 35, 40, 43, 44, 46, and 48. The selected trees are shaded in the following diagram.

Out of the 48 trees, 24 are selected. Therefore, the percentage of trees selected is $\frac{24}{48} \times 100\% = 50\%$. It takes 30 minutes per tree to do the inspection, so 2 trees can be inspected in one hour. In 12 hours, 2 trees per hour × 12 hours = 24 trees can be inspected. It costs $35 per hour. The total cost of the inspection is $35 per hour × 12 hours = $420.

**(b)** The random number in column 4 and row 135 is 50892. Use the first two digits and read down the column to randomly select 16 trees. The trees in the sample include 04, 40, 30, 39, 03, 10, 34, 23, 06, 37, 41, 11, 15, 09, 35, and 44. The selected trees are shaded in the following diagram.

Out of the 48 trees, 16 are selected. Therefore, the percentage of trees selected is $\frac{16}{48} \times 100\% = 33\frac{1}{3}\%$. It takes 30 minutes per tree to do the inspection, so 2 trees can be inspected in one hour. In 8 hours, 2 trees per hour × 8 hours = 16 trees can be inspected. It costs $35 per hour. The total cost of the inspection is $35 per hour × 8 hours = $280.

**(c)** The fourth digit in column 3 and row 142 is 1. Begin by including tree 01 in the sample and every third tree after that. The trees in the sample include 01, 04, 07, 10, 13, 16, 19, 22, 25, 28, 31, 34, 37, 40, 43, and 46. The selected trees are shaded in the following diagram.

Out of the 48 trees, 16 are selected. Therefore, the percentage of trees selected is $\frac{16}{48} \times 100\% \approx 33.3\%$. It takes 30 minutes per tree to do the inspection, so 2 trees can be inspected in one hour. In 8 hours, 2 trees per hour × 8 hours = 16 trees can be inspected. It costs $35 per hour. The total cost of the inspection is $35 per hour × 8 hours = $280.

**(d)** Answers will vary. Remember that independent samples do not have fixed sample sizes. The sample size is determined after the sample is taken. If money is limited, it would be best not to use a method that could put you over budget. With a simple random sample, it is possible that all the sampled trees could come from one area, so there is a chance that pockets of pests are missed in the sampling. Systematic sampling samples systematically throughout the orchard. The costs will not exceed a fixed amount. Systematic sampling may be the most appropriate in this case.

**37.** Answers will vary.
   **(a)** It is a cluster sample. The clusters are the cities. Once a city is selected, all children in each school in the cluster are interviewed. Interviewing all children in the schools in each selected city would help keep travel time and travel costs down. However, clusters may vary greatly in how many children get interviewed. It will cost to interview each child. With this technique, there is no way to fix the number of interviews in advance. There may be significant economic and social differences between cities. There is no guarantee that a representative sample of cities will be obtained.

**(b)** It is a cluster sample. The clusters are the schools. Once a school is selected, all children in the school are interviewed. With schools as the clusters, it is possible for the interviewer to have to go to a city and visit only one school in that city. Travel time and costs would increase, most likely, compared to the method in part (a). The number of interviews is not fixed due to differing enrollments in each school.

**(c)** It is a simple random sample. The sample size can be fixed in advance. The difficulty is that the interviewer would spend a lot of time traveling to different schools to interview possibly only one child in that school. This method would be the least biased.

**(d)** The state is divided into two nonoverlapping strata: urban and rural. A fixed number of schools are randomly selected from each stratum. This is a stratified random sample. The total number of interviews that will be done is unknown until the sample is taken. This method guarantees that there will be representative elements from both urban and rural areas.

**39.** Answers will vary.

**41.** Answers will vary.

## Solutions to Odd-Numbered Problems in Section 9.3

**1. (a) Calculate the mean:** There are 14 players. The sum of the heights is 1119 inches. The mean is $\frac{1119}{14} \approx 79.93$ inches.

**Calculate the median:** The heights must be ordered from least to greatest:
74, 74, 78, 79, 79, 79, 79, 80, 82, 82, 82, 83, 84, 84

Because there is an even number of heights, the median is the mean of the two middle heights. The median is $\frac{79+80}{2} = \frac{159}{2} = 79.5$ inches.

**Determine the mode:** The mode is the most frequently observed height. The height 79 inches occurs four times, and no other height appears more than three times in the list. The mode is 79 inches.

**(b) Calculate the mean:** There are 14 players. The sum of the weights is 3221 pounds. The mean is $\frac{3221}{14} \approx 230.07$ pounds.

**Calculate the median:** The weights must be ordered from least to greatest:
188, 195, 200, 215, 218, 220 230, 230, 230, 240, 250, 260, 265, 280

Because there is an even number of weights, the median is the mean of the two middle weights. The median is $\frac{230+230}{2} = \frac{460}{2} = 230$ pounds.

**Determine the mode:** The mode is the most frequently observed weight. The weight 230 pounds occurs three times and no other weight appears more than once in the list. The mode is 230 pounds.

**(c)** Marcus Banks has a height of 74 inches which is the smallest height.

**Recalculate the mean:** The sum of the heights without Marcus Banks' height is 1119 − 74 = 1045 inches. There are now 13 heights. The mean is $\frac{1045}{13} \approx 80.38$ inches.

**Recalculate the median:** The ordered heights without Marcus Banks' height are given next:
74, 78, 79, 79, 79, 79, 80, 82, 82, 82, 83, 84, 84

There is an odd number of heights, so the median is the middle height. The middle height is 80 inches. The median is 80 inches.

**Determine the new mode:** The mode is still 79 inches. It occurs four times.
The mode did not change. The deleted height was not one of the original most frequently observed heights.

(d) Kendrick Perkins has a weight of 280 pounds which is the largest weight.
**Recalculate the mean:** The sum of the weights without Kendrick Perkins' weight is $3221 - 280$ $= 2941$ pounds. There are now 13 weights. The mean is $\dfrac{2941}{13} \approx 226.23$ pounds.

**Recalculate the median:** The ordered weights without Kendrick Perkins' weight are given next:
$$188, 195, 200, 215, 218, 220 \ 230, 230, 230, 240, 250, 260, 265$$
There is an odd number of weights, so the median is the middle weight. The middle weight is 230 pounds. The median is 230 pounds.
**Determine the new mode:** The mode is 230 pounds.
The mean changed because it is calculated using all of the weights. If any one weight is deleted, then the mean will change.

3. (a) **Calculate the mean:** The sum of the homicide rates for the years 1980 through 1989 is 87.2. There are 10 rates. The mean is $\dfrac{87.2}{10} = 8.72$ homicides per 100,000.

**Calculate the median:** The rates must be ordered from smallest to largest:
$$7.9, 7.9, 8.3, 8.3, 8.4, 8.6, 8.7, 9.1, 9.8, 10.2$$
There is an even number of rates, so the median is the mean of the middle two rates. The median is $\dfrac{8.4+8.6}{2} = \dfrac{17}{2} = 8.5$ homicides per 100,000.

(b) **Calculate the mean:** The sum of the homicide rates for the years 1990 through 1999 is 81.4. There are 10 rates. The mean is $\dfrac{81.4}{10} = 8.14$ homicides per 100,000.

**Calculate the median:** The rates must be ordered from least to greatest:
$$5.7, 6.3, 6.8, 7.4, 8.2, 9.0, 9.3, 9.4, 9.5, 9.8$$
There is an even number of rates, so the median is the mean of the middle two rates. The median is $\dfrac{8.2+9.0}{2} = \dfrac{17.2}{2} = 8.6$ homicides per 100,000.

(c) To emphasize the idea that crime prevention policies are working, then we should report the statistics in such a way as to emphasize a decrease from the first time period to the next. For the earlier time period from 1980 to 1989, we should report the larger value which was the mean of 8.72 homicides per 100,000. In the later time period from 1990 to 1999, we should report the smaller value which was mean of 8.14 homicides per 100,000.

5. (a)

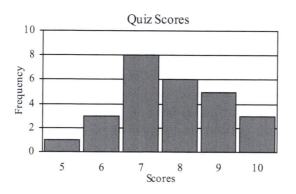

(b) **Calculate the mean:** The sum of the 26 scores is $(1 \times 5) + (3 \times 6) + (8 \times 7) + (6 \times 8) + (5 \times 9) +$ $(3 \times 10) = 5 + 18 + 56 + 48 + 45 + 30 = 202$. The mean is $\dfrac{202}{26} \approx 7.8$.

**Calculate the median:** Order the scores from least to greatest:

$$5, 6, 6, 6, 7, 7, 7, 7, 7, 7, 7, 7, 8, 8, 8, 8, 8, 8, 9, 9, 9, 9, 9, 10, 10, 10$$

There is an even number of scores, so the median is the mean of the middle two scores. The median is $\dfrac{8+8}{2} = \dfrac{16}{2} = 8$.

**Determine the mode:** Eight students earned a score of 7. This is the most frequently observed score. The mode is 7.

(c) From part (b), we see the mean is slightly less than the median. When this happens, the distribution is skewed left.

7. We will need the data values and must get them from the histogram. The data values are along the horizontal axis and their frequencies are represented by the heights of the bars. There are two 1s, four 3s, five 4s, two 5s, one 6, one 8, and one 9.

(a) **Calculate the mean:** The sum of the 16 data values is $(2 \times 1) + (4 \times 3) + (5 \times 4) + (2 \times 5) + (1 \times 6) + (1 \times 6) + (1 \times 8) + (1 \times 9) = 2 + 12 + 20 + 10 + 6 + 8 + 9 = 67$. The mean is $\dfrac{67}{16} \approx 4.2$.

**Calculate the median:** Order the data values from least to greatest:

$$1, 1, 3, 3, 3, 3, 4, 4, 4, 4, 4, 5, 5, 6, 8, 9$$

There is an even number of data values, so the median is the mean of the middle two values. The median is $\dfrac{4+4}{2} = \dfrac{8}{2} = 4$.

**Determine the mode:** The mode is the most frequently observed data value. It is easily read off of a histogram. The tallest bar represents the data value that occurs the most. The mode is 4.

(b) The mean is greater than the median. Notice the way the data values are more concentrated to the left and trail off to the right. This is a distribution that is skewed right.

(c) Adding a data value will change the mean because it uses all of the data values in its calculation. The median will not change because 4 will be the middle value no matter where the new value is placed. If a new value of 3 is added, then the distribution would be bimodal and would have two modes: 3 and 4.

9. Answers may vary.
(a) The median temperature may be the best representation of the normal body temperature because the median is not affected by extreme temperatures such as those a person has with a fever.
(b) The store wants to sell the shoes it orders, so the mode would be the most useful statistic. It represents the most often purchased shoe size.

11. The product of the per-capita income of a state and the population of the state gives the total income of the state. The per-capita personal income for Hawaii, Alaska, Washington, Oregon, and California can be calculated as the weighted mean of the per-capita personal incomes using the populations as the weights. The 2002 per-capita personal income is calculated as follows:

$$\frac{(1,234,514 \times 27,011) + (640,841 \times 28,947) + (6,067,146 \times 29,420) + (3,523,281 \times 25,867) + (34,988,261 \times 29,707)}{1,234,514 + 640,841 + 6,067,146 + 3,523,281 + 34,988,261}$$

$$\approx \$29,296.14$$

The 2002 personal per-capita income for the west coast states is $29,296.14.

13. (a) Many answers are possible. One solution will be given here, but you should try to create your own solution. We know the mode is 61, so it is the most frequent value. Out of the five values, if we include two values of 61, and make sure the other values occur one time each, then the mode will be 61. There are five values, so the median is the middle value. The middle value must be 55. So far, our data values include the following: Value 1, Value 2, 55, 61, and 61. The mean is 50, so we must have one or both of the remaining data values less than 50 because three values are

greater than 50. Suppose we let Value 2 be 45, then the data set includes Value 1, 45, 55, 61, and 61. If we were to calculate the mean from this information we would have the following:

$$\frac{\text{Value } 1 + 45 + 55 + 61 + 61}{5} = 50$$

$$\text{Value } 1 + 45 + 55 + 61 + 61 = 50 \times 5$$

$$\text{Value } 1 + 222 = 250$$

$$\text{Value } 1 = 28$$

One possible set of data values is 28, 45, 55, 61, and 61.

(b) Many answers are possible. One solution will be given here, but you should try to create your own solution. There is no mode, so the six values each occur one time. The range is 27, so we know the difference between the greatest and least data value is 27. The mean is 14, so data values must be selected so that some are less than 14 and some are greater than 14. Suppose the data value 1 is selected and is the smallest value in the data set, then we know the value 28 must also be included because the range is 27. If three more values are selected such as 10, 12, and 15, then there is one value left to include so the data set includes the following values: 1, 10, 12, 15, and 28. Let $x$ be the missing data value and calculate the mean.

$$\frac{x + 1 + 10 + 12 + 15 + 28}{6} = 14$$

$$x + 66 = 14 \times 6$$

$$x + 66 = 84$$

$$x = 18.$$

One possible set of data values is 1, 10, 12, 15, 18, and 28.

15. The ordered data set is 1, 3, 4, 5, 6, 8, 9, 10, 11, 12, and 15. There are 11 data values.
   (i) The range is the largest data value minus the smallest data value. Range = 15 − 1 = 14.
   (ii) The median is the middle value. In this case, the median is 8.
   (iii) The lower half of the data includes the values 1, 3, 4, 5, and 6. The first quartile is the median of the lower half of the data. There is an odd number of data values in the lower half of the data so the first quartile is 4.
   The upper half of the data includes the values 9, 10, 11, 12, and 15. The third quartile is the median of the upper half of the data. There is an odd number of data values in the upper half of the data so the third quartile is 11.
   (iv) The interquartile range is the difference between the third and first quartiles. The interquartile range is 11 − 4 = 7.
   (v) The five-number summary is 1, 4, 8, 11, 15.
   (vi)

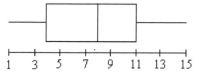

17. The ordered data set is 6, 9, 10, 12, 13, 14, 18, 21, 24, and 26. There are 10 data values.
   (i) The range is the largest data value minus the smallest data value. Range = 26 − 6 = 20.
   (ii) The median is the middle value. Because there is an even number of values, the median is the mean of the middle two values. In this case, the median is $\frac{13 + 14}{2} = \frac{27}{2} = 13.5$.
   (iii) The lower half of the data includes the values 6, 9, 10, 12, and 13. The first quartile is the median of the lower half of the data. There is an odd number of data values in the lower half of the data so the first quartile is 10.

The upper half of the data includes the values 14, 18, 21, 24, and 26. The third quartile is the median of the upper half of the data. There is an odd number of data values in the upper half of the data so the third quartile is 21.

    (iv) The interquartile range is the difference between the third and first quartiles. The interquartile range is $21 - 10 = 11$.

    (v) The five-number summary is 6, 10, 13.5, 21, 26.

(vi)

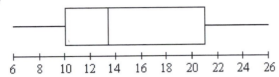

**19. (a)** The five-number summary is 39, 48, 57, 63, 69.

    **(b)** The location of the median is not in the center of the plot. The left whisker is longer than the right whisker. The distribution appears to be more concentrated to the right. This distribution is skewed to the left.

**21. (a)** Estimates may vary slightly. The five-number summary for Maryland Community Colleges is 68, 75, 80, 85, 100. The five-number summary for Oregon Community Colleges is 47, 51, 54, 60, 63.

    **(b)** Answers may vary. There is no overlap in the box plots for Maryland Community Colleges and Oregon Community Colleges. Every community college in Maryland charges more per credit than every community college in Oregon. The median cost per credit in Maryland is $26 more then the median cost per credit in Oregon. Notice the range of costs in Oregon is $63 – $47 = $16 while the range of costs in Maryland is $100 – $68 = $32. There is more variation in the cost per credit for community colleges in Maryland. Notice the plot for Oregon schools is slightly skewed to the right.

**23. (a)** Order the earthquake magnitudes from least to greatest: 2.0, 2.0, 2.0, 2.0, 2.0, 2.0, 2.0, 2.1, 2.1, 2.2, 2.2, 2.2, 2.2, 2.3, 2.3, 2.3, 2.4, 2.4, 2.5, 2.5, 2.7, 2.9, 3.2, and 3.5. The first quartile is $\dfrac{2.0 + 2.0}{2} = \dfrac{4.0}{2} = 2.0$. The median is $\dfrac{2.2 + 2.2}{2} = \dfrac{4.4}{2} = 2.2$. The third quartile is $\dfrac{2.4 + 2.5}{2} = \dfrac{4.9}{2} = 2.45$. The five-number summary is 2.0, 2.0, 2.2, 2.45, 3.5.

    **(b)**

```
 ┌──┬────────────────────────┐
 │  │                        │
 └──┴────────────────────────┘
├──┼──┼──┼──┼──┼──┼──┼──┤
2.0 2.2 2.4 2.6 2.8 3.0 3.2 3.4 3.6
        Earthquake Magnitudes
```

    **(c)** Notice the long right-hand whisker in the plot. Most of the data are concentrated to the left. This distribution is very skewed to the right.

**25.** The salary data are already ordered from least to greatest in the table. There are 16 salaries. The first quartile is $\dfrac{\$1,285,440 + \$1,325,000}{2} = \dfrac{\$2,610,440}{2} = \$1,305,220$. The median is $\dfrac{\$3,380,457 + \$3,416,520}{2} = \dfrac{\$6,796,977}{2} = \$3,398,488.50$. The third quartile is

$$\frac{\$5,955,000 + \$6,187,500}{2} = \frac{\$12,142,500}{2} = \$6,071,250.$$ The five-number summary is $366,931;
$1,305,220; $3,398,488.50; $6,071,250; $6,834,444.

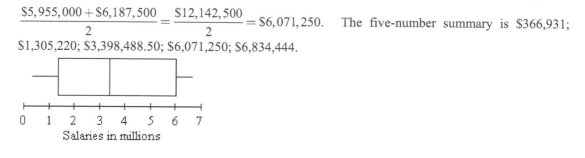

Salaries in millions

27. **(a)** The 14 ordered heights of the Denver Nuggets are 65, 74, 76, 76, 77, 80, 81, 81, 82, 82, 83, 83, 84, and 84. The first quartile is 76. The median is $\frac{81+81}{2} = \frac{162}{2} = 81$. The third quartile is 83. The five-number summary is 65, 76, 81, 83, 84.

 **(b)** The 13 ordered heights of the Houston Rockets are 71, 75, 75, 76, 76, 77, 79, 79, 81, 81, 81, 83, and 90. The first quartile is $\frac{75+76}{2} = \frac{151}{2} = 75.5$. The median is 79. The third quartile is $\frac{81+81}{2} = \frac{162}{2} = 81$. The five-number summary is 71, 75.5, 79, 81, 90.

 **(c)** Denver Nuggets' box-and-whisker plot:

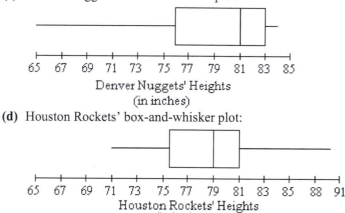

Denver Nuggets' Heights
(in inches)

 **(d)** Houston Rockets' box-and-whisker plot:

Houston Rockets' Heights
(in inches)

 **(e)** Observations will vary. The median of the heights for the Denver Nuggets is 81. The third quartile for the Houston Rockets heights is 81. Half of the Denver Nugget players are taller than three-fourths of the Houston Rocket players. The first quartile height for both teams is approximately 76 inches. For both teams one-quarter of the players are at most 76 inches tall. Notice that the smallest height for the Denver Nuggets is 65 inches while the smallest height for the Houston Rockets is 71 inches.

29. Answers will vary. One solution will be provided, but you should create a different one. If the mean, median, and mode are all the same, and the values must not all be the same, then a data set must be created that is symmetric. In a distribution that is skewed, the mean and median will be different. To make sure the mode is the same as the mean and median, the middle observation must be the most frequently observed value. One possible data set is 1, 2, 2, 2, and 3. For this data set, the mean, median, and mode are all 2.

**31.** Let $x$ represent the missing score. We know the mean is 80.7. Set up an equation and solve for the missing score.

$$\text{Mean Score} = \frac{x + 66 + 72 + 75 + 76 + 81 + 86 + 88 + 90 + 94}{10}$$

$$80.7 \times 10 = x + 728$$
$$807 = x + 728$$
$$807 - 728 = x$$
$$79 = x.$$

The missing score is 79.

**33. (a)** The mean of the data set is $\dfrac{4 + 10 + 7 + 1 + 5}{5} = \dfrac{27}{5} = 5.4$.

**(b)** To find the deviation from the mean for a value, subtract the mean from the data value:

$4 - 5.4 = -1.4$
$10 - 5.4 = 4.6$
$7 - 5.4 = 1.6$
$1 - 5.4 = -4.4$
$5 - 5.4 = -0.4$

The deviations from the mean are $-1.4$, $4.6$, $1.6$, $-4.4$, and $-0.4$.

**(c)** Square each of the deviations from the mean from part (b).

$(-1.4)^2 = 1.96$

$(4.6)^2 = 21.16$

$(1.6)^2 = 2.56$

$(-4.4)^2 = 19.36$

$(-0.4)^2 = 0.16$

**(d)** The sum of the squared deviations is $1.96 + 21.16 + 2.56 + 19.36 + 0.16 = 45.2$.

Divide the sum by the total number of data values: $\dfrac{45.2}{5} = 9.04$. This value is called the population variance.

**35. (a)** The sample mean is $\dfrac{4 + 6 + 7 + 10 + 13}{5} = \dfrac{40}{5} = 8$.

| Data Point | Deviation from the Mean | (Deviation from the Mean)$^2$ |
|---|---|---|
| 4 | $4 - 8 = -4$ | $(-4)^2 = 16$ |
| 6 | $6 - 8 = -2$ | $(-2)^2 = 4$ |
| 7 | $7 - 8 = -1$ | $(-1)^2 = 1$ |
| 10 | $10 - 8 = 2$ | $(2)^2 = 4$ |
| 13 | $13 - 8 = 5$ | $(5)^2 = 25$ |

The sum of the squared deviations from the mean is $16 + 4 + 1 + 4 + 25 = 50$. The sample variance is $\dfrac{50}{5-1} = \dfrac{50}{4} = 12.5$. The sample standard deviation is $\sqrt{12.5} \approx 3.54$.

**(b)** The sample mean is $\dfrac{2+2+1+2+4+12}{6} = \dfrac{23}{6} \approx 3.8333$.

| Data Point | Deviation from the Mean | (Deviation from the Mean)$^2$ |
|---|---|---|
| 2 | $2 - 3.8333 = -1.8333$ | $(-1.8333)^2 \approx 3.3610$ |
| 2 | $2 - 3.8333 = -1.8333$ | $(-1.8333)^2 \approx 3.3610$ |
| 1 | $1 - 3.8333 = -2.8333$ | $(-2.8333)^2 \approx 8.0276$ |
| 2 | $2 - 3.8333 = -1.8333$ | $(-1.8333)^2 \approx 3.3610$ |
| 4 | $4 - 3.8333 = 0.1667$ | $(0.1667)^2 \approx 0.0278$ |
| 12 | $12 - 3.8333 = 8.1667$ | $(8.1667)^2 \approx 66.6950$ |

The sum of the squared deviations from the mean is 84.8334. The sample variance is $\dfrac{84.8334}{6-1} = \dfrac{84.8334}{5} \approx 16.97$. The sample standard deviation is $\sqrt{\dfrac{84.8334}{5}} \approx 4.12$.

**(c)** The sample mean is $\dfrac{3+4+4+4+5+5+5+6}{8} = \dfrac{36}{8} = 4.5$.

| Data Point | Deviation from the Mean | (Deviation from the Mean)$^2$ |
|---|---|---|
| 3 | $3 - 4.5 = -1.5$ | $(-1.5)^2 = 2.25$ |
| 4 | $4 - 4.5 = -0.5$ | $(-0.5)^2 = 0.25$ |
| 4 | $4 - 4.5 = -0.5$ | $(-0.5)^2 = 0.25$ |
| 4 | $4 - 4.5 = -0.5$ | $(-0.5)^2 = 0.25$ |
| 5 | $5 - 4.5 = 0.5$ | $(0.5)^2 = 0.25$ |
| 5 | $5 - 4.5 = 0.5$ | $(0.5)^2 = 0.25$ |
| 5 | $5 - 4.5 = 0.5$ | $(0.5)^2 = 0.25$ |
| 6 | $6 - 4.5 = 1.5$ | $(1.5)^2 = 2.25$ |

The sum of the squared deviations from the mean is 6. The sample variance is $\dfrac{6}{8-1} = \dfrac{6}{7} \approx 0.86$.

The sample standard deviation is $\sqrt{\dfrac{6}{7}} \approx 0.93$.

**37. (a)** The original data are 3, 10, 9, 7, and 15.

The mean is $\dfrac{3+10+9+7+15}{5} = \dfrac{44}{5} = 8.8$.

$$
\begin{aligned}
s &= \sqrt{\frac{(3-8.8)^2 + (10-8.8)^2 + (9-8.8)^2 + (7-8.8)^2 + (15-8.8)^2}{5-1}} \\
&= \sqrt{\frac{(-5.8)^2 + (1.2)^2 + (0.2)^2 + (-1.8)^2 + (6.2)^2}{4}} \\
&= \sqrt{\frac{33.64 + 1.44 + 0.04 + 3.24 + 38.44}{4}} \\
&= \sqrt{\frac{76.8}{4}} \\
&= \sqrt{19.2}
\end{aligned}
$$

$s \approx 4.38$

The mean for the original data set is 8.8, and the sample standard deviation is approximately 4.38.

**(b)** Modify the data by adding 5 to each value. The modified data are 8, 15, 14, 12, and 20.

The mean is $\dfrac{8+15+14+12+20}{5} = \dfrac{69}{5} = 13.8$.

$$s = \sqrt{\frac{(8-13.8)^2 + (15-13.8)^2 + (14-13.8)^2 + (12-13.8)^2 + (20-13.8)^2}{5-1}}$$

$$= \sqrt{\frac{(-5.8)^2 + (1.2)^2 + (0.2)^2 + (-1.8)^2 + (6.2)^2}{4}}$$

$$= \sqrt{\frac{33.64 + 1.44 + 0.04 + 3.24 + 38.44}{4}}$$

$$= \sqrt{\frac{76.8}{4}}$$

$$= \sqrt{19.2}$$

$s \approx 4.38$

The mean for the modified data set is 13.8, and the sample standard deviation is approximately 4.38.

**(c)** The mean from part (a) is 8.8. The mean from part (b) is 13.8. Notice that we added 5 to each value and the mean of the modified data is 5 more than the original mean. In general whether we have a sample or a population, adding a constant to each data value will increase the mean by an amount equal to the constant. Consider the case in which we added 5 to each data value:

$$\text{Modified Mean} = \frac{(x_1+5)+(x_2+5)+\cdots+(x_N+5)}{N}$$

$$= \frac{x_1+x_2\cdots+x_N+5N}{N}$$

$$= \frac{x_1+x_2\cdots+x_N}{N} + \frac{5N}{N}$$

$$= \text{Original Mean} + 5$$

This argument will be true no matter what constant is added.

**(d)** The sample standard deviations for the original data and for the modified data are the same. Adding a constant to every data value in a data set does not change how spread out the data values are with respect to the mean. The sample standard deviation will not change.

**39. (a)** The original data are 9, 7, 3, 10, and 15.

The mean is $\dfrac{9+7+3+10+15}{5} = \dfrac{44}{5} = 8.8$.

$$s = \sqrt{\frac{(9-8.8)^2 + (7-8.8)^2 + (3-8.8)^2 + (10-8.8)^2 + (15-8.8)^2}{5-1}}$$

$$= \sqrt{\frac{(0.2)^2 + (-1.8)^2 + (-5.8)^2 + (1.2)^2 + (6.2)^2}{4}}$$

$$= \sqrt{\frac{0.04 + 3.24 + 33.64 + 1.44 + 38.44}{4}}$$

$$= \sqrt{\frac{76.8}{4}}$$

$$= \sqrt{19.2}$$

$s \approx 4.38$

The mean for the original data set is 8.8, and the sample standard deviation is approximately 4.38.

**(b)** Modify the data by multiplying each data value by 3. The modified data are 27, 21, 9, 30, and 45.

The mean is $\dfrac{27+21+9+30+45}{5} = \dfrac{132}{5} = 26.4$.

$$s = \sqrt{\dfrac{(27-26.4)^2 + (21-26.4)^2 + (9-26.4)^2 + (30-26.4)^2 + (45-26.4)^2}{5-1}}$$

$$= \sqrt{\dfrac{(0.6)^2 + (-5.4)^2 + (-17.4)^2 + (3.6)^2 + (18.6)^2}{4}}$$

$$= \sqrt{\dfrac{0.36 + 29.16 + 302.76 + 12.96 + 345.96}{4}}$$

$$= \sqrt{\dfrac{691.2}{4}}$$

$$= \sqrt{172.8}$$

$s \approx 13.15$

The mean for the modified data set is 26.4, and the sample standard deviation is approximately 13.15.

**(c)** The mean from part (a) is 8.8. The mean from part (b) is 26.4. Notice that we multiplied each original data value by 3. If we multiply the original mean by 3, we get 8.8 × 3 = 26.4 which is the modified mean. In general whether we have a sample or a population, the mean of a data set in which every value is multiplied by a constant is the constant multiplied by the original mean. Consider the case in which we multiplied each data value by 3:

$$\text{Modified mean} = \dfrac{(3x_1)+(3x_2)+\cdots+(3x_N)}{N}$$

$$= \dfrac{3(x_1+x_2\cdots+x_N)}{N}$$

$$= 3\left(\dfrac{x_1+x_2\cdots+x_N}{N}\right)$$

$$= 3(\text{Original mean})$$

This argument is true no matter what constant is multiplied.

**(d)** The sample standard deviation from part (a) is approximately 4.38. The sample standard deviation from part (b) is approximately 13.15. Multiplying every data value by 3 changes the spread of the data by a factor of 3. The sample standard deviation of the modified data set is three times the original sample standard deviation. Consider the case in which we multiplied each data value by 3:

$$\text{Original sample standard deviation} = \sqrt{\dfrac{(x_1-\overline{x})^2 + (x_2-\overline{x})^2 + \cdots + (x_n-\overline{x})^2}{n-1}}$$

$$\text{Modified sample standard deviation} = \sqrt{\frac{(3x_1 - 3\overline{x})^2 + (3x_2 - 3\overline{x})^2 + \cdots + (3x_n - 3\overline{x})^2}{n-1}}$$

$$= \sqrt{\frac{9(x_1 - \overline{x})^2 + 9(x_2 - \overline{x})^2 + \cdots + 9(x_n - \overline{x})^2}{n-1}}$$

$$= \sqrt{9\left(\frac{(x_1 - \overline{x})^2 + (x_2 - \overline{x})^2 + \cdots + (x_n - \overline{x})^2}{n-1}\right)}$$

$$= 3\sqrt{\frac{(x_1 - \overline{x})^2 + (x_2 - \overline{x})^2 + \cdots + (x_n - \overline{x})^2}{n-1}}$$

$$= 3(\text{Original sample standard deviation}).$$

This argument is true no matter what constant is multiplied.

**41.** The mean of the height data is $\dfrac{1119}{14} \approx 79.92857$.

| Data Value | Data Value − Mean | (Data Value − Mean)$^2$ |
|---|---|---|
| 74 | −5.92857 | 35.14794224 |
| 84 | 4.07143 | 16.57654224 |
| 79 | −0.92857 | 0.862242245 |
| 79 | −0.92857 | 0.862242245 |
| 74 | −5.92857 | 35.14794224 |
| 80 | 0.07143 | 0.005102245 |
| 83 | 3.07143 | 9.433682245 |
| 82 | 2.07143 | 4.290822245 |
| 84 | 4.07143 | 16.57654224 |
| 79 | −0.92857 | 0.862242245 |
| 82 | 2.07143 | 4.290822245 |
| 78 | −1.92857 | 3.719382245 |
| 82 | 2.07143 | 4.290822245 |
| 79 | −0.92857 | 0.862242245 |

The sum of the squared deviations from the mean is 132.9285714.

$$\text{Sample variance} = \frac{132.9285714}{14-1} = \frac{132.9285714}{13} \approx 10.2 \text{ square inches.}$$

$$\text{Sample variance} = \sqrt{\frac{132.9285714}{13}} \approx 3.2 \text{ inches.}$$

**43. (a)** Any set of five values such that the five values are identical will have a population variance of zero. There will be no deviation from the mean.

**(b)** It is impossible to obtain a population variance that is negative. All deviations from the mean are squared. This will always result in nonnegative numbers. The population variance will always be greater than or equal to zero.

# Solutions to Chapter 9 Review Problems

1. (a) The population is the set of voters in the city.
   (b) The sample is the set of adults who are selected to be surveyed who are shopping downtown on one afternoon.
   (c) The variable of interest is whether or not a voter is in favor of eliminating metered parking downtown.
   (d) If the survey includes only those adults who are shopping downtown, then they may be overwhelmingly in favor of eliminating metered parking. These are the people who have to pay for parking. Their opinions may not reflect the opinions of the population of voters in the city.

2. Answers will vary.
   (a) It is possible that students coming from the library may not be in favor of raising fees to add a new nonacademic feature to the university. The survey itself may contain misleading questions. We do not know how these students will be selected to be included in the survey. The selection process could be a source of bias.
   (b) To obtain a sample that is representative of all students at the university, a simple random sample of all students could be taken. The university has records of all students enrolled, so a random sample can be selected.

3. Answers may vary. The five people could be numbered 1, 2, 3, 4, and 5. A random-number table could be used to generate a sample of size three. Randomly select a random number in the random-number table, use the first digit, and read down the column eliminating repeats. The first three occurrences of the digits 1 through 5 will indicate who is included in the sample.

4. The random number in column 2 and row 109 is 24246. Use the last two digits, ignore duplicates, and read down the column. The people selected are those assigned the numbers 46, 01, 55, 96, 30, 71, 57, 82, 03, 83, 19, 14, 33, 25, 48, 43, 10, 00, 75, and 29.

5. (a) The population is the set of bags of potato chips produced in a single day. The sample is the set of 25 bags. The variable of interest is whether a bag contains at least 12.5 ounces of potato chips.
   (b) The sample size is fixed at 25. We know there are a total of 4993 bags produced in a single day. The first bag produced and the last bag produced during the day will be sampled. The 1-in-$k$ systematic sample must use a $k$-value so that when the sampling is done, there are 25 bags sampled. If the first bag produced is sampled, then $r$ must be 1. After that first bag is selected, then every $k$th bag will be selected so that bags 1, $1 + k$, $1 + 2k$, $1 + 3k$, ..., $1 + 24k$ are selected. The last bag in the sample is the last bag produced in the day, so we know $1 + 24k = 4993$. Thus, $24k = 4992$ and $k = \dfrac{4992}{24} = 208$.
   (c) The random number in column 2 and row 110 is 57601. Read the first three digits, ignore repeats, and read down the column. The crates that will be included in the sample have numbers 576, 933, 284, 911, 964, 324, 433, 736, 698, and 488.

6. (a) **Calculate the mean:** Add all the ages and divide by 43. The sum of the ages of the presidents at inauguration is 2357. Mean age at inauguration is $\dfrac{2357}{43} \approx 54.81$ years.

   **Calculate the median:** The ages must be ordered from smallest to largest. There are 43 ages in the data set. Because there is an odd number of data values, the median will be the middle value, which in this case will be the 22nd age. The first 22 ages in order are 42, 43, 46, 46, 47, 48, 49, 49, 50, 50, 51, 51, 51, 51, 52, 52, 54, 54, 54, 54, 54, and 55. The 22nd age is 55, so the median age is 55 years.

**Determine the mode:** The mode is the most frequently occurring value. In this case, the age that occurred most often is 54. The mode is 54 years.

**(b)** The smallest data value is 42. The first quartile is the median of the first 21 ages and is 51. The median was calculated in part (a) and found to be 55. The third quartile is the median of the last 21 ages and is 58. The largest data value is 69. The five-number summary is 42, 51, 55, 58, 69.

**(c)**

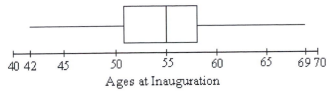

Ages at Inauguration

7. From problem 6(a) we know the mean of the presidents' ages at inauguration is $\dfrac{2357}{43} \approx 54.81395$ years. In order to calculate the population variance of the ages, find the sum of the squared deviations from the mean and divide by the number of data values.

$$\sigma^2 = \frac{(x_1 - \mu)^2 + \dots + (x_{43} - \mu)^2}{43}$$

$$\sigma^2 = \frac{(57 - 54.81395)^2 + \dots + (54 - 54.81395)^2}{43}$$

$$\sigma^2 \approx \frac{1632.512}{43}$$

$$\sigma^2 \approx 37.97$$

The population variance of ages is approximately 37.97 squared years. The population standard deviation is $\sigma = \sqrt{\sigma^2} = \sqrt{\dfrac{1632.512}{43}} \approx 6.16$ years.

8. **(a)** The digit 4 is randomly selected to start the systematic sample, so Madison is the first president to be selected into the sample. Every 6th president after Madison will also be selected into the sample. Select presidents 4, 10, 16, 22, 28, 34, and 40. Select Presidents Madison, Tyler, Lincoln, Cleveland, Wilson, Eisenhower, and Reagan.

**(b)** The ages of the presidents selected in the sample in part (a) are 57, 51, 52, 47, 56, 62, and 69. The sample mean is $\dfrac{57 + 51 + 52 + 47 + 56 + 62 + 69}{7} \approx 56.2857$.

| Data Value | Deviation from the Mean Data Value − 56.2857 | (Deviation from the Mean)$^2$ |
|---|---|---|
| 57 | 0.7143 | 0.5102 |
| 51 | −5.2857 | 27.9386 |
| 52 | −4.2857 | 18.3672 |
| 47 | −9.2857 | 86.2242 |
| 56 | −0.2857 | 0.0816 |
| 62 | 5.7143 | 32.6532 |
| 69 | 12.7143 | 161.6534 |

The sum of the squared deviations from the mean is 327.4284.

$$s = \sqrt{\frac{327.4284}{7-1}}$$

$$= \sqrt{\frac{327.4284}{6}}$$

$$s \approx 7.39$$

The sample mean is greater than the mean of the population. The sample included the oldest president at inauguration and did not include the youngest. The sample standard deviation is larger than the population standard deviation also.

9. The random number in row 120 and column 2 is 55519. Read across the row. For any of the first 43 digits that are a 1 or a 2, the corresponding president is included in the sample. A 1 or a 2 occurs in positions 4, 10, 14, 16, 18, 31, 32, 33, 34, and 40. The presidents included are Madison, Tyler, Pierce, Lincoln, Grant, Hoover, F. D. Roosevelt, Truman, Eisenhower, and Reagan. Ten out of the 43 presidents or $\frac{10}{43} \times 100\% \approx 23.3\%$ are included in the sample. A 20% independent sample does not mean that 20% of the population will be selected. It means that each element will have a 20% chance of being selected in the sample.

10. **(a)** The sum of the data values is 172. The mean is $\frac{172}{23} \approx 7.48$.

    The ordered data set is 6, 6, 6, 6, 7, 7, 7, 7, 7, 7, 7, 7, 7, 8, 8, 8, 8, 8, 8, 9, 9, 9, and 10. The median is 7. The mode is 7.

    **(b)** The mean of the data set is greater than the median, so this distribution is skewed right.

11. **(a)** The mean of the Week 1 exercise times is $\frac{40+20+27+20+30+28+45}{7} = \frac{210}{7} = 30$ minutes.

    Order the times as follows: 20, 20, 27, 28, 30, 40, 45. The median is the middle value. The median is 28 minutes.

    **(b)** We do not know the seventh exercise time during Week 2. The mean of the Week 2 exercise times is $\frac{30+22+25+25+30+27+x}{7}$. We want the mean exercise time during Week 1 to be the same as the mean exercise time for Week 2. The mean exercise time for Week 1 is 30 minutes. Therefore, we want:

    $$\text{Week 2 Mean} = \frac{30+22+25+25+30+27+x}{7}$$

    $$30 = \frac{159+x}{7}$$

    $$30 \times 7 = 159+x$$

    $$210 = 159+x$$

    $$210-159 = x$$

    $$51 = x.$$

    The person would have to exercise for 51 minutes on day seven of Week 2 so that the mean exercise times for both weeks are the same.

    **(c)** Order the known times for Week 2: 22, 25, 25, 27, 30, 30. We know the median exercise time for Week 1 is 28 minutes. For Week 2, we want to try to insert a time into the list, creating a list with seven times, so that the median is 28 minutes. If a time is inserted that is 25 minutes or less, the median time would be 25. If a time is inserted that is 27 minutes or greater, the median time would be 27. Finally, any time inserted between 25 and 27 would be the middle value in the data set, so the median would be that value that is between 25 and 27. There is no way to insert a value so that the median is 28.

**12.** The numbers in the data set are 3.2, 3.6, and 3.5. These data values are mean grades. The weights are the numbers of two-credit, three-credit, and four-credit classes taken.

$$\text{Weighted Mean} = \frac{(13 \times 3.2) + (22 \times 3.6) + (21 \times 3.5)}{13 + 22 + 21}$$

$$= \frac{41.6 + 79.2 + 73.5}{56}$$

$$= \frac{194.3}{56}$$

$$\approx 3.47$$

The grade point average for all classes taken is 3.47.

**13. (a)** The ordered data set is 1, 1, 2, 2, 3, 4, 4, 4, 5, 5, 6, 8, 9, and 10. There are 14 data values.

    (i)   The range is the largest data value minus the smallest data value. Range = 10 − 1 = 9.

    (ii)  The median is the mean of the two middle values. In this case, the median is $\dfrac{4+4}{2} = \dfrac{8}{2} = 4$.

    (iii) The lower half of the data includes the values 1, 1, 2, 2, 3, 4, and 4. The first quartile is the median of the lower half of the data. There is an odd number of data values in the lower half of the data so the first quartile is 2.
        The upper half of the data includes the values 4, 5, 5, 6, 8, 9, and 10. The third quartile is the median of the upper half of the data. There is an odd number of data values in the upper half of the data so the third quartile is 6.

    (iv) The interquartile range is the difference between the third and first quartiles. The interquartile range is 6 − 2 = 4.

    (v)  The five-number summary is 1, 2, 4, 6, 10.

    (vi)

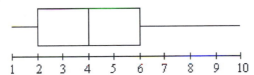

**(b)** The ordered data set is 22, 25, 26, 26, 29, 29, 30, 31, 31, 34, 38, and 40. There are 12 data values.

    (i)   The range is the largest data value minus the smallest data value. Range = 40 − 22 = 18.

    (ii)  The median is the mean of the two middle values. In this case, the median is $\dfrac{29+30}{2} = \dfrac{59}{2} = 29.5$.

    (iii) The lower half of the data includes the values 22, 25, 26, 26, 29, and 29. The first quartile is the median of the lower half of the data. There is an even number of data values in the lower half of the data so the first quartile is $\dfrac{26+26}{2} = \dfrac{52}{2} = 26$.

        The upper half of the data includes the values 30, 31, 31, 34, 38, and 40. The third quartile is the median of the upper half of the data. There is an even number of data values in the upper half of the data so the third quartile is $\dfrac{31+34}{2} = \dfrac{65}{2} = 32.5$.

    (iv) The interquartile range is the difference between the third and first quartiles. The interquartile range is 32.5 − 26 = 6.5.

    (v)  The five-number summary is 22, 26, 29.5, 32.5, 40.

(vi)

14. List the data values in the set. The heights of the bars in the histogram indicate how many of that particular data value there are. The data values are 15, 16, 16, 17, 17, 17, 17, 17, 18, 18, 18, 18, 18, 18, 19, 19, 19, 19, 20, 20, 20, 21, 21, 22, 22, 23, 24, 24, and 25. There are 29 data values.

   (a) The sum of the data values is 558. The mean is $\dfrac{558}{29} \approx 19.24$.

      The median is the middle data value. The median is 19.
      The mode is the most frequent value which corresponds to the data value with the tallest bar in the histogram. The mode is 18.

   (b) Looking at the histogram, we see more of the data values are concentrated to the left with fewer training off to the right. Also the mean is greater than the median. The distribution is skewed right.

   (c) We know the smallest value is 15. The greatest value is 25. The median from part (a) is 19. The first quartile is the median of the lower 14 values. The first quartile is $\dfrac{17+17}{2} = \dfrac{34}{2} = 17$. The third quartile is the median of the upper 14 values. The third quartile is $\dfrac{21+21}{2} = \dfrac{42}{2} = 21$. The five-number summary is 15, 17, 19, 21, 25.

   (d)

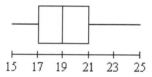

15. (a) Notice the average salaries are not listed in order from largest to smallest. There are 10 average salaries. Order the salaries:
      49,685; 51,076; 51,289; 51,424; 52,043; 52,600; 53,798; 54,158; 55,367; 56,283
      The smallest salary is $49,685. The first quartile is the median of the lower half of the data. The median of the lower half is the third salary. The first quartile is $51,289. The median is the mean of the middle two salaries. The median is $\dfrac{52,043+52,600}{2} = \dfrac{104,643}{2} = \$52,321.50$. The third quartile is the median of the upper half of the data. The median of the upper half is the eighth salary. The third quartile is $54,158. The largest salary is $56,283.
      The five-number summary is $49,685; $51,289; $52,321.50; $54,158; $56,283.

Top Ten Salaries
(in 1000s)

   (b) Notice the average salaries are not listed in order from largest to smallest. There are 10 average salaries. Order the salaries:
      32,416; 33,210; 34,555; 34,877; 35,754; 36,965; 37,300; 37,753; 37,896; 38,481
      The smallest salary is $32,416. The first quartile is the median of the lower half of the data. The median of the lower half is the third salary. The first quartile is $34,555. The median is the mean

of the middle two salaries. The median is $\dfrac{35,754+36,965}{2}=\dfrac{72,719}{2}=\$36,359.50$. The third

quartile is the median of the upper half of the data. The median of the upper half is the eighth salary. The third quartile is \$37,753. The largest salary is \$38,481.
The five-number summary is \$32,416; \$34,555; \$36,359.50; \$37,753; \$38,481.

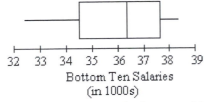

Bottom Ten Salaries
(in 1000s)

(c) Answers will vary. The lowest paid of the top ten group is approximately \$20,000 more than the highest paid of the bottom ten group. Notice the top ten average salary distribution is skewed right with more of the salaries clumped at the lower end of the scale. The bottom ten average salary distribution is skewed left with more of the salaries clumped at the upper end of the scale. The average salaries vary more in the top ten data.

16. (a) The sum of the top ten average salaries is \$527,723. The mean is $\dfrac{\$527,723}{10}=\$52,772.30$. The

range is the largest average salary – the lowest average salary. The range is \$56,283 – \$49,685 = \$6598.

(b) The sum of the bottom ten average salaries is \$359,207. The mean is $\dfrac{\$359,207}{10}=\$35,920.70$.

The range is the largest average salary – the lowest average salary. The range is \$38,481 – \$32,416 = \$6065.

17. (a) From the previous problem, we know the mean of the top ten average salaries.

| Data Value | Deviation from the Mean Data Value – 52,772.30 | (Deviation from the Mean)$^2$ |
|---|---|---|
| 56,283 | 3510.70 | 12,325,014.49 |
| 55,367 | 2594.70 | 6,732,468.09 |
| 54,158 | 1385.70 | 1,920,164.49 |
| 53,798 | 1025.70 | 1,052,060.49 |
| 52,600 | –172.30 | 29,687.29 |
| 51,424 | –1348.30 | 1,817,912.89 |
| 52,043 | –729.30 | 531,878.49 |
| 51,076 | –1696.30 | 2,877,433.69 |
| 51,289 | –1483.30 | 2,200,178.89 |
| 49,685 | –3087.30 | 9,531,421.29 |

The sum of the squared deviations from the mean is 39,018,220.10.
The population standard deviation for the top ten average salaries is calculated as follows:

$$\sqrt{\dfrac{39,018,220.10}{10}}=\sqrt{3,901,822.01}\approx\$1975.30.$$

**(b)** From the previous problem, we know the mean of the bottom ten average salaries.

| Data Value | Deviation from the Mean Data Value − 35,920.70 | (Deviation from the Mean)$^2$ |
|---|---|---|
| 37,753 | 1832.30 | 3,357,323.29 |
| 38,481 | 2560.30 | 6,555,136.09 |
| 36,965 | 1044.30 | 1,090,562.49 |
| 37,300 | 1379.30 | 1,902,468.49 |
| 37,896 | 1975.30 | 3,901,810.09 |
| 34,877 | −1043.70 | 1,089,309.69 |
| 35,754 | −166.70 | 27,788.89 |
| 34,555 | −1365.70 | 1,865,136.49 |
| 33,210 | −2710.70 | 7,347,894.49 |
| 32,416 | −3504.70 | 12,282,922.09 |

The sum of the squared deviations from the mean is 39,420,352.10.
The population standard deviation for the bottom ten average salaries is calculated as follows:

$$\sqrt{\frac{39,420,352.10}{10}} = \sqrt{3,942,035.21} \approx \$1985.46.$$

**18. (a)** Label the states with the top average salaries in the order they appear in the table with California being numbered 0 and Alaska being numbered 9. The random number in column 4 of row 120 is 55720. Read the digits along this row of numbers, ignore repeats, and sample five states. The first five states to be included in the sample are states numbered 5, 7, 2, 0, and 1. These states are Pennsylvania, Rhode Island, New Jersey, California, and Connecticut.

Label the states with the bottom average salaries in the order they appear in the table with Arkansas being numbered 0 and South Dakota being numbered 9. The random number in column 5 of row 125 is 81713. Read the digits along this row of numbers, ignore repeats, and sample five states. The first five states to be included in the sample are states numbered 8, 1, 7, 3, and 6. These states are North Dakota, West Virginia, Mississippi, Louisiana, and Montana.

**(b)** The average salaries for the five states in the sample are Pennsylvania: $51,424; Rhode Island: $51,076; New Jersey: $54,158; California: $56,283; and Connecticut: $55,367. The sum of the five salaries is $268,308. The mean of the sample of average salaries is $\dfrac{\$268,308}{5} = \$53,661.60.$

Use a table to organize the information and calculate the sample standard deviation.

| Data Value | Deviation from the Mean Data Value − 53,661.60 | (Deviation from the Mean)$^2$ |
|---|---|---|
| 56,283.00 | 2621.40 | 6,871,737.96 |
| 55,367.00 | 1705.40 | 2,908,389.16 |
| 54,158.00 | 496.40 | 246,412.96 |
| 51,424.00 | −2237.60 | 5,006,853.76 |
| 51,076.00 | −2585.60 | 6,685,327.36 |

The sum of the squared deviations from the mean is 21,718,721.20. The sample standard

deviation is $\sqrt{\dfrac{21,718,721.20}{5-1}} = \sqrt{\dfrac{21,718,721.20}{4}} = \sqrt{5,429,680.30} \approx \$2330.17.$

**(c)** The average salaries for the five states in the sample are West Virginia: $38,481; Louisiana: $37,300; Montana: $35,754; Mississippi: $34,555; and North Dakota: $33,210. The sum of the five salaries is $179,300. The mean of the sample of average salaries is $\dfrac{\$179,300}{5} = \$35,860.$

Use a table to organize the information and calculate the sample standard deviation.

| Data Value | Deviation from the Mean Data Value − 35,860 | (Deviation from the Mean)$^2$ |
|---|---|---|
| 38,481.00 | 2621.00 | 6,869,641 |
| 37,300.00 | 1440.00 | 2,073,600 |
| 35,754.00 | −106.00 | 11,236 |
| 34,555.00 | −1305.00 | 1,703,025 |
| 33,210.00 | −2650.00 | 7,022,500 |

The sum of the squared deviations from the mean is 17,680,002. The sample standard deviation is

$$\sqrt{\frac{17,680,002}{5-1}} = \sqrt{\frac{17,680,002}{4}} = \sqrt{4,420,000.50} \approx \$2102.38.$$

19. **(a)** The five-number summary for class 1 is 58, 69, 74, 82, 92.
The five-number summary for class 2 is 64, 72, 78, 84, 90.

**(b)** Answers will vary. Class 2 had higher first quartile, median, and third quartile scores and less variation in the scores compared to class 1.

# *Chapter 10:* Probability

## *Key Concepts*

**By the time you finish studying Section 10.1, you should be able to:**

☐ list the sample space for an experiment.

☐ find the experimental probability of an event.

☐ find the theoretical probability of an event with equally likely outcomes.

☐ find probabilities for experiments involving drawing with or without replacement.

☐ describe what it means for events to be mutually exclusive and find the probability of mutually exclusive events.

☐ describe in words and in symbols the union of two events and find the probability of the union of two events.

☐ describe in words and in symbols the intersection of two events and find the probability of the intersection of two events.

☐ describe the complement of an event and find the probability of the complement of an event.

**By the time you finish studying Section 10.2, you should be able to:**

☐ draw a one-stage or a two-stage tree diagram to represent the outcomes of an experiment.

☐ use the fundamental counting principal to determine the number of ways events can occur in succession.

☐ draw a probability tree diagram to represent probabilities in multistage experiments.

☐ determine the probability of mutually exclusive events by adding probabilities from a probability tree diagram.

☐ determine probabilities for experiments in which drawing is done with or without replacement.

☐ explain why probabilities of events can be found by multiplying probabilities along a series of branches in a probability tree diagram.

**By the time you finish studying Section 10.3, you should be able to:**

☐ describe the conditional sample space.

☐ calculate conditional probability using a table.

☐ calculate conditional probability using a tree diagram.

☐ give an example of two events that are independent.

☐ calculate the probability of independent events.

☐ determine whether two events are independent.

☐ find the expected value of an experiment.

☐ give the odds in favor of an event and against an event.

☐ given odds for an event, calculate the probability of the event.

☐ given the odds against an event, calculate the probability of the event.

## Skills Brush-Ups

Problems in Chapter 10 will require that you be comfortable adding, subtracting, multiplying, and dividing fractions. You will add or subtract fractions to find probabilities for more than one event, for mutually exclusive events, and for complementary events. You will multiply or divide fractions when finding probabilities for multi-stage experiments and for independent events as well as for conditional probabilities, expected value, and odds.

### FRACTIONS:

A fraction is a number written as $\frac{a}{b}$ where $a$ and $b$ are whole numbers and $b \neq 0$. The number $a$ is called the numerator and $b$ is called the denominator.

### Addition and Subtraction of Fractions with Common Denominators

Let $\frac{a}{b}$ and $\frac{c}{b}$ be fractions, then $\frac{a}{b} + \frac{c}{b} = \frac{a+c}{b}$ and $\frac{a}{b} - \frac{c}{b} = \frac{a-c}{b}$.

You can add or subtract fractions only if they have the same denominator. When the denominator is the same, add or subtract the numerators.

*Examples:* (a) $\dfrac{3}{7}+\dfrac{2}{7}=\dfrac{3+2}{7}=\dfrac{5}{7}$

(b) $\dfrac{7}{9}-\dfrac{5}{9}=\dfrac{7-5}{9}=\dfrac{2}{9}$

(c) $\dfrac{5}{8}+\dfrac{4}{8}-\dfrac{2}{8}=\dfrac{5+4-2}{8}=\dfrac{7}{8}$

### Addition and Subtraction of Fractions with Unlike Denominators

Let $\dfrac{a}{b}$ and $\dfrac{c}{d}$ be fractions, then

$$\dfrac{a}{b}+\dfrac{c}{d}=\dfrac{a\times d}{b\times d}+\dfrac{b\times c}{b\times d}=\dfrac{a\times d+b\times c}{b\times d} \quad \text{and} \quad \dfrac{a}{b}-\dfrac{c}{d}=\dfrac{a\times d}{b\times d}-\dfrac{b\times c}{b\times d}=\dfrac{a\times d-b\times c}{b\times d}.$$

When the denominators of fractions are different, create equivalent fractions with common denominators and then add or subtract the numerators.

*Examples:* (a) $\dfrac{3}{7}+\dfrac{2}{5}=\dfrac{3\times5}{7\times5}+\dfrac{7\times2}{7\times5}=\dfrac{15}{35}+\dfrac{14}{35}=\dfrac{29}{35}$

(b) $\dfrac{7}{9}-\dfrac{5}{8}=\dfrac{7\times8}{9\times8}-\dfrac{9\times5}{9\times8}=\dfrac{56}{72}-\dfrac{45}{72}=\dfrac{11}{72}$

(c) $1-\dfrac{5}{7}=\dfrac{7}{7}-\dfrac{5}{7}=\dfrac{2}{7}$

(d) $\dfrac{1}{8}+\dfrac{3}{4}-\dfrac{5}{16}=\dfrac{1\times2}{8\times2}+\dfrac{3\times4}{4\times4}-\dfrac{5}{16}=\dfrac{2}{16}+\dfrac{12}{16}-\dfrac{5}{16}=\dfrac{2+12-5}{16}=\dfrac{9}{16}$

Note: In this case, 16 is a multiple of each denominator, so we can create equivalent fractions with denominators of 16.

### Multiplication of Fractions

Let $\dfrac{a}{b}$ and $\dfrac{c}{d}$ be fractions, then $\dfrac{a}{b}\times\dfrac{c}{d}=\dfrac{a\times c}{b\times d}$.

Fractions do not need to have common denominators when they are multiplied.

*Examples:* (a) $\dfrac{8}{13}\times\dfrac{2}{3}=\dfrac{8\times2}{13\times3}=\dfrac{16}{39}$

(b) $\dfrac{4}{5}\times\dfrac{1}{6}=\dfrac{4\times1}{5\times6}=\dfrac{4}{30}=\dfrac{2}{15}$

Note: Express fractions in simplest form.

### Division of Fractions

Let $\dfrac{a}{b}$ and $\dfrac{c}{d}$ be fractions with $c\neq0$, then $\dfrac{a}{b}\div\dfrac{c}{d}=\dfrac{a}{b}\times\dfrac{d}{c}=\dfrac{a\times d}{b\times c}$.

*Examples:* (a) $\dfrac{7}{9}\div\dfrac{2}{9}=\dfrac{7}{9}\times\dfrac{9}{2}=\dfrac{7\times9}{9\times2}=\dfrac{7}{2}$

(b) $\dfrac{2}{13}\div\dfrac{4}{5}=\dfrac{2}{13}\times\dfrac{5}{4}=\dfrac{2\times5}{13\times4}=\dfrac{10}{52}=\dfrac{5}{26}$

Note: Express fractions in simplest form.

## Skills Practice

1. Add and express in simplest form.

(a) $\dfrac{3}{11}+\dfrac{4}{11}$      (b) $\dfrac{2}{15}+\dfrac{4}{5}$      (c) $\dfrac{2}{7}+\dfrac{4}{9}$      (d) $\dfrac{3}{4}+\dfrac{1}{6}$

2. Add or subtract and express in simplest form.

(a) $\dfrac{13}{14}-\dfrac{5}{7}$      (b) $1-\dfrac{7}{17}$      (c) $\dfrac{2}{3}+\dfrac{1}{5}-\dfrac{4}{15}$      (d) $\dfrac{1}{5}+\dfrac{5}{6}-\dfrac{3}{10}$

3. Multiply or divide and express in simplest form.

(a) $\dfrac{4}{7}\times\dfrac{2}{5}$      (b) $\dfrac{3}{5}\div\dfrac{2}{5}$      (c) $\dfrac{2}{3}\div\dfrac{4}{15}$      (d) $\dfrac{5}{6}\times\dfrac{3}{10}$

*Answers to Skills Practice:*

1. (a) $\dfrac{7}{11}$      (b) $\dfrac{14}{15}$      (c) $\dfrac{46}{63}$      (d) $\dfrac{11}{12}$

2. (a) $\dfrac{3}{14}$      (b) $\dfrac{10}{17}$      (c) $\dfrac{3}{5}$      (d) $\dfrac{11}{15}$

3. (a) $\dfrac{8}{35}$      (b) $\dfrac{3}{2}$      (c) $\dfrac{5}{2}$      (d) $\dfrac{1}{4}$

## Hints for Odd-Numbered Problems

### Section 10.1

1. (a) Remember that you need a common denominator to add fractions.
   (c) Remember that you do not need a common denominator to multiply fractions.

3. (b) Multiply first then add the fractions.

5. Refer to Example 10.1 and the discussion about interpreting probability.

7. See Example 10.2. Many events are possible.

9. See Example 10.2.

11. There will be 16 outcomes in the sample space.

13. You will have 8 outcomes in the array.

15. (a) See Example 10.3.
    (b) Use the experimental results.

17. See Example 10.4.

19. See Example 10.5.

21. See Example 10.5.

23. (a) How many equally-spaced sectors are there and how many are colored yellow?

25. (c) A prime number is a number greater than 1 with no positive integer divisors other than 1 and the number itself.

27. List the sample space.

29. Write out all the months of the year. How many correspond to event $T$ and how many correspond to event $Y$?

31. (a) How many children are there altogether?
    (b) When calculating $P(G \cap D)$ be sure not to count students twice.

33. (a) Compare the angle of each sector.

35. (a) The probability a person has blood type O is 0.45.

**37. (a)** Is it possible for both events *B* and *C* to occur together?

**39. (a)** For example, *GCG* is the event that the car is behind curtain number 2 and goats are behind curtains number 1 and 3.

**41.** See Example 10.9.

## Section 10.2

**1.** The branches in the tree diagram should represent the outcomes of the experiment.
 **(b)** How many outcomes are possible?

**3. (a)** In the first stage, a coin is tossed. In the second stage, a coin is tossed.
 **(b)** Select paint colors in the first stage. Select trim colors in the second stage.

**5.** Once a red ball is chosen, no more branches are needed at that point in the tree.

**7. (a)** There are two stages corresponding to the trip to New York City and the trip to London.

**9.** There are three stages for this experiment.

**11. (b)** There are four marbles, but you can combine branches as is done in Example 10.15.

**13.** Refer to Example 10.17.

**15.** Notice the central angles of each sector are the same.

**17. (a)** What must the sum of the probabilities be?
 **(b)** You are given the probability of event *R* and the probability of event *RG*.
 **(c)** Count the number of branches in the second stage of the probability tree diagram.

**19. (a)** There are four stages in this experiment. How many outcomes are possible for each stage?
 **(b)** How many outcomes in the sample space correspond to getting four heads?

**21. (c)** This will be a two-stage probability tree diagram. Probabilities will change in the second stage depending on what was selected in the first stage.
 **(f)** It might be easier to use the complement of the event.

**23. (e)** First find the probability of getting, for example, two identical 9s. There are eight 9s in the deck. Once one of those is selected, how many ways can the identical 9 be selected on the next draw?

**25. (b)** Use the probability tree diagram. Multiply the probabilities along the branches that correspond to getting defective bulbs.

**27. (b)** The player must lose on the first try and win on the second try.
 **(c)** The player must lose on the first two tries and win on the third try.

**29. (b)** There is no way to select M twice. Find the probabilities of selecting two Is, two Ss, and two Ps.

**31. (c)** List the ways to get two odd numbers.

## Section 10.3

**1. (f)** Use the definition of conditional probability and the probabilities calculated in part (e).

**3.** Create a probability tree diagram.

**5. (d)** Use the probabilities calculated in part (a).
 **(e)** Use the probabilities calculated in part (c).

**7.** Count the dots to determine the number of outcomes in the sample space.
 **(a)** Be sure to count all of the dots in the circle that represents event *A*.
 **(c)** Count the dots in the region that corresponds to events *A* and *B* happening at the same time.

9. Remember that $\overline{L}$ and $\overline{M}$ are the complements of the events $L$ and $M$ respectively.

11. See Example 10.23.

13. See Example 10.25.

15. See Example 10.25.
   (c) Assume that if the person does not have a college degree, that person also does not have a graduate degree.

17. It might be helpful to create a probability tree diagram. See Example 10.24.

19. Multiply probabilities along the branches.

21. Refer to the probability tree diagram from problem 19.

23. Refer to the probability tree diagram from problem 19.

25. Determine the probability a student randomly selected from the first class speaks Spanish. Do the same for the second class. Does the selection of a student in the first class influence the selection of a student in the second class?

27. (b) Create a probability tree diagram.

29. (a) If events $A$ and $B$ are independent, then $P(A \cap B) = P(A) \times P(B)$.
   (b) If events $A$ and $C$ are independent, then $P(A \cap C) = P(A) \times P(C)$.
   (c) If events $B$ and $C$ are independent, then $P(B \cap C) = P(B) \times P(C)$.

31. See Example 10.28.

33. See Example 10.29.

35. See Example 10.29.

37. (a) What is the probability you will win a quarter and the probability you lose a dollar?

39. See Example 10.31.

41. See Example 10.32.

43. (a) First find the probability of getting all heads.
   (b) First find the probability of getting one head.
   (c) First find the probability of getting exactly two tails.

## Solutions to Odd-Numbered Problems in Section 10.1

1. (a) $\dfrac{1}{8} + \dfrac{3}{4} = \dfrac{1}{8} + \dfrac{6}{8} = \dfrac{7}{8}$

   (b) $\dfrac{1}{3} + 1 - \dfrac{11}{15} = \dfrac{5}{15} + \dfrac{15}{15} - \dfrac{11}{15} = \dfrac{9}{15} = \dfrac{3}{5}$

   (c) $\dfrac{3}{10} \cdot \dfrac{2}{9} = \dfrac{6}{90} = \dfrac{1}{15}$

   (d) $\dfrac{\frac{4}{7}}{\frac{3}{2}} = \dfrac{4}{7} \cdot \dfrac{2}{3} = \dfrac{8}{21}$

3.  **(a)** $\dfrac{\dfrac{8}{15}}{\dfrac{11}{45}} = \dfrac{8}{15} \cdot \dfrac{45}{11} = \dfrac{8}{1} \cdot \dfrac{3}{11} = \dfrac{24}{11} = 2\dfrac{2}{11}$

   **(b)** $\dfrac{1}{4} \cdot \dfrac{2}{5} + \dfrac{3}{4} \cdot \dfrac{3}{5} = \dfrac{2}{20} + \dfrac{9}{20} = \dfrac{11}{20}$

   **(c)** $\dfrac{1}{3}(240) + \dfrac{2}{5}(-50) = 80 - 20 = 60$

5.  Select option (iii). Of past days when conditions were similar, 1 of 5 had some snow in the country.

7.  Events will vary.
    **(a)** Sample space: {H, T}
    Event: A head is tossed.
    **(b)** Sample space: {*A, B, C, D, E, F*}
    Event: An *E* is rolled.
    **(c)** Sample space: {0, 1, 2, 3, 4, 5, 6, 7, 8, 9}
    Event: A 5 is observed.

9.  **(a)** The sample space is {*A, B, C, D, E, F, G, H, I, J*}.
    **(b)** There are 3 vowels in the sample space: {*A, E, I*}.
    **(c)** There are 6 consonants in the sample space: {*B, C, D, F, G, H, J*}.
    **(d)** There are 4 letters between *B* and *G*: {*C, D, E, F*}.
    **(e)** The sample space contains only 1 letter in the word ZOOLOGY: {*G*}.

11. **(a)** There are 16 outcomes in the sample space: {HHHH, HHHT, HHTH, HTHH, THHH, HHTT, HTHT, HTTH, TTHH, THTH, THHT, HTTT, THTT, TTHT, TTTH, TTTT}.
    **(b)** Half of the outcomes in the sample space list the first coin as showing a head: {HHHH, HHHT, HHTH, HTHH, HHTT, HTHT, HTTH, HTTT}.
    **(c)** Four of the outcomes in the sample space show three heads: {HHHT, HHTH, HTHH, THHH}.
    **(d)** Half of the outcomes in the sample space list the fourth coin as showing a tail: {HHHT, HHTT, HTHT, THHT, HTTT, THTT, TTHT, TTTT}.
    **(e)** Four of the outcomes in the sample space list the second coin as showing a head and the third coin as showing a tail: {HHTH, HHTT, THTH, THTT}.

13.

```
      H │ (1, H)  (2, H)  (3, H)  (4, H)
Coin    │
      T │ (1, T)  (2, T)  (3, T)  (4, T)
        └──────────────────────────────
           1      2      3      4
              Four-Sided Die
```

The sample space of the experiment is {(1, H), (2, H), (3, H), (4, H), (1, T), (2, T), (3, T), (4, T)}.

15. **(a)** The experimental probability of an event is the number of times the event occurred divided by the total number of events.
    **(1)** The experimental probability of getting a 4 is $\dfrac{12}{60} = \dfrac{1}{5}$.
    **(2)** The odd numbers are 1, 3, and 5, and odd numbers were rolled a total of $10 + 10 + 8 = 28$ times out of the 60 rolls. The experimental probability of getting an odd number is $\dfrac{28}{60} = \dfrac{7}{15}$.

(3) Numbers greater than 3 include 4, 5, and 6, and these were observed a total of $12 + 8 + 11 = 31$ times out of the 60 rolls. The experimental probability of getting a number greater than 3 is $\dfrac{31}{60}$.

**(b)** An even number was observed $9 + 12 + 11 = 32$ times out of the 60 rolls so the experimental probability of getting an even number is $\dfrac{32}{60} = \dfrac{8}{15}$. If the die is rolled 250 times, we would expect approximately $\dfrac{8}{15}$ of the rolls to show an even number. We expect that $\dfrac{8}{15} \times 250 \approx 133.33$ or approximately 133 rolls show an even number.

**17. (a)** The sample space contains 8 outcomes {HHH, HHT, HTH, THH, TTH, THT, HTT, TTT}. Each outcome is equally likely.

| Outcome | HHH | HHT | HTH | THH | TTH | THT | HTT | TTT |
|---|---|---|---|---|---|---|---|---|
| Theoretical Probability | $\dfrac{1}{8}$ | $\dfrac{1}{8}$ | $\dfrac{1}{8}$ | $\dfrac{1}{8}$ | $\dfrac{1}{8}$ | $\dfrac{1}{8}$ | $\dfrac{1}{8}$ | $\dfrac{1}{8}$ |

**(b)** The theoretical probability of the event of getting at least 1 head is the probability of observing any one of the following 7 outcomes {HHH, HHT, HTH, THH, TTH, THT, HTT}. Therefore,

$$P(\text{getting at least 1 head}) = \frac{\text{number of outcomes in the event}}{\text{number of outcomes in sample space}} = \frac{7}{8}.$$

**(c)** The theoretical probability of the event of getting exactly 2 heads is the probability of observing any one of the following 3 outcomes {HHT, HTH, THH}. Therefore, $P(\text{getting exactly 2 heads}) = $

$$\frac{\text{number of outcomes in the event}}{\text{number of outcomes in sample space}} = \frac{3}{8}.$$

**19. (a)** There are 6 outcomes out of the 36 outcomes in the sample space that correspond to getting a 4 on the second die. Therefore, $P(\text{getting a 4 on the second die}) = \dfrac{6}{36} = \dfrac{1}{6}$.

**(b)** There are 9 outcomes out of the 36 outcomes in the sample space that correspond to getting an even number on each die. Therefore, $P(\text{getting an even number on each die}) = \dfrac{9}{36} = \dfrac{1}{4}$.

**(c)** There are 21 outcomes out of the 36 outcomes in the sample space that correspond to getting a total of at least 7 dots. Therefore, $P(\text{getting a total of at least 7 dots}) = \dfrac{21}{36} = \dfrac{7}{12}$.

**(d)** There are 0 outcomes out of the 36 outcomes in the sample space that correspond to getting a total of 15 dots. Therefore, $P(\text{getting a total of 15 dots}) = \dfrac{0}{36} = 0$.

**21. (a)** The sample space contains the following 36 outcomes:

(1, 1)  (1, 2)  (1, 3)  (1, 4)  (1, 5)  (1, 6)
(2, 1)  (2, 2)  (2, 3)  (2, 4)  (2, 5)  (2, 6)
(3, 1)  (3, 2)  (3, 3)  (3, 4)  (3, 5)  (3, 6)
(4, 1)  (4, 2)  (4, 3)  (4, 4)  (4, 5)  (4, 6)
(5, 1)  (5, 2)  (5, 3)  (5, 4)  (5, 5)  (5, 6)
(6, 1)  (6, 2)  (6, 3)  (6, 4)  (6, 5)  (6, 6)

**(b)** There are 27 outcomes out of the 36 outcomes in the sample space that correspond to getting even product. Therefore, $P(\text{getting an even product}) = \dfrac{27}{36} = \dfrac{3}{4}$.

**(c)** There are 9 outcomes out of the 36 outcomes in the sample space that correspond to getting odd product. Therefore, $P(\text{getting an odd product}) = \dfrac{9}{36} = \dfrac{1}{4}$.

**(d)** There are 11 outcomes out of the 36 outcomes in the sample space that correspond to getting a product that is a multiple of 5. Therefore, $P(\text{getting a product that is a multiple of 5}) = \dfrac{11}{36}$.

**23.** The spinner is divided into 8 equal sectors. Two sectors are white, 3 are yellow, 2 are blue, and 1 is red.

**(a)** $P(\text{Yellow}) = \dfrac{3}{8}$

**(b)** Two of the sectors are white and 2 are blue. The central angles are the same for both colors, so the probabilities are the same.

**25.** The sample space will have $12 \times 12 = 144$ outcomes. For each outcome, add the numbers showing on the dice. Each die can show any of the numbers 1 through 12. The outcomes for each die are listed down the far-left column and across the top row. The numbers showing on each die are added to get the totals listed in the body of the table.

| Total | 1 | 2 | 3 | 4 | 5 | 6 | 7 | 8 | 9 | 10 | 11 | 12 |
|---|---|---|---|---|---|---|---|---|---|---|---|---|
| 1 | 2 | 3 | 4 | 5 | 6 | 7 | 8 | 9 | 10 | 11 | 12 | 13 |
| 2 | 3 | 4 | 5 | 6 | 7 | 8 | 9 | 10 | 11 | 12 | 13 | 14 |
| 3 | 4 | 5 | 6 | 7 | 8 | 9 | 10 | 11 | 12 | 13 | 14 | 15 |
| 4 | 5 | 6 | 7 | 8 | 9 | 10 | 11 | 12 | 13 | 14 | 15 | 16 |
| 5 | 6 | 7 | 8 | 9 | 10 | 11 | 12 | 13 | 14 | 15 | 16 | 17 |
| 6 | 7 | 8 | 9 | 10 | 11 | 12 | 13 | 14 | 15 | 16 | 17 | 18 |
| 7 | 8 | 9 | 10 | 11 | 12 | 13 | 14 | 15 | 16 | 17 | 18 | 19 |
| 8 | 9 | 10 | 11 | 12 | 13 | 14 | 15 | 16 | 17 | 18 | 19 | 20 |
| 9 | 10 | 11 | 12 | 13 | 14 | 15 | 16 | 17 | 18 | 19 | 20 | 21 |
| 10 | 11 | 12 | 13 | 14 | 15 | 16 | 17 | 18 | 19 | 20 | 21 | 22 |
| 11 | 12 | 13 | 14 | 15 | 16 | 17 | 18 | 19 | 20 | 21 | 22 | 23 |
| 12 | 13 | 14 | 15 | 16 | 17 | 18 | 19 | 20 | 21 | 22 | 23 | 24 |

**(a)** Four outcomes out of the 144 outcomes in the sample space correspond to getting a total of 5. Therefore, $P(\text{getting a total of 5}) = \dfrac{4}{144} = \dfrac{1}{36}$.

**(b)** Totals that are perfect squares are 4, 9, and 16. Twenty outcomes out of the 144 outcomes in the sample space correspond to getting a total that is a perfect square. Therefore, $P(\text{getting a total that is a perfect square}) = \dfrac{20}{144} = \dfrac{5}{36}$.

**(c)** Totals that are prime numbers are 2, 3, 5, 7, 11, 13, 17, 19, and 23. There are 51 outcomes out of the 144 outcomes in the sample space that correspond to getting a total that is a prime number. Therefore, $P(\text{getting a total that is a prime number}) = \dfrac{51}{144} = \dfrac{17}{48}$.

**27.** Assume the die is fair.

   **(a)** Two of the 6 outcomes correspond to getting a 2, so $P(\text{getting a 2}) = \dfrac{2}{6} = \dfrac{1}{3}$.

   **(b)** Four of the 6 outcomes correspond to not getting a 2, so $P(\text{not getting a 2}) = \dfrac{4}{6} = \dfrac{2}{3}$.

   **(c)** Four of the 6 outcomes correspond to getting an odd number, so $P(\text{getting an odd number}) = \dfrac{4}{6} = \dfrac{2}{3}$.

   **(d)** Two of the 6 outcomes correspond to not getting an odd number, so $P(\text{not getting an odd number}) = \dfrac{2}{6} = \dfrac{1}{3}$.

**29.** The sample space contains 12 months. List the number of outcomes in each event:
   $T = \{\text{April, June, September, November}\}$
   $Y = \{\text{January, February, May, July}\}$
   $T \cup Y = \{\text{January, February, April, May, June, July, September, November}\}$

   $P(T) = \dfrac{4}{12} = \dfrac{1}{3}$ and is the probability the month is 30 days long.

   $P(Y) = \dfrac{4}{12} = \dfrac{1}{3}$ and is the probability the month ends in "y".

   $P(T \cup Y) = P(T) + P(Y) = \dfrac{4}{12} + \dfrac{4}{12} = \dfrac{8}{12} = \dfrac{2}{3}$ and is the probability the month is 30 days long or ends in "y".

**31. (a)** There are a total of $14 + 11 = 25$ students in the class. $P(G) = \dfrac{14}{25}$ and $P(D) = \dfrac{13}{25}$.

   **(b)** $G \cup D$ is the set of students who are girls or who have homework done. Out of the 25 students, a total of 21 are girls (with or without homework done) or boys (with homework done), so $P(G \cup D) = \dfrac{21}{25}$ and is the probability the student is a girl or has homework done.

   $G \cap D$ is the set of students who are girls and who have homework done. Out of the 25 students, a total of 6 are girls with homework done, so $P(G \cap D) = \dfrac{6}{25}$ and is the probability the student is a girl and has her homework done.

   **(c)** Events $G$ and $D$ are not mutually exclusive. There are 6 students who are girls and who have their homework done. Therefore, $G \cap D \neq \emptyset$.

**33. (a)** The central angles of each of the four sectors are congruent. Therefore, $P(A) = \dfrac{1}{4}$ and $P(B) = \dfrac{1}{4}$.

   There are 16 possible outcomes in the sample space: $\{BB, BR, BY, BG, RB, RR, RY, RG, YB, YR, YY, YG, GB, GR, GY, GG\}$. The event $A \cap B$ corresponds to the outcome $GY$; that is getting green on the first spin and yellow on the second spin. Therefore, $P(A \cap B) = \dfrac{1}{16}$. The event $A \cup B$ corresponds to the subset of outcomes $\{GB, GR, GY, GG, YB, YR, YY\}$, that is getting green on the first spin or yellow on the second spin. Therefore, $P(A \cup B) = \dfrac{7}{16}$.

**(b)** $P(A \cup B) = P(A) + P(B) - P(A \cap B) = \frac{1}{4} + \frac{1}{4} - \frac{1}{16} = \frac{7}{16}$.

**35. (a)** $\overline{E}$ is the event that a person does not have type O blood and $P(\overline{E}) = 1 - 0.45 = 0.55$.

**(b)** $\overline{F}$ is the event that a baby is not left-handed and $P(\overline{F}) = 1 - 0.08 = 0.92$.

**37. (a)** Events $B$ and $C$ are mutually exclusive. In event $B$ the second marble must be white, and in event $C$ the second marble must be red. Both of these cannot happen at the same time.

**(b)** Events $A$ and $C$ are not mutually exclusive. In event $A$ the first marble must be green, and in event $C$ the second marble must be red. These two events can happen at the same time.

**(c)** The complement of event $A$ is the event that the first marble is not green.

**(d)** $P(A) = \frac{1}{4}$, and $P(\overline{A}) = \frac{3}{4}$

**(e)** $P(\overline{A}) = \frac{3}{4}$ and $1 - P(A) = 1 - \frac{1}{4} = \frac{3}{4}$. The equation holds for the probabilities from part (d).

**39. (a)** The sample space can be written as: $\{CGG, GCG, GGC\}$.
**(b)** One outcome corresponds to event $E$: $CGG$.
**(c)** $\overline{E}$ is the event that the car is not hidden behind door number 1. The outcomes in $\overline{E}$ are $GCG$ and $GGC$.
**(d)** $P(E) = \frac{1}{3}$ and $P(\overline{E}) = 1 - \frac{1}{3} = \frac{2}{3}$.

**41.** There are 17 outcomes in the sample space.

**(a)** $P(A) = \frac{5}{17}$      **(b)** $P(B) = \frac{4}{17}$      **(c)** $P(C) = \frac{6}{17}$

**(d)** $P(S) = \frac{17}{17} = 1$      **(e)** $P(A \cup B) = \frac{9}{17}$      **(f)** $P(B \cap C) = \frac{3}{17}$

**(g)** $P(A \cap C) = \frac{0}{17} = 0$      **(h)** $P(\overline{A}) = \frac{12}{17}$      **(i)** $P(\overline{B}) = \frac{13}{17}$

**(j)** $P(\overline{C}) = \frac{11}{17}$

## Solutions to Odd-Numbered Problems in Section 10.2

**1. (a)** The coin can land heads up or tails up.

**(b)** There are 10 digits.

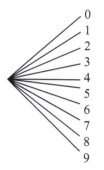

**3. (a)** There are two outcomes at each stage.

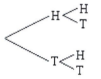

**(b)** There are four outcomes in the first stage and three outcomes in the second.

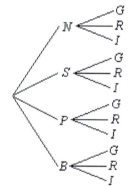

**5.**

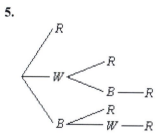

7. **(a)** The primary branches correspond to the travel options to New York. The secondary branches correspond to the travel options to London.

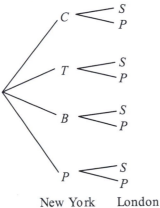

New York    London

**(b)** Eight routes are possible.

**(c)** There are 4 options in the first stage and 2 options in the second stage, so there are $4 \times 2 = 8$ routes possible. This agrees with the number of routes from part (b).

9. **(a)** The coin can land heads up or tails up, so there are 2 outcomes.

**(b)** The first die can land with any 1 of the 6 numbers facing up, so there are 6 outcomes.

**(c)** The second die can land with any 1 of the 6 numbers facing up, so there are 6 outcomes.

**(d)** There are $2 \times 6 \times 6 = 72$ outcomes for the experiment.

11. **(a)** $P(B) = \dfrac{1}{3}$

**(b)** The event of drawing a blue marble and the event of drawing a yellow marble are mutually exclusive, so probabilities can be added: $P(B) = \dfrac{1}{4}$, $P(Y) = \dfrac{1}{2}$, and $P(B \text{ or } Y) = \dfrac{1}{4} + \dfrac{1}{2} = \dfrac{3}{4}$.

13. (a)

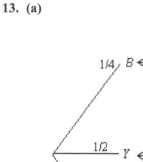

(b) Multiply the probabilities along the $YY$ branches:  $P(YY) = \dfrac{1}{2} \times \dfrac{1}{2} = \dfrac{1}{4}$.

(c) Multiply the probabilities along the $BB$, $YB$, and $GB$ branches and then add:
$$P(BB \cup YB \cup GB) = \frac{1}{4} \times \frac{1}{4} + \frac{1}{2} \times \frac{1}{4} + \frac{1}{4} \times \frac{1}{4} = \frac{1}{16} + \frac{1}{8} + \frac{1}{16} = \frac{4}{16} = \frac{1}{4}.$$

(d) Multiply the probabilities along the $BB$, $YY$, and $GG$ branches and then add:
$$P(BB \cup YY \cup GG) = \frac{1}{4} \times \frac{1}{4} + \frac{1}{2} \times \frac{1}{2} + \frac{1}{4} \times \frac{1}{4} = \frac{1}{16} + \frac{1}{4} + \frac{1}{16} = \frac{6}{16} = \frac{3}{8}.$$

(e)

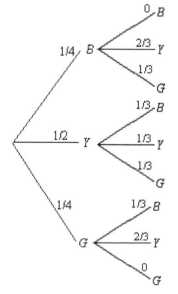

Multiply the probabilities along the $YY$ branches:  $P(YY) = \dfrac{1}{2} \times \dfrac{1}{3} = \dfrac{1}{6}$.

Multiply the probabilities along the $BB$, $YB$, and $GB$ branches and then add:
$$P(BB \cup YB \cup GB) = \frac{1}{4} \times 0 + \frac{1}{2} \times \frac{1}{3} + \frac{1}{4} \times \frac{1}{3} = 0 + \frac{1}{6} + \frac{1}{12} = \frac{3}{12} = \frac{1}{4}.$$ Multiply the probabilities along the $BB$, $YY$, and $GG$ branches and then add:
$$P(BB \cup YY \cup GG) = \frac{1}{4} \times 0 + \frac{1}{2} \times \frac{1}{3} + \frac{1}{4} \times 0 = 0 + \frac{1}{6} + 0 = \frac{1}{6}.$$

15. (a)

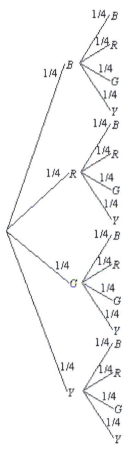

(b) The event the spinner lands on yellow both times is $YY$. Therefore, $P(YY) = \frac{1}{4} \times \frac{1}{4} = \frac{1}{16}$.

(c) The event the spinner lands on red on the second spin corresponds to outcomes $BR$, $RR$, $GR$, and $YR$:
$$P(BR \cup RR \cup GR \cup YR) = \frac{1}{4} \times \frac{1}{4} + \frac{1}{4} \times \frac{1}{4} + \frac{1}{4} \times \frac{1}{4} + \frac{1}{4} \times \frac{1}{4} = \frac{1}{16} + \frac{1}{16} + \frac{1}{16} + \frac{1}{16} = \frac{4}{16} = \frac{1}{4}.$$

(d) The event the spinner lands on blue then green or green then blue corresponds to outcomes $BG$ or $GB$: $P(BG \cup GB) = \frac{1}{4} \times \frac{1}{4} + \frac{1}{4} \times \frac{1}{4} = \frac{1}{16} + \frac{1}{16} = \frac{2}{16} = \frac{1}{8}.$

**17. (a)** The probabilities for events $B$, $R$, and $G$ must add to 1, so $P(G) = 1 - \dfrac{1}{7} - \dfrac{2}{7} = \dfrac{7}{7} - \dfrac{1}{7} - \dfrac{2}{7} = \dfrac{4}{7}$.

**(b)** The probability of event $RB$ is found by multiplying the probabilities along the branches, so
$P(RB) = \dfrac{2}{7} \times \dfrac{1}{7} = \dfrac{2}{49}$.

**(c)** The probability of event $RR$ is found by multiplying the probabilities along the branches, so
$P(RR) = \dfrac{2}{7} \times \dfrac{2}{7} = \dfrac{4}{49}$.

**(d)** The probabilities for events $B$, $R$, and $G$ must add to 1, so $P(G) = 1 - \dfrac{1}{7} - \dfrac{2}{7} = \dfrac{7}{7} - \dfrac{1}{7} - \dfrac{2}{7} = \dfrac{4}{7}$.

**(e)** From part (a), we know $P(G) = \dfrac{4}{7}$. The probability of event $GB$ is found by multiplying the

probabilities along the branches, so $P(GB) = \dfrac{4}{7} \times \dfrac{1}{7} = \dfrac{4}{49}$.

**(f)** From part (a), we know $P(G) = \dfrac{4}{7}$. The probability of event $GR$ is found by multiplying the

probabilities along the branches, so $P(GR) = \dfrac{4}{7} \times \dfrac{2}{7} = \dfrac{8}{49}$.

**(g)** There are 9 outcomes in the sample space.

**(h)** The marbles were drawn with replacement. The probabilities do not change for a color in the second stage after that color has been drawn.

**(i)** There is no way to tell how many marbles were originally in the box. Probabilities are reduced to lowest form when given in fraction form so the original numbers are not given.

**(j)** The probability of the event of getting two marbles that are the same color corresponds to outcomes $BB$, $RR$, and $GG$:
$$P(BB \cup RR \cup GG) = \dfrac{1}{7} \times \dfrac{1}{7} + \dfrac{2}{7} \times \dfrac{2}{7} + \dfrac{4}{7} \times \dfrac{4}{7} = \dfrac{1}{49} + \dfrac{4}{49} + \dfrac{16}{49} = \dfrac{21}{49} = \dfrac{3}{7}.$$

**19. (a)** There are 4 stages in this experiment. In each stage, there are two outcomes, so there are $2 \times 2 \times 2 \times 2 = 16$ outcomes in the experiment.

**(b)** There is one way to get 4 heads, HHHH. Therefore, $P(\text{getting 4 heads}) = \dfrac{1}{16}$.

**(c)** There are 6 ways to get exactly 2 heads {HHTT, HTHT, THHT, THTH, TTHH, HTTH}.
Therefore, $P(\text{getting exactly 2 heads}) = \dfrac{6}{16} = \dfrac{3}{8}$.

**(d)** There are 4 ways to get exactly 3 tails {HTTT, THTT, TTHT, TTTH}. Therefore,
$P(\text{getting exactly 3 tails}) = \dfrac{4}{16} = \dfrac{1}{4}$.

**21. (a)** There are 2 stages, one stage for the selection of each chocolate.

**(b)** There are $0.10 \times 20 = 2$ nut-filled, $0.3 \times 20 = 6$ caramel, and $0.60 \times 20 = 12$ nougats.

**(c)** In the first stage of the experiment, there are 20 chocolates in the box. There are, for example, 2

nut-filled chocolates, so the probability of getting a nut-filled chocolate is $\dfrac{2}{20} = \dfrac{1}{10}$. Other

probabilities in the first stage are calculated similarly. In the second stage, there are only 19 chocolates, so the denominator of the probabilities changes to 19.

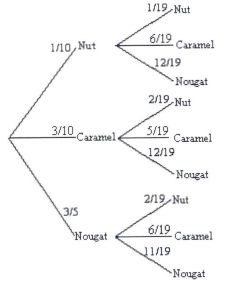

**(d)** Multiply probabilities along the branches in the probability tree diagram corresponding to getting a nut-filled chocoate at each stage: $P(\text{getting 2 nut-filled chocolates}) = \dfrac{1}{10} \times \dfrac{1}{19} = \dfrac{1}{190}$.

**(e)** Multiply probabilities along the branches in the probability tree diagram corresponding to getting a caramel and a nut-filled or a nut-filled and a caramel then add:

$$P(\text{getting a caramel or a nut-filled}) = \dfrac{3}{10} \times \dfrac{2}{19} + \dfrac{1}{10} \times \dfrac{6}{19} = \dfrac{6}{190} + \dfrac{6}{190} = \dfrac{12}{190} = \dfrac{6}{95}.$$

**(f)** It will be easier to find the probability of the complement of the event. The complement of getting a nougat or a caramel is the event of getting 2 nut-filled chocolates:

$$P(\text{getting a caramel or a nougat}) = 1 - P(\text{getting 2 nut-filled}) = 1 - \dfrac{1}{10} \times \dfrac{1}{19} = 1 - \dfrac{1}{190} = \dfrac{189}{190}.$$

**23. (a)** For the first draw, there are 48 cards to choose from. The drawing is done without replacement, so for the second draw, there are 47 cards to choose from. Therefore there are 48 × 47 = 2256 outcomes.

**(b)** For the first draw, there are (4 suits) × (3 face cards) × (2 of each) = 24 face cards to choose from. If a face card is selected in the first draw, then there will be only 23 face cards left to choose from in the second draw. Therefore, there are 24 × 23 = 552 outcomes.

**(c)** From parts (a) and (b), we can determine the probability both cards are face cards:

$$P(\text{getting 2 face cards}) = \dfrac{552}{2256} = \dfrac{23}{94}.$$

**(d)** We can get a pair of 9s, 10s, jacks, queens, kings, or aces. For each of the 6 different types of cards, there are 8 × 7 = 56 ways to select a pair when selecting cards without replacement. Therefore,

$$P(\text{getting a pair}) = 6 \times \left( \dfrac{56}{2256} \right) = \dfrac{336}{2256} = \dfrac{7}{47}.$$

**(e)** In a pinochle deck there are 2 identical 9s, 10s, jacks, queens, kings, and aces for each of the 4 suits. There are 8 ways to select an identical pair of 9s, for example, because there are eight 9s in the deck, and once one of the 9s is selected, there is only one way to select its identical match from the deck. Therefore, $P(\text{getting an identical pair}) = 6 \times \left( \dfrac{8}{2256} \right) = \dfrac{48}{2256} = \dfrac{1}{47}.$

25. (a)

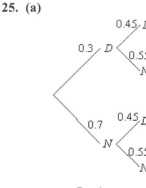

Box 1    Box 2

  (b) Multiply the probabilities along the branches of the tree corresponding to getting a defective bulb each time: $P(DD) = 0.3 \times 0.45 = 0.135$.

  (c) Multiply the probabilities along the branches of the tree corresponding to getting a defective bulb followied by a nondefective bulb: $P(DN) = 0.3 \times 0.55 = 0.165$.

27. (a) Let $W$ represent the event that a player wins. There are 2 stars behind 8 balloons, so
$$P(W) = \frac{2}{8} = \frac{1}{4}.$$

  (b) Balloons are replaced when they are popped, so in order to win in exactly two shots, you must lose after throwing the first dart and win after throwing the second dart: $P(LW) = \frac{6}{8} \times \frac{2}{8} = \frac{12}{64} = \frac{3}{16}$.

  (c) Balloons are replaced when they are popped, so in order to win in exactly three shots, you must lose after throwing the first dart and the second dart, and win after throwing the third dart:
$$P(LLW) = \frac{6}{8} \times \frac{6}{8} \times \frac{2}{8} = \frac{72}{512} = \frac{9}{64}.$$

  (d) Balloons are replaced when they are popped, so in order not win in three shots, you must lose every time you throw a dart: $P(LLL) = \frac{6}{8} \times \frac{6}{8} \times \frac{6}{8} = \frac{216}{512} = \frac{27}{64}$.

29. (a) Letters are selected without replacement. There are 11 letters and 2 of them are Ps. Therefore,
$$P(PP) = \frac{2}{11} \times \frac{1}{10} = \frac{2}{110} = \frac{1}{55}.$$

  (b) To select the same letter both times means two of the letter I, two of the letter S, or two of the letter P must be selected: $P(II) + P(SS) + P(PP) = \frac{4}{11} \times \frac{3}{10} + \frac{4}{11} \times \frac{3}{10} + \frac{2}{11} \times \frac{1}{10} = \frac{26}{110} = \frac{13}{55}$.

  (c) There are 7 consonants. Therefore, $P(\text{selecting 2 consonants}) = \frac{7}{11} \times \frac{6}{10} = \frac{42}{110} = \frac{21}{55}$.

31. Probabilities will not change from roll to roll. The probability a 1 is rolled is $\frac{1}{6}$. The probability a 2 is rolled is $\frac{2}{6} = \frac{1}{3}$. The probability a 3 is rolled is $\frac{3}{6} = \frac{1}{2}$.

  (a) $P(\text{two 3s are rolled}) = \frac{1}{2} \times \frac{1}{2} = \frac{1}{4}$

  (b) You could get two 1s, two 2s, or two 3s:

$$P(\text{two 1s}) + P(\text{two 2s}) + P(\text{two 3s}) = \frac{1}{6}\times\frac{1}{6}+\frac{1}{3}\times\frac{1}{3}+\frac{1}{2}\times\frac{1}{2}$$
$$= \frac{1}{36}+\frac{1}{9}+\frac{1}{4}$$
$$= \frac{14}{36}$$
$$= \frac{7}{18}$$

(c) There are four ways to get two odd numbers. You can get two 1s, two 3s, a 1 and a 3, and a 3 and a 1.

$$P(\text{two odd numbers}) = P(\text{two 1s}) + P(\text{two 3s}) + P(\text{a 1 and a 3}) + P(\text{a 3 and a 1})$$
$$= \frac{1}{6}\times\frac{1}{6}+\frac{1}{2}\times\frac{1}{2}+\frac{1}{6}\times\frac{1}{2}+\frac{1}{2}\times\frac{1}{6}$$
$$= \frac{1}{36}+\frac{1}{4}+\frac{1}{12}+\frac{1}{12}$$
$$= \frac{16}{36}$$
$$= \frac{4}{9}$$

## Solutions to Odd-Numbered Problems in Section 10.3

1. (a) Since there are 6 different outcomes for each die, there are $6 \times 6 = 36$ outcomes in the sample space.

(b) The numbers 3 and 6 are multiples of 3. As long as the first die shows a 3 or a 6, the second die can show any of the numbers 1 through 6: $M = \{(3, 1), (3, 2), (3, 3), (3, 4), (3, 5), (3, 6), (6, 1), (6, 2), (6, 3), (6, 4), (6, 5), (6, 6)\}$.

(c) The first die can show any of the number 1 through 6. The second die must show 1, 3, or 5. $O = \{(1, 1), (1, 3), (1, 5), (2, 1), (2, 3), (2, 5), (3, 1), (3, 3), (3, 5), (4, 1), (4, 3), (4, 5), (5, 1), (5, 3), (5, 5), (6, 1), (6, 3), (6, 5)\}$

(d) The event $M \cap O$ is the event the first die shows a multiple of 3 and the second die shows an odd number. Therefore, $M \cap O = \{(3, 1), (3, 3), (3, 5), (6, 1), (6, 3), (6, 5)\}$.

(e) $P(M) = \frac{12}{36} = \frac{1}{3}$, $P(O) = \frac{18}{36} = \frac{1}{2}$, and $P(M \cap O) = \frac{6}{36} = \frac{1}{6}$.

(f) $P(M|O) = \frac{P(M \cap O)}{P(O)} = \frac{\frac{1}{6}}{\frac{1}{2}} = \frac{2}{6} = \frac{1}{3}$ and represents the probability that the first die shows a multiple of 3 given the second die shows an odd number.

$$P(O|M) = \frac{P(O \cap M)}{P(M)} = \frac{\frac{1}{6}}{\frac{1}{3}} = \frac{3}{6} = \frac{1}{2}$$ and represents the probability that the second die shows an odd number given that the first shows a multiple of 3.

3. Construct a probability tree diagram.

First Draw    Second Draw

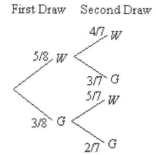

(a)
$P$(the second ball is green) = $P$(first white and second green) + $P$(first green and second green)

$$= \frac{5}{8} \times \frac{3}{7} + \frac{3}{8} \times \frac{2}{7}$$

$$= \frac{21}{56}$$

$$= \frac{3}{8}$$

(b) $P$(the first ball is white) $= \dfrac{5}{8}$

(c) $P$(the first ball is white and the second ball is green) $= \dfrac{5}{8} \times \dfrac{3}{7} = \dfrac{15}{56}$

(d)
$$P\text{(second ball is green given first ball is white)} = \frac{P\text{(second ball is green and first ball is white)}}{P\text{(first ball is white)}}$$

$$= \frac{\frac{3}{7} \times \frac{5}{8}}{\frac{5}{8}}$$

$$= \frac{15}{56} \times \frac{8}{5}$$

$$= \frac{3}{7}$$

**(e)**

$$P(\text{first ball is white given second is green}) = \frac{P(\text{first ball is white and second ball is green})}{P(\text{second ball is green})}$$

$$= \frac{\dfrac{5}{8} \times \dfrac{3}{7}}{\dfrac{5}{8} \times \dfrac{3}{7} + \dfrac{3}{8} \times \dfrac{2}{7}}$$

$$= \frac{\dfrac{15}{56}}{\dfrac{3}{8}}$$

$$= \frac{15}{56} \times \frac{8}{3}$$

$$= \frac{5}{7}$$

**(f)**

$$P(\text{first ball is green given second ball is green}) = \frac{P(\text{first ball is green and second ball is green})}{P(\text{second ball is green})}$$

$$= \frac{\dfrac{3}{8} \times \dfrac{2}{7}}{\dfrac{5}{8} \times \dfrac{3}{7} + \dfrac{3}{8} \times \dfrac{2}{7}}$$

$$= \frac{\dfrac{6}{56}}{\dfrac{3}{8}}$$

$$= \frac{6}{56} \times \frac{8}{3}$$

$$= \frac{2}{7}$$

5.  **(a)** The probabilities in the first stage of the probability tree diagram must add to 1. Therefore the missing probability (i) is $1 - \dfrac{4}{5} = \dfrac{1}{5}$. In the second stage of the probability tree diagram, $P(R) + P(S) = 1$ and we know $P(S) = \dfrac{5}{8}$, therefore probability (ii) is $\dfrac{3}{8}$.

**(b)** Use the probabilities from part (a). To find $P(R \cap P)$ multiply the probabilities along the branches of the tree diagram from $P$ to $R$: $P(R \cap P) = \dfrac{4}{5} \times \dfrac{3}{8} = \dfrac{12}{40} = \dfrac{3}{10}$. To find $P(S \cap P)$ multiply the probabilities along the branches of the tree diagram from $P$ to $S$: $P(S \cap P) = \dfrac{4}{5} \times \dfrac{5}{8} = \dfrac{20}{40} = \dfrac{1}{2}$.

Event $P$ is given in the first stage of the probability tree diagram: $P(P) = \dfrac{4}{5}$.

**(c)** Use the probabilities from part (a). To find $P(R \cap Q)$ multiply the probabilities along the branches of the tree diagram from $R$ to $Q$: $P(R \cap Q) = \dfrac{1}{5} \times \dfrac{1}{3} = \dfrac{1}{15}$. To find $P(S \cap Q)$ multiply the probabilities along the branches of the tree diagram from $Q$ to $S$: $P(S \cap Q) = \frac{1}{5} \times \frac{2}{3} = \frac{2}{15}$. Event $R$ is given in the second stage of the probability tree diagram, so multiply probabilities along the branches leading to $R$ and add the products: $P(R) = \frac{4}{5} \times \frac{3}{8} + \frac{1}{5} \times \frac{1}{3} = \frac{12}{40} + \frac{1}{15} = \frac{3}{10} + \frac{1}{15} = \frac{11}{30}$.

**(d)** $P(S|P)$ represents the probability that event $S$ occurs given that event $P$ occurred.

$$P(S|P) = \frac{P(S \cap P)}{P(P)} = \frac{\frac{1}{2}}{\frac{4}{5}} = \frac{1}{2} \times \frac{5}{4} = \frac{5}{8}.$$

**(e)** $P(Q|R)$ represents the probability that event $Q$ occurs given that event $R$ occurred.

$$P(Q|R) = \frac{P(Q \cap R)}{P(R)} = \frac{\frac{1}{15}}{\frac{11}{30}} = \frac{1}{15} \times \frac{30}{11} = \frac{2}{11}.$$

**7.** There are 15 outcomes in the sample space.

**(a)** There are 8 outcomes that correspond to event $A$. Therefore, $P(A) = \frac{8}{15}$.

**(b)** There are 6 outcomes that correspond to event $B$. Therefore, $P(B) = \frac{6}{15} = \frac{2}{5}$.

**(c)** The area in the overlapping circles that represent events $A$ and $B$ contains 3 outcomes. Therefore,
$$P(A \cap B) = \frac{3}{15} = \frac{1}{5}.$$

**(d)** $P(A|B) = \frac{P(A \cap B)}{P(B)} = \frac{3}{6} = \frac{1}{2}$

**(e)** $P(B|A) = \frac{P(B \cap A)}{P(A)} = \frac{3}{8}$

**9.** **(a)** $P(L|M)$ is the probability that a person is left-handed given that they are a male.

**(b)** $P(M|L)$ is the probability that a person is a male given that they are left-handed.

**(c)** $P(L \cap M)$ is the probability that a person is left-handed and a male.

**(d)** $P(\overline{L} \cap \overline{M})$ is the probability that a person is not left-handed and not a male.

**(e)** $P(\overline{L}|M)$ is the probability that a person is not left-handed given that they are a male.

**(f)** $P(\overline{L}|\overline{M})$ is the probability that a person is not left-handed given that they are not a male.

**11.** **(a)** $P(\text{a girl is selected}) = \frac{1}{2} \times \frac{18}{29} + \frac{1}{2} \times \frac{10}{27} = \frac{18}{58} + \frac{10}{54} = \frac{486}{1566} + \frac{290}{1566} = \frac{776}{1566} = \frac{388}{783}$

**(b)** $P(\text{a boy is selected}) = \frac{1}{2} \times \frac{11}{29} + \frac{1}{2} \times \frac{17}{27} = \frac{11}{58} + \frac{17}{54} = \frac{297}{1566} + \frac{493}{1566} = \frac{790}{1566} = \frac{395}{783}$

**(c)**

$$P\big(\text{a girl is selected given that the student came from class II}\big) = \frac{P(\text{girl} \cap \text{class II})}{P(\text{class II})}$$

$$= \frac{\dfrac{1}{2} \times \dfrac{10}{27}}{\dfrac{1}{2}}$$

$$= \frac{10}{27}$$

**(d)**

$$P\big(\text{a boy is selected given that the student came from class I}\big) = \frac{P(\text{boy} \cap \text{class I})}{P(\text{class I})}$$

$$= \frac{\dfrac{1}{2} \times \dfrac{11}{29}}{\dfrac{1}{2}}$$

$$= \frac{11}{29}$$

**13. (a)** We want to find the conditional probability that a fetus has Down syndrome given that the test result was positive, that is we want $P(\text{fetus has Down syndrome} \mid \text{a positive test result})$. From the table, we see that the total number of tests given was $5 + 1 + 208 + 2386 = 2600$, so $P(\text{fetus has Down syndrome} \cap \text{positive test result}) = \dfrac{5}{2600}$ and $P(\text{positive test result}) = \dfrac{213}{2600}$.

Therefore, we have the following:

$$P(\text{fetus has Down syndrome} \mid \text{positive test}) = \frac{P(\text{fetus has Down syndrome} \cap \text{positive test})}{P(\text{positive test})}$$

$$= \frac{\dfrac{5}{2600}}{\dfrac{213}{2600}}$$

$$= \frac{5}{213}$$

**(b)** We want to find the conditional probability that the fetus does not have Down syndrome given that the test result was positive, that is we want to find $P(\text{fetus does not have Down syndrome} \mid \text{a positive test result})$. From the table, we see that 2600 tests were given, so $P(\text{fetus does not have Down syndrome} \cap \text{positive test result}) = \dfrac{208}{2600}$ and $P(\text{positive test result}) = \dfrac{213}{2600}$.

Therefore, we have the following:

$P(\text{fetus does not have Down syndrome}|\text{positive test})$

$$= \frac{P(\text{fetus does not have Down syndrome} \cap \text{positive test})}{P(\text{positive test})}$$

$$= \frac{\dfrac{208}{2600}}{\dfrac{213}{2600}}$$

$$= \frac{208}{213}$$

(c) We want to find the conditional probability that the test is negative given that the fetus has Down syndrome; that is we want to find $P(\text{negative test result}|\text{fetus has Down syndrome})$. From the table, we see that 2600 tests were given, so $P(\text{negative test result} \cap \text{fetus has Down syndrome}) = \dfrac{1}{2600}$. Also, from the table, we see that a total of 6 fetuses have Down syndrome, so $P(\text{fetus has Down syndrome}) = \dfrac{6}{2600}$.

Therefore, we have the following:

$P(\text{negative test result}|\text{fetus has Down syndrome})$

$$= \frac{P(\text{negative test result} \cap \text{fetus has Down syndrome})}{P(\text{fetus has Down syndrome})}$$

$$= \frac{\dfrac{1}{2600}}{\dfrac{6}{2600}}$$

$$= \frac{1}{6}$$

**15. (a)** We want to find the conditional probability that the person is female given that the person has a graduate degree, that is we want to find $P(\text{female}|\text{graduate degree})$. From the table, we see that 400 adults were classified, so $P(\text{female} \cap \text{graduate degree}) = \dfrac{14}{400} = \dfrac{7}{200}$. Also, from the table, we see that a total of $14 + 16 = 30$ people have a graduate degree, so $P(\text{graduate degree}) = \dfrac{30}{400} = \dfrac{3}{40}$.

Therefore, we have the following:

$P(\text{female}|\text{graduate degree})$

$$= \frac{P(\text{female} \cap \text{graduate degree})}{P(\text{graduate degree})}$$

$$= \frac{\frac{7}{200}}{\frac{3}{40}}$$

$$= \frac{7}{200} \times \frac{40}{3}$$

$$= \frac{280}{600}$$

$$= \frac{7}{15}$$

**(b)** We want to find the conditional probability that the person is male given that the person has at most a high school diploma, that is we want to find $P(\text{male}|\text{high school diploma})$. From the table, we see that 400 adults were classified, so $P(\text{male} \cap \text{high school diploma}) = \frac{87}{400}$. Also, from the table, we see that a total of $99 + 87 = 186$ people have a high school diploma as their highest level of education completed, so $P(\text{high school diploma}) = \frac{186}{400} = \frac{93}{200}$.

Therefore, we have the following:

$P(\text{male}|\text{high school diploma})$

$$= \frac{P(\text{male} \cap \text{high school diploma})}{P(\text{high school diploma})}$$

$$= \frac{\frac{87}{400}}{\frac{93}{200}}$$

$$= \frac{87}{400} \times \frac{200}{93}$$

$$= \frac{29}{62}$$

**(c)** We want to find the conditional probability that the person does not have a college degree given that the person is female. In other words, we want to find the probability that the person has graduated from elementary school or has a high school diploma given that the person is female, that is we want to find $P(\text{elementary or high school diploma}|\text{female})$. From the table, we see that 400 adults were classified, so $P(\text{elementary school or high school diploma} \cap \text{female}) = \frac{163}{400}$.

Also, from the table, we see that a total of $64 + 99 + 28 + 14 = 205$ females were classified, so $P(\text{female}) = \frac{205}{400} = \frac{41}{80}$.

Therefore, we have the following:

$$P(\text{elementary or high school diploma}|\text{female})$$

$$= \frac{P(\text{elementary or high school diploma} \cap \text{female})}{P(\text{female})}$$

$$= \frac{\dfrac{163}{400}}{\dfrac{41}{80}}$$

$$= \frac{163}{400} \times \frac{80}{41}$$

$$= \frac{163}{205}$$

17. Create a probability tree diagram. We are told that the screening test is 95% accurate for both infected and uninfected persons. Therefore, of those who are infected, 95% will have a positive test result, and of those who are uninfected, 95% will have a negative test result.

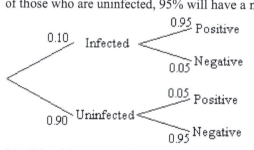

(a) $P(\text{positive test result}) = 0.10 \times 0.95 + 0.90 \times 0.05 = 0.14$.

(b) A false positive result occurs when a person is uninfected and has a positive test result. From the probability tree diagram, we see $P(\text{uninfected} \cap \text{positive}) = 0.9 \times 0.05 = 0.045$.

(c) Find the conditional probability that a person is infected given that the person has a positive test result, that is find $P(\text{infected}|\text{positive})$.

$$P(\text{infected}|\text{positive}) = \frac{P(\text{infected} \cap \text{positive})}{P(\text{positive})}$$

$$= \frac{0.10 \times 0.95}{0.10 \times 0.95 + 0.90 \times 0.05}$$

$$= \frac{0.095}{0.14}$$

$$\approx 0.6786$$

(d) Find the conditional probability that the test result is positive given that the person is infected, that is find $P(\text{positive}|\text{infected})$.

$$P(\text{positive}|\text{infected}) = \frac{P(\text{positive} \cap \text{infected})}{P(\text{infected})}$$

$$= \frac{0.10 \times 0.95}{0.10}$$

$$= 0.95$$

**19.** Multiply probabilities along the branches.

(a) $P(\text{car is behind door 1 and host chooses door 1}) = \dfrac{1}{3} \times 0 = 0$

(b) $P(\text{car is behind door 1 and host chooses door 2}) = \dfrac{1}{3} \times \dfrac{1}{2} = \dfrac{1}{6}$

(c) $P(\text{car is behind door 1 and host chooses door 3}) = \dfrac{1}{3} \times \dfrac{1}{2} = \dfrac{1}{6}$

(d) $P(\text{car is behind door 2 and host chooses door 1}) = \dfrac{1}{3} \times 0 = 0$

(e) $P(\text{car is behind door 2 and host chooses door 2}) = \dfrac{1}{3} \times 0 = 0$

(f) $P(\text{car is behind door 2 and host chooses door 3}) = \dfrac{1}{3} \times 1 = \dfrac{1}{3}$

(g) $P(\text{car is behind door 3 and host chooses door 1}) = \dfrac{1}{3} \times 0 = 0$

(h) $P(\text{car is behind door 3 and host chooses door 2}) = \dfrac{1}{3} \times 1 = \dfrac{1}{3}$

(i) $P(\text{car is behind door 3 and host chooses door 3}) = \dfrac{1}{3} \times 0 = 0$

**21.** Refer to the probability tree diagram in problem 19.

(a)

$$P(\text{car is behind door 1} \mid \text{host opens door 2}) = \frac{P(\text{car behind door 1} \cap \text{host opens door 2})}{P(\text{host opens door 2})}$$

$$= \frac{\dfrac{1}{3} \times \dfrac{1}{2}}{\dfrac{1}{3} \times \dfrac{1}{2} + \dfrac{1}{3} \times 0 + \dfrac{1}{3} \times 1}$$

$$= \frac{\dfrac{1}{6}}{\dfrac{1}{6} + \dfrac{1}{3}}$$

$$= \frac{\dfrac{1}{6}}{\dfrac{1}{2}}$$

$$= \frac{1}{6} \times \frac{2}{1}$$

$$= \frac{1}{3}$$

**(b)**

$$P(\text{car is behind door 3}|\text{host opens door 2}) = \frac{P(\text{car behind door 3} \cap \text{host opens door 2})}{P(\text{host opens door 2})}$$

$$= \frac{\dfrac{1}{3} \times 1}{\dfrac{1}{3} \times \dfrac{1}{2} + \dfrac{1}{3} \times 0 + \dfrac{1}{3} \times 1}$$

$$= \frac{\dfrac{1}{3}}{\dfrac{1}{6} + \dfrac{1}{3}}$$

$$= \frac{\dfrac{1}{3}}{\dfrac{1}{2}}$$

$$= \frac{1}{3} \times \frac{2}{1}$$

$$= \frac{2}{3}$$

**(c)** The contestant should switch doors. Switching doors doubles the probability of winning the car.

**23.** Refer to the probability tree diagram in problem 19.

**(a)**

$$P(\text{car is behind door 1}|\text{host opens door 3}) = \frac{P(\text{car behind door 1} \cap \text{host opens door 3})}{P(\text{host opens door 3})}$$

$$= \frac{\dfrac{1}{3} \times \dfrac{1}{2}}{\dfrac{1}{3} \times \dfrac{1}{2} + \dfrac{1}{3} \times 1 + \dfrac{1}{3} \times 0}$$

$$= \frac{\dfrac{1}{6}}{\dfrac{1}{6} + \dfrac{1}{3}}$$

$$= \frac{\dfrac{1}{6}}{\dfrac{1}{2}}$$

$$= \frac{1}{6} \times \frac{2}{1}$$

$$= \frac{1}{3}$$

**(b)**

$$P(\text{car is behind door 2}\,|\,\text{host opens door 3}) = \frac{P(\text{car behind door 2} \cap \text{host opens door 3})}{P(\text{host opens door 3})}$$

$$= \frac{\frac{1}{3} \times 1}{\frac{1}{3} \times \frac{1}{2} + \frac{1}{3} \times 1 + \frac{1}{3} \times 0}$$

$$= \frac{\frac{1}{3}}{\frac{1}{6} + \frac{1}{3}}$$

$$= \frac{\frac{1}{3}}{\frac{1}{2}}$$

$$= \frac{1}{3} \times \frac{2}{1}$$

$$= \frac{2}{3}$$

**(c)** The contestant should switch doors. Switching doors doubles the probability of winning the car.

**25.** The probability a student from the first class speaks Spanish is $\dfrac{20}{25} = \dfrac{4}{5}$. The probability a student from the second class speaks Spanish is $\dfrac{12}{18} = \dfrac{2}{3}$. The probability that both students speak Spanish is $\dfrac{4}{5} \times \dfrac{2}{3} = \dfrac{8}{15}$. The selections are random and the selection of a student from the first class has no influence on the selection of a student from the second class. The events are independent, so we can multiply the probabilities.

**27. (a)** Consider a probability tree diagram. There are three songs that are your favorites out of 11. Let $F$ represent the event that a song played is one of your favorites.

First    Second
Song    Song

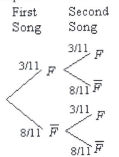

Therefore, $P(\text{second song is one of your favorites}) = \dfrac{3}{11} \times \dfrac{3}{11} + \dfrac{8}{11} \times \dfrac{3}{11} = \dfrac{9}{121} + \dfrac{24}{121} = \dfrac{33}{121} = \dfrac{3}{11}.$

(b) Use the probability tree diagram from part (a).

$P(\text{second song is a favorite}|\text{first song is a favorite})$

$$= \frac{P(\text{second song is a favorite} \cap \text{first song is a favorite})}{P(\text{first song is a favorite})}$$

$$= \frac{\dfrac{3}{11} \times \dfrac{3}{11}}{\dfrac{3}{11}}$$

$$= \frac{3}{11}$$

(c) Song selections are independent since the probability that the second song was one of your favorites was equal to the conditional probability that the second song was one of your favorites given the first song was one of your favorites.

29. (a) In one roll of the standard die, there are 6 outcomes in the sample space. In two rolls of a standard die, there are 36 outcomes in the sample space.

In the first roll, a 3 can occur in 1 way. Therefore, $P(A) = \dfrac{1}{6}$

In two rolls of the die, a sum of 7 can occur in 6 ways. Therefore, $P(B) = \dfrac{6}{36} = \dfrac{1}{6}$.

In two rolls of the die, there is one outcome that corresponds to getting a 3 on the first roll and getting a sum of 7 in two rolls. Therefore, $P(A \cap B) = \dfrac{1}{36}$

Events $A$ and $B$ are independent because $P(A \cap B) = \dfrac{1}{36} = \dfrac{1}{6} \times \dfrac{1}{6} = P(A) \times P(B)$.

(b) In the first roll, a 3 can occur in 1 way. Therefore, $P(A) = \dfrac{1}{6}$

In two rolls, the same number can be rolled in 6 ways. Therefore, $P(C) = \dfrac{6}{36} = \dfrac{1}{6}$.

Events $A$ and $C$ are independent because $P(A \cap C) = \dfrac{1}{36} = \dfrac{1}{6} \times \dfrac{1}{6} = P(A) \times P(C)$..

(c) In two rolls of the die, a sum of 7 can occur in 6 ways. Therefore, $P(B) = \dfrac{6}{36} = \dfrac{1}{6}$.

In two rolls, the same number can be rolled in 6 ways. Therefore, $P(C) = \dfrac{6}{36} = \dfrac{1}{6}$.

There is no way for the sum of two rolls to be 7 and the same number rolled each time, so $P(B \cap C) = 0$.

Events $B$ and $C$ are not independent because $P(B \cap C) = 0 \neq \dfrac{1}{6} \times \dfrac{1}{6} = P(B) \times P(C)$.

**31.** Use the definition of expected value.

| Number of Girls | Probability |
|:---:|:---:|
| 0 | $\dfrac{1}{16}$ |
| 1 | $\dfrac{4}{16}$ |
| 2 | $\dfrac{6}{16}$ |
| 3 | $\dfrac{4}{16}$ |
| 4 | $\dfrac{1}{16}$ |

The expected number of girls is calculated as follows:

$$0 \times \frac{1}{6} + 1 \times \frac{4}{16} + 2 \times \frac{6}{16} + 3 \times \frac{4}{16} + 4 \times \frac{1}{16} = \frac{0}{16} + \frac{4}{16} + \frac{12}{16} + \frac{12}{16} + \frac{4}{16} = \frac{32}{16} = 2$$

Therefore, the expected number of girls is 2.

**33. (a)** Find the expected value of the lottery to determine a fair price for one ticket. The probability of winning a \$10 prize is $\dfrac{50}{1000}$. The probability of winning a \$15 prize is $\dfrac{10}{1000}$. The probability of winning a \$30 prize is $\dfrac{5}{1000}$. The probability of winning a \$50 prize is $\dfrac{1}{1000}$. Finally, the probability of winning \$0 is $\dfrac{934}{1000}$. The expected value is calculated as follows:

$$\begin{aligned}
\text{Expected Value} &= \$10 \times \frac{50}{1000} + \$15 \times \frac{10}{1000} + \$30 \times \frac{5}{1000} + \$50 \times \frac{1}{1000} + \$0 \times \frac{934}{1000} \\
&= \frac{\$500}{1000} + \frac{\$150}{1000} + \frac{\$150}{1000} + \frac{\$50}{1000} + \frac{\$0}{1000} \\
&= \frac{\$850}{1000} \\
&= \$0.85
\end{aligned}$$

On average, you expect to win \$0.85 with a ticket. A fair price to pay for one ticket would be \$0.85.

**(b)** From part (a), we know with a price of one ticket set at \$0.85, a person expects to break even. We need to adjust the price of one ticket if we want people to lose \$0.50 on average. If we want people to lose \$0.50 on average, then the price of a ticket should be increased \$0.50 to \$0.85 + \$0.50 = \$1.35.

**(c)** We know the lottery will pay out $50 \times \$10 + 10 \times \$15 + 5 \times \$30 + 1 \times \$50 = \$850$. This can be verified by the expected value from part (a). The expected loss to the lottery is \$0.85 per ticket, so the lottery expected to lose $1000 \times \$0.85 = \$850$. At \$2 per ticket, the lottery takes in \$2000 for a net gain of $\$2000 - \$850 = \$1150$.

**35.** The expected value of the gift is calculated as follows:

$$\frac{9272 \times 1}{52,000} + \frac{44.95 \times 25,736}{52,000} + \frac{2500 \times 1}{52,000} + \frac{729.95 \times 3}{52,000} + \frac{26.99 \times 25,736}{52,000} + \frac{1000 \times 3}{52,000} + \frac{44.99 \times 180}{52,000} + \frac{63.98 \times 180}{52,000} + \frac{25 \times 160}{52,000}$$

$$= \frac{1,892,024.29}{52,000}$$

$$\approx 36.39$$

Therefore, the expected value of the gift is $36.39.

**37. (a)** There are 36 outcomes in the sample space. The table contains the 36 outcomes that occur from multiplying the numbers of dots showing on each die.

|   | 1 | 2 | 3 | 4 | 5 | 6 |
|---|---|---|---|---|---|---|
| 1 | 1 | 2 | 3 | 4 | 5 | 6 |
| 2 | 2 | 4 | 6 | 8 | 10 | 12 |
| 3 | 3 | 6 | 9 | 12 | 15 | 18 |
| 4 | 4 | 8 | 12 | 16 | 20 | 24 |
| 5 | 5 | 10 | 15 | 20 | 25 | 30 |
| 6 | 6 | 12 | 18 | 24 | 30 | 36 |

The probability of getting an even product is $\frac{27}{36} = \frac{3}{4}$. The probability of getting an odd product

is $\frac{9}{36} = \frac{1}{4}$. For you, the exepected value is calculated as follows:

$$\text{Expected Value} = \$0.25 \times \frac{3}{4} - \$1.00 \times \frac{1}{4} = \frac{\$0.75}{4} - \frac{\$1.00}{4} = \frac{-\$0.25}{4} \approx -\$0.06.$$

**(b)** Your friend wins $1.00 when an odd product occurs and loses $0.25 when an even product occurs. For your friend, the expected value is calculated as follows:

$$\text{Expected Value} = -\$0.25 \times \frac{3}{4} + \$1.00 \times \frac{1}{4} = \frac{-\$0.75}{4} + \frac{\$1.00}{4} = \frac{\$0.25}{4} \approx \$0.06.$$

**(c)** If an even product occurs, you win $0.25. We want to determine how much you should lose when an odd product occurs so that the expected value of the game for you is $0.05. Let $P$ represent the amount you must pay your friend if an odd product occurs.

$$\$0.05 = \$0.25 \times \frac{3}{4} - P \times \frac{1}{4}$$

$$\$0.05 = \frac{\$0.75}{4} - \frac{P}{4}$$

$$\$0.20 = \$0.75 - P$$

$$\$0.55 = P$$

You should pay the friend $0.55.

**39.** The probability of an event is $\frac{1}{5}$. Therefore, the probability of the complement of the event is $\frac{4}{5}$.

**(a)** The odds in favor of the event are $\dfrac{\frac{1}{5}}{\frac{4}{5}} = \frac{1}{5} \times \frac{5}{4} = \frac{1}{4}$ or 1:4.

**(b)** The odds against the event are $\dfrac{\frac{4}{5}}{\frac{1}{5}} = \frac{4}{5} \times \frac{5}{1} = \frac{4}{1}$ or 4:1.

**41.** If the odds against an event are 2 to 1 or 2:1, then we know the probability against the event is $\dfrac{2}{2+1} = \dfrac{2}{3}$. Therefore, the probability of the event is $1 - \dfrac{2}{3} = \dfrac{1}{3}$.

**43. (a)** There are 8 outcomes in the sample space, 1 of which corresonds to getting 3 heads. The probability of getting 3 heads when 3 coins are tossed is $\dfrac{1}{8}$. The odds in favor of getting 3 heads are $\dfrac{\frac{1}{8}}{\frac{7}{8}} = \dfrac{1}{8} \times \dfrac{8}{7} = \dfrac{1}{7}$ or 1:7.

**(b)** There are 8 outcomes in the sample space, 3 of which correspond to getting exactly 1 head. The probability of getting 1 head when 3 coins are tossed is $\dfrac{3}{8}$. Therefore, the probability of the complement of the event is $\dfrac{5}{8}$. The odds against getting 1 head are $\dfrac{\frac{5}{8}}{\frac{3}{8}} = \dfrac{5}{8} \times \dfrac{8}{3} = \dfrac{5}{3}$ or 5:3.

**(c)** There are 8 outcomes in the sample space, 3 of which correspond to event $T$, getting exactly 2 tails. The probability of $T$ is $\dfrac{3}{8}$. Therefore, the probability of $\overline{T}$ is $\dfrac{5}{8}$. The odds for $\overline{T}$ are $\dfrac{\frac{5}{8}}{\frac{3}{8}} = \dfrac{5}{8} \times \dfrac{8}{3} = \dfrac{5}{3}$ or 5:3.

## Solutions to Chapter 10 Review Problems

**1. (a)** For any event, the probability of the event is at least 0 and no more than 1.
  **(b)** A sure thing has a probability of 1.
  **(c)** An impossible event has a probability of 0.
  **(d)** $A$ and $B$ must be mutually exclusive events.
  **(e)** $A$ and $B$ must be independent events.

**2. (a)** The probabilities of events $A$, $B$, and $C$ must add to 1. The probability of $B$ is $1 - 0.2 - 0.5 = 0.3$.
  **(b)** The probabilities of events $D$, $E$, and $F$ must add to 1. The probability of $F$ is $1 - 0.3 - 0.1 = 0.6$.
  **(c)** The probability of $D$ is $1 - 0.5 - 0.2 = 0.3$.
  **(d)** The probability of $E$ is $1 - 0.6 - 0.3 = 0.1$.

**3. (a)** Multiply the probabilities along the branches: $P(A \cap D) = (0.2)(0.3) = 0.06$.
  **(b)** Use multiplicative and additive properties of probability tree diagrams:
  $$P(E) = P(AE \cup BE \cup CE) = 0.2 \times 0.1 + 0.3 \times 0.5 + 0.5 \times 0.1 = 0.22$$
  **(c)** $P(B|F) = \dfrac{P(B \cap F)}{P(F)} = \dfrac{0.3 \times 0.2}{0.2 \times 0.6 + 0.3 \times 0.2 + 0.5 \times 0.3} = \dfrac{0.06}{0.33} \approx 0.18$
  **(d)** $P(\overline{C}) = 1 - P(C) = 1 - 0.5 = 0.5$
  **(e)** $P(E|A) = \dfrac{P(E \cap A)}{P(A)} = \dfrac{0.2 \times 0.1}{0.2} = 0.1$

4. **(a)** The sample space can be written {HHH, HHT, HTH, THH, HTT, THT, TTH, TTT}.

   **(b)** The outcomes that correspond to getting a tail on the second toss are {HTH, HTT, TTH, TTT}.

   **(c)** There is 1 outcome out of the 8 outcomes in the sample space that corresponds to getting three tails. Therefore, $P(\text{TTT}) = \dfrac{1}{8}$.

   **(d)** There are 2 outcomes, THH or TTH, out of the 8 outcomes in the sample space that correspond to getting a tail on the first coin and a head on the third coin. Therefore, the probability of getting a tail on the first coin and a head on the third coin is $\dfrac{2}{8} = \dfrac{1}{4}$.

   **(e)** The complement of the event of getting at least 1 head is the event of getting 3 tails. We know there is 1 outcome out of 8 that corresponds to getting 3 tails, so $P(\text{TTT}) = \dfrac{1}{8}$. Therefore, the probability of getting at least 1 head is $1 - \dfrac{1}{8} = \dfrac{7}{8}$.

5. **(a)** The sample space can be written {HHHH, HHHT, HHTH, HTHH, THHH, HHTT, HTHT, HTTH, THHT, THTH, TTHH, HTTT, THTT, TTHT, TTTH, TTTT}.

   **(b)** The outcomes that correspond to getting more heads than tails are {HHHH, HHHT, HHTH, HTHH, THHH}.

   **(c)** Five out of the 16 outcomes in the sample space correspond to getting more heads than tails, so $P(\text{getting more heads than tails}) = \dfrac{5}{16}$.

   **(d)** Getting at most 1 tail is the same as getting 0 tails or exactly 1 tail. There are 5 outcomes out of the 16 outcomes in the sample space that correspond to getting at most 1 tail, so $P(\text{getting at most one tail}) = \dfrac{5}{16}$.

6. The student observed the dryer broken 12 times out of 30. The experimental probability that the clothes dryer will be broken the next time the student uses the laundry room is $\dfrac{12}{30} = \dfrac{2}{5}$.

7. **(a)** The sample space can be written {PN, PD, PQ, NP, ND, NQ, DP, DN, DQ, QP, QN, QD}.

   **(b)**

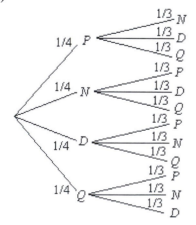

(c) The events $A$ and $C$ are mutually exclusive. The outcomes for event $A$ are $PQ, NQ, DQ, QP, QN,$ and $QD$ while the outcomes for event $C$ are $PN, PD, NP,$ and $DP$. Therefore, since the events have no outcomes in common, they are mutually exclusive.

(d) $P(A) = \dfrac{6}{12} = \dfrac{1}{2}$, $P(C) = \dfrac{4}{12} = \dfrac{1}{3}$, $P(\overline{B}) = \dfrac{6}{12} = \dfrac{1}{2}$.

(e) From the sample space in part (a), we see there are 10 ways for you to remove the quarter or the dime, so $P(A \cup B) = \dfrac{10}{12} = \dfrac{5}{6}$ and is the probability you remove the quarter or the dime. There are 8 ways for you to remove the dime or remove less than 12 cents, so $P(B \cup C) = \dfrac{8}{12} = \dfrac{2}{3}$ and is the probability you remove the dime or you remove less than 12 cents.

8.  (a) You purchase one box of cereal. The sample space can be written as {grand prize, prize worth $20, prize worth $0.25}.

(b) There is 1 box out of 100,000 that contains the grand prize. The probability you will win the grand prize is $\dfrac{1}{100,000}$.

(c) There are 100 boxes out of 100,000 that contain prizes worth $20. The probability you will win one of the prizes worth $20 is $\dfrac{100}{100,000} = \dfrac{1}{1000}$.

(d) To win a prize worth at least $20 means you could win a prize worth $20 or the grand prize worth $10,000. There are 101 ways this can happen. Therefore, the probability you will win a prize worth at least $20 is $\dfrac{101}{100,000}$.

(e) The probability of winning the grand prize is $\dfrac{1}{100,000}$. The probability of winning a prize worth $20 is $\dfrac{100}{100,000}$. The probability of winning a prize worth $0.25 is $\dfrac{99,899}{100,000}$. The cost of one box of cereal is $2.45 so subtract the cost from the expected value. The expected value is calculated as follows:

$$\text{Expected Value} = \$10,000 \times \dfrac{1}{100,000} + \$20 \times \dfrac{100}{100,000} + \$0.25 \times \dfrac{99,899}{100,000} - \$2.45$$

$$= \dfrac{\$10,000}{100,000} + \dfrac{\$2000}{100,000} + \dfrac{\$24,974.75}{100,000} - \$2.45$$

$$= \dfrac{\$36,974.75}{100,000} - \$2.45$$

$$\approx \$0.37 - \$2.45$$

$$= -\$2.08$$

The expected value of 1 box of cereal is –$1.90.

9.  (a) There are 3 colors for shirts and 4 colors for slacks, so there are $3 \times 4 = 12$ uniform combinations possible.

(b) Suppose the employee randomly select the slacks first. The first stage of the probability tree diagram has 4 branches. For each branch in the first stage, there are 3 shirt options in the second stage.

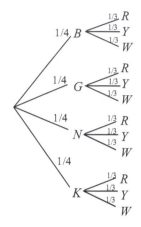

(c) There are two ways an employee can select black slacks and a shirt that is not white. The employee can select black slacks and a red shirt or black pants and a yellow shirt. Therefore,

$$P(BR \cup BY) = P(BR) + P(BY) = \frac{1}{4} \times \frac{1}{3} + \frac{1}{4} \times \frac{1}{3} = \frac{1}{12} + \frac{1}{12} = \frac{2}{12} = \frac{1}{6}.$$

10. (a) Selections are made with replacement, so each time a card is selected, the deck contains 52 cards. There are 4 aces in a standard deck of cards. In one draw, the probability of selecting an ace is $\frac{4}{52} = \frac{1}{13}$. Card selections are independent, so the probability of drawing 3 aces is $\left(\frac{1}{13}\right)^3 = \frac{1}{2197}$ or approximately 0.00046.

(b) This time, the cards are drawn without replacement, so the probability changes with each draw. The number of aces available in each draw changes from 4 to 3 to 2 and the number of cards in the deck changes from 52 to 51 to 50. Therefore, the probability of drawing 3 aces in a row is $\frac{4}{52} \times \frac{3}{51} \times \frac{2}{50} = \frac{24}{132,600}$ or approximately 0.00018.

11. (a) There are 200 patients. A total of $69 + 34 + 22 + 3 = 128$ have at least one cavity. The probability a patient selected at random has at least one cavity is $\frac{128}{200} = \frac{16}{25}$.

(b) A total of $69 + 2 = 71$ patients out of the 200 patients classified brush only. The probability a patient selected at random brushes only is $\frac{71}{200}$.

(c) We want to find the conditional probability that a patient has no cavities given that he or she brushes, flosses, and has tooth sealants.

$$P(\text{no cavities}|\text{brush, floss, and sealants}) = \frac{P(\text{no cavities} \cap \text{brush, floss, and sealants})}{P(\text{brush, floss, and sealants})}$$

$$= \frac{46}{3 + 46}$$

$$= \frac{46}{49}$$

Therefore, the probability the patient has no cavities given that he or she brushes, flosses, and has tooth sealants is approximately 0.94.

(d) We want to find the conditional probability that a patient brushes only given that he or she has at least one cavity.

$$P(\text{brushes only} \,|\, \text{at least one cavity}) = \frac{P(\text{brushes only} \cap \text{at least one cavity})}{P(\text{at least one cavity})}$$

$$= \frac{69}{69 + 34 + 22 + 3}$$

$$= \frac{69}{128}$$

Therefore, the probability the patient brushes only given that he or she has at least one cavity is approximately 0.54.

12. Let $R$ represent the event that a red marble is drawn and $B$ represent the event that a blue marble is drawn. Marbles are drawn without replacement until two of the same color have been drawn.

(a)

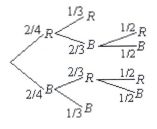

(b) There are 2 blue and 2 red marbles in the jar: $P(B) = \frac{2}{4} = \frac{1}{2}$.

(c) There are two ways to draw marbles so that the second marble is blue: $RB$ or $BB$. Therefore,

$$P(RB \cup BB) = \frac{2}{4} \times \frac{2}{3} + \frac{2}{4} \times \frac{1}{3} = \frac{4}{12} + \frac{2}{12} = \frac{6}{12} = \frac{1}{2}.$$

(d) Find the probability of events $RBB$ or $BRB$.

$$P(RBB \cup BRB) = \frac{2}{4} \times \frac{2}{3} \times \frac{1}{2} + \frac{2}{4} \times \frac{2}{3} \times \frac{1}{2} = \frac{4}{24} + \frac{4}{24} = \frac{8}{24} = \frac{1}{3}.$$

(e) For only two drawings to be necessary, 2 red marbles must be drawn or 2 blue marbles must be drawn. Find the probability of events $RR$ or $BB$.

$$P(RR \cup BB) = \frac{2}{4} \times \frac{1}{3} + \frac{2}{4} \times \frac{1}{3} = \frac{2}{12} + \frac{2}{12} = \frac{4}{12} = \frac{1}{3}.$$

13. Two of the urns are selected at random and their contents are mixed. There are three equally-likely ways this can be done; that is, urns I and II can be mixed, urns II and III can be mixed, or urns I and III can be mixed. The first stage of the probability tree diagram is shown next.

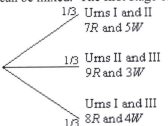

All four balls selected can be red and come from the mixture of urns I and II or all four balls can be red and come from the mixture of urns II and III, or all four balls can be red and come from the mixture of urns I and III. The balls are selected without replacement:

$$P(4\ \text{red}) = \frac{1}{3} \times \frac{7}{12} \times \frac{6}{11} \times \frac{5}{10} \times \frac{4}{9} + \frac{1}{3} \times \frac{9}{12} \times \frac{8}{11} \times \frac{7}{10} \times \frac{6}{9} + \frac{1}{3} \times \frac{8}{12} \times \frac{7}{11} \times \frac{6}{10} \times \frac{5}{9}$$

$$= \frac{840}{35,640} + \frac{3024}{35,640} + \frac{1680}{35,640}$$

$$= \frac{5544}{35,640}$$

$$= \frac{7}{45}$$

14. There are 36 possible outcomes when two fair dice are rolled. Six outcomes correspond to getting a 4 on the first die. Half of the outcomes, 18 of them, have an odd number showing on the second die. Three outcomes correspond to the event $F \cap O$ which is the event of getting a 4 on the first die and an odd number on the second die. $P(F \cap O) = \frac{3}{36} = \frac{1}{12}$ and is the probability of getting a 4 on the first die and an odd number on the second. $F \cup O$ is the event of getting a 4 on the first die or an odd number on the second. $P(F \cup O) = P(F) + P(O) - P(F \cap O) = \frac{6}{36} + \frac{18}{36} - \frac{3}{36} = \frac{21}{36} = \frac{7}{12}$ and is the probability of getting a 4 on the first die or an odd number on the second. Events $F$ and $O$ are independent because $P(F \cap O) = \frac{1}{12} = \frac{1}{6} \times \frac{1}{2} = P(F) \times P(O)$.

15. Create a probability tree diagram. The class is randomly selected, so in the first stage of the probability tree diagram each class has probability $\frac{1}{2}$ of being selected. In class I, there are 15 females out of 25, so the probability of a female being selected from class I is $\frac{15}{25} = \frac{3}{5}$. Once a female is selected in class I, there are 14 females left out of 24 students. In class II, there are 8 females out of 18, so the probability of a female being selected from class II is $\frac{8}{18} = \frac{4}{9}$. Once a female is selected in class II, there are 7 females left out of 17 students.

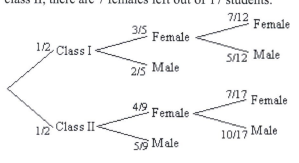

We want to find the conditional probability that both students came from class I given that they are both female.

$$P(\text{both from class I}|\text{both are female}) = \frac{P(\text{both from class I} \cap \text{both female})}{P(\text{both female})}$$

$$= \frac{\dfrac{1}{2} \times \dfrac{3}{5} \times \dfrac{7}{12}}{\dfrac{1}{2} \times \dfrac{3}{5} \times \dfrac{7}{12} + \dfrac{1}{2} \times \dfrac{4}{9} \times \dfrac{7}{17}}$$

$$= \frac{\dfrac{21}{120}}{\dfrac{21}{120} + \dfrac{28}{306}}$$

$$= \frac{\dfrac{21}{120}}{\dfrac{1071}{6120} + \dfrac{560}{6120}}$$

$$= \frac{\dfrac{21}{120}}{\dfrac{1631}{6120}}$$

$$= \frac{21}{120} \times \frac{6120}{1631}$$

$$= \frac{153}{233}$$

16. Let $S$ represent the event the student studies and let $A$ represent the event the student earns an A.

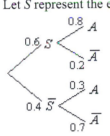

The student can study and earn an A or not study and earn an A. The probability the student earns an A is $P(SA \cup \overline{S}A) = 0.6 \times 0.8 + 0.4 \times 0.3 = 0.6$. We also want to find the probability the student studied given that she earned an A.

$$P(S|A) = \frac{P(S \cap A)}{P(A)}$$

$$= \frac{0.6 \times 0.8}{0.6 \times 0.8 + 0.4 \times 0.3}$$

$$= \frac{0.48}{0.6}$$

$$= 0.8$$

Therefore, the probability the student studied given that she earned an A is 0.8.

17. (a) There are 52 cards in a standard deck, 13 of which are hearts. The cards are drawn without replacement. Once a card is selected, the number of cards in the deck decreases. Create a probability tree diagram. Let $H$ represent the event a heart is drawn.

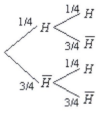

$$P(A) = \frac{13}{52} = \frac{1}{4}.$$

$$P(B) = \frac{1}{4} \times \frac{12}{51} + \frac{3}{4} \times \frac{13}{51} = \frac{12}{204} + \frac{39}{204} = \frac{51}{204} = \frac{1}{4}.$$

The probability both cards are hearts is $P(A \cap B) = \frac{1}{4} \times \frac{12}{51} = \frac{12}{204} = \frac{1}{17}$. Events $A$ and $B$ are not

independent because $P(A \cap B) = \frac{1}{17} \ne \frac{1}{4} \times \frac{1}{4} = P(A) \times P(B)$.

(b) There are 52 cards in a standard deck, 13 of which are hearts. The cards are drawn with replacement. Let $H$ represent the event a heart is drawn.

$$P(A) = \frac{13}{52} = \frac{1}{4}.$$

$$P(B) = \frac{1}{4} \times \frac{1}{4} + \frac{3}{4} \times \frac{1}{4} = \frac{1}{16} + \frac{3}{16} = \frac{4}{16} = \frac{1}{4}.$$

The probability both cards are hearts is $P(A \cap B) = \frac{1}{4} \times \frac{1}{4} = \frac{1}{16}$. Events $A$ and $B$ are independent

because $P(A \cap B) = \frac{1}{16} = \frac{1}{4} \times \frac{1}{4} = P(A) \times P(B)$.

18. (a) There are 15 bills in the box.

Expected Value = $\$1 \times \frac{5}{15} + \$5 \times \frac{4}{15} + \$10 \times \frac{3}{15} + \$20 \times \frac{2}{15} + \$100 \times \frac{1}{15} = \frac{\$195}{15} = \$13.$

(b) On average, you expect to win $13. If you pay $20 for a chance to pull one bill out of the box, then you expect to lose $7 on average..

19. The company wants to make \$10 per claim on average. We do not know the premium or cost the company should charge for the towing portion of a policy. Let $P$ represent the premium that is paid to the insurance company. When the company must pay money on a claim, it is a loss, so subtract.

Expected Value $= -\$30 \times 0.12 - \$55 \times 0.08 + P$

$$\$10 = -\$8 + P$$
$$\$18 = P$$

The insurance company should charge an \$18 premium.

20. The event that at least one of the four units has a defect is the complement of the event that none of the units have a defect. There are 16 computers out of 20 that are not defective, and there are 10 printers out of 12 that are not defective. The computers and printers are selected without replacement.

$P(\text{at least one defective}) = 1 - P(\text{no computers defective and no printers defective})$

$$= 1 - \frac{16}{20} \times \frac{15}{19} \times \frac{14}{18} \times \frac{13}{17} \times \frac{10}{12} \times \frac{9}{11} \times \frac{8}{10} \times \frac{7}{9}$$

$$= 1 - \frac{5096}{31,977}$$

$$= \frac{26,881}{31,977}$$

$$\approx 0.84$$

21. **(a)** You purchased 1 out of 2000 tickets, so the probability of winning is $\dfrac{1}{2000}$.

**(b)** Not winning the prize is the complement of the event that you win the prize. The probability of not winning is $1 - \dfrac{1}{2000} = \dfrac{1999}{2000}$.

**(c)** The odds in favor of winning the prize are $\dfrac{\frac{1}{2000}}{\frac{1999}{2000}} = \dfrac{1}{2000} \times \dfrac{2000}{1999} = \dfrac{1}{1999}$ or 1:1999.

**(d)** The odds against winning the prize are $\dfrac{\frac{1999}{2000}}{\frac{1}{2000}} = \dfrac{1999}{2000} \times \dfrac{2000}{1} = \dfrac{1999}{1}$ or 1999:1.

22. Let $R$ represent the event that a block of cheese must be rewrapped. Create a probability tree diagram. The first stage will represent the selection of a block of cheese from the morning shift. The second stage will represent the selection of a block of cheese from the afternoon shift.

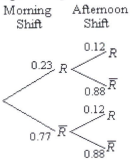

**(a)** The probability that neither block of cheese must be rewrapped is $P(\overline{R}\,\overline{R}) = 0.77 \times 0.88 = 0.6776$.

**(b)** The outcomes associated with exactly one block of cheese needing to be rewrapped are $\overline{R}R$ and $R\overline{R}$. The probability that exactly one block must be rewrapped is $P(\overline{R}R) + P(R\overline{R}) = 0.77 \times 0.12 + 0.23 \times 0.88 = 0.2948$.

**(c)** Create a new probability tree diagram. Let $M$ represent the event that the block came from the morning shift.

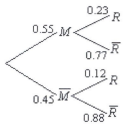

We want to find the probability the cheese came from the morning shift given that it needs to be rewrapped.

$$P(M|R) = \frac{P(M \cap R)}{P(R)} = \frac{0.55 \times 0.23}{0.55 \times 0.23 + 0.45 \times 0.12} = \frac{0.1265}{0.1805} \approx 0.7.$$

Therefore, if a block of cheese is selected and needs to be rewrapped there is a 70% chance it came from the morning shift.

**23.** The probability for bad weather is 0.35 so the proabaility for fair weather is $1 - 0.35 = 0.65$. The expected value is $\$900 \times 0.65 - \$250 \times 0.35 = \$497.50$.

# *Chapter 11:* Inferential Statistics

## *Key Concepts*

**By the time you finish studying Section 11.1, you should be able to:**

☐ calculate the area under the standard normal curve between two points using a table.

☐ be able to look at graphs of normal distributions and determine which has the largest (or smallest) mean and which has the largest (or smallest) standard deviation.

☐ discuss the features of the standard normal distribution.

**By the time you finish studying Section 11.2, you should be able to:**

☐ explain how any normal distribution is related to the standard normal distribution.

☐ use the 68-95-99.7 rule to determine approximate percentages of measurements that lie within one, two, or three standard deviations of the mean.

☐ calculate the *z*-score for a measurement.

☐ find the percentage of values in an interval for a given normal distribution using the *z*-score and areas from the standard normal distribution.

**By the time you finish studying Section 11.3, you should be able to:**

☐ determine the population proportion and the sample proportion for a given situation.

☐ find the mean and the standard deviation of the set of all of the sample proportions.

☐ discuss under what conditions we are able to use the normal distribution to approximate the distribution of sample proportions.

☐ calculate the approximate percentage of samples for which the sample proportion is between two given values, greater than a given value, or less than a given value.

☐ find and discuss the meaning of the standard error.

☐ calculate a 95% confidence interval for the population proportion.

☐ find and discuss the meaning of the margin of error.

## Skills Brush-Ups

Problems in Chapter 11 will require that you be comfortable working with percents and correctly applying the order of operations.

### PERCENTS:

Percent means "per 100". Therefore, 10% means 10 per 100 or 10 out of 100.

To convert a number from a percent to a decimal, divide the number by 100 and eliminate the percent symbol. In general, you can drop the percent symbol and move the decimal point two places to the left.

*Examples:* (a) $3.4\% = \dfrac{3.4}{100} = 0.034$ (b) $160\% = \dfrac{160}{100} = 1.6$ (c) $0.035\% = \dfrac{0.035}{100} = 0.00035$

To convert a number to a percent, multiply the number by 100%. In general, you can shift the decimal point two places to the right and insert a % symbol.

*Example:* 0.0219 is $0.0219 \times 100\% = 2.19\%$.

### ORDER OF OPERATIONS:

When more than one operation occurs in an expression, it matters which operation is performed first. Different results may arise depending on the order in which the operations are carried out. To avoid such problems, remember the correct order of operations. First, simplify expressions in parentheses, evaluate exponents (this includes radicals), evaluate multiplications and divisions in order from left to right, and evaluate additions and subtractions in order from left to right. The following example shows the type of expression you will be asked to simplify in Chapter 11.

*Example:* Simplify inside parentheses first. Notice that the radical (square root) acts like parentheses so simplify the expression under the radical as shown.

$$0.32 - 2\sqrt{\frac{(0.32)(1-0.32)}{32}} = 0.32 - 2\sqrt{\frac{(0.32)(0.68)}{32}} \qquad \text{Simplify inside parentheses}$$

$$= 0.32 - 2\sqrt{\frac{0.2176}{32}} \qquad \text{Multiply under the radical}$$

$$= 0.32 - 2\sqrt{0.0068} \qquad \text{Divide under the radical}$$

$$\approx 0.32 - 2(0.082462113) \qquad \text{Evaluate the radical}$$

$$\approx 0.32 - 0.164924225 \qquad \text{Multiply}$$

$$\approx 0.155075775 \qquad \text{Subtract}$$

## Skills Practice

1. Convert each of the following from percent notation to decimal notation.
   (a) 2%  (b) 4.5%  (c) 0.69%  (d) 1.05%

2. Convert each of the following from decimal notation to percent notation.
   (a) 0.09  (b) 0.0786  (c) 0.0013  (d) 0.155

3. Simplify each of the following expressions using the correct order of operations.
   (a) $0.1326 + 2 \times 0.056$  (b) $\sqrt{3 - 2.11} + 8.4 \times 1.7 - 5$

(c) $\dfrac{2.5 \times 5.8}{3} + 4 \times 3^2$      (d) $4.2(3.1 - 0.4)^2 + 5.2$

4. Simplify each of the following expressions using the correct order of operations.

  (a) $0.25 + 3\sqrt{\dfrac{0.21 \times 0.79}{40}}$    (b) $0.13 + 2\sqrt{\dfrac{0.65(1 - 0.65)}{53}}$   (c) $0.968 + 2\sqrt{\dfrac{0.43(1 - 0.43)}{100}}$

*Answers to Skills Practice:*
1. (a) 0.02    (b) 0.045    (c) 0.0069    (d) 0.0105
2. (a) 9%     (b) 7.86%    (c) 0.13%     (d) 15.5%
3. (a) 0.2246   (b) 10.2234   (c) $40.8\overline{3}$   (d) 35.818
4. (a) 0.4432   (b) 0.2610   (c) 1.067

---

## Hints for Odd-Numbered Problems

### Section 11.1

1. If a smooth curve is sketched through the tops of the rectangles in each figure, which smooth curve would most accurately estimate the area of the rectangles?

3. The area under the distribution curve between the given weights can be used to approximate the percentage of dogs in the population who have weights between 67 and 74 pounds.

5. The area under the distribution curve between the given weights can be used to approximate the percentage of dogs in the population who have weights between 50 and 67 pounds.

7. The area under the distribution curve between the given lengths can be used to approximate the percentage of hamsters in the population who have lengths between 3.5 and 4 inches.

9. The area under the distribution curve between the given lengths can be used to approximate the percentage of hamsters in the population who have lengths between 3 and 5.5 inches.

11. The peak of a normal distribution is located at the mean. Compare the location of the peak of each curve.

13. The standard deviation determines the height of the peak and the spread, or width, of the curve. Compare the heights and widths of each curve.

15. How will the distributions compare if the means are the same? Can two distributions with different standard deviations have the same height?

17. The entries in Table 11.3 represent the areas under the standard normal curve in the interval from 0 to $z$.

19. To find the area under the standard normal curve in an interval between two nonzero points, split the interval up into two pieces. Find the area from −2 to 0 then find the area from 0 to 1.7.

21. Due to symmetry in the standard normal distribution, finding the area under the curve in the interval from −0.8 to −0.2 is the same as finding the area under the curve in the interval from 0.2 to 0.8. Table 11.3 will give the area in the interval from 0 to 0.8 and also in the interval from 0 to 0.2. How can you use those two areas to find the desired area?

23. Shade the region under the standard normal curve in the interval from 0 to 1.5.

25. Shade the region under the standard normal curve in the interval from −2.1 to 1.9.

27. The percentage of the data in a standard normal distribution that has a value between two points is the same as the area under the

standard normal curve in the interval between the two points.

29. The percentage of the data in a standard normal distribution that has a value between two points is the same as the area under the standard normal curve in the interval between the two points.

31. The percentage of measurements that are between two values in a standard normal distribution is the same as the area under the standard normal curve in the interval between the two values.

33. The percentage of measurements that are between two values in a standard normal distribution is the same as the area under the standard normal curve in the interval between the two values.

35. The mean in a standard normal distribution is 0. The standard deviation is 1.
   (a) One standard deviation above the mean will occur at $z = 1$.
   (b) Two standard deviations below the mean will occur at $z = -2$.

37. The mean in a standard normal distribution is 0. The standard deviation is 1.
   (a) "Within one standard deviation of the mean" refers to an interval extending from one standard deviation below the mean to one standard deviation above the mean.

39. (c) The percentage of measurements within two standard deviations of the mean in this normal distribution is the same as the percentage in the standard normal distribution.
   (d) The percentage of measurements above the mean in any normal distribution is the same.

41. The percentage of measurements above what point in the standard normal distribution is 0.02?

43. (b) What percentage of values are within 1.5 standard deviations of the mean in the standard normal distribution?

## Section 11.2

1. Subtract to determine how far from the mean the data value is. Then compare the difference to the standard deviation.

3. Find each data value by adding a multiple of the standard deviation to the mean or subtracting a multiple of the standard deviation from the mean.

5. See Example 11.7.

7. Use the 68-95-99.7 rule and determine how many standard deviations above or below the mean the given data values are.

9. Use the 68-95-99.7 rule and refer to Figure 11.15.

11. (b) Compare the given graph to Figure 11.15.

13. Use the definition of the $z$-score and the given values of the mean and standard deviation.

15. Use the definition of the $z$-score and the given values of the mean and standard deviation.

17. Use the definition of the $z$-score and the given values of the mean and standard deviation.

19. Use the definition of the $z$-score and substitute in the values you know.

21. Use the definition of the $z$-score and substitute in the values you know.

23. We are assuming the gross Saturday sales are approximately normally distributed. Use the standard normal distribution areas in Table 11.3 and the $z$-scores to find the percentages related to sales.

25. We are assuming the weights are normally distributed. Use the standard normal distribution areas in Table 11.3 and the $z$-scores of the weights to find the percentages related to the weights.

27. We are assuming the SAT scores are normally distributed. Use the standard normal distribution areas in Table 11.3 and the $z$-scores to find the desired percentages.

29. We are assuming the SAT scores are normally distributed. Use the standard normal distribution areas in Table 11.3 and the $z$-scores to find the desired percentages.

31. We are assuming the SAT scores are normally distributed. Use the standard normal distribution areas in Table 11.3 and the $z$-scores to find the desired percentages.

33. The annual total rainfall in Guatemala is approximately normally distributed. Find the $z$-score and use Table 11.3.

35. Flea counts on the dog and in the surroundings are assumed to be normally distributed. Find the $z$-scores and use Table 11.3.

## Section 11.3

1. Round as the last step. Allow your calculator to keep as many digits as possible during the calculations.

3. Use the definition of the mean and standard deviation for the set of all sample proportions.

5. (b) What fraction of the total population of cars had a problem needing correction?
   (c) What fraction of the sample of cars had a problem needing correction?

7. (b) How many of the registered voters were involved in the pre-election canvassing?

9. (c) The following two conditions must be met before we can assume the distribution of sample proportions is approximately normal:

$$p - 3\sqrt{\frac{p(1-p)}{n}} > 0$$

and

$$p + 3\sqrt{\frac{p(1-p)}{n}} < 1$$

11. (d) Compare the means and the standard deviations from parts (b) and (c). Does the mean or the standard deviation depend on the sample size?

13. One is related to the population and one is related to the sample.

15. (a) The population is the set of the five required courses.
    (b) The table contains all possible samples of three courses selected from the five.
    (c) There are ten samples of size three. To find the mean, add up the sample proportions and divide by 10.

17. Write each sample proportion as a decimal and use the definition of the standard error.

19. Write the sample proportion as a decimal and use the definition of the standard error.

21. Out of 500 students selected, 265 are females. Calculate the sample proportion of females.

23. Out of 1003 adults selected, 361 thought it was morally acceptable. Calculate the sample proportion of adults who thought it was morally acceptable.

25. (b) Use the definition of a 95% confidence interval and the value for the sample proportion and standard error from part (a).
    (c) Is there any reason to suspect that a larger (or smaller) proportion of left-handed students would sign up for biology compared to the general student body?

27. How are the margin of error and the standard error related? Use the definition of the standard error and the given sample proportion to solve for the unknown sample size.

**29. (c)** If many samples were taken and 95% confidence intervals created, what percentage of the intervals is expected to contain the true proportion?

**31.** See Example 11.20.

**33.** Find the sample proportion and the standard error from the given information. The 95% confidence interval is the interval within two standard errors of the sample proportion.

**35.** The margin of error is twice the standard error.

**37.** To find the sample size, use the definition for the standard error and let 0.15 be the sample proportion.

**39. (b)** Compare the sample size required to obtain a margin of error of 10% to the sample sizes required to achieve one tenth the margin of error or one hundredth the margin of error.

**41. (b)** Compare the margins of error across each row and notice how the margins of error change as the sample size increases. Also compare the margins of error down each column and notice how the margins of error change as the sample proportion changes.

## Solutions to Odd-Numbered Problems in Section 11.1

**1.** Histogram (II) could best be approximated by the region under a smooth curve. The rectangles are narrower in this histogram, so there are smaller gaps compared to a smooth curve.

**3.** The percentage of dogs that weigh between 67 and 74 pounds can be approximated by the area under the distribution curve between those two weights. The area under the curve from 67 to 74 is listed as 20% of the total area. Therefore, approximately 20% of dogs in the population weigh between 67 and 74 pounds.

**5.** The percentage of dogs that weigh between 50 and 67 pounds can be approximated by the area under the distribution curve between those two weights. The area under the curve from 50 to 67 can be broken up into three intervals: 50 to 57, 57 to 62, and 62 to 67. The area under the curve in each of the three intervals is listed as 20%. Add the areas. Therefore, approximately 20% + 20% + 20% = 60% of dogs in the population weigh between 50 and 67 pounds.

**7.** The percentage of hamsters that have lengths between 3.5 and 4 inches can be approximated by the area under the distribution curve between those two lengths. The area under the curve from 3.5 to 4 is listed as 16% of the total area. Therefore, approximately 16% of hamsters in the population have lengths between 3.5 and 4 inches.

**9.** The percentage of hamsters that have lengths between 3 and 5.5 inches can be approximated by the area under the distribution curve between those two lengths. The area under the curve from 3 to 5.5 is broken up into four intervals: 3 to 3.5, 3.5 to 4, 4 to 5, and 5 to 5.5. Add the area under the curve in each of the four intervals: 14% + 16% + 30% + 9% = 69%. Therefore, approximately 69% of hamsters in the population have lengths between 3 and 5.5 inches.

**11.** The peak of the curve is located at its mean. The data set with the largest mean is data set III. Data set III has its peak located farthest to the right compared to the other two distributions. The data set with the smallest mean is data set I. Data set I has its peak located farthest to the left compared to the other distributions.

**13.** The standard deviation determines the height of the peak and spread (or width) of the curve. The data set with the largest standard deviation is data set II. Data set II is the shortest and the widest. The data set with the smallest standard deviation is data set I. Data set I is the tallest and thinnest.

**15.** The two normal distributions shown next have the same mean. As a result, their peaks are centered over the same point. The standard deviations are not the same, however, so the heights and widths of the distributions are different. The one with the smaller standard deviation is taller and thinner. The one with the larger standard deviation is shorter and wider.

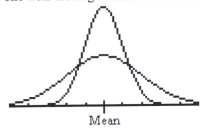

**17.** Table 11.3 gives areas under the standard normal curve in intervals from 0 to $z$. Let $z = 1.8$ and locate 1.8 in the table. The area is 0.4641. Therefore the area under the standard normal curve in the interval from 0 to 1.8 is 0.4641. We can conclude that 46.41% of the data in a standard normal distribution lies in the interval from 0 to 1.8.

**19.** Table 11.3 gives areas under the standard normal curve in the interval from 0 to $z$. If we want to find the area under the standard normal curve between a negative value and a positive value, then it will be necessary to split the interval up into two pieces. In order to find the area under the curve in the interval from $z = -2$ to $z = 1.7$, find the area in the interval from $z = -2$ to $z = 0$ and also in the interval from $z = 0$ to $z = 1.7$. The desired area is shaded in the following graph.

**Interval from −2 to 0:** The area under the curve from $z = -2$ to $z = 0$ is the same as the area under the curve from $z = 0$ to $z = 2$ due to symmetry in the standard normal distribution. Let $z = 2$ and locate 2 in Table 11.3. The area is 0.4772 and is shaded in the following graph.

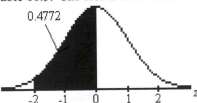

**Interval from 0 to 1.7:** Let $z = 1.7$ and locate 1.7 in Table 11.3. The area is 0.4554.

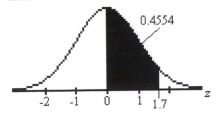

The total area under the curve from $z = -2$ to $z = 1.7$ is found by adding the two areas 0.4772 + 0.4554 = 0.9326. We can conclude that 93.26% of the data in a standard normal distribution lies in the interval from −2 to 1.7.

**21.** Table 11.3 gives areas under the standard normal curve in the interval from 0 to $z$. We want to find the area under the standard normal curve between two negative values. Due to the symmetry in the standard normal curve, the area under the curve in the interval from $z = -0.8$ to $z = -0.2$ is the same as the area under the curve in the interval from $z = 0.2$ to $z = 0.8$ (shaded in solid black). First, find the area under the curve in the interval from $z = 0$ to $z = 0.8$. This area is more than we need. Next, find the area under the curve in the interval from $z = 0$ to $z = 0.2$ (shaded in stripes). The area we want can be found by subtracting these two areas as shown in the following graph.

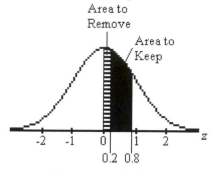

**Interval from 0 to 0.8:** Let $z = 0.8$ and locate 0.8 in Table 11.3. The area is 0.2881.
**Interval from 0 to 0.2:** Let $z = 0.2$ and locate 0.2 in Table 11.3. The area is 0.0793.
The total area under the curve in the interval from $z = 0.2$ to $z = 0.8$ (and also in the interval from $-0.8$ to $-0.2$) is found by subtracting the two areas: $0.2881 - 0.0793 = 0.2088$. We can conclude that 20.88% of the data in a standard normal distribution lies in the interval from $-0.8$ to $-0.2$.

**23.** Let $z = 1.5$ and locate 1.5 in Table 11.3. The area given is 0.4332. Therefore, 43.32% of the area under the standard normal curve is in the interval from 0 to 1.5. This region is shaded in the following graph of the standard normal curve.

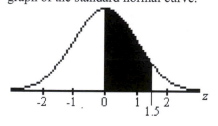

**25.** Table 11.3 gives areas under the standard normal curve in the interval from 0 to $z$. If we want to find the area under the standard normal curve between a negative value and a positive value, then it will be necessary to split the interval up into two pieces. In order to find the area under the curve in the interval from $z = -2.1$ to $z = 1.9$, find the area under the curve in the interval from $z = -2.1$ to $z = 0$ and also in the interval from $z = 0$ to $z = 1.9$.

**Interval from $-2.1$ to 0:** The area under the curve in the interval from $z = -2.1$ to $z = 0$ is the same as the area under the curve in the interval from $z = 0$ to $z = 2.1$ due to symmetry in the standard normal distribution. Let $z = 2.1$ and locate 2.1 in Table 11.3. The area is 0.4821.
**Interval from 0 to 1.9:** Let $z = 1.9$ and locate 1.9 in Table 11.3. The area is 0.4713.
The total area under the curve in the interval from $z = -2.1$ to $z = 1.9$ is found by adding the two areas $0.4821 + 0.4713 = 0.9534$. This region is shaded in the following graph of the standard normal curve.

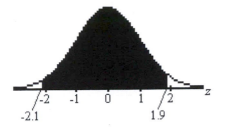

27. The percentage of data values in a standard normal distribution between two points is the same as the percentage of the area under the standard normal curve in the interval between the two points.
    (a) The area under the standard normal curve in the interval from $z = 0$ to $z = 3$ is found from Table 11.3 as 0.4987. The area under the standard normal curve in the interval from $z = 0$ to $z = 1$ is found from Table 11.3 as 0.3413. Subtract to obtain the area in the interval from $z = 1$ to $z = 3$: $0.4987 - 0.3413 = 0.1574$. Therefore, the percentage of the data that lie between 1 and 3 is 15.74%.
    (b) The area under the standard normal curve in the interval from $z = 0$ to $z = 2$ is found from Table 11.3 as 0.4772. The area under the standard normal curve greater than $z = 0$ is 0.5. Subtract to find the area greater than $z = 2$: $0.5 - 0.4772 = 0.0228$. Therefore, the percentage of the data that is greater than 2 is 2.28%.
    (c) From Table 11.3, we know the area in the interval from $z = 0$ to $z = 1$ is 0.3413. The area under the standard normal curve in the interval from $z = -1$ to $z = 1$ is twice as large as the area in the interval from $z = 0$ to $z = 1$ due to symmetry in the standard normal curve. Therefore, the area in the interval from $z = -1$ to $z = 1$ is $2 \times 0.3413 = 0.6826$. The percentage of data that lie between $-1$ and 1 is 68.26%. The percentage of data that is not between $-1$ and 1 is $100\% - 68.26\% = 31.74\%$.

29. The percentage of the data in a standard normal distribution that has a value in an interval is the same as the area under the standard normal curve in the interval. Use Table 11.3.
    (a) The area under the curve in the interval from $z = 0$ to $z = 3$ is 0.4987. The area under the curve in the interval from $z = 0$ to $z = 2$ is 0.4772. Subtract to find the area under the curve from $z = 2$ to $z = 3$: $0.4987 - 0.4772 = 0.0215$. Therefore, the percentage of data that has a value between 2 and 3 is 2.15%.
    (b) The area under the curve less than $z = 0$ is 0.5. The area under the curve in the interval from $z = 0$ to $z = 2$ is 0.4772. Add to find the area under the curve less than $z = 2$: $0.5 + 0.4772 = 0.9772$. Therefore, the percentage of data that has a value less than 2 is 97.72%.
    (c) The total area under the standard normal curve is 1. The area that is *not* between $z = -2$ and $z = 2$ is shaded in the following graph.

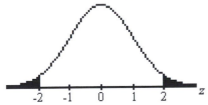

    To find the area that is *not* between $z = -2$ and $z = 2$, find the area that *is* in the interval from $z = -2$ to $z = 2$ and then subtract that area from 1. The area in the interval from $z = -2$ to $z = 2$ is twice the area in the interval from $z = 0$ to $z = 2$ due to the symmetry of the standard normal curve. Therefore the area in the interval from $z = -2$ to $z = 2$ is $2 \times 0.4772 = 0.9544$. Finally, the area that is not in the interval from $z = -2$ to $z = 2$ is $1 - 0.9544 = 0.0456$. The percentage of the data that is not between $-2$ and 2 is 4.56%.

**31.** To find the percentage of measurements that are between two values in a standard normal distribution, find the area under the standard normal curve in the interval between the two values. Use Table 11.3.

   **(a)** The area under the curve in the interval from $z = 0$ to $z = 1.2$ is 0.3849. Therefore, the percentage of measurements that are between 0 and 1.2 is 38.49%.

   **(b)** The area under the curve in the interval from $z = -2.3$ to $z = 0$ is the same as the area under the curve in the interval from $z = 0$ to $z = 2.3$. The area is 0.4893. Therefore, the percentage of measurements that are between $-2.3$ and 0 is 48.93%.

   **(c)** Add the area under the curve in the interval from $z = -0.7$ to $z = 0$ to the area under the curve in the interval from $z = 0$ to $z = 1.8$. The area under the curve in the interval from $z = -0.7$ to $z = 0$ is the same as the area in the interval from $z = 0$ to $z = 0.7$. This area is 0.2580. The area in the interval from $z = 0$ to $z = 1.8$ is 0.4641. Add the areas: $0.2580 + 0.4641 = 0.7221$. Therefore the percentage of measurements that are between $-0.7$ and 1.8 is 72.21%.

**33.** To find the percentage of measurements that are between two values in a standard normal distribution, find the area under the standard normal curve in the interval between the two values. Use Table 11.3.

   **(a)** From Table 11.3, we know the area under the curve in the interval from $z = 0$ to $z = 1.8$ is 0.4641. The area greater than $z = 0$ is 0.5. We want the area greater than $z = 1.8$. Subtract to find the area: $0.5 - 0.4641 = 0.0359$. Therefore the percentage of measurements greater than 1.8 is 3.59%.

   **(a)** From part (b), we know the area under the curve greater than $z = 1.8$ is 0.0359. The total area under the standard normal curve is 1. Subtract to find the area under the curve less than $z = 1.8$: $1 - 0.0359 = 0.9641$. Therefore the percentage of measurements less than 1.8 is 96.41%.

**35.** The standard normal distribution has a mean of 0 and a standard deviation of 1.

   **(a)** One standard deviation above the mean occurs at $z = 1$. Find the area under the curve in the interval from $z = 0$ to $z = 1$. This area is 0.3413. Therefore, 34.13% of the measurements lie between 0 and 1 standard deviation above the mean.

   **(b)** Two standard deviations below the mean occurs at $z = -2$. Find the area under the curve in the interval from $z = -2$ to $z = 0$. This area is the same as the area in the interval from $z = 0$ to $z = 2$ and is found in Table 11.3 as 0.4772. Therefore, 47.72% of the measurements lie between 2 standard deviations below the mean and 0.

   **(c)** Two standard deviations below the mean occurs at $z = -2$. One standard deviation above the mean occurs at $z = 1$. Find the area under the curve in the interval from $z = -2$ to $z = 1$. The area from $z = -2$ to $z = 0$ is 0.4772 and the area in the interval from $z = 0$ to $z = 1$ is 0.3413. Add the areas: $0.4772 + 0.3413 = 0.8185$. Therefore, 81.85% of the measurements lie between 2 standard deviations below the mean and 1 standard deviation above the mean.

**37.** The standard normal distribution has a mean of 0 and a standard deviation of 1.

   **(a)** The statement "within one standard deviation of the mean" refers to the interval that extends from one standard deviation below the mean to one standard deviation above the mean as shown in the following graph.

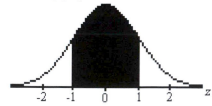

   Find the area under the curve in the interval from $z = -1$ to $z = 1$. This area can be found by doubling the area from $z = 0$ to $z = 1$ due to the symmetry in the standard normal distribution. The area in the interval from $z = 0$ to $z = 1$ is 0.3413. Therefore the area in the interval from $z = -1$ to $z = 1$ is $2 \times 0.3413 = 0.6826$. Finally, we conclude 68.26% of measurements are within one standard deviation of the mean.

(b) Within three standard deviations of the mean refers to the interval that extends from three standard deviation below the mean to three standard deviation above the mean as shown in the following graph. We want the area that is not in this region.

Find the area under the curve in the interval from $z = -3$ to $z = 3$ and subtract the area from 1. The shaded area can be found by doubling the area from $z = 0$ to $z = 3$ due to the symmetry in the standard normal distribution. The area in the interval from $z = 0$ to $z = 3$ is 0.4987. Therefore the area in the interval from $z = -3$ to $z = 3$ is $2 \times 0.4987 = 0.9974$. Finally, we conclude $1 - 0.9974 = 0.0026$ so 0.26% of measurements are not within three standard deviations of the mean.

39. The mean of the normal distribution is 16 inches, and the standard deviation is 2 inches.
   (a) The mean + one standard deviation $= 16 + 2 = 18$. Therefore, 18 inches is the measurement that is one standard deviation above the mean.
   (b) The mean – two standard deviations $= 16 - (2 \times 2) = 12$. Therefore, 12 inches is the measurement that is two standard deviations below the mean.
   (c) The percentage of measurements within two standard deviations of the mean is the same for the standard normal distribution as it is for any normal distribution. From problem 38, part (a), we know this value to be 95.44%.
   (d) The mean is 16 inches. The percentage of measurements greater than the mean is 50% which is the same in any normal distribution.

41. In order to be accepted into Mensa, a person must have an IQ in the top 2% of all IQ values. Graphically, we are looking for the IQ score above which the area under the normal curve is approximately 0.02. On a standard normal distribution, the following shaded region represents the area we are talking about.

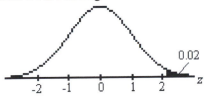

If we let $z = 2$, then the area under the standard normal curve greater than $z = 2$ is $0.5 - 0.4772 = 0.0228$. Alternately, if we let $z = 2.1$, then the area under the standard normal curve greater than $z = 2.1$ is $0.5 - 0.4821 = 0.0179$. The point above which 2% of the area is found is between 2 and 2.1 standard deviations above the mean. The IQ score that is 2 standard deviations above the mean is $100 + (2 \times 15) = 130$. The IQ score that is 2.1 standard deviations above the mean is $100 + (2.1 \times 15) = 131.5$. Therefore, the lowest IQ accepted for membership into Mensa is approximately 130.

43. (a) Women with normal serum total cholesterol levels will be included in the study. Normal is defined as a level within 1.5 standard deviations of the mean. The range of normal serum total cholesterol levels is $197 - (1.5 \times 43.1) = 132.35$ to $197 + (1.5 \times 43.1) = 261.65$. Therefore, women with levels from 132.35 mg/dL to 261.65 mg/dL will be included in the study.
   (b) In any normal distribution, the percentage of data within 1.5 standard deviations of the mean is the same. In the standard normal distribution, the area under the curve from $z = -1.5$ to $z = 1.5$ is twice the area from $z = 0$ to $z = 1.5$. This area is $2 \times 0.4332 = 0.8664$. Out of 5000 women, approximately 86.64% would be eligible to participate. Therefore, 86.64% of 5000 = $0.8664 \times 5000 = 4332$ women would be eligible to participate.

## Solutions to Odd-Numbered Problems in Section 11.2

**1.** For the given data set, the mean is 25 and the standard deviation is 3.

  (a) We know $28 - 25 = 3$, and $\frac{3}{3} = 1$. Therefore, the data value 28 is one standard deviation above the mean.

  (b) We know $31 - 25 = 6$, and $\frac{6}{3} = 2$. Therefore, the data value 31 is two standard deviations above the mean.

  (c) We know $22 - 25 = -3$, and $\frac{-3}{3} = -1$. Therefore, the data value 22 is one standard deviation below the mean.

  (d) We know $26.5 - 25 = 1.5$, and $\frac{1.5}{3} = 0.5$. Therefore, the data value 26.5 is 0.5 standard deviations above the mean.

  (e) We know $20.5 - 25 = -4.5$, and $\frac{-4.5}{3} = -1.5$. Therefore, the data value 20.5 is 1.5 standard deviations below the mean.

  (f) We know $16 - 25 = -9$, and $\frac{-9}{3} = -3$. Therefore, the data value 16 is three standard deviations below the mean.

**3.** (a) The value that is 3 standard deviations above the mean is $52 + (3 \times 10) = 82$.
  (b) The value that is 2 standard deviations below the mean is $52 - (2 \times 10) = 32$.
  (c) The value that is 1.5 standard deviations below the mean is $52 - (1.5 \times 10) = 37$.
  (d) The value that is 2.5 standard deviations above the mean is $52 + (2.5 \times 10) = 77$.
  (e) The value that is $\frac{1}{4}$ of a standard deviation above the mean is $52 + \left(\frac{1}{4} \times 10\right) = 54.5$.

**5.** (a) The data value that is $\frac{1}{4}$ of a standard deviation above the mean is $84 + \left(\frac{1}{4} \times 8\right) = 86$.

  (b) The data value that is $\frac{1}{4}$ of a standard deviation below the mean is $84 - \left(\frac{1}{4} \times 8\right) = 82$.

  (c) Because the interval from 82 to 86 represents values within $\frac{1}{4}$ of a standard deviation from the mean, 20% of the measurements lie between 82 and 86.

**7.** The mean of the population of turkeys is 12 pounds and the standard deviation is 2.5 pounds.
  (a) Notice that $12 - 2.5 = 9.5$, so 9.5 is one standard deviation below the mean. Notice also that $12 + 2.5 = 14.5$, so 14.5 is one standard deviation above the mean. By the 68-95-99.7 rule, we know approximately 68% of the turkeys have weights within one standard deviation of the mean.
  (b) Notice that $12 - (3 \times 2.5) = 4.5$, so 4.5 is three standard deviations below the mean. Notice also that $12 + (3 \times 2.5) = 19.5$, so 19.5 is three standard deviations above the mean. By the 68-95-99.7 rule, we know approximately 99.7% of the turkeys have weights within three standard deviations of the mean.

(c) Notice that $12 - (2 \times 2.5) = 7$, so 7 is two standard deviations below the mean. The value 12 is exactly the same as the mean. By the 68-95-99.7 rule, we know approximately 95% of the turkeys have weights within two standard deviation of the mean. Therefore approximately 47.5% of the turkeys have weights ranging from 7 to 12 pounds.

9. Refer to Figure 11.15.
   (a) Add the areas between $\mu - \sigma$ and $\mu + 2\sigma$ to obtain $0.34 + 0.34 + 0.135 = 0.815$. Therefore, 81.5% of measurements lie between one standard deviation below the mean and two standard deviations above the mean.
   (b) Add the areas between $\mu - 3\sigma$ and $\mu + \sigma$ to obtain $0.0235 + 0.135 + 0.34 + 0.34 = 0.8385$. Therefore, 83.85% of measurements lie between three standard deviations below the mean and one standard deviation above the mean.
   (c) According to the 68-95-99.7 rule, approximately 68% of measurements in any normal distribution lie within one standard deviation of the mean. Since the total area under the curve is 100%, the area that is *not* between $\mu - \sigma$ and $\mu + \sigma$ is $100\% - 68\% = 32\%$.

11. (a) The mean of the population is 15. The peak of the curve is centered above 15.
    (b) By the 68-95-99.7 rule, we know approximately 68% of the measurements in any normal distribution lie within one standard deviation of the mean. From the graph provided, we can see that 68% of the area under the curve lies between 11 and 19. Therefore, 11 must be one standard deviation below the mean and 19 must be one standard deviation above the mean. The standard deviation is $19 - 15 = 4$.
    (c) Add the areas under the curve between the values provided: $0.34 + 0.34 + 0.135 = 0.815$. Therefore, approximately 81.5% of the measurements lie between 11 and 23.
    (d) Add the areas under the curve between the values provided: $0.135 + 0.34 + 0.34 = 0.815$. Therefore, approximately 81.5% of the measurements lie between 7 and 19.
    (e) The area under the curve below 7 is 0.0235. The area under the curve above 23 is 0.0235. The total area that is not between 7 and 23 is $0.0235 + 0.0235 = 0.047$. Therefore, 4.7% of measurements are *not* between 7 and 23.

13. Use the definition of the z-score and let $\mu = 10$ and $\sigma = 2$.
    $$z = \frac{x - \mu}{\sigma} = \frac{x - 10}{2}.$$

| Measurement $x$ | z-score |
|---|---|
| 9 | $z = \dfrac{9-10}{2} = \dfrac{-1}{2} = -0.5$ |
| 10 | $z = \dfrac{10-10}{2} = \dfrac{0}{2} = 0$ |
| 11 | $z = \dfrac{11-10}{2} = \dfrac{1}{2} = 0.5$ |
| 14 | $z = \dfrac{14-10}{2} = \dfrac{4}{2} = 2$ |
| 17 | $z = \dfrac{17-10}{2} = \dfrac{7}{2} = 3.5$ |

**15.** Use the definition of the z-score and let $\mu = 100$ and $\sigma = 15$.

$$z = \frac{x-\mu}{\sigma} = \frac{x-100}{15}.$$

| Measurement $x$ | z-score |
|---|---|
| 64 | $z = \dfrac{64-100}{15} = \dfrac{-36}{15} = -2.4$ |
| 80 | $z = \dfrac{80-100}{15} = \dfrac{-20}{15} \approx -1.33$ |
| 96 | $z = \dfrac{96-100}{15} = \dfrac{-4}{15} \approx -0.27$ |
| 111 | $z = \dfrac{111-100}{15} = \dfrac{11}{15} \approx 0.73$ |
| 136 | $z = \dfrac{136-100}{15} = \dfrac{36}{15} = 2.4$ |
| 145 | $z = \dfrac{145-100}{15} = \dfrac{45}{15} = 3$ |

**17.** Use the definition of the z-score and let $\mu = 55.6$ and $\sigma = 11.3$.

$$z = \frac{x-\mu}{\sigma} = \frac{x-55.6}{11.3}.$$

| Measurement $x$ | z-score |
|---|---|
| 45.16 | $z = \dfrac{45.16-55.6}{11.3} = \dfrac{-10.44}{11.3} \approx -0.92$ |
| 49.82 | $z = \dfrac{49.82-55.6}{11.3} = \dfrac{-5.78}{11.3} \approx -0.51$ |
| 55.20 | $z = \dfrac{55.2-55.6}{11.3} = \dfrac{-0.4}{11.3} \approx -0.04$ |
| 58.63 | $z = \dfrac{58.63-55.6}{11.3} = \dfrac{3.03}{11.3} \approx 0.27$ |

**19. (a)** Use the definition of the z-score and let $\sigma = 4.25$, $z = 1.9$, and $x = 52.1$.

$$z = \frac{x-\mu}{\sigma}$$
$$1.9 = \frac{52.1-\mu}{4.25}$$
$$1.9 \times 4.25 = 52.1-\mu$$
$$8.075 = 52.1-\mu$$
$$\mu = 52.1-8.075$$
$$\mu = 44.025$$

The mean is 44.025.

**(b)** Use the definition of the z-score and let $\sigma = 0.6$, $z = -2.3$, and $x = 2$.

$$z = \frac{x - \mu}{\sigma}$$

$$-2.3 = \frac{2 - \mu}{0.6}$$

$$-2.3 \times 0.6 = 2 - \mu$$

$$-1.38 = 2 - \mu$$

$$\mu = 2 + 1.38$$

$$\mu = 3.38$$

The mean is 3.38.

**21. (a)** Use the definition of the z-score and let $\mu = 9.8$, $z = 2$, and $x = 10.3$.

$$z = \frac{x - \mu}{\sigma}$$

$$2 = \frac{10.3 - 9.8}{\sigma}$$

$$2 \times \sigma = 0.5$$

$$\sigma = \frac{0.5}{2}$$

$$\sigma = 0.25$$

The standard deviation is 0.25.

**(b)** Use the definition of the z-score and let $\mu = 577$, $z = -0.5$, and $x = 533$.

$$z = \frac{x - \mu}{\sigma}$$

$$-0.5 = \frac{533 - 577}{\sigma}$$

$$-0.5 \times \sigma = -44$$

$$\sigma = \frac{-44}{-0.5}$$

$$\sigma = 88$$

The standard deviation is 88.

**23. (a)** Use the definition of the z-score and let $\mu = 4610$ and $\sigma = 370$.

| Measurement $x$ | z-score |
|---|---|
| 3870 | $z = \dfrac{3870 - 4610}{370} = \dfrac{-740}{370} = -2$ |
| 4425 | $z = \dfrac{4425 - 4610}{370} = \dfrac{-185}{370} = -0.5$ |
| 5535 | $z = \dfrac{5535 - 4610}{370} = \dfrac{925}{370} = 2.5$ |

**(b)** The percentage of the franchises that had gross Saturday sales between $4425 and $5535 is the same as the percentage of the area under the standard normal curve that lies between $z = -0.5$ and $z = 2.5$. We know the area in the interval from $z = -0.5$ to $z = 0$ is the same as the area in the interval from $z = 0$ to $z = 0.5$. From Table 11.3, this area is 0.1915. The area in the interval from $z = 0$ to $z = 2.5$ is 0.4938. The total area is $0.1915 + 0.4938 = 0.6853$. Therefore, approximately 68.53% of the franchises had gross Saturday sales between $4425 and $5535.

(c) The percentage of the franchises that had gross Saturday sales between $3870 and $5535 is the same as the percentage of the area under the standard normal curve that lies between $z = -2$ and $z = 2.5$. We know the area in the interval from $z = -2$ to $z = 0$ is the same as the area in the interval from $z = 0$ to $z = 2$. From Table 11.3, this area is 0.4772. The area in the interval from $z = 0$ to $z = 2.5$ is 0.4938. The total area is $0.4772 + 0.4938 = 0.9710$. Therefore, approximately 97.10% of the franchises had gross Saturday sales between $3870 and $5535.

(d) The percentage of the franchises that had gross Saturday sales less than $5535 is the same as the percentage of the area under the standard normal curve that is less than $z = 2.5$. We know the area under the standard normal curve that is less than $z = 0$ is 0.5. The area in the interval from $z = 0$ to $z = 2.5$ is 0.4938. The total area is $0.5 + 0.4938 = 0.9938$. Therefore, approximately 99.38% of the franchises had gross Saturday sales less than $5535.

25. (a) Find the $z$-scores for the given weights. The percentage of dogs that will weigh between 5.75 and 16.25 pounds is the same as the percentage of the area under the standard normal curve between

$$z = \frac{5.75 - 11}{3.5} = -1.5 \text{ and } z = \frac{16.25 - 11}{3.5} = 1.5.$$ The area in the interval from $z = -1.5$ to $z = 1.5$

is twice the area from $z = 0$ to $z = 1.5$ which is $2 \times 0.4332 = 0.8664$. Therefore, 86.64% of dogs will weigh between 5.75 and 16.25 pounds.

(b) Find the $z$-scores for the given weights. The percentage of dogs that will weigh between 11 and 18 pounds is the same as the percentage of the area under the standard normal curve between

$$z = \frac{11 - 11}{3.5} = 0 \text{ and } z = \frac{18 - 11}{3.5} = 2.$$ The area in the interval from $z = 0$ to $z = 2$ is 0.4772.

Therefore, 47.72% of dogs will weigh between 11 and 18 pounds.

(c) The percentage of dogs that will weigh more than 15.55 pounds is the same as the percentage of

the area under the standard normal curve greater than $z = \dfrac{15.55 - 11}{3.5} = \dfrac{4.55}{3.5} = 1.3$. The area in

the interval from $z = 0$ to $z = 1.3$ is 0.4032. The area less than $z = 0$ is 0.5. Thus the area less than $z = 1.3$ is $0.4032 + 0.5 = 0.9032$. Therefore, 90.32% of dogs will weigh less than 15.55 pounds and $100\% - 90.32\% = 9.68\%$ of dogs will weigh more than 15.55 pounds.

(d) The percentage of dogs that will weigh less than 1.2 pounds is the same as the percentage of the

area under the standard normal curve less than $z = \dfrac{1.2 - 11}{3.5} = \dfrac{-9.8}{3.5} = -2.8$. The area in the

interval from $z = -2.8$ to $z = 0$ is 0.4974 due to symmetry in the curve. The area less than $z = 0$ is 0.5. Therefore, the area less than $z = -2.8$ is $0.5 - 0.4974 = 0.0026$. Thus, 0.26% of dogs will weigh less than 1.2 pounds.

27. (a) The percentage of students who had a math score between 300 and 500 points is the same as the

percentage of the area under the standard normal curve between $z = \dfrac{300 - 507}{111} = \dfrac{-207}{111} \approx -1.9$

and $z = \dfrac{500 - 507}{111} = \dfrac{-7}{111} \approx -0.1$. Due to symmetry this is the same as the area in the interval

from $z = 0.1$ to $z = 1.9$. Find the area by subtracting. The area is $0.4713 - 0.0398 = 0.4315$. Therefore, 43.15% of students scored between 300 and 500 points on the math portion of the SAT.

(b) The percentage of students who had a math score less than 200 points is the same as the

percentage of the area under the standard normal curve less than $z = \dfrac{200 - 507}{111} = \dfrac{-307}{111} \approx -2.8$.

The area in the interval from $z = -2.8$ to $z = 0$ is 0.4974 due to symmetry. The area less than $z = 0$ is 0.5. Therefore, the area less than $z = -2.8$ is $0.5 - 0.4974 = 0.0026$. Thus, 0.26% of students scored less than 200 points on the math portion of the SAT.

(c) The percentage of students who had a math score of at least 400 points is the same as the percentage of the area under the standard normal curve greater than $z = \dfrac{400-507}{111} = \dfrac{-107}{111} \approx -1$.

The area in the interval from $z = -1$ to $z = 0$ is 0.3413 due to symmetry. The area greater than $z = 0$ is 0.5. Therefore, the area greater than $z = -1$ is 0.3413 + 0.5 = 0.8413. Thus, 84.13% of students scored at least 400 points on the math portion of the SAT. Out of all students who took the SAT, $0.8413 \times 1{,}406{,}324 \approx 1{,}183{,}140$ students scored at least 400 on the math portion.

29. (a) The percentage of students who had a math score of at least 650 points is the same as the percentage of the area under the standard normal curve greater than $z = \dfrac{650-507}{111} = \dfrac{143}{111} \approx 1.3$.

The area in the interval from $z = 0$ to $z = 1.3$ is 0.4032. The area greater than $z = 0$ is 0.5. Therefore, the area greater than $z = 1.3$ is 0.5 − 0.4032 = 0.0968. Thus, 9.68% of students scored at least 650 points on the math portion of the SAT.

(b) The percentage of students who had a verbal score of at least 650 points is the same as the percentage of the area under the standard normal curve greater than $z = \dfrac{650-519}{115} = \dfrac{131}{115} \approx 1.1$.

The area in the interval from $z = 0$ to $z = 1.1$ is 0.3643. The area greater than $z = 0$ is 0.5. Therefore, the area greater than $z = 1.1$ is 0.5 − 0.3643 = 0.1357. Thus, 13.57% of students scored at least 650 points on the verbal portion of the SAT.

31. (a) The percentage of students who had a verbal score between 500 and 600 points is the same as the percentage of the area under the standard normal curve between $z = \dfrac{500-413}{100} = \dfrac{87}{100} \approx 0.9$ and $z = \dfrac{600-413}{100} = \dfrac{187}{100} \approx 1.9$. Find the area by subtracting. The area is 0.4713 − 0.3159 = 0.1554.

Therefore, 15.54% of students whose parents did not earn a high school diploma scored between 500 and 600 points on the verbal portion of the SAT.

(b) The percentage of students who had a math score between 500 and 600 points is the same as the percentage of the area under the standard normal curve between $z = \dfrac{500-443}{114} = \dfrac{57}{114} = 0.5$ and $z = \dfrac{600-443}{114} = \dfrac{157}{114} \approx 1.4$. Find the area by subtracting. The area is 0.4192 − 0.1915 = 0.2277.

Therefore, 22.77% of students whose parents did not earn a high school diploma scored between 500 and 600 points on the math portion of the SAT.

(c) The percentage of students who had a verbal score less than 300 points is the same as the percentage of the area under the standard normal curve less than $z = \dfrac{300-413}{100} = \dfrac{-113}{100} \approx -1.1$.

The area in the interval from $z = -1.1$ to $z = 0$ is 0.3643 due to symmetry. The area less than $z = 0$ is 0.5. Therefore, the area less than $z = -1.1$ is 0.5 − 0.3643 = 0.1357. Thus, 13.57% of students whose parents did not earn a high school diploma scored less than 300 points on the verbal portion of the SAT.

(d) The percentage of students who had a math score less than 300 points is the same as the percentage of the area under the standard normal curve less than $z = \dfrac{300-443}{114} = \dfrac{-143}{114} \approx -1.3$.

The area in the interval from $z = -1.3$ to $z = 0$ is 0.4032 due to symmetry. The area less than $z = 0$ is 0.5. Therefore, the area less than $z = -1.3$ is 0.5 − 0.4032 = 0.0968. Thus, 9.68% of students whose parents did not earn a high school diploma scored less than 300 points on the math portion of the SAT.

33. The percentage of years during which Guatemala would suffer from drought conditions, that is, receive less than 600 mm of rain, is the same as the percentage of the area under the standard normal curve less than $z = \dfrac{600-955}{257} = \dfrac{-355}{257} \approx -1.4$. The area in the interval from $z = -1.4$ to $z = 0$ is 0.4192 due to symmetry. The area less than $z = 0$ is 0.5. Therefore, the area less than $z = -1.4$ is $0.5 - 0.4192 = 0.0808$. Therefore, during 8.08% of the years Guatemala would suffer drought conditions.

35. **(a)** The percentage of dogs that had at most 1 flea on them on day 14 is approximately the same as the percentage of the area under the standard normal curve less than $z = \dfrac{1-1.4}{2.5} = \dfrac{-0.4}{2.5} \approx -0.2$. The area in the interval from $z = -0.2$ to $z = 0$ is 0.0793. The area less than $z = -0.2$ is $0.5 - 0.0793 = 0.4207$. Therefore, approximately 42.07% of dogs had at most 1 flea on them on day 14.

    The percentage of dogs that had at most 1 flea on them on day 28 is approximately the same as the percentage of the area under the standard normal curve less than $z = \dfrac{1-1.7}{4.7} = \dfrac{-0.7}{4.7} \approx -0.1$. The area in the interval from $z = -0.1$ to $z = 0$ is 0.0398. The area less than $z = -0.1$ is $0.5 - 0.0398 = 0.4602$. Therefore, approximately 46.02% of dogs had at most 1 flea on them on day 28.

    It appears that more dogs have at most 1 flea on their body on day 28 compared to day 14. This might suggest that the Imidacloprid has not worn off by day 28 and continues to be effective.

    **(b)** The percentage of dogs that had at most 1 flea in their surroundings on day 14 is approximately the same as the percentage of the area under the standard normal curve less than $z = \dfrac{1-12.7}{14} = \dfrac{-11.7}{14} \approx -0.8$. The area in the interval from $z = -0.8$ to $z = 0$ is 0.2881. The area less than $z = -0.8$ is $0.5 - 0.2881 = 0.2119$. Therefore, approximately 21.19% of dogs had at most 1 flea in their surroundings on day 14.

    The percentage of dogs that had at most 1 flea in their surroundings on day 28 is approximately the same as the percentage of the area under the standard normal curve less than $z = \dfrac{1-1}{1.7} = \dfrac{0}{1.7} = 0$. This area is 0.5. Therefore, approximately 50% of dogs had at most 1 flea in their surroundings on day 28.

    It appears that significantly more dogs have at most 1 flea in their surroundings on day 28 compared to day 14. This might suggest that the Imidacloprid is more effective at controlling fleas in a dog's surroundings after 28 days.

    **(c)** After 28 days, the percentage of dogs with at most 5 fleas on their bodies is the same as the percentage of the area under the standard normal curve less than $z = \dfrac{5-1.7}{4.7} = \dfrac{3.3}{4.7} \approx 0.7$. This area is $0.5 + 0.2580 = 0.7580$. Therefore, of the 50 dogs treated, we would expect $0.7580 \times 50 \approx 38$ of dogs have at most 5 fleas on their bodies after 28 days.

    After 28 days, the percentage of dogs with at least 11 fleas on their bodies is the same as the percentage of the area under the standard normal curve greater than $z = \dfrac{11-1.7}{4.7} = \dfrac{9.3}{4.7} \approx 2$. This area is $0.5 - 0.4772 = 0.0228$. Therefore, of the 50 dogs treated, we would expect $0.0228 \times 50 \approx 1$ dog to have at least 11 fleas on its body after 28 days.

## Solutions to Odd-Numbered Problems in Section 11.3

1. Round to the nearest thousandth in the last step.

   (a) $\sqrt{\dfrac{(0.2)(0.8)}{50}} = \sqrt{\dfrac{0.16}{50}} = \sqrt{0.0032} \approx 0.057$

   (b) $\sqrt{\dfrac{(0.14)(1-0.14)}{634}} = \sqrt{\dfrac{(0.14)(0.86)}{634}} = \sqrt{\dfrac{0.1204}{634}} \approx 0.014$

   (c) $0.49 - 3\sqrt{\dfrac{(0.49)(1-0.49)}{470}} = 0.49 - 3\sqrt{\dfrac{(0.49)(0.51)}{470}} = 0.49 - 3\sqrt{\dfrac{0.2499}{470}} \approx 0.421$

3. We know $p = 0.58$ and $n = 250$.

   (a) The mean of the set of sample proportions is 0.58.

   (b) The standard deviation of the set of sample proportions is
   $\sqrt{\dfrac{p(1-p)}{n}} = \sqrt{\dfrac{0.58(1-0.58)}{250}} = \sqrt{\dfrac{0.58(0.42)}{250}} = \sqrt{\dfrac{0.2436}{250}} \approx 0.031$.

5. (a) The population is the set of 6000 cars produced in a factory in a certain week. The sample is the set of 60 cars selected for a detailed inspection.

   (b) The population proportion of cars having problems is $p = \dfrac{300}{6000} = 0.05$.

   (c) The sample proportion of cars having problems is $\hat{p} = \dfrac{5}{60} \approx 0.083$.

7. (a) The population is the set of registered voters in a certain city. The sample is the set of canvassed voters.

   (b) The total number of canvassed voters is $185 + 210 + 25 = 420$. The sample size is 420.

   (c) The population proportion of registered Republicans is $p = \dfrac{3250}{7140} \approx 0.455$.

   (d) The sample proportion of registered Republicans is $\hat{p} = \dfrac{210}{420} = 0.5$.

9. (a) The population proportion of boxes of cereal that are significantly under filled is
   $p = \dfrac{300}{6000} = 0.05$.

   (b) The mean of the set of all sample proportions of under filled boxes is 0.05. The standard deviation of the set of all sample proportions of under filled boxes is
   $\sqrt{\dfrac{p(1-p)}{n}} = \sqrt{\dfrac{0.05(1-0.05)}{55}} = \sqrt{\dfrac{0.05(0.95)}{55}} = \sqrt{\dfrac{0.0475}{55}} \approx 0.029$.

   (c) We can conclude the distribution of sample proportions is approximately normal if both of the following conditions are true:
   $$p - 3\sqrt{\dfrac{p(1-p)}{n}} > 0 \text{ and } p + 3\sqrt{\dfrac{p(1-p)}{n}} < 1$$
   Substitute the known values for $p$ and $n$ into the inequalities and see if the conditions are met.

$$0.05 - 3\sqrt{\frac{0.05(1-0.05)}{55}} > 0 \quad \text{and} \quad 0.05 + 3\sqrt{\frac{0.05(1-0.05)}{55}} < 1$$

$$0.05 - 3\sqrt{\frac{0.05(0.95)}{55}} > 0 \quad \text{and} \quad 0.05 + 3\sqrt{\frac{0.05(0.95)}{55}} < 1$$

$$0.05 - 3\sqrt{\frac{0.0475}{55}} > 0 \quad \text{and} \quad 0.05 + 3\sqrt{\frac{0.0475}{55}} < 1$$

$$-0.038 > 0 \quad \text{and} \quad 0.138 < 1$$

Notice that −0.038 is not greater than zero. Therefore, we cannot assume the distribution of sample proportions is approximately normal.

11. (a) The population proportion of people who support the continuation of the manned space shuttle program is $p = \dfrac{22,048}{27,560} = 0.8$.

(b) Samples of size 150 are taken from the population. The mean of the set of all sample proportions of people who support the continuation of the manned space shuttle program is 0.8. The standard deviation of the set of all sample proportions of people who support the continuation of the manned space shuttle program is $\sqrt{\dfrac{p(1-p)}{n}} = \sqrt{\dfrac{0.8(1-0.8)}{150}} = \sqrt{\dfrac{0.8(0.2)}{150}} = \sqrt{\dfrac{0.16}{150}} \approx 0.033$.

(c) Samples of size 3000 are taken from the population. The mean of the set of all sample proportions of people who support the continuation of the manned space shuttle program is 0.8. The standard deviation of the set of all sample proportions of people who support the continuation of the manned space shuttle program is $\sqrt{\dfrac{p(1-p)}{n}} = \sqrt{\dfrac{0.8(1-0.8)}{3000}} = \sqrt{\dfrac{0.8(0.2)}{3000}} = \sqrt{\dfrac{0.16}{3000}} \approx 0.007$.

(d) As the sample size increases, the standard deviation of the set of all population sample proportions decreases.

13. The population proportion is $p$. If $p$ is unknown, then we estimate it with the sample proportion $\hat{p}$.

15. (a) Out of the five courses, three are humanities courses. The population proportion of required humanities courses he can take is $\dfrac{3}{5} = 0.6$.

(b) For each sample, find the proportion $\dfrac{\text{number of humanities courses in the sample}}{\text{number of courses in the sample}}$.

| Course Selections | Sample Proportion of Humanities Courses |
|---|---|
| SA, SB, HA | $\dfrac{1}{3}$ |
| SA, SB, HB | $\dfrac{1}{3}$ |
| SA, SB, HC | $\dfrac{1}{3}$ |
| SA, HA, HB | $\dfrac{2}{3}$ |
| SA, HA, HC | $\dfrac{2}{3}$ |

| SA, HB, HC | $\dfrac{2}{3}$ |
|---|---|
| SB, HA, HB | $\dfrac{2}{3}$ |
| SB, HA, HC | $\dfrac{2}{3}$ |
| SB, HB, HC | $\dfrac{2}{3}$ |
| HA, HB, HC | $\dfrac{3}{3}$ |

(c) To find the mean of the sample proportions, add up the sample proportions and divide by 10. The

mean is $\dfrac{\dfrac{1}{3}+\dfrac{1}{3}+\dfrac{1}{3}+\dfrac{2}{3}+\dfrac{2}{3}+\dfrac{2}{3}+\dfrac{2}{3}+\dfrac{2}{3}+\dfrac{2}{3}+\dfrac{3}{3}}{10} = \dfrac{\dfrac{18}{3}}{10} = \dfrac{6}{10} = 0.6.$ The mean of all possible sample proportions is the same as the population proportion.

17. Use the definition of the standard error. If $\hat{p}$ is the sample proportion, then the standard error is

approximately $\hat{s} = \sqrt{\dfrac{\hat{p}(1-\hat{p})}{n}}$ .

(a) In this case, $\hat{p} = 0.1$. $\hat{s} = \sqrt{\dfrac{\hat{p}(1-\hat{p})}{n}} = \sqrt{\dfrac{0.1(1-0.1)}{40}} = \sqrt{\dfrac{0.1(0.9)}{40}} \approx 0.047.$

(b) In this case, $\hat{p} = 0.25$. $\hat{s} = \sqrt{\dfrac{\hat{p}(1-\hat{p})}{n}} = \sqrt{\dfrac{0.25(1-0.25)}{40}} = \sqrt{\dfrac{0.25(0.75)}{40}} \approx 0.068.$

(c) In this case, $\hat{p} = 0.5$. $\hat{s} = \sqrt{\dfrac{\hat{p}(1-\hat{p})}{n}} = \sqrt{\dfrac{0.5(1-0.5)}{40}} = \sqrt{\dfrac{0.5(0.5)}{40}} \approx 0.079.$

(d) In this case, $\hat{p} = 0.8$. $\hat{s} = \sqrt{\dfrac{\hat{p}(1-\hat{p})}{n}} = \sqrt{\dfrac{0.8(1-0.8)}{40}} = \sqrt{\dfrac{0.8(0.2)}{40}} \approx 0.063.$

19. Use the definition of the standard error. If $\hat{p} = 0.45$, then the standard error is approximately

$\hat{s} = \sqrt{\dfrac{\hat{p}(1-\hat{p})}{n}}$ .

(a) In this case, $n = 50$. $\hat{s} = \sqrt{\dfrac{\hat{p}(1-\hat{p})}{n}} = \sqrt{\dfrac{0.45(1-0.45)}{50}} = \sqrt{\dfrac{0.45(0.55)}{50}} \approx 0.070.$

(b) In this case, $n = 100$. $\hat{s} = \sqrt{\dfrac{\hat{p}(1-\hat{p})}{n}} = \sqrt{\dfrac{0.45(1-0.45)}{100}} = \sqrt{\dfrac{0.45(0.55)}{100}} \approx 0.050.$

(c) In this case, $n = 500$. $\hat{s} = \sqrt{\dfrac{\hat{p}(1-\hat{p})}{n}} = \sqrt{\dfrac{0.45(1-0.45)}{500}} = \sqrt{\dfrac{0.45(0.55)}{500}} \approx 0.022.$

(d) In this case, $n = 1000$. $\hat{s} = \sqrt{\dfrac{\hat{p}(1-\hat{p})}{n}} = \sqrt{\dfrac{0.45(1-0.45)}{1000}} = \sqrt{\dfrac{0.45(0.55)}{1000}} \approx 0.016.$

**21.** The sample size is 500. It is observed that 265 are females. The sample proportion of females is $\hat{p} = \dfrac{265}{500} = 0.53$. The standard error for the proportion of females is calculated as follows:

$\hat{s} = \sqrt{\dfrac{\hat{p}(1-\hat{p})}{n}} = \sqrt{\dfrac{0.53(1-0.53)}{500}} = \sqrt{\dfrac{0.53(0.47)}{500}} \approx 0.022.$ The true value of the population proportion of females is unknown. The sample proportion is our best guess of the true proportion. The standard error is an approximation to the true standard deviation of the set of all sample proportions and is an indication of how good the sample proportion is as an estimate of the population proportion.

**23.** The sample size is 1003. It is observed that 361 adults thought it was morally acceptable. The sample proportion of adults who thought it was morally acceptable is $\hat{p} = \dfrac{361}{1003} \approx 0.36$. The standard error for the proportion of adults who thought it was morally acceptable is calculated as follows:

$\hat{s} = \sqrt{\dfrac{\hat{p}(1-\hat{p})}{n}} = \sqrt{\dfrac{0.36(1-0.36)}{1003}} = \sqrt{\dfrac{0.36(0.64)}{1003}} \approx 0.015.$ The true value of the population proportion of adults nationwide who thought it was morally acceptable is unknown. The sample proportion of 0.36 is our best guess of the true proportion. The standard error is an approximation to the true standard deviation of the set of all sample proportions and is an indication of how good the sample proportion is as an estimate of the population proportion.

**25. (a)** The sample size is 42. It is observed that 7 students are left-handed. The sample proportion of students who are left-handed is $\hat{p} = \dfrac{7}{42} \approx 0.167$. The standard error for the proportion of students who are left-handed is calculated as follows:

$\hat{s} = \sqrt{\dfrac{\hat{p}(1-\hat{p})}{n}} = \sqrt{\dfrac{0.167(1-0.167)}{42}} = \sqrt{\dfrac{0.167(0.833)}{42}} \approx 0.058.$

**(b)** The 95% confidence interval for the population proportion of left-handed students is calculated as follows:

$$\hat{p} - 2\hat{s} \text{ to } \hat{p} + 2\hat{s}$$

$$0.167 - 2(0.058) \text{ to } 0.167 + 2(0.058)$$

$$0.051 \text{ to } 0.283$$

With a confidence level of 95%, we conclude the true population proportion of left-handed students is between 0.051 and 0.283.

**(c)** The students were randomly assigned to the class and there is no reason to think that students who are left-handed are more likely to sign up for biology. The assumption is probably reasonable.

**27.** The margin of error is defined as $2\hat{s}$, and we know that $2\hat{s} = 4\%$. Therefore, $\hat{s} = 2\% = 0.02$. In order to determine the sample size, recall the definition of the standard error and substitute in the value of the sample proportion.

$$\widehat{s} = \sqrt{\frac{\widehat{p}(1-\widehat{p})}{n}}$$

$$0.02 = \sqrt{\frac{0.39(0.61)}{n}}$$

$$(0.02)^2 = \left(\sqrt{\frac{0.2379}{n}}\right)^2$$

$$0.0004 = \frac{0.2379}{n}$$

$$n = \frac{0.2379}{0.0004}$$

$$n = 594.75$$

The sample size was approximately 595.

29. **(a)** The sample size is 280. It is observed that 172 students are satisfied with the new procedures. The sample proportion is $\widehat{p} = \dfrac{172}{280} \approx 0.614$. The standard error is calculated as follows:

$$\widehat{s} = \sqrt{\frac{\widehat{p}(1-\widehat{p})}{n}} = \sqrt{\frac{0.614(1-0.614)}{280}} = \sqrt{\frac{0.614(0.386)}{280}} \approx 0.029.$$

**(b)** The 95% confidence interval for the percentage of students who are satisfied with the new registration procedures is calculated as follows:

$$\widehat{p} - 2\widehat{s} \quad \text{to} \quad \widehat{p} + 2\widehat{s}$$

$$0.614 - 2(0.029) \quad \text{to} \quad 0.614 + 2(0.029)$$

$$0.556 \quad \text{to} \quad 0.672$$

With a confidence level of 95%, we conclude the true percentage of students who are satisfied with the new registration procedures is 55.6% to 67.2%.

**(c)** If all possible samples of size 280 were taken, and confidence intervals were created, then we expect 95% of those to contain the true population proportion or percentage. We do not know for sure if our confidence interval contains the true percentage of students who are satisfied with the new registration procedures, but we are 95% confident that it does.

31. **(a)** We know $\widehat{p} = 0.54$. The standard error is $\widehat{s} = \sqrt{\dfrac{\widehat{p}(1-\widehat{p})}{n}} = \sqrt{\dfrac{0.54(0.46)}{1385}} \approx 0.013.$

The 95% confidence interval for the proportion of adults who favor President Bush's proposal is calculated as follows:

$$\widehat{p} - 2\widehat{s} \quad \text{to} \quad \widehat{p} + 2\widehat{s}$$

$$0.54 - 2(0.013) \quad \text{to} \quad 0.54 + 2(0.013)$$

$$0.514 \quad \text{to} \quad 0.566$$

With a confidence level of 95%, we conclude the true proportion of adults who favor President Bush's proposal is 0.514 to 0.566.

**(b)** The margin of error is $2\widehat{s} = 2(0.013) = 0.026$. We could say that the percentage of people who approve of Bush's proposal is 54% ± 2.6%. The margin of error gives an estimate as to how close our sample proportion is to the actual population proportion.

**33.** We know $\hat{p} = \dfrac{105}{240} = 0.4375$. The standard error is $\hat{s} = \sqrt{\dfrac{\hat{p}(1-\hat{p})}{n}} = \sqrt{\dfrac{0.4375(0.5625)}{240}} \approx 0.032$.

The 95% confidence interval for the percentage of full-time students who are working 10 or more hours each week is calculated as follows:

$$\hat{p} - 2\hat{s} \quad \text{to} \quad \hat{p} + 2\hat{s}$$

$$0.4375 - 2(0.032) \quad \text{to} \quad 0.4375 + 2(0.032)$$

$$0.3735 \quad \text{to} \quad 0.5015$$

We are 95% confident that the percentage of full-time students who are working 10 or more hours each week is between 37.35% and 50.15%.

**35.** We know $\hat{p} = 0.68$. The standard error is $\hat{s} = \sqrt{\dfrac{\hat{p}(1-\hat{p})}{n}} = \sqrt{\dfrac{0.68(0.32)}{500}} \approx 0.021$.

The 95% confidence interval for the proportion of people who approve of Title IX is calculated as follows:

$$\hat{p} - 2\hat{s} \quad \text{to} \quad \hat{p} + 2\hat{s}$$

$$0.68 - 2(0.021) \quad \text{to} \quad 0.68 + 2(0.021)$$

$$0.638 \quad \text{to} \quad 0.722$$

We are 95% confident that the proportion of people who approve of Title IX is between 0.638 and 0.722. The margin of error is $2\hat{s} = 2(0.021) = 0.042$ or 4.2%.

**37.** Our best guess for the proportion of the population who would favor reinstating the draft is 0.15.

    **(a)** If we want the margin of error to be 5%, then $2\hat{s} = 0.05$ and $\hat{s} = 0.025$. To determine the sample size needed, use the definition of the standard error and let $\hat{p} = 0.15$.

$$\hat{s} = \sqrt{\dfrac{\hat{p}(1-\hat{p})}{n}}$$

$$0.025 = \sqrt{\dfrac{0.15(0.85)}{n}}$$

$$(0.025)^2 = \left(\sqrt{\dfrac{0.1275}{n}}\right)^2$$

$$0.000625 = \dfrac{0.1275}{n}$$

$$n = \dfrac{0.1275}{0.000625}$$

$$n = 204$$

Therefore, the sample size was 204.

    **(b)** If we want the margin of error to be 1%, then $2\hat{s} = 0.01$ and $\hat{s} = 0.005$. To determine the sample size needed, use the definition of the standard error and let $\hat{p} = 0.15$.

$$\hat{s} = \sqrt{\frac{\hat{p}(1-\hat{p})}{n}}$$

$$0.005 = \sqrt{\frac{0.15(0.85)}{n}}$$

$$(0.005)^2 = \left(\sqrt{\frac{0.1275}{n}}\right)^2$$

$$0.000025 = \frac{0.1275}{n}$$

$$n = \frac{0.1275}{0.000025}$$

$$n = 5100$$

Therefore, the sample size was 5100.

**39. (a)** Let $\hat{p} = 0.5$.

| Margin of Error | $\hat{s} = \frac{1}{2}$ (margin of error) | Sample Size $n = \frac{\hat{p}(1-\hat{p})}{(\hat{s})^2}$ | 95% Confidence Interval $\hat{p} - 2\hat{s}$ to $\hat{p} + 2\hat{s}$ |
|---|---|---|---|
| 10% | 5% or 0.05 | $n = \frac{0.5(0.5)}{(0.05)^2} = 100$ | $0.5 - 2(0.05)$ to $0.5 + 2(0.05)$<br>0.4 to 0.6 |
| 5% | 2.5% or 0.025 | $n = \frac{0.5(0.5)}{(0.025)^2} = 400$ | $0.5 - 2(0.025)$ to $0.5 + 2(0.025)$<br>0.45 to 0.55 |
| 1% | 0.5% or 0.005 | $n = \frac{0.5(0.5)}{(0.005)^2} = 10,000$ | $0.5 - 2(0.005)$ to $0.5 + 2(0.005)$<br>0.49 to 0.51 |
| 0.1% | 0.05% or 0.0005 | $n = \frac{0.5(0.5)}{(0.0005)^2} = 1,000,000$ | $0.5 - 2(0.0005)$ to $0.5 + 2(0.0005)$<br>0.499 to 0.501 |

**(b)** In order to decrease the margin of error, the sample size must increase. For a small margin of error, the confidence interval is very narrow. The tradeoff for increased accuracy is a dramatic increase in the necessary sample size.

**41. (a)** The margin of error is $2\hat{s}$ and $\hat{s} = \sqrt{\frac{\hat{p}(1-\hat{p})}{n}}$. Therefore, the margin of error $= 2\sqrt{\frac{\hat{p}(1-\hat{p})}{n}}$.

For each combination of sample proportion and sample size, calculate the margin of error.

| | $n = 30$ | $n = 50$ | $n = 100$ | $n = 500$ |
|---|---|---|---|---|
| $\hat{p} = 0.25$ | $2\sqrt{\frac{0.25(0.75)}{30}}$ $\approx 0.158$ | $2\sqrt{\frac{0.25(0.75)}{50}}$ $\approx 0.122$ | $2\sqrt{\frac{0.25(0.75)}{100}}$ $\approx 0.087$ | $2\sqrt{\frac{0.25(0.75)}{500}}$ $\approx 0.039$ |
| $\hat{p} = 0.3$ | $2\sqrt{\frac{0.3(0.7)}{30}}$ $\approx 0.167$ | $2\sqrt{\frac{0.3(0.7)}{50}}$ $\approx 0.130$ | $2\sqrt{\frac{0.3(0.7)}{100}}$ $\approx 0.092$ | $2\sqrt{\frac{0.3(0.7)}{500}}$ $\approx 0.041$ |
| $\hat{p} = 0.5$ | $2\sqrt{\frac{0.5(0.5)}{30}}$ $\approx 0.183$ | $2\sqrt{\frac{0.5(0.5)}{50}}$ $\approx 0.141$ | $2\sqrt{\frac{0.5(0.5)}{100}}$ $= 0.1$ | $2\sqrt{\frac{0.5(0.5)}{500}}$ $\approx 0.045$ |

(b) Consider a column. For a sample size of 30, for example, the sample proportion changes from 0.25 to 0.5. The resulting margins of error do not change very much however. On the other hand, consider a row. For a sample proportion of 0.25, for example, the sample sizes change from 30 to 500. The resulting margins of error change dramatically. It appears that the sample size is what causes the greatest change in the margin of error.

## Solutions to Chapter 11 Review Problems

1. (a) The statement "within one standard deviation of the mean" refers to the interval that extends from one standard deviation below the mean to one standard deviation above the mean as shown in the following graph of the standard normal curve.

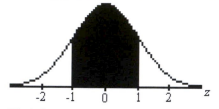

Find the area under the curve in the interval from $z = -1$ to $z = 1$. This area can be found by doubling the area from $z = 0$ to $z = 1$ due to the symmetry in the standard normal distribution. The area in the interval from $z = 0$ to $z = 1$ is 0.3413. Therefore the area in the interval from $z = -1$ to $z = 1$ is $2 \times 0.3413 = 0.6826$. Finally, we conclude 68.26% of measurements are within one standard deviation of the mean.

(b) Find the area under the standard normal curve in the interval from $z = -2$ to $z = 2$. The desired region is shaded in the following graph.

This area can be found by doubling the area from $z = 0$ to $z = 2$ due to the symmetry in the standard normal distribution. The area in the interval from $z = 0$ to $z = 2$ is 0.4772. Therefore the area in the interval from $z = -2$ to $z = 2$ is $2 \times 0.4772 = 0.9544$. Finally, we conclude 95.44% of values are between −2 and 2.

2. (a) According to the 68-95-99.7Rule, approximately 95% of the measurements in any normal distribution lie within two standard deviations of the mean.

(b) In any normal distribution, half of the curve lies above the mean and half of the curve lies below the mean. Therefore, 50% of the values in any normal distribution lie above the mean.

(c) According to the 68-95-99.7 rule, approximately $0.34 + 0.34 + 0.135 = 0.815$ or 81.5% of measurements in any standard normal distribution lie from one standard deviation below the mean to up to two standard deviations above the mean.

3. (a) Table 11.3 gives areas under the standard normal curve in the interval from 0 to $z$. If we want to find the area under the standard normal curve between a negative value and a positive value, then it will be necessary to split the interval up into two pieces. In order to find the area under the curve in the interval from $z = -1$ to $z = 3$, find the area in the interval from $z = -1$ to $z = 0$ and also in the interval from $z = 0$ to $z = 3$. The total desired area is shaded in the following graph.

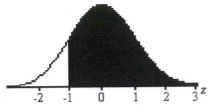

**Interval from −1 to 0:** The area under the curve from z = −1 to z = 0 is the same as the area under the curve from z = 0 to z = 1 due to symmetry in the standard normal distribution. The area is 0.3413.

**Interval from 0 to 3:** The area is 0.4987.

The total area under the curve from z = −1 to z = 3 is found by adding the two areas 0.3413 + 0.4987 = 0.84. Therefore, 84% of the values in a standard normal distribution are between −1 and 3.

**(b)** The area under the standard normal curve in the interval from z = 0 to z = 3 is found in Table 11.3 as 0.4987. The area under the curve greater than z = 0 is 0.5. To find the area greater than z = 3, subtract the two areas: 0.5 − 0.4987 = 0.0013. Therefore 0.13% of the values in a standard normal population are greater than 3.

**(c)** The area under the curve less than z = −1 is shaded in the following graph.

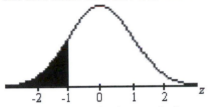

Due to symmetry in the standard normal curve, finding the area under the curve less than z = −1 is the same as finding the area under the curve greater than z = 1.

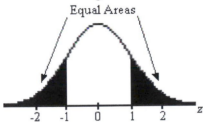

We know that the area under the curve greater than z = 0 is 0.5. We also know the area under the curve in the interval from z = 0 to z = 1 is 0.3413. Subtract to find the area under the curve greater than z = 1 (and equivalently less than z = −1): 0.5 − 0.3413 = 0.1587. Therefore, 15.87% of the values in a standard normal population are less than −1.

**(d)** We have accounted for the area between −1 and 3, the area greater than 3, and the area less than −1. All of the area under the standard normal curve has been accounted for. All of the percentages must add to 100%.

4. **(a)** Table 11.3 gives areas under the standard normal curve in the interval from 0 to z. If we want to find the area under the standard normal curve between a negative value and a positive value, then it will be necessary to split the interval up into two pieces. In order to find the area under the curve in the interval from z = −1.2 to z = 2.7, find the area in the interval from z = −1.2 to z = 0 and also in the interval from z = 0 to z = 2.7. The desired area is shaded in the following graph.

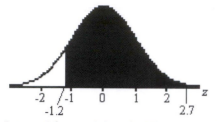

**Interval from −1.2 to 0:** The area under the curve from $z = -1.2$ to $z = 0$ is the same as the area under the curve from $z = 0$ to $z = 1.2$ due to symmetry in the standard normal distribution. The area is 0.3849.

**Interval from 0 to 2.7:** The area is 0.4965.

The total area under the curve from $z = -1.2$ to $z = 2.7$ is found by adding the two areas $0.3849 + 0.4965 = 0.8814$. Therefore, 88.14% of the values in a standard normal population are between $-1.2$ and 2.7.

(b) Table 11.3 gives areas under the standard normal curve in the interval from 0 to $z$. We want to find the area under the standard normal curve between two positive values, $z = 0.7$ and $z = 2.1$. First, find the area under the curve in the interval from $z = 0$ to $z = 2.1$. This area is more than we need. Next, find the area under the curve in the interval from $z = 0$ to $z = 0.7$. The area we want is shaded in the following graph.

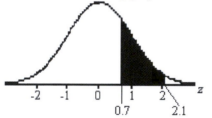

**Interval from 0 to 2.1:** The area is 0.4821.

**Interval from 0 to 0.7:** The area is 0.2580.

The total area under the curve in the interval from $z = 0.7$ to $z = 2.1$ is found by subtracting the two areas: $0.4821 - 0.2580 = 0.2241$. Therefore, 22.41% of the values in a standard normal population are between 0.7 and 2.1.

(c) The area under the standard normal curve in the interval from $z = 0$ to $z = 1.6$ is found in Table 11.3 as 0.4452. The area under the curve greater than $z = 0$ is 0.5. To find the area greater than $z = 1.6$, subtract the two areas: $0.5 - 0.4452 = 0.0548$. Therefore 5.48% of the values in a standard normal population are greater than 1.6.

5. Use the definition of the $z$-score and let $\mu = 17.9$ and $\sigma = 1.4$.

$$z = \frac{x - \mu}{\sigma} = \frac{x - 17.9}{1.4}.$$

| Measurement $x$ | $z$-score |
|---|---|
| 16.1 | $z = \dfrac{16.1 - 17.9}{1.4} = \dfrac{-1.8}{1.4} \approx -1.3$ |
| 21.9 | $z = \dfrac{21.9 - 17.9}{1.4} = \dfrac{4}{1.4} \approx 2.9$ |
| 22.3 | $z = \dfrac{22.3 - 17.9}{1.4} = \dfrac{4.4}{1.4} \approx 3.1$ |
| 18.6 | $z = \dfrac{18.6 - 17.9}{1.4} = \dfrac{0.7}{1.4} = 0.5$ |

**6.** **(a)** Find the percent of measurements under the normal curve between 9.5 and 12.125 as shown next.

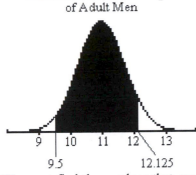

Distribution of Foot Lengths
of Adult Men

We must find the z-values that correspond to foot lengths of 9.5 and 12.125. The z-values are

$$z = \frac{x-\mu}{\sigma} = \frac{9.5-11}{0.75} = \frac{-1.5}{0.75} = -2 \quad \text{and} \quad z = \frac{x-\mu}{\sigma} = \frac{12.125-11}{0.75} = \frac{1.125}{0.75} = 1.5. \quad \text{To find the}$$

shaded area, we need to find the area under the standard normal curve between $z = -2$ and $z = 1.5$. Table 11.3 gives areas under the standard normal curve in the interval from 0 to z. If we want to find the area under the standard normal curve between a negative value and a positive value, then it will be necessary to split the interval up into two pieces. In order to find the area under the curve in the interval from $z = -2$ to $z = 1.5$, find the area under the curve in the interval from $z = -2$ to $z = 0$ and also in the interval from $z = 0$ to $z = 1.5$.

**Interval from −2 to 0:** The area under the curve in the interval from $z = -2$ to $z = 0$ is the same as the area under the curve in the interval from $z = 0$ to $z = 2$ due to symmetry in the standard normal distribution. The area is 0.4772.

**Interval from 0 to 1.5:** The area is 0.4332.

The total area under the curve in the interval from $z = -2$ to $z = 1.5$ is found by adding the two areas $0.4772 + 0.4332 = 0.9104$. Therefore, approximately 91.04% of measurements lie between 9.5 and 12.125 inches.

**(b)** Find the percent of measurements under the normal curve between 6 and 9.6 as shown next.

Distribution of Foot Lengths
of Adult Women

We must find the z-values that correspond to foot lengths of 6 and 9.6. The z-values are

$$z = \frac{x-\mu}{\sigma} = \frac{6-9}{1.2} = \frac{-3}{1.2} = -2.5 \quad \text{and} \quad z = \frac{x-\mu}{\sigma} = \frac{9.6-9}{1.2} = \frac{0.6}{1.2} = 0.5. \quad \text{To find the shaded area,}$$

we need to find the area under the standard normal curve between $z = -2.5$ and $z = 0.5$.

Table 11.3 gives areas under the standard normal curve in the interval from 0 to z. If we want to find the area under the standard normal curve between a negative value and a positive value, then it will be necessary to split the interval up into two pieces. In order to find the area under the curve in the interval from $z = -2.5$ to $z = 0.5$, find the area under the curve in the interval from $z = -2.5$ to $z = 0$ and also in the interval from $z = 0$ to $z = 0.5$.

**Interval from −2.5 to 0:** The area under the curve in the interval from $z = -2.5$ to $z = 0$ is the same as the area under the curve in the interval from $z = 0$ to $z = 2.5$ due to symmetry in the standard normal distribution. The area is 0.4938.

**Interval from 0 to 0.5:** The area is 0.1915.
The total area under the curve in the interval from $z = -2.5$ to $z = 0.5$ is found by adding the two areas $0.4938 + 0.1915 = 0.6853$. Therefore, approximately 68.53% of measurements lie between 6 and 9.6 inches.

(c) The normal distribution curve for men's foot lengths has a smaller standard deviation, so it is taller and thinner. The normal distribution curve for men's foot lengths is centered at a mean of 11 inches, so the curve would be centered to the right of the curve for women's foot lengths which has a mean of 9 inches.

7. The mean of the distribution is 517. Notice that 99.74% of measurements lie between 421 and 613 and 421 is $421 - 517 = 96$ units below the mean while 613 is $613 - 517 = 96$ units above the mean. Therefore half of 99.74%, or 49.87%, of measurements are from 517 to 613. Compare this percentage to the area under a standard normal curve from $z = 0$ to $z = 3$. The area given is 0.4987. This means that the measurement 613 is 3 standard deviations above the mean.

$$z = \frac{x - \mu}{\sigma}$$

$$3 = \frac{613 - 517}{\sigma}$$

$$3\sigma = 96$$

$$\sigma = 32$$

Therefore, the standard deviation is 32.

8. (a) Find the $z$-scores for the given values. The area of the shaded region is the same as the area under the standard normal curve between $z = \frac{62.4 - 72}{8} = \frac{-9.6}{8} = -1.2$ and $z = \frac{84 - 72}{8} = \frac{12}{8} = 1.5$.

The area in the interval from $z = -1.2$ to $z = 0$ is the same as the area from $z = 0$ to $z = 1.2$ due to symmetry. This area is 0.3849. The area from $z = 0$ to $z = 1.5$ is 0.4332. The total area is $0.3849 + 0.4332 = 0.8181$.

(b) Find the $z$-scores for the given values. The area of the shaded region is the same as the area under the standard normal curve between $z = \frac{33.7 - 31}{3} = \frac{2.7}{3} = 0.9$ and $z = \frac{38.8 - 31}{3} = \frac{7.8}{3} = 2.6$.

The area in the interval from $z = 0$ to $z = 2.6$ is 0.4953, but we want to remove the area from $z = 0$ to $z = 0.9$ which is 0.3159. The area of the shaded region is $0.4953 - 0.3159 = 0.1794$.

9. (a) The percentage of students who take fewer than 40 exams is approximately the same as the percentage of the area under the standard normal curve less than $z = \frac{40 - 54.5}{8.4} = \frac{-14.5}{8.4} \approx -1.7$.

The area from $z = -1.7$ to $z = 0$ is 0.4554 due to symmetry in the normal curve. The area less than $z = 0$ is 0.5. Therefore, the area less than $z = -1.7$ is $0.5 - 0.4554 = 0.0446$. Approximately 4.46% of students take fewer than 40 exams.

(b) The percentage of students who take between 50 and 60 exams is approximately the same as the percentage of the area under the standard normal curve between $z = \frac{50 - 54.5}{8.4} = \frac{-4.5}{8.4} \approx -0.5$

and $z = \frac{60 - 54.5}{8.4} = \frac{5.5}{8.4} \approx 0.7$. The area from $z = -0.5$ to $z = 0$ is 0.1915 due to symmetry in the normal curve. The area from $z = 0$ to $z = 0.7$ is 0.2580. Therefore, the total area is $0.1915 + 0.2580 = 0.4495$. Approximately 44.95% of students take between 50 and 60 exams.

(c) The percentage of students who take more than 60 exams is approximately the same as the percentage of the area under the standard normal curve greater than $z = \dfrac{60 - 54.5}{8.4} = \dfrac{5.5}{8.4} \approx 0.7$.

The area from $z = 0$ to $z = 0.7$ is 0.2580. The area greater than $z = 0$ is 0.5. Therefore, the area greater than $z = 0.7$ is $0.5 - 0.2580 = 0.2420$. Approximately 24.20% of students take more than 60 exams.

10. (a) The percentage of children who can read fewer than 25 sight words per minute is approximately the same as the percentage of the area under the standard normal curve less than $z = \dfrac{25 - 41}{9.19} = \dfrac{-16}{9.19} \approx -1.7$. The area from $z = -1.7$ to $z = 0$ is 0.4554 due to symmetry in the normal curve. The area less than $z = 0$ is 0.5. Therefore, the area less than $z = -1.7$ is $0.5 - 0.4554 = 0.0446$. Approximately 4.46% of children can read fewer than 25 sight words per minute.

(b) The percentage of children who can read more than 50 sight words per minute is approximately the same as the percentage of the area under the standard normal curve greater than $z = \dfrac{50 - 41}{9.19} = \dfrac{9}{9.19} \approx 1$. The area from $z = 0$ to $z = 1$ is 0.3413. The area greater than $z = 0$ is 0.5.

Therefore, the area greater than $z = 1$ is $0.5 - 0.3413 = 0.1587$. Approximately 15.87% of children can read more than 50 sight words per minute.

(c) The probability that a child selected at random will be able to read between 45 and 65 sight words per minute is approximately the same as the area under the standard normal curve between $z = \dfrac{45 - 41}{9.19} = \dfrac{4}{9.19} \approx 0.4$ and $z = \dfrac{65 - 41}{9.19} = \dfrac{24}{9.19} \approx 2.6$. The area from $z = 0$ to $z = 2.6$ is 0.4953. The area from $z = 0$ is 0.4 is 0.1554. Therefore, the area between $z = 0.4$ and 2.6 is $0.4953 - 0.1554 = 0.3399$. The probability that a child selected at random will be able to read between 45 and 65 sight words per minute is 0.3399.

(d) The probability that a child selected at random will be able to read between 10 and 35 sight words per minute is approximately the same as the area under the standard normal curve between $z = \dfrac{10 - 41}{9.19} = \dfrac{-31}{9.19} \approx -3.4$ and $z = \dfrac{35 - 41}{9.19} = \dfrac{-6}{9.19} \approx -0.7$. The area from $z = -3.4$ to $z = 0$ is approximately 0.5. The area from $z = -0.7$ to $z = 0$ is 0.2580 due to symmetry in the curve. Therefore, the area between $z = -3.4$ and $z = -0.7$ is $0.5 - 0.2580 = 0.2420$. The probability that a child selected at random will be able to read between 10 and 35 sight words per minute is 0.2420.

11. (a) The percentage of students from the 2003 group who have an ACT composite score of at least 18 is approximately the same as the percentage of the area under the standard normal curve greater than $z = \dfrac{18 - 20.8}{4.8} = \dfrac{-2.8}{4.8} \approx -0.6$. The area from $z = -0.6$ to $z = 0$ is 0.2257 due to symmetry in the normal curve. The area greater than $z = 0$ is 0.5. Therefore, the area greater than $z = -0.6$ is $0.5 + 0.2257 = 0.7257$. Approximately 72.57% of students in the 2003 group would be preferred by the University of Virginia's College at Wise.

(b) The percentage of students from the 2003 group who have an ACT composite score of at least 24 is approximately the same as the percentage of the area under the standard normal curve greater than $z = \dfrac{24 - 20.8}{4.8} = \dfrac{3.2}{4.8} \approx 0.7$. The area from $z = 0$ to $z = 0.7$ is 0.2580. The area greater than $z = 0$ is 0.5. Therefore, the area greater than $z = 0.7$ is $0.5 - 0.2580 = 0.2420$. Approximately 24.20% of students in the 2003 group could apply at the University of Missouri-Columbia.

(c) For the **mathematics** portion of the ACT, the percentage of students who scored between 20 and 24 is approximately the same as the percentage of the area under the standard normal curve between $z = \dfrac{20 - 20.6}{5.1} = \dfrac{-0.6}{5.1} \approx -0.1$ and $z = \dfrac{24 - 20.6}{5.1} = \dfrac{3.4}{5.1} \approx 0.7$. The area from $z = -0.1$ to $z = 0$ is 0.0398 due to symmetry. The area from $z = 0$ to $z = 0.7$ is 0.2580. Therefore, the total area is $0.0398 + 0.2580 = 0.2978$. Approximately 29.78% of students scored between 20 and 24 on the mathematics portion.

For the **reading** portion of the ACT, the percentage of students who scored between 20 and 24 is approximately the same as the percentage of the area under the standard normal curve between $z = \dfrac{20 - 21.2}{6.1} = \dfrac{-1.2}{6.1} \approx -0.2$ and $z = \dfrac{24 - 21.2}{6.1} = \dfrac{2.8}{6.1} \approx 0.5$. The area from $z = -0.2$ to $z = 0$ is 0.0793 due to symmetry. The area from $z = 0$ to $z = 0.5$ is 0.1915. Therefore, the total area is $0.0793 + 0.1915 = 0.2708$. Approximately 27.08% of students scored between 20 and 24 on the reading portion.

For the **science** portion of the ACT, the percentage of students who scored between 20 and 24 is approximately the same as the percentage of the area under the standard normal curve between $z = \dfrac{20 - 20.8}{4.6} = \dfrac{-0.8}{4.6} \approx -0.2$ and $z = \dfrac{24 - 20.8}{4.6} = \dfrac{3.2}{4.6} \approx 0.7$. The area from $z = -0.2$ to $z = 0$ is 0.0793 due to symmetry. The area from $z = 0$ to $z = 0.7$ is 0.2580. Therefore, the total area is $0.0793 + 0.2580 = 0.3373$. Approximately 33.73% of students scored between 20 and 24 on the science portion.

For the **English** portion of the ACT, the percentage of students who scored between 20 and 24 is approximately the same as the percentage of the area under the standard normal curve between $z = \dfrac{20 - 20.3}{5.8} = \dfrac{-0.3}{5.8} \approx -0.1$ and $z = \dfrac{24 - 20.3}{5.8} = \dfrac{3.7}{5.8} \approx 0.6$. The area from $z = -0.1$ to $z = 0$ is 0.0398 due to symmetry. The area from $z = 0$ to $z = 0.6$ is 0.2257. Therefore, the total area is $0.0398 + 0.2257 = 0.2655$. Approximately 26.55% of students scored between 20 and 24 on the English portion.

Consider the section with the smallest percentage. In the reading portion, approximately 27.08% scored between 20 and 24. This means that $100\% - 27.08\% = 72.92\%$ did not score between 20 and 24 and that $\dfrac{72.92\%}{2} = 36.46\%$ scored above 24. Out of the four sections, a larger percentage scored above 24 in reading. The largest percentage of students scored between 20 and 24 on the science portion.

12. **(a)** For the science portion of the ACT, the percentage of students who scored at least 24 is approximately the same as the percentage of the area under the standard normal curve greater than $z = \dfrac{24 - 20.8}{4.6} = \dfrac{3.2}{4.6} \approx 0.7$. The area from $z = 0$ to $z = 0.7$ is 0.2580. Therefore, the area greater than $z = 0.7$ is $0.5 - 0.2580 = 0.2420$. Approximately 24.20% of students scored at least 24 on the science portion.

    **(b)** For the mathematics portion of the ACT, the percentage of students who scored at least 22 is approximately the same as the percentage of the area under the standard normal greater than $z = \dfrac{22 - 20.6}{5.1} = \dfrac{1.4}{5.1} \approx 0.3$. The area from $z = 0$ to $z = 0.3$ is 0.1179. Therefore, the area greater than $z = 0.3$ is $0.5 - 0.1179 = 0.3821$. Approximately 38.21% of students scored at least 22 on the mathematics portion.

    **(c)** A larger percentage of students who took the ACT in 2003 were ready for college algebra than were ready for college biology.

13. The proportion of chocolate candies in the jar is unknown. Our best guess for the unknown proportion is to use the sample proportion of chocolate candies. The sample proportion is $\frac{9}{30} = 0.3$. Out of 200 candies in the entire jar, we would expect $0.3 \times 200$ or 60 of them to be chocolate.

14. (a) The population is the set of 20,000 jelly beans. The sample is the set of 48 jelly beans taken from the jar.

(b) The population proportion of cherry jelly beans is $\frac{5000}{20,000} = 0.25$.

(c) The mean of the set of all the sample proportions of cherry jelly beans is the same as the population proportion of cherry jelly beans. The mean is 0.25. The standard deviation of the set of all the sample proportions of cherry jelly beans is found as follows with $p = 0.25$.

$$\sqrt{\frac{p(1-p)}{n}} = \sqrt{\frac{0.25(0.75)}{48}} = \sqrt{\frac{0.1875}{48}} = 0.0625$$

The standard deviation is 0.0625.

(d) The following two conditions must be met before we can assume the distribution of sample proportions is approximately normal:

$$p - 3\sqrt{\frac{p(1-p)}{n}} > 0 \text{ and } p + 3\sqrt{\frac{p(1-p)}{n}} < 1$$

$$0.25 - 3\sqrt{\frac{0.25(0.75)}{48}} > 0 \text{ and } 0.25 + 3\sqrt{\frac{0.25(0.75)}{48}} < 1$$

$$0.25 - 3(0.0625) > 0 \text{ and } 0.25 + 3(0.0625) < 1$$

$$0.0625 > 0 \text{ and } 0.4375 < 1$$

Because it is true that $0.0625 > 0$, and at the same time, $0.4375 < 1$, we conclude the sample size is large enough.

(e) We know $\hat{p} = \frac{9}{48} = 0.1875$. The standard error is $\hat{s} = \sqrt{\frac{\hat{p}(1-\hat{p})}{n}} = \sqrt{\frac{0.1875(0.8125)}{48}} \approx 0.056$.

The 95% confidence interval for the proportion of cherry flavored jelly beans in the jar is calculated as follows:

$$\hat{p} - 2\hat{s} \text{ to } \hat{p} + 2\hat{s}$$

$$0.1875 - 2(0.056) \text{ to } 0.1875 + 2(0.056)$$

$$0.0755 \text{ to } 0.2995$$

We are 95% confident that the proportion of cherry flavored jelly beans in the jar is between 0.0755 and 0.2995.

15. The following two conditions must be met before we can assume the distribution of sample proportions is approximately normal. The sample size in this case is 15.

$$p - 3\sqrt{\frac{p(1-p)}{n}} > 0 \text{ and } p + 3\sqrt{\frac{p(1-p)}{n}} < 1$$

$$0.25 - 3\sqrt{\frac{0.25(0.75)}{15}} > 0 \text{ and } 0.25 + 3\sqrt{\frac{0.25(0.75)}{15}} < 1$$

$$0.25 - 3(0.1118) > 0 \text{ and } 0.25 + 3(0.1118) < 1$$

$$-0.0854 > 0 \text{ and } 0.5854 < 1$$

Because it is not true that $-0.0854 > 0$, we conclude the sample size is not large enough.

**16. (a)** The sample proportion of families with school-age children who send their children to private school is $\dfrac{34}{300} \approx 0.1133$.

**(b)** Use the definition of the standard error. If $\hat{p} \approx 0.1133$, then the standard error is approximately

$\hat{s} = \sqrt{\dfrac{\hat{p}(1-\hat{p})}{n}} = \sqrt{\dfrac{0.1133(0.8867)}{300}} \approx 0.018$. The standard error is a good approximation to the true, unknown standard deviation of the set of all sample proportions.

**(c)** We know $\hat{p} \approx 0.1133$. The standard error is $\hat{s} \approx 0.018$.

The 95% confidence interval for the proportion of families with school-age children who send their children to private school is calculated as follows:

$$\hat{p} - 2\hat{s} \text{ to } \hat{p} + 2\hat{s}$$

$$0.1133 - 2(0.018) \text{ to } 0.1133 + 2(0.018)$$

$$0.0773 \text{ to } 0.1493$$

We are 95% confident that the proportion is between 0.0773 and 0.1493.

**17. (a)** The mean is 0.11 and the standard deviation is $\sqrt{\dfrac{p(1-p)}{n}} = \sqrt{\dfrac{0.11(0.89)}{100}} = \sqrt{\dfrac{0.0979}{100}} \approx 0.031$.

**(b)** The percentage of samples that would have a sample proportion less than 8% is approximately the same as the area under the standard normal curve less than $z = \dfrac{0.08 - 0.11}{0.031} = \dfrac{-0.03}{0.031} \approx -1$. The area between $z = -1$ and $z = 0$ is 0.3413 due to symmetry in the curve. The area less than $z = 0$ is 0.5. Therefore, the area less than $z = -1$ is 0.5 − 0.3413 = 0.1587. We conclude 15.87% of samples would have a sample proportion less than 8%.

**(c)** The percent of samples that would have a sample proportion greater than 15% is approximately the same as the area under the standard normal curve greater than $z = \dfrac{0.15 - 0.11}{0.031} = \dfrac{0.04}{0.031} \approx 1.3$.

The area between $z = 0$ and $z = 1.3$ is 0.4032. The area greater than $z = 0$ is 0.5. Therefore, the area greater than $z = 1.3$ is 0.5 − 0.4032 = 0.0968. We conclude 9.68% of samples would have a sample proportion greater than 15%.

**(d)** The percentage of samples that would have a sample proportion of at least 21% is approximately the same as the area under the standard normal curve greater than $z = \dfrac{0.21 - 0.11}{0.031} = \dfrac{0.1}{0.031} \approx 3.2$.

A $z$-value of 3.2 is not listed in Table 11.3. The area between $z = 0$ and $z = 3$ is 0.4987, so the area greater than $z = 3$ is 0.5 − 0.4987 = 0.0013. There would be even less area under the standard normal curve greater than $z = 3.2$. We conclude less than 0.13% of samples would have a sample proportion of at least 21%. This is a very unlikely sample to occur.

**(e)** We know $\hat{p} = \dfrac{8}{100} = 0.08$. The standard error is $\sqrt{\dfrac{\hat{p}(1-\hat{p})}{n}} = \sqrt{\dfrac{0.08(0.92)}{100}} = \sqrt{\dfrac{0.0736}{100}} \approx 0.027$.

The 95% confidence interval for the proportion computer users who have internet service through a cable modem is calculated as follows:

$$\hat{p} - 2\hat{s} \text{ to } \hat{p} + 2\hat{s}$$

$$0.08 - 2(0.027) \text{ to } 0.08 + 2(0.027)$$

$$0.026 \text{ to } 0.134$$

We are 95% confident that the proportion is between 0.026 and 0.134.

**(f)** We know $\hat{p} = \dfrac{10}{100} = 0.1$. The standard error is $\sqrt{\dfrac{\hat{p}(1-\hat{p})}{n}} = \sqrt{\dfrac{0.1(0.9)}{100}} = \sqrt{\dfrac{0.09}{100}} = 0.03$.

The 95% confidence interval for the proportion computer users who have internet service through a cable modem is calculated as follows:

$$\hat{p} - 2\hat{s} \ \text{ to } \ \hat{p} + 2\hat{s}$$

$$0.1 - 2(0.03) \ \text{ to } \ 0.1 + 2(0.03)$$

$$0.04 \ \text{ to } \ 0.16$$

We are 95% confident that the proportion is between 0.04 and 0.16.

**(g)** The confidence interval from part (e) is narrower than the one from part (f) due to the smaller standard error. Notice that both confidence intervals do contain the true population proportion of computer users who have internet service through a cable modem. The true value is 0.11.

18. **(a)** We are given the formula for the standard deviation for the distribution of the set of all the sample proportions. Solve for $n$ as shown next.

$$s = \sqrt{\dfrac{p(1-p)}{n}}$$

$$s^2 = \dfrac{p(1-p)}{n} \qquad \text{square both sides}$$

$$ns^2 = p(1-p) \qquad \text{multiply both sides by } n$$

$$n = \dfrac{p(1-p)}{s^2} \qquad \text{divide both sides by } s^2$$

**(b)** Use the formula from part (a) with $s = 0.01$ and $p = 0.5$.

$$n = \dfrac{p(1-p)}{s^2}$$

$$n = \dfrac{0.5(0.5)}{0.01^2}$$

$$n = \dfrac{0.25}{0.0001}$$

$$n = 2500$$

Therefore, a sample size of at least 2500 would be required to obtain a value of $s = 0.01$.

**(c)** Use the formula from part (a) with $s = 0.01$ and $p = 0.2$.

$$n = \dfrac{p(1-p)}{s^2}$$

$$n = \dfrac{0.2(0.8)}{0.01^2}$$

$$n = \dfrac{0.16}{0.0001}$$

$$n = 1600$$

Therefore, a sample size of at least 1600 would be required to obtain a value of $s = 0.01$.

19. In the first sample, $n = 100$ and $\hat{p} = 0.62$. In the second sample, $n = 1000$ and $\hat{p} = 0.60$.

**(a)** The margin of error in the first sample is $2\sqrt{\dfrac{\hat{p}(1-\hat{p})}{n}} = 2\sqrt{\dfrac{0.62(0.38)}{100}} \approx 0.097$.

The margin of error in the second sample is $2\sqrt{\dfrac{\hat{p}(1-\hat{p})}{n}} = 2\sqrt{\dfrac{0.60(0.40)}{1000}} \approx 0.031$.

The margin of error for the sample with the larger sample size is much smaller. In general, a better estimate of the true, unknown proportion will be found by taking a larger sample.

**(b)** For the sample with a sample size of 100, the 95% confidence interval for the proportion of people who feel they spend too much money on taxes is calculated as follows:

$$\hat{p} - 2\hat{s} \text{ to } \hat{p} + 2\hat{s}$$

$$0.62 - 0.097 \text{ to } 0.62 + 0.097$$

$$0.523 \text{ to } 0.717$$

We are 95% confident that the proportion is between 0.523 and 0.717.

For the sample with a sample size of 1000, the 95% confidence interval for the proportion of people who feel they spend too much money on taxes is calculated as follows:

$$\hat{p} - 2\hat{s} \text{ to } \hat{p} + 2\hat{s}$$

$$0.60 - 0.031 \text{ to } 0.60 + 0.031$$

$$0.569 \text{ to } 0.631$$

We are 95% confident that the proportion is between 0.569 and 0.631.

The confidence interval obtained for the sample of size 1000 is narrower than the one obtained for a sample of size 100.

**(c)** For a population proportion of 0.61 and a margin of error of 0.01, use the definition of the margin of error.

$$\text{margin of error} = 2\sqrt{\frac{\hat{p}(1-\hat{p})}{n}}$$

$$0.01 = 2\sqrt{\frac{0.61(0.39)}{n}}$$

$$\frac{0.01}{2} = \sqrt{\frac{0.61(0.39)}{n}}$$

$$\left(\frac{0.01}{2}\right)^2 = \frac{0.61(0.39)}{n}$$

$$n = \frac{0.61(0.39)}{\left(\frac{0.01}{2}\right)^2}$$

$$n = 9516$$

For a sample size of 1000, the margin of error was 0.031. Increasing the sample size by a factor of 10 caused the margin of error to be only one third as large. If sampling is expensive, then the added cost may not be worth the gain in accuracy.

**20. (a)** We know $\hat{p} = \dfrac{8973}{146{,}545} \approx 0.061$.

The standard error is $\sqrt{\dfrac{\hat{p}(1-\hat{p})}{n}} = \sqrt{\dfrac{0.061(0.939)}{146{,}545}} = \sqrt{\dfrac{0.057279}{146{,}545}} \approx 0.000625.$

The 95% confidence interval for the unemployment rate is calculated as follows:

$$\hat{p} - 2\hat{s} \text{ to } \hat{p} + 2\hat{s}$$

$$0.061 - 2(0.000625) \text{ to } 0.061 + 2(0.000625)$$

$$0.05975 \text{ to } 0.06225$$

We are 95% confident that the U.S. unemployment rate is between 5.975% and 6.225%.

**(b)** The following two conditions must be met before we can assume the distribution of sample proportions is approximately normal. The sample size in this case is 100. Use the sample proportion and the standard error since the population proportion and standard deviation are unknown.

$$p - 3\sqrt{\frac{p(1-p)}{n}} > 0 \text{ and } p + 3\sqrt{\frac{p(1-p)}{n}} < 1$$

$$0.061 - 3(0.000625) > 0 \text{ and } 0.061 + 3(0.000625) < 1$$

$$0.059125 > 0 \text{ and } 0.062875 < 1$$

Both conditions are true, so we conclude the sample size 100 would have been large enough.

**21. (a)** The population is the set of all Americans age 18 and older. The sample is the set of 1006 Americans surveyed.

**(b)** The sample proportion is $\dfrac{533}{1006} \approx 0.530$.

The margin of error is $2\sqrt{\dfrac{\widehat{p}(1-\widehat{p})}{n}} = 2\sqrt{\dfrac{0.530(0.467)}{1006}} \approx 0.031$. In reporting the results of the poll, we could say that the percentage of Americans who approve of the way George W. Bush is handling his job as president is 53% with a margin of error of 3.1%. The margin of error is an indicator of how good an estimate we have for the true, unknown percentage.

**(c)** The 95% confidence interval is calculated as follows:

$$\widehat{p} - 2\widehat{s} \text{ to } \widehat{p} + 2\widehat{s}$$

$$0.53 - 0.031 \text{ to } 0.53 + 0.031$$

$$0.499 \text{ to } 0.561$$

We are 95% certain that the true percentage of Americans who approve is between 49.9% and 56.1%.

**22. (a)** The sample proportion is $\dfrac{605}{840} \approx 0.72$.

The margin of error is $2\sqrt{\dfrac{\widehat{p}(1-\widehat{p})}{n}} = 2\sqrt{\dfrac{0.72(0.28)}{840}} \approx 0.031$.

The 95% confidence interval is calculated as follows:

$$\widehat{p} - 2\widehat{s} \text{ to } \widehat{p} + 2\widehat{s}$$

$$0.72 - 0.031 \text{ to } 0.72 + 0.031$$

$$0.689 \text{ to } 0.751$$

We are 95% certain that the true percentage of students who have at least two cards is between 68.9% and 75.1%.

**(b)** The sample proportion is $\dfrac{218}{840} \approx 0.26$.

The margin of error is $2\sqrt{\dfrac{\widehat{p}(1-\widehat{p})}{n}} = 2\sqrt{\dfrac{0.26(0.74)}{840}} \approx 0.03$.

The 95% confidence interval is calculated as follows:

$$\widehat{p} - 2\widehat{s} \text{ to } \widehat{p} + 2\widehat{s}$$

$$0.26 - 0.03 \text{ to } 0.26 + 0.03$$

$$0.23 \text{ to } 0.29$$

We are 95% certain that the true percentage of students who have missed payments is between 23% and 29%.

**23. (a)** The sample proportion in survey 1 is $\dfrac{120}{200} = 0.6.$ The margin of error is

$$2\sqrt{\dfrac{\widehat{p}(1-\widehat{p})}{n}} = 2\sqrt{\dfrac{0.6(0.4)}{200}} \approx 0.069.$$

The sample proportion in survey 2 is $\dfrac{480}{800} = 0.6.$ The margin of error is

$$2\sqrt{\dfrac{\widehat{p}(1-\widehat{p})}{n}} = 2\sqrt{\dfrac{0.6(0.4)}{800}} \approx 0.035.$$

The sample proportion in survey 3 is $\dfrac{1080}{1800} = 0.6.$ The margin of error is

$$2\sqrt{\dfrac{\widehat{p}(1-\widehat{p})}{n}} = 2\sqrt{\dfrac{0.6(0.4)}{1800}} \approx 0.023.$$

**(b)** The number polled in survey 2 is $\dfrac{800}{200} = 4$ times greater than the number polled in survey 1.

The margin of error in survey 2 is $\dfrac{0.035}{0.069} \approx 0.51$ times the margin of error in survey 1. The sample size had to be increased by a factor of 4 to obtain a margin of error half the size.

**(c)** The number polled in survey 3 is $\dfrac{1800}{200} = 9$ times greater than the number polled in survey 1.

The margin of error in survey 3 is $\dfrac{0.023}{0.069} \approx 0.33$ times the margin of error in survey 1. The sample size had to be increased by a factor of 9 to obtain a margin of error of one third the size.

**24. (a)** The sample proportion is of females who passed the test is 0.705.

The margin of error is $2\sqrt{\dfrac{\widehat{p}(1-\widehat{p})}{n}} = 2\sqrt{\dfrac{0.705(0.295)}{400}} \approx 0.046.$

The 95% confidence interval for the proportion of females who would pass is calculated as follows:

$$\widehat{p} - 2\widehat{s} \text{ to } \widehat{p} + 2\widehat{s}$$

$$0.705 - 0.046 \text{ to } 0.705 + 0.046$$

$$0.659 \text{ to } 0.751$$

We are 95% certain that the true percentage of females who would pass the test is between 65.9% and 75.1%.

The sample proportion is of males who passed the test is 0.622.

The margin of error is $2\sqrt{\dfrac{\widehat{p}(1-\widehat{p})}{n}} = 2\sqrt{\dfrac{0.622(0.378)}{360}} \approx 0.051.$

The 95% confidence interval for the proportion of males who would pass is calculated as follows:

$$\widehat{p} - 2\widehat{s} \text{ to } \widehat{p} + 2\widehat{s}$$

$$0.622 - 0.051 \text{ to } 0.622 + 0.051$$

$$0.571 \text{ to } 0.673$$

We are 95% certain that the true percentage of males who would pass the test is between 57.1% and 67.3%.

(b) For each interval, we are 95% certain that it contains the true, unknown percentage of females or males who would pass the test. Notice that the confidence intervals overlap. Each confidence interval contains a range of percentages that is the same. It is possible that the true percentage for females and the true percentage for males are in this common region. It is possible for the true percentages to be the same. The teacher is making a reasonable conclusion.

# *Chapter 12:* Growth and Decay

## *Key Concepts*

**By the time you finish studying Section 12.1, you should be able to:**

☐ use the Malthusian population growth model to predict the total population.

☐ calculate the annual rate of growth for a population.

☐ determine the minimum number of people needed to invest in a Ponzi scheme in a certain period.

☐ determine the amount of money required to payoff investors involved in a Ponzi scheme.

☐ calculate the payoff in rising from the bottom to the top of the list of a chain letter.

☐ calculate the number of people who must join for an individual to rise from the bottom of the list to the top of the list of a chain letter.

**By the time you finish studying Section 12.2, you should be able to:**

☐ calculate the annual rate of decline for a population.

☐ calculate the amount of a radioactive substance that will be left after a certain period of time.

☐ use half-life to determine the annual decay rate of a radioactive substance.

☐ determine the approximate half-life of a radioactive substance given the annual decay rate.

☐ estimate the age of a fossil given the percentage of $^{14}C$ present.

**By the time you finish studying Section 12.3, you should be able to:**

☐ differentiate between Malthusian growth and logistic growth.

☐ use the logistic growth law to calculate population totals.

☐ calculate the steady-state population.

☐ calculate and interpret the population fraction.

☐ use Verhulst's equation to calculate the population fractions for many breeding seasons.

☐ calculate and interpret the steady-state population fraction.

## Skills Brush-Ups

Problems in Chapter 12 will require that you be comfortable working with percents, using scientific notation, and evaluating expressions involving exponents by correctly applying rules for order of operations.

### PERCENT:

Percent means "per 100". Therefore, 9% means 9 per 100, 9 out of 100, or $\dfrac{9}{100}$.

To convert a number from a percent to a decimal, eliminate the % symbol and divide the number by 100. In general, you can drop the percent symbol and move the decimal point two places to the left.

*Examples:*  (a) $3.4\% = \dfrac{3.4}{100} = 0.034$  (b) $160\% = \dfrac{160}{100} = 1.6$  (c) $0.035\% = \dfrac{0.035}{100} = 0.00035$

To convert a number to a percent, multiply the number by 100 and write the % symbol on the right side of the number. In general, you can shift the decimal point two places to the right and insert a % symbol.

*Example:*  0.0219 is $0.0219 \times 100\% = 2.19\%$.

### SCIENTIFIC NOTATION:

Very large or very small numbers can be expressed using a shorthand method known as scientific notation. A number in scientific notation is made up of the product of two parts, a number $a$ where $1 \le a < 10$ and an integer power of 10. A number expressed in scientific notation has the form $a \times 10^n$.

**How to convert a number from standard notation to scientific notation:**
Place the decimal point after the first nonzero digit from the left, and count the number of places the decimal has moved. The number of places the decimal has moved becomes the exponent on the 10. If the decimal has moved to the right, then the exponent on the 10 is negative. If the decimal has moved to the left, then the exponent on the 10 is positive. Trailing zeros can be dropped when using scientific notation.

*Examples:*
(a) Express 2,410,000,000,000 in scientific notation. The first nonzero digit is 2 so the decimal point is placed after the 2 and all trailing zeros are dropped. Compared to the original number, the decimal has been moved 12 places to the left, so the exponent on the 10 is 12. The number written in scientific notation is $2.41 \times 10^{12}$.
(b) Express 0.00000000098355 in scientific notation. The first nonzero digit is 9 so the decimal point is placed after the 9. Compared to the original number, the decimal has been moved 10 places to the right so the exponent on the 10 is −10. The number written in scientific notation is $9.8355 \times 10^{-10}$.

**How to convert a number from scientific notation to standard notation:**
If the exponent on the 10 is positive, then move the decimal that many places to the right. Fill the empty places with zeros. If the exponent on the 10 is negative, then move the decimal that many places to the left. Fill the empty places with zeros.

*Examples:*
(a) Express $8.401 \times 10^{-5}$ in standard notation. The exponent on the 10 is −5, so move the decimal point 5 places to the left and fill the empty places with zeros. The number in standard notation is 0.00008401.

(b) Express $6.2 \times 10^7$ in standard notation. The exponent on the 10 is 7, so move the decimal point 7 places to the right and fill the empty places with zeros. The number in standard notation is 62,000,000.

## *ORDER OF OPERATIONS*

When more than one operation occurs in an expression, it matters which operation is performed first. Different results may arise depending on the order in which the operations are carried out. To avoid such problems, remember the correct order of operations. First, simplify expressions in parentheses, evaluate exponents (this includes radicals), evaluate multiplications and divisions in order from left to right, and evaluate additions and subtractions in order from left to right. The following example shows the type of expression you will be asked to simplify in Chapter 12.

*Example*:

$$(1.0388)(546) - \left(\frac{1.0388}{20,000}\right)(546)^2 = (1.0388)(546) - (0.00005194)(546)^2 \quad \text{Simplify in parentheses}$$

$$= (1.0388)(546) - (0.00005194)(298,116) \quad \text{Evaluate the exponent}$$

$$= 567.1848 - (0.00005194)(298,116) \quad \text{Multiply}$$

$$= 567.1848 - 15.48414504 \quad \text{Multiply}$$

$$\approx 551.7001 \quad \text{Subtract}$$

## *Skills Practice*

1. Convert each of the following from percent notation to decimal notation.
   (a) 3%  (b) 3.1%  (c) 0.742%  (d) 4.05%

2. Convert each of the following from decimal notation to percent notation.
   (a) 0.06  (b) 0.0305  (c) 0.0098  (d) 0.212

3. Each of the following numbers is in scientific notation. Write each in standard notation.
   (a) $8.902 \times 10^7$  (b) $6.5862 \times 10^{-5}$  (c) $1.6 \times 10^5$  (d) $1.005 \times 10^{-11}$

4. Write each of the following in scientific notation.
   (a) 56,730,000,000  (b) 301,000,000  (c) 0.000000509  (d) 0.00000000000002334

5. Simplify each of the following and round your answers to the nearest thousandth.

   (a) $650(1-0.0385)^5$  (b) $\left(\frac{250}{98}\right)^{\frac{1}{15}} - 1$  (c) $3.5\left(1+\frac{0.03}{40}\right)^{\frac{10}{2.5}}$  (d) $1.2(1-0.0347)(0.0347)$

*Answers to Skills Practice:*
1. (a) 0.03  (b) 0.031  (c) 0.00742  (d) 0.0405
2. (a) 6%  (b) 3.05%  (c) 0.98%  (d) 21.2%
3. (a) 89,020,000  (b) 0.000065862  (c) 160,000  (d) 0.00000000001005
4. (a) $5.673 \times 10^{10}$  (b) $3.01 \times 10^8$  (c) $5.09 \times 10^{-9}$  (d) $2.334 \times 10^{-14}$
5. (a) 534.146  (b) 0.064  (c) 3.511  (d) 0.040

# Hints for Odd-Numbered Problems

## Section 12.1

1. The growth rate is constant, so you will be continuing to use the formula for Malthusian population growth.

3. Compare the given expression to the formula for Malthusian population growth.

5. The growth rate is constant so use the formula for Malthusian population growth. The initial population, $P_0$, is 50,000. The rate of growth, $r$, is 0.03. You will find the population for three different years.

7. Consider the formula for Malthusian population growth. Which values have been given?

9. Use the Malthusian population growth formula to find the populations in the year 2005. Then subtract to find the difference.

11. Be careful when expressing the rate as a decimal. The rate of growth is 0.007. Use the formula for Malthusian population growth. If 1990 is the starting year, what would m be in the year 2005? 2010?

13. (b) If the growth rate remains constant from 2003 to 2011, then $m = 8$. Find the population total in 2011 and use that as the initial population for the next 9 years when the growth rate is different.

15. This is not a population, but the formula for Malthusian growth can be used because the rate of growth is constant. The initial value is $24.

17. When $m = 1$, 20 minutes have passed. When $m = 2$, 40 minutes have passed. The population doubles every 20 minutes.

19. You are given an initial value and a value after 5 years.
    (a) Substitute what you know into the formula for Malthusian growth and solve for the growth rate.

(b) Use the growth rate from part (a) and the initial value of $215,000 to predict the value in five more years.

21. Use the Malthusian population growth formula with the value of $r$ unknown. In the formula, the value of $n$ is the time beginning when the initial population value is $P$. In this case, we know how the population changed over the past 10 years, so let $m = 10$. We want to know the population total in 20 *more* years. Therefore, $n = 30$.

23. Use the Malthusian population growth formula with the value of $r$ unknown. In the formula, the value of $n$ is the time beginning when the initial population value is $P$. In this case, we know how the population changed over the past 8 years, so let $m = 8$. We want to know the population total in 22 *more* years. Therefore, $n = 30$.

25. Use the formula for the annual growth rate to determine the growth rate for each year. In each case, $m = 1$.

27. (a) Use the formula for the annual growth rate with $m = 23$.
    (c) Use dimensional analysis to convert square feet to square miles and multiply by the population.

29. Write the Malthusian population growth formula given the growth rate. If the population begins at an initial value of $P$, then it will be doubled when the final value is $2P$. Guess and test values for $m$.

31. Use the relationship $S_{m+1} \geq (1.4)^m \times S_1$ with $S_1 = 5$. Find $S_2$, $S_3$, and $S_4$.

33. Use the relationship $S_{m+1} \geq (1.4)^m \times S_1$. In the first quarter there are 10 investors. Guess and test values for $m$ so that in the $(m + 1)^{st}$ quarter there are 290,000,000 investors.

35. (a) Use the payoff expression $P \times n^l$. What are the values of each of the variables?

**(b)** Use the expression $\dfrac{n^{l+1}-1}{n-1}$.

**37.** Use the payoff expression $P \times n^l$. You are given values for $n$ and $l$. For each part of the problem, substitute the given dollar amounts for $P$ and simplify.

**39.** You are given only two parts of the payoff expression: the payoff and the price to participate. Try a value for the number of levels, for example, and solve for the number of letters that must be sent out. Many answers are possible.

**41.** Look for a pattern between the number of layers and the number of folds. Once you know the number of folds, how do you determine the total thickness? Create an equation relating the total thickness and the number of folds. How many millimeters are in 480 meters?

**45.** Create a systematic way to label the dogs according to their age. Construct a table and list the number of dogs by age and keep track of the total number of dogs. Find the approximate annual growth rate using the formula for the annual growth rate in this section.

## Section 12.2

**1. (a)** Use the formula for the annual growth rate from section 12.1.
**(b)** Use the formula for Malthusian population growth. Even though the population is declining, the formula still applies.

**3. (c)** Guess and test until you find a value for $m$ so that $P_m = 1$.

**5.** Use the radioactive decay formula: $A_m = (1-d)^m \times A_0$. Let $A_0 = 27$ and $d = 0.03$.

**7. (a)** Compare the given equation to the radioactive decay formula.
**(c)** When you input a value for $m$ and find the corresponding value for $A_m$, you are finding the amount of the radioactive substance that is still present. How will you determine how much of the substance has decayed?

**9.** Use the radioactive decay formula. The decay rate and the initial amount are given. $d = 0.025$ and $A_0 = 300$.

**11.** Use the decay rate formula. Let $h$ be the given half-life.

**13. (a)** Use the decay rate formula. Let $h$ be the given half-life.

**15. (b)** The year 2035 is 35 years after the year 2000, so $m = 35$.

**17.** It is fine to use the decay rate formula and let $h = 18$ minutes even though the formula specifies that $h$ is in years. The rate will be a rate per minute in this case rather than an annual rate. We can still use it to make a prediction a certain number of minutes into the future. Use the radioactive decay formula and express time in minutes.

**19.** Find the rate of decay by using the decay rate formula then use the radioactive decay formula. Let $m = 2003 - 1950 = 53$ years.

**21.** Use the half-life radioactive decay formula with $A_0 = 500$ grams and $m = 100$ years.

**23.** Use the half-life radioactive decay formula with $A_0 = 100$ grams and $h = 2$ years.

**25.** Use the half-life approximation formula with $d = 0.02445$.

**27.** Compare the given percentage to those given in Table 12.6 to determine the age of the sample in half-lives. Use the age of the sample in half-lives and the half-life of $^{14}C$ to determine the age of the fossil.

**29.** Compare the given percentage to those given in Table 12.6 to determine the age of the sample in half-lives. Use the age of the sample in half-lives and the half-life of $^{14}C$ to determine the age of the fossil.

## Section 12.3

**1.** **(b)** Find the change in the population during each breeding season by subtracting: (Population Total in season $n$) − (Population Total in season $n-1$).

**(d)** Calculate the growth rate from one season to the next. Growth rate is found by dividing:

$$\frac{\text{Population Total in season } n}{\text{Population Total in season } n-1} - 1$$

If the population is governed by the Malthusian growth model, what would you expect to see in the growth rates from season to season? What would you expect to see if the population followed a logistic model?

**3.** Use the given formula. Notice $P_1 = P_{0+1}$, so when you find $P_1$, you will have to let $m = 0$ in the formula. The initial population, $P_0$, is 300.

**5.** Use the logistic growth law and find $P_1$, $P_2$, $P_3$, $P_4$, and $P_5$. When you plot the data on a graph, put breeding seasons along the horizontal axis and the population totals along the vertical axis.

**7.** **(a)** Use the logistic growth law and find $P_1$ through $P_{10}$. When you plot the data on a graph, put breeding seasons along the horizontal axis and the population totals along the vertical axis.

**(b)** $P_{steady} = \dfrac{r \times c}{1+r}$

**9.** **(a)** Use the logistic growth law and find $P_1$ through $P_{15}$. When you plot the data on a graph, put breeding seasons along the horizontal axis and the population totals along the vertical axis.

**(b)** $P_{steady} = \dfrac{r \times c}{1+r}$

**(c)** Find the change in population during each breeding season.

**11.** **(c)** The population fraction is found from the formula $p_m = \dfrac{P_m}{c}$.

**13.** **(a)** $P_0 = 100{,}000$

**(b)** The steady-state population has been given to you.

**(c)** Find $p_0$ and $p_{15}$.

**15.** **(a)** To find $p_0$, use the definition of the population fraction, $p_m = \dfrac{P_m}{c}$, with $m = 0$. Then use Verhulst's equation to find the population fractions for breeding seasons 1, 2, 3, and 4.

**(b)** Use the formula: $p_{steady} = \dfrac{\tilde{r}-1}{\tilde{r}}$.

**(c)** If some event caused the population to increase above the steady-state value, then what should happen to the population over time?

**17 and 19.**
To find $p_0$, use the definition of the population fraction, $p_m = \dfrac{P_m}{c}$, with $m = 0$. Then use Verhulst's equation to find the population fractions for the first ten breeding seasons.

**(a)** Calculate the steady-state population fraction value using the formula: $p_{steady} = \dfrac{\tilde{r}-1}{\tilde{r}}$.

**(b)** For each breeding season, $n$, calculate $p_n - p_{steady}$.

**21.** **(a)** Find the number of square meters in Nepal. If every person is allotted 1000 square meters, then divide the number of square meters in Nepal by 1000 to find the carrying capacity.

**(b)** Assume each year is one breeding season.

**23.** **(a)** Since 2000 is 0.1% of carrying capacity, create an equation and solve for carrying capacity.

## *Solutions to Odd-Numbered Problems in Section 12.1*

**1.** The total number of rodents can be found by using the formula for Malthusian population growth: $P_m = (1+r)^m \times P_0$. The first two years have been set up for you in the table. Continue using 200 for $P_0$ and 0.95 for $r$. Each year the value for $m$ changes.

| Time, in Years | Total Number of Rodents |
|----------------|-------------------------|
| 0 | $(1+0.95)^0 \times 200 = 200$ |
| 1 | $(1+0.95)^1 \times 200 = 390$ |
| 2 | $(1+0.95)^2 \times 200 \approx 761$ |
| 3 | $(1+0.95)^3 \times 200 \approx 1483$ |
| $m$ | $(1+0.95)^m \times 200 = (1.95)^m \times 200$ |

**3.** Compare the expression $(1.45)^m \times 8500$ to the Malthusian population growth formula, $(1+r)^m \times P_o$.

    **(a)** The initial population, $P_o$, of the island is 8500.

    **(b)** The annual growth rate, $r$, is 0.45 or 45% per year.

    **(c)** The population after 2 years will be $(1.45)^2 \times 8500 \approx 17,871$.

    **(d)** The population after 15 years will be $(1.45)^{15} \times 8500 \approx 2,238,407$.

**5.** Use the Malthusian population growth formula with $r = 0.03$ and $P_o = 50,000$.

The population after 5 years will be $P_5 = (1+0.03)^5 \times 50,000 \approx 57,964$.

The population after 20 years will be $P_{20} = (1+0.03)^{20} \times 50,000 \approx 90,306$.

The population after 45 years will be $P_{45} = (1+0.03)^{45} \times 50,000 \approx 189,080$.

**7.** The initial population and the population of Limon after 20 years are both given. Assuming a Malthusian population model applies in this situation, let $P_o = 168,000$ and $P_{20} = 380,000$.

$$P_m = (1+r)^m \times P_o$$

$$380,000 = (1+r)^{20} \times 168,000$$

$$\frac{380,000}{168,000} = (1+r)^{20}$$

$$\sqrt[20]{\frac{380,000}{168,000}} = 1+r$$

$$\sqrt[20]{\frac{380,000}{168,000}} - 1 = r$$

$$0.0416546 \approx r$$

The rate of growth is approximately 4.17%.

9. The initial population for both cities was 35,000. Use the Malthusian population growth formula to predict the population of each city in 10 years. For one city $r = 0.02$, and for the other, $r = 0.08$.

City with 2% annual growth rate:      City with 8% annual growth rate:

$P_m = (1+r)^m \times P_o$                               $P_m = (1+r)^m \times P_o$

$P_{10} = (1+0.02)^{10} \times 35,000$              $P_{10} = (1+0.08)^{10} \times 35,000$

$P_{10} \approx 42,665$                              $P_{10} \approx 75,562$

Their populations will differ by approximately 32,897 people.

11. Assuming the growth rate remains constant, use the formula for Malthusian population growth. The population in the year 1990 is the initial population, so $P_0 = 249,000,000$. The rate is 0.007. Find the population in 2005, when $m = 15$, and in 2010, when $m = 20$.

Population in 2005:                     Population in 2010:

$P_m = (1+r)^m \times P_o$                               $P_m = (1+r)^m \times P_o$

$P_{15} = (1+0.007)^{15} \times 249,000,000$      $P_{20} = (1+0.007)^{20} \times 249,000,000$

$P_{15} \approx 276,465,794$                       $P_{20} \approx 286,278,517$

In 2005, the U.S. population will be approximately 276 million and 286 million in the year 2010.

13. Assuming the growth rate remains constant, use the formula for Malthusian population growth. The population in the year 2003 is the initial population, so $P_0 = 40,000,000$. The rate is 0.0016.

   (a) Find the population in 2020, when $m = 17$.

   $P_m = (1+r)^m \times P_o$

   $P_{17} = (1+0.0016)^{17} \times 40,000,000$

   $P_{15} \approx 41,102,038$

   In the year 2020, the population of Spain will be approximately 41,102,000.

   (b) The growth rate remains 0.16% from 2003 to 2011 at which time the growth rate changes. Find the population in 2011.

   $P_m = (1+r)^m \times P_o$

   $P_8 = (1+0.0016)^8 \times 40,000,000$

   $P_8 \approx 40,514,876$

   From 2011 to 2020, the growth rate is 0.32%. Use the population in 2011 as the initial population and adjust the rate in the formula for Malthusian population growth. Predict the population in 2020.

   $P_m = (1+r)^m \times P_o$

   $P_9 = (1+0.0032)^9 \times 40,514,876$

   $P_9 \approx 41,696,752$

   The population in 2020 is approximately 41,697,000.

15. Since we assume the growth was constant, we can use the formula for Malthusian growth. The initial value is $24, the rate is 0.03, and the time is $2005 - 1626 = 379$ years.

   $P_m = (1+r)^m \times P_o$

   $P_{379} = (1+0.03)^{379} \times \$24$

   $P_{379} \approx 1,760,027$

   If the value of the goods increased at 3% per year, then they would be worth about $1,760,027 in the year 2005. If the value of the goods increased at 4% per year, then change the rate in the formula.

$$P_m = (1+r)^m \times P_o$$

$$P_{379} = (1+0.04)^{379} \times \$24$$

$$P_{379} \approx 68,524,652$$

At a rate of 4%, they would be worth about $68,524,652.

17. **(a)** After one 20-minute time period, $m = 1$. After 40 minutes, or two 20-minute time periods, $m = 2$. Every 20-minutes, each bacterium duplicates itself, so every 20-minutes, the population doubles.

| Number of 20-Minute Time Periods, $m$ | Total Number of E.coli |
|---|---|
| 0 | 1 |
| 1 (20 minutes) | 2 |
| 2 (40 minutes) | 4 |
| 3 | 8 |
| 4 | 16 |

**(b)** Use the formula for Malthusian population growth. Substitute in values you know and solve for the growth rate. If the initial population is 1 bacterium, then one 20-minute time period later, $m = 1$ and $P_1 = 2$.

$$P_m = (1+r)^m \times P_o$$

$$2 = (1+r)^1 \times 1$$

$$2 = 1+r$$

$$1 = r$$

The value of $r$ in this case is 1, or we would say the growth rate is 100%.

**(c)** Substitute the values you know into the formula and simplify.

$$P_m = (1+1)^m \times 1$$

$$P_m = (2)^m \times 1$$

**(d)** Since there are twelve 20-minute intervals in 4 hours, let $m = 12$.

$$P_m = (2)^m \times 1$$

$$P_{12} = (2)^{12} \times 1$$

$$P_{12} = 4096$$

So, the total number of E. coli present after 4 hours is 4096. Since there are seventy two 20-minute intervals in 24 hours (1 day), let $m = 72$.

$$P_m = (2)^m \times 1$$

$$P_{72} = (2)^{72} \times 1$$

$$P_{72} \approx 4.7 \times 10^{21}$$

So, the total number of E. coli present after one day is approximately $4.7 \times 10^{21}$.

19. Assume the rate of growth was constant.

**(a)** Substitute the values into the formula for the annual growth rate with $P = \$145,000$, $Q = \$215,000$, and $m = 5$.

$$r = \left(\frac{Q}{P}\right)^{\frac{1}{m}} - 1$$

$$r = \left(\frac{\$215,000}{\$145,000}\right)^{\frac{1}{5}} - 1$$

$$r \approx 0.081967$$

The rate of growth is approximately 8.2% annually.

**(b)** Use the rate of growth from part (a). The value of the home has already grown to $215,000 so that is the initial value needed to make the prediction.

$$P_m = (1+r)^m \times P_o$$

$$P_m = (1+0.082)^5 \times \$215,000$$

$$P_m \approx \$318,841$$

The value of the home after 5 more years is approximately $319,000.

**21.** In order to make a prediction when the rate is unknown, use the formula for Malthusian population growth with $r$ unknown. Let $P = 20,000$ and $Q = 33,000$. Because the population changed from 20,000 to 33,000 in 10 years and we want to predict 20 more years into the future, $m = 10$, and $n = 30$.

$$P = \left(\frac{Q}{P}\right)^{\frac{n}{m}}$$

$$= 20,000\left(\frac{33,000}{20,000}\right)^{\frac{30}{10}}$$

$$\approx 89,843$$

In 20 more years the population of the city will be approximately 89,843.

**23.** In order to make a prediction when the rate is unknown, use the formula for Malthusian population growth with $r$ unknown. Let $P = 28,000$ and $Q = 35,500$. Because the population changed from 28,000 to 35,500 in 8 years and we want to predict 22 more years into the future, $m = 8$, and $n = 30$.

$$P = \left(\frac{Q}{P}\right)^{\frac{n}{m}}$$

$$= 28,000\left(\frac{35,500}{28,000}\right)^{\frac{30}{8}}$$

$$\approx 68,182$$

In 22 more years the population of Greenville will be approximately 68,182.

**25. (a)** Use the formula for the annual growth rate to determine the growth rate for each year.

From 1999 to 2000: From 2000 to 2001: From 2001 to 2002:

$$r = \left(\frac{Q}{P}\right)^{\frac{1}{m}} - 1 \qquad r = \left(\frac{Q}{P}\right)^{\frac{1}{m}} - 1 \qquad r = \left(\frac{Q}{P}\right)^{\frac{1}{m}} - 1$$

$$r = \left(\frac{11,697,600}{11,527,900}\right)^{\frac{1}{1}} - 1 \qquad r = \left(\frac{11,894,900}{11,697,600}\right)^{\frac{1}{1}} - 1 \qquad r = \left(\frac{12,068,300}{11,894,900}\right)^{\frac{1}{1}} - 1$$

$$r \approx 0.0147208 \qquad r \approx 0.0168667 \qquad r \approx 0.0145777$$

Therefore, the growth rates for each pair of consecutive years are approximately 1.47%, 1.69%, and 1.46%.

**(b)** A Malthusian population model can be assumed since the rate of growth from year to year is roughly constant.

**27. (a)** Use the formula for the annual growth rate to determine the growth rate.

$$r = \left(\frac{Q}{P}\right)^{\frac{1}{m}} - 1$$

$$r = \left(\frac{6,300,000,000}{4,400,000,000}\right)^{\frac{1}{23}} - 1$$

$$r \approx 0.0157287$$

The rate of growth is approximately 1.57%.

**(b)** Assume the rate remains a constant 1.57287% so that the formula for Malthusian population growth can be used. In the year 2010, $m = 30$.

$$P_m = (1+r)^m \times P_o$$

$$P_{30} = (1+0.0157287)^{30} \times 4,400,000,000$$

$$P_2 \approx 7,027,233,630$$

The population will be approximately 7.0 billion.

**(c)** If each person in the world stood on a 2-foot by 2-foot square, then each person would have 4 square feet of area. Convert 4 square feet to square miles and multiply by the population to obtain the number of square miles required for the world's population.

$$\frac{6,300,000,000}{1} \cdot \frac{4 \text{ ft}^2}{1} \cdot \frac{1 \text{ mi}^2}{(5280)^2 \text{ ft}^2} \approx 904 \text{ mi}^2$$

The state of Rhode Island has a total land area of approximately 1045 square miles. The 6.3 billion people would require a total area of less than one state the size of Rhode Island. Only about 86.5% of the state of Rhode Island would be needed.

**29.** Use the formula for Malthusian population growth. If the initial population is $P$, then the population that is doubled is written $2P$.

$$P_m = (1+r)^m \times P_o$$

$$2P = (1+0.0113)^m \times P$$

$$\frac{2P}{P} = (1.0113)^m$$

$$2 = (1.0113)^m$$

Guess and test values for $m$ until you find one that produces a true equation. It would take between 61 and 62 years.

**31.** The investors are guaranteed a 40% rate of return so we know that the number of investors in the each quarter must be at least 1.40 times the number of investors in the previous quarter. Use the relationship $S_{m+1} \geq (1.4)^m \times S_1$ where $S_{m+1}$ represents the $(m+1)^{st}$ quarter and $S_1 = 5$. Find the number of investors in the second, third, and fourth quarters. That is, find $S_2$, $S_3$, and $S_4$.

$$S_2: \quad S_{1+1} \geq (1.4)^1 \times 5 = 7$$

$$S_3: \quad S_{2+1} \geq (1.4)^2 \times 5 = 9.8 \approx 10$$

$$S_4: \quad S_{3+1} \geq (1.4)^3 \times 5 = 13.72 \approx 14$$

In the second quarter there would need to be at least 7 new investors, in the third quarter there would need to be at least 10 new investors, and in the fourth quarter there would need to be at least 14 new investors.

**33.** Use the relationship $S_{m+1} \geq (1.4)^m \times S_1$ where $S_{m+1}$ represents the $(m+1)^{st}$ quarter, $S_1 = 10$ and $S_{m+1} = 290{,}000{,}000$. Guess and test values for $m$ until the number of people required in the $(m+1)^{st}$ quarter is approximately 290,000,000.

$$S_{m+1} \approx (1.4)^m \times S_1$$

$$290{,}000{,}000 \approx (1.4)^m \times 10$$

$$\frac{290{,}000{,}000}{10} \approx (1.4)^m$$

$$29{,}000{,}000 \approx (1.4)^m$$

It would take approximately 52 quarters or about 13 years.

**35. (a)** For a chain letter, the payoff in rising from the bottom to the top is $P \times n^l$. In this case, there are $l = 5$ levels, the price to participate is $P = \$2$, and you are required to send out $n = 10$ new letters. The payoff in rising from the bottom of the list to the top is $P \times n^l = \$2 \times (10)^5 = \$200{,}000$.

**(b)** The number of people who must join for you to rise from the bottom of the list to the top is

$$\frac{n^{l+1} - 1}{n - 1} = \frac{10^{5+1} - 1}{10 - 1} = \frac{10^6 - 1}{9} = 111{,}111.$$

**37.** For a chain letter, the payoff in rising from the bottom to the top is $P \times n^l$. In this case, there are $l = 4$ levels, and you are required to send out $n = 5$ new letters so the payoff is $P \times n^l = P \times 5^4$ and depends on the price to participate.

**(a)** When $P = \$1$, the payoff is $P \times 5^4 = \$1 \times 5^4 = \$625$.

**(b)** When $P = \$5$, the payoff is $P \times 5^4 = \$5 \times 5^4 = \$3125$.

**(c)** When $P = \$10$, the payoff is $P \times 5^4 = \$10 \times 5^4 = \$6250$.

**(d)** When $P = \$50$, the payoff is $P \times 5^4 = \$50 \times 5^4 = \$31{,}250$.

**39.** Answers may vary. You are given the payoff and the price to participate. Use the payoff expression $P \times n^l$. Try different values for the number of levels and then figure out the number of letters you are required to send out. For example, since the number of levels is not given, assume that there are 5 levels. So, we want $3 \times n^5 \approx 90{,}000$. Solve for $n$. If $n = 8$, then the payoff is $\$98{,}304$. So a good scheme would be to have 5 levels and have each person send out 8 letters.

## *Solutions to Odd-Numbered Problems in Section 12.2*

**1.** Population values are in millions.

**(a)** Use the formula for the annual growth rate from section 12.1. It does not matter that the population decreased or that the population rate is declining rather than growing, the same formula applies.

$$r = \left(\frac{Q}{P}\right)^{\frac{1}{m}} - 1$$

$$r = \left(\frac{144}{148.7}\right)^{\frac{1}{10}} - 1$$

$$r \approx -0.0032066$$

The rate of decline of the population of Russia is $r \approx -0.0032066$. So, the population decreased at an annual rate of about 0.32% per year between 1992 and 2002.

**(b)** Use the formula for Malthusian population growth from section 12.1.

$$P_m = (1+r)^m \times P_o$$

$$P_{11} = (1-0.0032066)^{11} \times 148.7$$

$$P_{11} \approx 143.54$$

The population in the year 2003 was approximately 143.54 million.

(c) The predicted value was about 0.64 million more than the actual population. The difference could be a result of the population not declining at a constant rate or perhaps there was a large amount of emigration in Russia. There are many possible reasons for the difference.

3. (a) Use the formula for the annual growth rate from section 12.1.

$$r = \left(\frac{Q}{P}\right)^{\frac{1}{m}} - 1$$

$$r = \left(\frac{9000}{30,000}\right)^{\frac{1}{6}} - 1$$

$$r \approx -0.1818112$$

The rate of decline of the population of Black Rhinoceros is $r \approx -0.1818112$. So, the population decreased about 18.18% per year.

(b) Use the formula for Malthusian population growth: $P_m = (0.8181888)^m \times 30,000$.

The year 2010 is 32 years after the year 1978. The year 2020 is 42 years after 1978.

Population in 2010:                     Population in 2020:

$$P_m = (0.8181888)^m \times 30,000 \qquad P_m = (0.8181888)^m \times 30,000$$

$$P_{32} = (0.8181888)^{32} \times 30,000 \qquad P_{42} = (0.8181888)^{42} \times 30,000$$

$$P_{32} \approx 49 \qquad\qquad\qquad\qquad P_{42} \approx 7$$

In 2010, the population will decline to about 49 rhinos. In 2020, the population will decline to about 7 rhinos.

(c) Use the Malthusian population growth formula and find the value $m$ such that $P_m = 1$.

$$P_m = (0.8181888)^m \times 30,000$$

$$1 = (0.8181888)^m \times 30,000$$

$$\frac{1}{30,000} = (0.8181888)^m$$

$$0.0000333 \approx (0.8181888)^m$$

Guess and test values for $m$. At the given rate of decline, the Black Rhinoceros would become extinct in the year 2030 when $m = 52$.

5. Use the radioactive decay formula: $A_m = (1-d)^m \times A_0$. Let $A_0 = 27$ and $d = 0.03$.

| Time, in Years | Amount of Substance Present, in Kilograms |
|---|---|
| 0 | $(1-0.03)^0 \times 27 = 27$ |
| 1 | $(1-0.03)^1 \times 27 = 26.19$ |
| 2 | $(1-0.03)^2 \times 27 = 25.4043$ |
| 3 | $(1-0.03)^3 \times 27 = 24.642171$ |
| $m$ | $(1-0.03)^m \times 27 = (0.97)^m \times 27$ |

7. Compare the given equation, $A_m = (0.955)^m \times 1400$, to the radioactive decay formula, $A_m = (1-d)^m \times A_0$
   (a) There were initially 1400 grams of the radioactive substance.
   (b) The decay rate is $d$. Notice $1-d = 0.955$ so $d = 1 - 0.955 = 0.045$ or 4.5% per year.
   (c) Let $m = 1$. $A_1 = (0.955)^1 \times 1400 = 1337$. In the first year, 63 grams decayed.
   (d) Let $m = 20$. $A_{20} = (0.955)^{20} \times 1400 \approx 557.4$. After 20 years, approximately 557.4 grams remain.
   (e) Let $m = 51$. $A_{51} = (0.955)^{51} \times 1400 \approx 133.8$. After 51 years, approximately 133.8 grams remain.

9. Use the radioactive decay formula. The decay rate is 2.5%, so $d = 0.025$. We know 300 grams of the substance was initially present, so $A_0 = 300$.
$$A_m = (1-d)^m \times A_0$$
$$A_m = (1-0.025)^m \times 300$$
$$A_m = (0.975)^m \times 300$$
After 5 years, 10 years, and 15 years, $m = 5$, $m = 10$, and $m = 15$ respectively.

After 5 years:   After 10 years:   After 15 years:
$A_5 = (0.975)^5 \times 300$   $A_{10} = (0.975)^{10} \times 300$   $A_{15} = (0.975)^{15} \times 300$
$A_5 \approx 264.3287$   $A_{10} \approx 232.8989$   $A_{15} \approx 205.2062$

After 5 years, there would be approximately 264.3 grams remaining. After 10 years, there would be approximately 232.9 grams remaining. After 15 years, there would be approximately 205.2 grams of the radioactive substance remaining.

11. Use the decay rate formula. The half-life is 28 years, so $h = 28$.
$$d = 1 - \left(\frac{1}{2}\right)^{\frac{1}{h}}$$
$$d = 1 - \left(\frac{1}{2}\right)^{\frac{1}{28}}$$
$$d \approx 0.02445$$
The annual decay rate for Strontium-90 is approximately 2.445%

13. (a) Use the decay rate formula. The half-life is 13 years, so $h = 13$.
$$d = 1 - \left(\frac{1}{2}\right)^{\frac{1}{h}}$$
$$d = 1 - \left(\frac{1}{2}\right)^{\frac{1}{13}}$$
$$d \approx 0.0519225$$
The annual decay rate for Plutonium-241 is approximately 5.19%
   (b) Use the radioactive decay formula with $d = 0.0519225$ and $m = 6.5$.
$$A_m = (1-d)^m \times A_0$$
$$A_{6.5} = (1-0.0519225)^{6.5} \times 50$$
$$A_{6.5} = (0.9480775)^{6.5} \times 50$$
$$A_{6.5} \approx 35.36$$
In 6.5 years there will be approximately 35.36 grams of the radioactive sample remaining.

**(c)** Use the radioactive decay formula from part (b).

| After 13 years: | After 26 years: | After 39 years: |
|---|---|---|
| $A_{13} = (0.9480775)^{13} \times 50$ | $A_{26} = (0.9480775)^{26} \times 50$ | $A_{39} = (0.9480775)^{39} \times 50$ |
| $A_{13} \approx 25$ | $A_{26} \approx 12.5$ | $A_{39} \approx 6.25$ |

In 13 years there will be approximately 25 grams; in 26 years there will be approximately 12.5 grams; and in 39 years there will be approximately 6.25 grams.

**15. (a)** Use the decay rate formula. The half-life is 2.6 years, so $h = 2.6$.

$$d = 1 - \left(\frac{1}{2}\right)^{\frac{1}{h}}$$

$$d = 1 - \left(\frac{1}{2}\right)^{\frac{1}{2.6}}$$

$$d \approx 0.2340168$$

The annual decay rate for Sodium-22 is approximately 23.4%

**(b)** Use the radioactive decay formula with $d = 0.2340168$ and $m = 35$.

$$A_m = (1-d)^m \times A_0$$

$$A_{35} = (1 - 0.2340168)^{35} \times 200$$

$$A_{35} = (0.7659832)^{35} \times 200$$

$$A_{35} \approx 0.0177298$$

In the year 2035 there will be approximately 0.02 gram of the radioactive sample remaining.

**17.** Use the decay rate formula. In the formula, the half-life is in years, but we can put $h = 18$ minutes in the formula as long as we understand that we get a rate per minute rather than an annual rate.

$$d = 1 - \left(\frac{1}{2}\right)^{\frac{1}{h}}$$

$$d = 1 - \left(\frac{1}{2}\right)^{\frac{1}{18}}$$

$$d \approx 0.0377762$$

Use the radioactive decay formula to predict how much of the radioactive element will be left 4 hours later. Be sure to express time in minutes. Four hours is $4 \times 60 = 240$ minutes.

$$A_m = (1-d)^m \times A_0$$

$$A_{240} = (1 - 0.0377762)^{240} \times 10$$

$$A_{240} = (0.9622238)^{240} \times 10$$

$$A_{240} \approx 0.0009689$$

In 4 hours there will be approximately 0.00097 gram remaining.

**19.** Use the decay rate formula. Let $h = 13$.

$$d = 1 - \left(\frac{1}{2}\right)^{\frac{1}{h}}$$

$$d = 1 - \left(\frac{1}{2}\right)^{\frac{1}{13}}$$

$$d \approx 0.0519225$$

Use the radioactive decay formula to predict how much of the Plutonium-241 will be left 53 years later.

$$A_m = (1-d)^m \times A_0$$

$$A_{53} = (1-0.0519225)^{53} \times 100$$

$$A_{53} = (0.9480775)^{53} \times 100$$

$$A_{53} \approx 5.9254797$$

In 2003, there will be approximately 5.9 grams remaining.

21. Use the half-life radioactive decay formula. In each case $A_0 = 500$ grams and $m = 100$ years. The only change will be the half-life.

$$A_m = \left(\frac{1}{2}\right)^{\frac{m}{h}} \times A_0$$

$$A_{100} = \left(\frac{1}{2}\right)^{\frac{100}{h}} \times 500$$

(a) If the half-life is 100 years, then $h = 100$.

$$A_{100} = \left(\frac{1}{2}\right)^{\frac{100}{100}} \times 500 = \left(\frac{1}{2}\right)^{1} \times 500 = 250$$

After 100 years, there will be 250 grams remaining.

(b) If the half-life is 50 years, then $h = 50$.

$$A_{100} = \left(\frac{1}{2}\right)^{\frac{100}{50}} \times 500 = \left(\frac{1}{2}\right)^{2} \times 500 = 125$$

After 100 years, there will be 125 grams remaining.

(c) If the half-life is 10 years, then $h = 10$.

$$A_{100} = \left(\frac{1}{2}\right)^{\frac{100}{10}} \times 500 = \left(\frac{1}{2}\right)^{10} \times 500 = 0.48828125$$

After 100 years, there will be approximately 0.49 grams remaining.

(d) If the half-life is 1 year, then $h = 1$.

$$A_{100} = \left(\frac{1}{2}\right)^{\frac{100}{1}} \times 500 = \left(\frac{1}{2}\right)^{100} \times 500 \approx 3.9 \times 10^{-28}$$

After 100 years, there will be approximately $3.9 \times 10^{-28}$ grams remaining. That number is so close to zero that we could say after 100 years none of the radioactive substance remains.

23. Use the half-life radioactive decay formula. In each case $A_0 = 100$ grams and $h = 2$ years. The only change will be the time.

$$A_m = \left(\frac{1}{2}\right)^{\frac{m}{h}} \times A_0$$

$$A_m = \left(\frac{1}{2}\right)^{\frac{m}{2}} \times 100$$

(a) In 1 year, $m = 1$.

$$A_1 = \left(\frac{1}{2}\right)^{\frac{1}{2}} \times 100 \approx 70.7106781$$

There will be approximately 70.7 grams of the substance remaining.

**(b)** In 18 months, $m = 1.5$.

$$A_{1.5} = \left(\frac{1}{2}\right)^{\frac{1.5}{2}} \times 100 = \left(\frac{1}{2}\right)^{0.75} \times 100 \approx 59.4603558$$

There will be approximately 59.46 grams of the substance remaining.

**(c)** In 2 years, $m = 2$.

$$A_2 = \left(\frac{1}{2}\right)^{\frac{2}{2}} \times 100 = \left(\frac{1}{2}\right)^{1} \times 100 = 50$$

There will be 50 grams of the substance remaining.

**(d)** In 4 years, $m = 4$.

$$A_4 = \left(\frac{1}{2}\right)^{\frac{4}{2}} \times 100 = \left(\frac{1}{2}\right)^{2} \times 100 = 25$$

There will be 25 grams of the substance remaining.

**(e)** In 8 years, $m = 8$.

$$A_8 = \left(\frac{1}{2}\right)^{\frac{8}{2}} \times 100 = \left(\frac{1}{2}\right)^{4} \times 100 = 6.25$$

There will be 6.25 grams of the substance remaining.

**25.** Use the half-life approximation formula:  $h = \dfrac{0.693}{\left(d + \dfrac{d^2}{2}\right)}$.

**(a)** The annual decay rate is 2.445% so $d = 0.02445$.

$$h = \frac{0.693}{\left(0.02445 + \dfrac{0.02445^2}{2}\right)} = \frac{0.693}{\left(0.02445 + \dfrac{0.0005978025}{2}\right)} \approx \frac{0.693}{0.0247489} \approx 28$$

The approximate half-life for this substance is 28 years.

**(b)** The element could be Strontium-90.

**27.** The fossil contains approximately 90% of its original amount of $^{14}C$. From Table 12.6, if 90% remains, then the age of the sample in half-lives is 0.15. Since the half-life of $^{14}C$ is 5730 years, the age of the fossil is $0.15 \times 5730 = 859.5$ years.

**29.** If 40% of the $^{14}C$ is still present, the bones are approximately 1.32 half-lives old or about $1.32 \times 5730 = 7563.6$ years. This is approximately 8000 years, so the archeologist's belief is reasonable.

## Solutions to Odd-Numbered Problems in Section 12.3

**1.** Study the given graph of population versus breeding season.

  **(a)** The initial population is approximately 100 mice. It occurs in breeding season 0. The point is the vertical intercept.

  **(b)** Answers may vary depending on the approximations made using the graph. To find the change in the population during season $m$, subtract the population total in the $(m-1)^{st}$ breeding season from the population total in the $m$th breeding season. For example, to find the change in population during season 4, subtract the population total in the 3rd breeding season from the population total in the $4^{th}$ breeding season: $290 - 210 = 80$.

| Breeding Season $m$ | Population at the End of Breeding Season $m$ | Change in Population During Season $m$ |
|---|---|---|
| 0 | 100 | |
| 1 | 120 | 20 |
| 2 | 150 | 30 |
| 3 | 210 | 60 |
| 4 | 290 | 80 |
| 5 | 400 | 110 |
| 6 | 540 | 140 |
| 7 | 700 | 160 |
| 8 | 900 | 200 |
| 9 | 1190 | 290 |
| 10 | 1580 | 390 |
| 11 | 2050 | 470 |
| 12 | 2680 | 630 |

(c) From the table in part (b), we see the change in the population increases with the passing of each breeding season. The change in the 12th breeding season is the greatest, so the population increase is the greatest during the 12th breeding season.

(d) In order to determine the type of growth that is exhibited by this population, it is necessary to calculate the rate of growth from one breeding season to the next. Find the growth rate by dividing the population in one breeding season by the population in the previous season and then subtract 1.

| Breeding Seasons | Growth Rate |
|---|---|
| 0 to 1 | $\frac{120}{100}-1=0.2$ |
| 1 to 2 | $\frac{150}{120}-1=0.25$ |
| 2 to 3 | $\frac{210}{150}-1=0.4$ |
| 3 to 4 | $\frac{290}{210}-1\approx0.38$ |
| 4 to 5 | $\frac{400}{290}-1\approx0.38$ |
| 5 to 6 | $\frac{540}{400}-1=0.35$ |
| 6 to 7 | $\frac{700}{540}-1\approx0.30$ |
| 7 to 8 | $\frac{900}{700}-1\approx0.29$ |
| 8 to 9 | $\frac{1190}{900}-1\approx0.32$ |
| 9 to 10 | $\frac{1580}{1190}-1\approx0.33$ |
| 10 to 11 | $\frac{2050}{1580}-1\approx0.30$ |
| 11 to 12 | $\frac{2680}{2050}-1\approx0.31$ |

The graph represents Malthusian growth since the population appears to have roughly a constant growth rate.

3. Initially there are 300 animals so $P_0 = 300$. Use the following formula to find $P_1$, $P_2$, and $P_3$:

$P_{m+1} = (1+0.75)P_m - \left(\dfrac{1+0.75}{1500}\right)P_m^2$. Round values to the nearest whole number.

$P_1 = P_{0+1} = (1+0.75)P_0 - \left(\dfrac{1+0.75}{1500}\right)P_0^2 = (1.75)(300) - \left(\dfrac{1.75}{1500}\right)(300)^2 = 420$

$P_2 = P_{1+1} = (1+0.75)P_1 - \left(\dfrac{1+0.75}{1500}\right)P_1^2 = (1.75)(420) - \left(\dfrac{1.75}{1500}\right)(420)^2 \approx 529$

$P_3 = P_{2+1} = (1+0.75)P_2 - \left(\dfrac{1+0.75}{1500}\right)P_2^2 = (1.75)(529) - \left(\dfrac{1.75}{1500}\right)(529)^2 \approx 599$

Therefore, $P_1 = 420$, $P_2 \approx 529$, and $P_3 \approx 599$. These values represent the population after 1 year, 2 years and 3 years if the initial population is 300, the natural growth rate of the population is 75% and the carrying capacity is 1500 when using the logistic growth law.

5. Use the logistic growth law with $P_0 = 500$, $r = 0.3$, and $c = 4000$.

Logistic Growth Law: $P_{m+1} = (1+r)P_m - \left(\dfrac{1+r}{c}\right)P_m^2$

| Breeding Season $m$ | Population at the End of Breeding Season $m$ |
|---|---|
| 1 | $P_1 = (1.3)(500) - \left(\dfrac{1.3}{4000}\right)(500)^2 \approx 569$ |
| 2 | $P_2 = (1.3)(569) - \left(\dfrac{1.3}{4000}\right)(569)^2 \approx 634$ |
| 3 | $P_3 = (1.3)(634) - \left(\dfrac{1.3}{4000}\right)(634)^2 \approx 694$ |
| 4 | $P_4 = (1.3)(694) - \left(\dfrac{1.3}{4000}\right)(694)^2 \approx 746$ |
| 5 | $P_5 = (1.3)(746) - \left(\dfrac{1.3}{4000}\right)(746)^2 \approx 789$ |

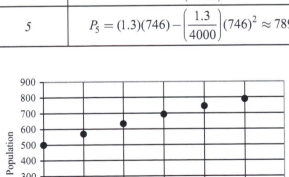
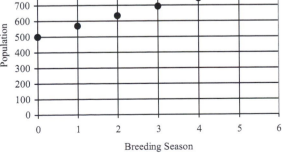

7. **(a)** Use the logistic growth law with $P_0 = 10$, $r = 0.8$, and $c = 700$.

Logistic Growth Law: $P_{m+1} = (1+r)P_m - \left(\dfrac{1+r}{c}\right)P_m^2$

| Breeding Season $m$ | Population at the End of Breeding Season $m$ |
|---|---|
| 0 | $P_0 = 10$ |
| 1 | $P_1 = (1.8)(10) - \left(\dfrac{1.8}{700}\right)(10)^2 \approx 18$ |
| 2 | $P_2 = (1.8)(18) - \left(\dfrac{1.8}{700}\right)(18)^2 \approx 32$ |
| 3 | $P_3 = (1.8)(32) - \left(\dfrac{1.8}{700}\right)(32)^2 \approx 55$ |
| 4 | $P_4 = (1.8)(55) - \left(\dfrac{1.8}{700}\right)(55)^2 \approx 91$ |
| 5 | $P_5 = (1.8)(91) - \left(\dfrac{1.8}{700}\right)(91)^2 \approx 143$ |
| 6 | $P_6 = (1.8)(143) - \left(\dfrac{1.8}{700}\right)(143)^2 \approx 205$ |
| 7 | $P_7 = (1.8)(205) - \left(\dfrac{1.8}{700}\right)(205)^2 \approx 261$ |
| 8 | $P_8 = (1.8)(261) - \left(\dfrac{1.8}{700}\right)(261)^2 \approx 295$ |
| 9 | $P_9 = (1.8)(295) - \left(\dfrac{1.8}{700}\right)(295)^2 \approx 307$ |
| 10 | $P_{10} = (1.8)(307) - \left(\dfrac{1.8}{700}\right)(307)^2 \approx 310$ |

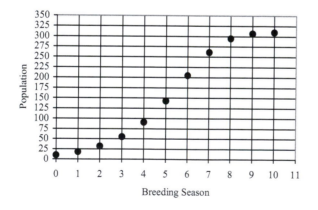

**(b)** The steady-state population is $P_{steady} = \dfrac{r \times c}{1+r} = \dfrac{0.80 \times 700}{1+0.80} = \dfrac{560}{1.8} \approx 311$. So, once the population reaches 311 mammals, it will stay the same from one breeding season to the next.

**(c)** The table contains the population totals for the next five breeding seasons. Notice the population levels off to the steady-state population.

| Breeding Season $m$ | Population at the End of Breeding Season $m$ |
|---|---|
| 11 | $P_{11} = (1.8)(310) - \left(\dfrac{1.8}{700}\right)(310)^2 \approx 311$ |
| 12 | $P_{12} = (1.8)(311) - \left(\dfrac{1.8}{700}\right)(311)^2 \approx 311$ |
| 13 | $P_{13} = (1.8)(311) - \left(\dfrac{1.8}{700}\right)(311)^2 \approx 311$ |
| 14 | $P_{14} = (1.8)(311) - \left(\dfrac{1.8}{700}\right)(311)^2 \approx 311$ |
| 15 | $P_{15} = (1.8)(311) - \left(\dfrac{1.8}{700}\right)(311)^2 \approx 311$ |

**9. (a)** Use the logistic growth law with $P_0 = 1000$, $r = 0.2$, and $c = 8000$.

Logistic Growth Law: $P_{m+1} = (1+r)P_m - \left(\dfrac{1+r}{c}\right)P_m^2$

| Breeding Season $m$ | Population at the End of Breeding Season $m$ |
|---|---|
| 0 | $P_0 = 1000$ |
| 1 | $P_1 = (1.2)(1000) - \left(\dfrac{1.2}{8000}\right)(1000)^2 = 1050$ |
| 2 | $P_2 = (1.2)(1050) - \left(\dfrac{1.2}{8000}\right)(1050)^2 \approx 1095$ |
| 3 | $P_3 = (1.2)(1095) - \left(\dfrac{1.2}{8000}\right)(1095)^2 \approx 1134$ |
| 4 | $P_4 = (1.2)(1134) - \left(\dfrac{1.2}{8000}\right)(1134)^2 \approx 1168$ |
| 5 | $P_5 = (1.2)(1168) - \left(\dfrac{1.2}{8000}\right)(1168)^2 \approx 1197$ |
| 6 | $P_6 = (1.2)(1197) - \left(\dfrac{1.2}{8000}\right)(1197)^2 \approx 1221$ |
| 7 | $P_7 = (1.2)(1221) - \left(\dfrac{1.2}{8000}\right)(1221)^2 \approx 1242$ |
| 8 | $P_8 = (1.2)(1242) - \left(\dfrac{1.2}{8000}\right)(1242)^2 \approx 1259$ |
| 9 | $P_9 = (1.2)(1259) - \left(\dfrac{1.2}{8000}\right)(1259)^2 \approx 1273$ |
| 10 | $P_{10} = (1.2)(1273) - \left(\dfrac{1.2}{8000}\right)(1273)^2 \approx 1285$ |
| 11 | $P_{11} = (1.2)(1285) - \left(\dfrac{1.2}{8000}\right)(1285)^2 \approx 1294$ |

| 12 | $P_{12} = (1.2)(1294) - \left(\dfrac{1.2}{8000}\right)(1294)^2 \approx 1302$ |
| 13 | $P_{13} = (1.2)(1302) - \left(\dfrac{1.2}{8000}\right)(1302)^2 \approx 1308$ |
| 14 | $P_{14} = (1.2)(1308) - \left(\dfrac{1.2}{8000}\right)(1308)^2 \approx 1313$ |
| 15 | $P_{15} = (1.2)(1313) - \left(\dfrac{1.2}{8000}\right)(1313)^2 \approx 1317$ |

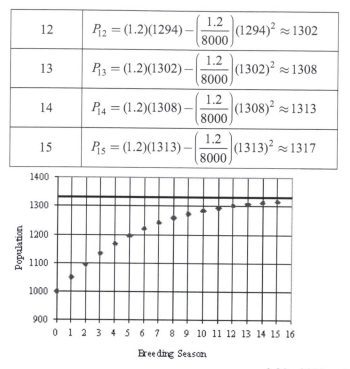

**(b)** The steady-state population is $P_{steady} = \dfrac{r \times c}{1+r} = \dfrac{0.20 \times 8000}{1+0.20} = \dfrac{1600}{1.2} \approx 1333$.

**(c)** The population growth appears to begin to slow during the second breeding season as shown in the following table. The growth of 50 in the first breeding season is a baseline value. In the next season, the increase is smaller.

| Breeding Season $m$ | Population at the End of Breeding Season $m$ | Change in Population During Season $m$ |
|---|---|---|
| 0 | 1000 | |
| 1 | 1050 | 50 |
| 2 | 1095 | 45 |
| 3 | 1134 | 39 |
| 4 | 1168 | 34 |
| 5 | 1197 | 29 |
| 6 | 1221 | 24 |
| 7 | 1242 | 21 |
| 8 | 1259 | 17 |
| 9 | 1273 | 14 |
| 10 | 1285 | 12 |
| 11 | 1294 | 9 |
| 12 | 1302 | 8 |
| 13 | 1308 | 6 |
| 14 | 1313 | 5 |
| 15 | 1317 | 4 |

**11. (a)** Use the logistic growth law with $P_0 = 546$, $r = 0.0388$, and $c = 20,000$.

Logistic Growth Law: $P_{m+1} = (1+r)P_m - \left(\dfrac{1+r}{c}\right)P_m^2$.

| Breeding Season $m$ | One-Horned Rhinoceros Population at the End of Breeding Season $m$ |
|---|---|
| 0 | $P_0 = 546$ |
| 1 | $P_1 = (1.0388)(546) - \left(\dfrac{1.0388}{20,000}\right)(546)^2 \approx 552$ |
| 2 | $P_2 = (1.0388)(552) - \left(\dfrac{1.0388}{20,000}\right)(552)^2 \approx 558$ |
| 3 | $P_3 = (1.0388)(558) - \left(\dfrac{1.0388}{20,000}\right)(558)^2 \approx 563$ |
| 4 | $P_4 = (1.0388)(563) - \left(\dfrac{1.0388}{20,000}\right)(563)^2 \approx 568$ |
| 5 | $P_5 = (1.0388)(568) - \left(\dfrac{1.0388}{20,000}\right)(568)^2 \approx 573$ |
| 6 | $P_6 = (1.0388)(573) - \left(\dfrac{1.0388}{20,000}\right)(573)^2 \approx 578$ |
| 7 | $P_7 = (1.0388)(578) - \left(\dfrac{1.0388}{20,000}\right)(578)^2 \approx 583$ |
| 8 | $P_8 = (1.0388)(583) - \left(\dfrac{1.0388}{20,000}\right)(583)^2 \approx 588$ |
| 9 | $P_9 = (1.0388)(588) - \left(\dfrac{1.0388}{20,000}\right)(588)^2 \approx 593$ |
| 10 | $P_{10} = (1.0388)(593) - \left(\dfrac{1.0388}{20,000}\right)(593)^2 \approx 598$ |

**(b)** The steady-state population is $P_{steady} = \dfrac{r \times c}{1 + r} = \dfrac{0.0388 \times 20,000}{1 + 0.0388} = \dfrac{776}{1.0388} \approx 747$.

**(c)** The population fraction is found from the formula $p_m = \dfrac{P_m}{c}$. In the 10th breeding season

$p_{10} = \dfrac{P_{10}}{c} = \dfrac{598}{20,000} = 0.0299$. After ten breeding cycles, the population is about 2.99% of the maximum population.

**13. (a)** Use the logistic growth law with $P_0 = 100,000$, $r = 0.47$, and $c = 500,000$.

Logistic Growth Law: $P_{m+1} = (1 + r)P_m - \left(\dfrac{1 + r}{c}\right)P_m^2$.

| Breeding Season $m$ | Goat Population at the End of Breeding Season $m$ |
|---|---|
| 0 | $P_0 = 100,000$ |
| 1 | $P_1 = (1.47)(100,000) - \left(\dfrac{1.47}{500,000}\right)(100,000)^2 = 117,600$ |
| 2 | $P_2 = (1.47)(117,600) - \left(\dfrac{1.47}{500,000}\right)(117,600)^2 \approx 132,213$ |

| | |
|---|---|
| 3 | $P_3 = (1.47)(132,213) - \left(\dfrac{1.47}{500,000}\right)(132,213)^2 \approx 142,961$ |
| 4 | $P_4 = (1.47)(142,961) - \left(\dfrac{1.47}{500,000}\right)(142,961)^2 \approx 150,065$ |
| 5 | $P_5 = (1.47)(150,065) - \left(\dfrac{1.47}{500,000}\right)(150,065)^2 \approx 154,388$ |
| 6 | $P_6 = (1.47)(154,388) - \left(\dfrac{1.47}{500,000}\right)(154,388)^2 \approx 156,874$ |
| 7 | $P_7 = (1.47)(156,874) - \left(\dfrac{1.47}{500,000}\right)(156,874)^2 \approx 158,253$ |
| 8 | $P_8 = (1.47)(158,253) - \left(\dfrac{1.47}{500,000}\right)(158,253)^2 \approx 159,003$ |
| 9 | $P_9 = (1.47)(159,003) - \left(\dfrac{1.47}{500,000}\right)(159,003)^2 \approx 159,405$ |
| 10 | $P_{10} = (1.47)(159,405) - \left(\dfrac{1.47}{500,000}\right)(159,405)^2 \approx 159,620$ |
| 11 | $P_{11} = (1.47)(159,620) - \left(\dfrac{1.47}{500,000}\right)(159,620)^2 \approx 159,734$ |
| 12 | $P_{12} = (1.47)(159,734) - \left(\dfrac{1.47}{500,000}\right)(159,734)^2 \approx 159,795$ |
| 13 | $P_{13} = (1.47)(159,795) - \left(\dfrac{1.47}{500,000}\right)(159,795)^2 \approx 159,827$ |
| 14 | $P_{14} = (1.47)(159,827) - \left(\dfrac{1.47}{500,000}\right)(159,827)^2 \approx 159,844$ |
| 15 | $P_{15} = (1.47)(159,844) - \left(\dfrac{1.47}{500,000}\right)(159,844)^2 \approx 159,853$ |

**(b)** The steady-state population is $P_{steady} = \dfrac{r \times c}{1+r} = \dfrac{0.47 \times 500,000}{1+0.47} = \dfrac{235,000}{1.47} \approx 159,864$ . The population of goats will be within 100 of the steady-state value in the last 4 breeding seasons shown in part (a).

**(c)** The population fraction is found from the formula $p_m = \dfrac{P_m}{c}$. Initially, the population fraction is

$p_0 = \dfrac{P_0}{c} = \dfrac{100,000}{500,000} = 0.2.$ After the 15th breeding season, the population fraction is

$p_{15} = \dfrac{P_{15}}{c} = \dfrac{159,853}{500,000} \approx 0.3197.$ Initially the population is 20% of the maximum population.

After the 15th breeding season, the population is approximately 32% of the maximum population.

**15. (a)** Verhulst's equation uses the growth parameter $\tilde{r} = 1 + r = 1 + 2 = 3$. The population fraction after $m$ breeding seasons is $p_m$. Verhulst's equation, $p_{m+1} = \tilde{r}(1 - p_m)p_m$ , gives the population fraction after $m+1$ breeding seasons.

| Breeding Season $m$ | Population Fraction |
|---|---|
| 0 | $p_0 = \dfrac{P_0}{c} = \dfrac{400}{2500} = 0.16$ |
| 1 | $p_1 = p_{0+1} = 3(1-p_0)p_0 = 3(1-0.16)(0.16) = 0.4032$ |
| 2 | $p_2 = p_{1+1} = 3(1-p_1)p_1 = 3(1-0.4032)(0.4032) \approx 0.7218893$ |
| 3 | $p_3 = p_{2+1} = 3(1-p_2)p_2 = 3(1-0.7218893)(0.7218893) \approx 0.6022954$ |
| 4 | $p_4 = p_{3+1} = 3(1-p_3)p_3 = 3(1-0.6022954)(0.6022954) \approx 0.7186070$ |

(b) The steady-state population fraction is $p_{steady} = \dfrac{\tilde{r}-1}{\tilde{r}} = \dfrac{3-1}{3} = \dfrac{2}{3} \approx 0.6666667$.

(c) During the second breeding season the population fraction exceeded the steady-state population fraction. Therefore, the population was too high and the population decreased during the third breeding season to move the population fraction back toward the steady-state population fraction.

17. (a) Calculate the population fraction for the first 10 breeding seasons. The growth parameter is $\tilde{r} = 1+r = 1+0.5 = 1.5$. Verhulst's equation, $p_{m+1} = \tilde{r}(1-p_m)p_m$, gives the population fraction after $m+1$ breeding seasons.

| Breeding Season $m$ | Population Fraction |
|---|---|
| 0 | $p_0 = \dfrac{P_0}{c} = \dfrac{10}{100} = 0.1$ |
| 1 | $p_1 = 1.5(1-0.1)(0.1) = 0.135$ |
| 2 | $p_2 = 1.5(1-0.135)(0.135) = 0.1751625$ |
| 3 | $p_3 = 1.5(1-0.1751625)(0.1751625) \approx 0.2167209$ |
| 4 | $p_4 = 1.5(1-0.2167209)(0.2167209) \approx 0.2546294$ |
| 5 | $p_5 = 1.5(1-0.2546294)(0.2546294) \approx 0.2846899$ |
| 6 | $p_6 = 1.5(1-0.2846899)(0.2846899) \approx 0.3054623$ |
| 7 | $p_7 = 1.5(1-0.3054623)(0.3054623) \approx 0.3182326$ |
| 8 | $p_8 = 1.5(1-0.3182326)(0.3182326) \approx 0.3254409$ |
| 9 | $p_9 = 1.5(1-0.3254409)(0.3254409) \approx 0.3292937$ |
| 10 | $p_{10} = 1.5(1-0.3292937)(0.3292937) \approx 0.3312890$ |

(b) The steady-state population fraction is $p_{steady} = \dfrac{\tilde{r}-1}{\tilde{r}} = \dfrac{1.5-1}{1.5} = \dfrac{0.5}{1.5} = \dfrac{1}{3} \approx 0.333$. A horizontal line has been sketched through the steady-state population fraction in the following graph.

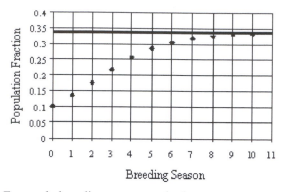

**(c)** For each breeding season calculate $p_m - p_{steady} = p_m - 0.333$. The population fractions are approaching the steady-state value.

| Breeding Season $m$ | Difference Between Population Fraction and the Steady-State Population Fraction |
|---|---|
| 0 | −0.233 |
| 1 | −0.198 |
| 2 | −0.158 |
| 3 | −0.117 |
| 4 | −0.079 |
| 5 | −0.049 |
| 6 | −0.028 |
| 7 | −0.015 |
| 8 | −0.008 |
| 9 | −0.004 |
| 10 | −0.002 |

**19. (a)** Calculate the population fraction for the first 10 breeding seasons. The growth parameter is $\tilde{r} = 1 + r = 1 + 2.5 = 3.5$. Verhulst's equation, $p_{m+1} = \tilde{r}(1 - p_m)p_m$, gives the population fraction after $m+1$ breeding seasons.

| Breeding Season $m$ | Population Fraction |
|---|---|
| 0 | $p_0 = \dfrac{P_0}{c} = \dfrac{10}{100} = 0.1$ |
| 1 | $p_1 = 3.5(1 - 0.1)(0.1) = 0.315$ |
| 2 | $p_2 = 3.5(1 - 0.315)(0.315) = 0.7552125$ |
| 3 | $p_3 = 3.5(1 - 0.7552125)(0.7552125) \approx 0.6470330$ |
| 4 | $p_4 = 3.5(1 - 0.6470330)(0.6470330) \approx 0.7993345$ |
| 5 | $p_5 = 3.5(1 - 0.7993345)(0.7993345) \approx 0.5613960$ |
| 6 | $p_6 = 3.5(1 - 0.5613960)(0.5613960) \approx 0.8618069$ |
| 7 | $p_7 = 3.5(1 - 0.8618069)(0.8618069) \approx 0.4168352$ |
| 8 | $p_8 = 3.5(1 - 0.4168352)(0.4168352) \approx 0.8507927$ |
| 9 | $p_9 = 3.5(1 - 0.8507927)(0.8507927) \approx 0.4443057$ |
| 10 | $p_{10} = 3.5(1 - 0.4443057)(0.4443057) \approx 0.8641435$ |

**(b)** The steady-state population fraction is $p_{steady} = \dfrac{\tilde{r}-1}{\tilde{r}} = \dfrac{3.5-1}{3.5} = \dfrac{2.5}{3.5} = \dfrac{5}{7} \approx 0.7142857$. A horizontal line has been sketched through the steady-state population fraction in the following graph.

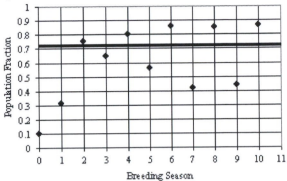

**(c)** For each breeding season calculate $p_m - p_{steady} = p_m - 0.714$. The difference between the population fraction and the steady-state population fraction starts to alternate between positive and negative and does not seem to be settling down.

| Breeding Season $m$ | Difference Between Population Fraction and the Steady-State Population Fraction |
|---|---|
| 0 | −0.614 |
| 1 | −0.399 |
| 2 | +0.041 |
| 3 | −0.067 |
| 4 | +0.085 |
| 5 | −0.153 |
| 6 | +0.148 |
| 7 | −0.297 |
| 8 | +0.137 |
| 9 | −0.270 |
| 10 | +0.150 |

**21. (a)** The land area of Nepal is 136,800 km². Convert to square meters:

$$\frac{136,800 \text{ km}^2}{1} \cdot \frac{1,000,000 \text{ m}^2}{1 \text{ km}^2} = 136,800,000,000 \text{ m}^2$$

Based on the square meters available in Nepal, the carrying capacity would be

$$\frac{136,800,000,000 \text{ m}^2}{1000 \text{ m}^2} = 136,800,000 \text{ people.}$$

The population fraction is $p_0 = \dfrac{P_0}{c} = \dfrac{23,000,000}{136,800,000} \approx 0.1681287$. The steady-state population

fraction is $p_{steady} = \dfrac{\tilde{r}-1}{\tilde{r}} = \dfrac{1.0225-1}{1.0225} = \dfrac{9}{409} \approx 0.0220049$.

**(b)** Calculate the population fraction for the first 10 years (assume a year is a breeding season). The growth parameter is $\tilde{r} = 1+r = 1+0.0225 = 1.0225$. Verhulst's equation, $p_{m+1} = \tilde{r}(1-p_m)p_m$, gives the population fraction after $m+1$ breeding seasons.

| Breeding Season $m$ | Population Fraction |
|---|---|
| 0 | $p_0 = \dfrac{P_0}{c} = \dfrac{23{,}000{,}000}{136{,}800{,}000} \approx 0.1681287$ |
| 1 | $p_1 = 1.0225(1-0.1681287)(0.1681287) \approx 0.1430083$ |
| 2 | $p_2 = 1.0225(1-0.1430083)(0.1430083) \approx 0.1253145$ |
| 3 | $p_3 = 1.0225(1-0.1253145)(0.1253145) \approx 0.1120770$ |
| 4 | $p_4 = 1.0225(1-0.1120770)(0.1120770) \approx 0.1017549$ |
| 5 | $p_5 = 1.0225(1-0.1017549)(0.1017549) \approx 0.0934574$ |
| 6 | $p_6 = 1.0225(1-0.0934574)(0.0934574) \approx 0.0866294$ |
| 7 | $p_7 = 1.0225(1-0.0866294)(0.0866294) \approx 0.0809051$ |
| 8 | $p_8 = 1.0225(1-0.0809051)(0.0809051) \approx 0.0760326$ |
| 9 | $p_9 = 1.0225(1-0.0760326)(0.0760326) \approx 0.0718323$ |
| 10 | $p_{10} = 1.0225(1-0.0718323)(0.0718323) \approx 0.0681726$ |

Verhulst's equation predicts that the population of Nepal will decrease each year for the next ten years.

**23. (a)** If 0.1% of the carrying capacity is 2000, then set up an equation and solve for the carrying capacity.

$$0.1\% \times c = 2000$$
$$0.001 \times c = 2000$$
$$c = \frac{2000}{0.001}$$
$$c = 2{,}000{,}000$$

The carrying capacity is 2,000,000 lice.
Calculate the population fraction for the first few breeding seasons. The growth parameter is $\tilde{r} = 1 + r = 1 + 3 = 4$. Verhulst's equation, $p_{m+1} = \tilde{r}(1 - p_m)p_m$, gives the population fraction after $m+1$ breeding seasons.

| Breeding Season $m$ | Population Fraction |
|---|---|
| 0 | $p_0 = \dfrac{P_0}{c} = \dfrac{2000}{2{,}000{,}000} = 0.001$ |
| 1 | $p_1 = 4(1-0.001)(0.001) = 0.003996$ |
| 2 | $p_2 = 4(1-0.003996)(0.003996) \approx 0.0159201$ |
| 3 | $p_3 = 4(1-0.0159201)(0.0159201) \approx 0.0626666$ |
| 4 | $p_4 = 4(1-0.0626666)(0.0626666) \approx 0.2349580$ |
| 5 | $p_5 = 4(1-0.2349580)(0.2349580) \approx 0.7190110$ |
| 6 | $p_6 = 4(1-0.7190110)(0.7190110) \approx 0.8081367$ |

In 6 breeding seasons the lice population will be at approximately 81% of the carrying capacity.

**(b)** The steady-state population fraction is $p_{steady} = \dfrac{\tilde{r}-1}{\tilde{r}} = \dfrac{4-1}{4} = \dfrac{3}{4} = 0.75$. A horizontal line has been sketched through the steady-state population fraction in the following graph.

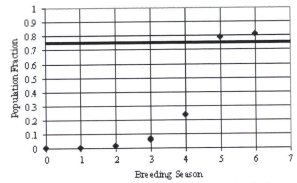

Breeding Season

**(c)** Find the steady-state population under the logistic growth law.

$$P_{steady} = \frac{r \times c}{1+r} = \frac{3 \times 2,000,000}{1+3} = \frac{6,000,000}{4} = 1,500,000$$

There would be 1,500,000 lice on a sheep if the population of lice reached the steady-state value. Some factors that could limit the size of the lice population on a sheep would be the size of the sheep, any medicine/treatment that the sheep is given, the weather, etc.

## Solutions to Chapter 12 Review Problems

**1.** The annual growth rate of 11.5% is constant year after year, so a Malthusian population model can be assumed.

**(a)** The population total in years 5, 9, and 17 can be found by using the formula for Malthusian population growth: $P_m = (1+r)^m \times P_0$. The initial population is given, so $P_0 = 850$. Substituting the known values into the formula, we have $P_m = (1+0.115)^m \times 850$.

| Time, in Years | Population Total |
|---|---|
| 5 | $P_5 = (1+0.115)^5 \times 850 = (1.115)^5 \times 850 \approx 1465$ |
| 9 | $P_9 = (1+0.115)^9 \times 850 = (1.115)^9 \times 850 \approx 2264$ |
| 17 | $P_{17} = (1+0.115)^{17} \times 850 = (1.115)^{17} \times 850 \approx 5409$ |

**(b)** Guess and test to find the number of years, $m$, it will take for $P_m = 20,000$.

$$P_m = (1+0.115)^m \times 850$$

$$20,000 = (1.115)^m \times 850$$

$$\frac{20,000}{850} = (1.115)^m$$

$$23.5294118 \approx (1.115)^m$$

It will take just over 29 years for the population to reach 20,000.

**2.** The annual growth of 80% is constant year after year, so a Malthusian population model can be assumed.

**(a)** The population totals for the next 15 years can be found by using the formula for Malthusian population growth: $P_m = (1+r)^m \times P_0$. Substitute the known values into the formula. Let $r = 0.8$ and $P_0 = 6500$.

| Year $m$ | Population Total |
|----------|------------------|
| 1 | $P_1 = (1.8)^1 \times 6500 = 11,700$ |
| 2 | $P_2 = (1.8)^2 \times 6500 = 21,060$ |
| 3 | $P_3 = (1.8)^3 \times 6500 = 37,908$ |
| 4 | $P_4 = (1.8)^4 \times 6500 \approx 68,234$ |
| 5 | $P_5 = (1.8)^5 \times 6500 \approx 122,822$ |
| 6 | $P_6 = (1.8)^6 \times 6500 \approx 221,079$ |
| 7 | $P_7 = (1.8)^7 \times 6500 \approx 397,943$ |
| 8 | $P_8 = (1.8)^8 \times 6500 \approx 716,297$ |
| 9 | $P_9 = (1.8)^9 \times 6500 \approx 1,289,335$ |
| 10 | $P_{10} = (1.8)^{10} \times 6500 \approx 2,320,804$ |
| 11 | $P_{11} = (1.8)^{11} \times 6500 \approx 4,177,447$ |
| 12 | $P_{12} = (1.8)^{12} \times 6500 \approx 7,519,404$ |
| 13 | $P_{13} = (1.8)^{13} \times 6500 \approx 13,534,927$ |
| 14 | $P_{14} = (1.8)^{14} \times 6500 \approx 24,362,869$ |
| 15 | $P_{15} = (1.8)^{15} \times 6500 \approx 43,853,164$ |

**Total Population**

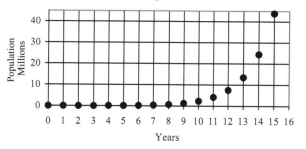

In year 15, there will be approximately 43,853,164 cats in a town with an area of 70 square miles. Calculate the number of square feet in the town:

$$\frac{70 \text{ mi}^2}{1} \cdot \frac{(5280)^2 \text{ ft}^2}{1 \text{ mi}^2} = 1,951,488,000 \text{ ft}^2$$

Calculate the number of square feet available per cat. Divide the number of square feet by the number of cats in year 15.

$$\frac{1,951,488,000 \text{ ft}^2}{43,853,164 \text{ cats}} \approx 44.5 \text{ ft}^2 \text{ per cat}$$

Therefore, each cat in year 15 could have approximately 44.5 square feet of space.

**(b)** Assuming there are limits to the growth of the population with a carrying capacity of 250,000, use the logistic growth law to calculate the population totals for the next 15 years. Let $P_0 = 6500$, $r = 0.8$, and $c = 250,000$.

Logistic Growth Law: $P_{m+1} = (1+r)P_m - \left(\dfrac{1+r}{c}\right)P_m^2$.

| Breeding Season $m$ | Population at the End of Breeding Season $m$ |
|---|---|
| 0 | $P_0 = 6500$ |
| 1 | $P_1 = (1.8)(6500) - \left(\dfrac{1.8}{250,000}\right)(6500)^2 \approx 11,396$ |
| 2 | $P_2 = (1.8)(11,396) - \left(\dfrac{1.8}{250,000}\right)(11,396)^2 \approx 19,578$ |
| 3 | $P_3 = (1.8)(19,578) - \left(\dfrac{1.8}{250,000}\right)(19,578)^2 \approx 32,481$ |
| 4 | $P_4 = (1.8)(32,481) - \left(\dfrac{1.8}{250,000}\right)(32,481)^2 \approx 50,870$ |
| 5 | $P_5 = (1.8)(50,870) - \left(\dfrac{1.8}{250,000}\right)(50,870)^2 \approx 72,934$ |
| 6 | $P_6 = (1.8)(72,934) - \left(\dfrac{1.8}{250,000}\right)(72,934)^2 \approx 92,982$ |
| 7 | $P_7 = (1.8)(92,982) - \left(\dfrac{1.8}{250,000}\right)(92,982)^2 \approx 105,119$ |
| 8 | $P_8 = (1.8)(105,119) - \left(\dfrac{1.8}{250,000}\right)(105,119)^2 \approx 109,654$ |
| 9 | $P_9 = (1.8)(109,654) - \left(\dfrac{1.8}{250,000}\right)(109,654)^2 \approx 110,804$ |
| 10 | $P_{10} = (1.8)(110,804) - \left(\dfrac{1.8}{250,000}\right)(110,804)^2 \approx 111,049$ |
| 11 | $P_{11} = (1.8)(111,049) - \left(\dfrac{1.8}{250,000}\right)(111,049)^2 \approx 111,099$ |
| 12 | $P_{12} = (1.8)(111,099) - \left(\dfrac{1.8}{250,000}\right)(111,099)^2 \approx 111,109$ |
| 13 | $P_{13} = (1.8)(111,109) - \left(\dfrac{1.8}{250,000}\right)(111,109)^2 \approx 111,111$ |
| 14 | $P_{14} = (1.8)(111,111) - \left(\dfrac{1.8}{250,000}\right)(111,111)^2 \approx 111,111$ |
| 15 | $P_{15} = (1.8)(111,111) - \left(\dfrac{1.8}{250,000}\right)(111,111)^2 \approx 111,111$ |

**Total Population**

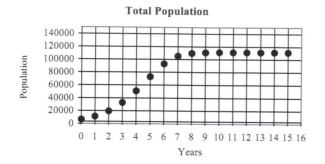

The steady-state population is $P_{steady} = \dfrac{r \times c}{1+r} = \dfrac{0.80 \times 250{,}000}{1+0.80} = \dfrac{200{,}000}{1.8} \approx 111{,}111$.

3. The growth is constant and is not limited, so assume a Malthusian population model.
   (a) Use the formula for the annual growth rate with $P = 250$, $Q = 473$, and $m = 4$.

   $$r = \left(\frac{Q}{P}\right)^{\frac{1}{m}} - 1$$

   $$r = \left(\frac{473}{250}\right)^{\frac{1}{4}} - 1$$

   $$r \approx 0.1728171$$

   The annual growth rate is approximately 0.1728 or 17.28%.

   (b) The initial population of 250 occurred in 2003. Find the population totals in 2010, 2020, and 2050 when $m = 7$, $m = 17$, and $m = 47$ respectively. Use the formula for Malthusian population growth:
   $P_m = (1+r)^m \times P_0$.

   | Year | $m$ | Population Total |
   |------|-----|------------------|
   | 2010 | 7 | $P_7 = (1.1728)^7 \times 250 \approx 763$ |
   | 2020 | 17 | $P_{17} = (1.1728)^{17} \times 250 \approx 3756$ |
   | 2050 | 47 | $P_{47} = (1.1728)^{47} \times 250 \approx 448{,}194$ |

4. The annual growth rate is a constant 3%. Use the formula for Malthusian population growth. If the initial population is 1800, then the population that is doubled is 3600.

   $$P_m = (1+r)^m \times P_o$$

   $$3600 = (1+0.03)^m \times 1800$$

   $$2 = (1.03)^m$$

   Guess and test values for $m$ until you find one that produces a true equation. It would take between 23 and 24 years.

5. (a) In 6 months each investor will get paid $1600.
   (b) The investors are guaranteed a 60% rate of return so we know that the number of investors in the each half year must be at least 1.60 times the number of investors in the previous half year. Use the relationship $S_{m+1} \geq (1.6)^m \times S_1$ where $S_{m+1}$ represents the $(m+1)^{st}$ half year and $S_1 = 20$. Find the number of investors required in each half year time period 2 through 8. That is, find $S_2$, $S_3$, $S_4$, $S_5$, $S_6$, $S_7$, and $S_8$.

$S_2: \quad S_{1+1} \geq (1.6)^1 \times 20 = 32$

$S_3: \quad S_{2+1} \geq (1.6)^2 \times 20 = 51.2 \approx 52$

$S_4: \quad S_{3+1} \geq (1.6)^3 \times 20 = 81.92 \approx 82$

$S_5: \quad S_{4+1} \geq (1.6)^4 \times 20 = 131.072 \approx 132$

$S_6: \quad S_{5+1} \geq (1.6)^5 \times 20 = 209.7152 \approx 210$

$S_7: \quad S_{6+1} \geq (1.6)^6 \times 20 = 335.54432 \approx 336$

$S_8: \quad S_{7+1} \geq (1.6)^7 \times 20 = 536.870912 \approx 537$

The minimum number of people needed to invest in each of the next seven half years is 32, 52, 82, 132, 210, 336, and 537 respectively.

(c) In the 8th time period, there would be 537 new investors. They would have each invested $1000 for a total of 537× $1000 = $537,000. Each investor expects a 60% return, so the payout would be the original $537,000 plus 60% interest which means the payout would be $537,000 + $537,000 × 0.60 = $859,200.

6. (a) The price to participate is $P$, so $P$ = $10. The number of levels is $m$, so $m = 8$. The number of new participants for each letter is $n$, so $n = 5$.

   (b) The payoff in rising from the bottom of the list to the top of the list is
   $$P \times n^m = \$10 \times 5^8 = \$10 \times 390,625 = \$3,906,250.$$

   (c) In order for you to rise from the bottom of the list to the top of the list and earn the promised amount of money, many new people must join. The total number of people who must join is $\dfrac{n^{m+1}-1}{n-1} = \dfrac{5^{8+1}-1}{5-1} = \dfrac{5^9-1}{4} = \dfrac{1,953,124}{4} = 488,281.$

7. Use the radioactive decay formula. The decay rate is 3.5%, so $d = 0.035$. We know 85 grams of the substance was initially present, so $A_0 = 85$.

   $A_m = (1-d)^m \times A_0$

   $A_m = (1-0.035)^m \times 85$

   $A_m = (0.965)^m \times 85$

   After 5 years, 14 years, 50 years, and 100 years, $m = 5$, $m = 14$, $m = 50$, and $m = 100$ respectively.

   | After 5 years: | After 14 years: | After 50 years: | After 100 years: |
   |---|---|---|---|
   | $A_5 = (0.965)^5 \times 85$ | $A_{14} = (0.965)^{14} \times 85$ | $A_{50} = (0.965)^{50} \times 85$ | $A_{100} = (0.965)^{100} \times 85$ |
   | $A_5 \approx 71.13$ | $A_{14} \approx 51.62$ | $A_{50} \approx 14.31$ | $A_{100} \approx 2.41$ |

   After 5 years, there would be approximately 71.13 grams remaining. After 14 years, there would be approximately 51.62 grams remaining. After 50 years, there would be approximately 14.31 grams remaining. After 100 years, there would be approximately 2.41 grams remaining.

8. Use the decay rate formula. The half-life is 2.5 years, so $h = 2.5$.

   $d = 1 - \left(\dfrac{1}{2}\right)^{\frac{1}{h}}$

   $d = 1 - \left(\dfrac{1}{2}\right)^{\frac{1}{2.5}}$

   $d \approx 0.2421417$

   The annual decay rate of the radioactive substance is approximately 24.21%

9. **(a)** Use the half-life radioactive decay formula. In this case $A_0 = 500$ grams and $h = 13$ years.

$$A_m = \left(\frac{1}{2}\right)^{\frac{m}{h}} \times A_0$$

$$A_{22} = \left(\frac{1}{2}\right)^{\frac{22}{13}} \times 500$$

$$A_{22} \approx 154.72$$

There will be approximately 154.72 grams of Plutonium-241 remaining.

**(b)** Find $A_0$ when $m = 17$ and $A_m = 7.63$ grams.

$$A_m = \left(\frac{1}{2}\right)^{\frac{m}{h}} \times A_0$$

$$7.63 = \left(\frac{1}{2}\right)^{\frac{17}{13}} \times A_0$$

$$\frac{7.63}{\left(\frac{1}{2}\right)^{\frac{17}{13}}} = A_0$$

$$18.89 \approx A_0$$

Initially there were approximately 18.9 grams of Plutonium-241.

10. **(a)** Use the half-life approximation formula: $h = \dfrac{0.693}{\left(d + \dfrac{d^2}{2}\right)}$. The decay rate is given, so $d = 0.12$.

$$h = \frac{0.693}{\left(d + \dfrac{d^2}{2}\right)} = \frac{0.693}{\left(0.12 + \dfrac{0.12^2}{2}\right)} = \frac{0.693}{(0.12 + 0.0072)} = \frac{0.693}{0.1272} \approx 5.45$$

Therefore, the approximate half-life of the radioactive substance is 5.45 years.

**(b)** Use the radioactive decay formula. The decay rate is 12%, so $d = 0.12$. Initially there are $A_0$ grams present, so find $m$ such that 10% of the initial amount remains. That is, find $m$ so that $A_m = 0.10A_0$. Use guess and test.

$$A_m = (1-d)^m \times A_0$$

$$0.10A_0 = (1-0.12)^m \times A_0$$

$$\frac{0.10A_0}{A_0} = (0.88)^m$$

$$0.10 = 0.88^m$$

It will take between 18 and 19 years.

11. The bone contains 80% of its original amount of $^{14}C$. From Table 12.6, if 80% remains, then the age of the bone in half-lives is 0.32. Since the half-life of $^{14}C$ is 5730 years, the age of the fossil is $0.32 \times 5730 \approx 1833.6$ years.

12. Use the logistic growth law with $P_0 = 20,000$, $r = 0.04$, and $c = 400,000$.

Logistic Growth Law: $P_{m+1} = (1+r)P_m - \left(\dfrac{1+r}{c}\right)P_m^2$

| Year $m$ | Population at the End of Year $m$ |
|---|---|
| 1 | $P_1 = (1.04)(20,000) - \left(\dfrac{1.04}{400,000}\right)(20,000)^2 = 19,760$ |
| 2 | $P_2 = (1.04)(19,760) - \left(\dfrac{1.04}{400,000}\right)(19,760)^2 \approx 19,535$ |
| 3 | $P_3 = (1.04)(19,535) - \left(\dfrac{1.04}{400,000}\right)(19,535)^2 \approx 19,324$ |
| 4 | $P_4 = (1.04)(19,324) - \left(\dfrac{1.04}{400,000}\right)(19,324)^2 \approx 19,126$ |
| 5 | $P_5 = (1.04)(19,126) - \left(\dfrac{1.04}{400,000}\right)(19,126)^2 \approx 18,940$ |

13. (a) Use the logistic growth law with $P_0 = 28,000$, $r = 0.0725$, and $c = 250,000$.

Logistic Growth Law: $P_{m+1} = (1+r)P_m - \left(\dfrac{1+r}{c}\right)P_m^2$

| Breeding Season $m$ | Population at the End of Breeding Season $m$ |
|---|---|
| 0 | $P_0 = 28,000$ |
| 1 | $P_1 = (1.0725)(28,000) - \left(\dfrac{1.0725}{250,000}\right)(28,000)^2 \approx 26,667$ |
| 2 | $P_2 = (1.0725)(26,667) - \left(\dfrac{1.0725}{250,000}\right)(26,667)^2 \approx 25,550$ |
| 3 | $P_3 = (1.0725)(25,550) - \left(\dfrac{1.0725}{250,000}\right)(25,550)^2 \approx 24,602$ |
| 4 | $P_4 = (1.0725)(24,602) - \left(\dfrac{1.0725}{250,000}\right)(24,602)^2 \approx 23,789$ |
| 5 | $P_5 = (1.0725)(23,789) - \left(\dfrac{1.0725}{250,000}\right)(23,789)^2 \approx 23,086$ |
| 6 | $P_6 = (1.0725)(23,086) - \left(\dfrac{1.0725}{250,000}\right)(23,086)^2 \approx 22,473$ |
| 7 | $P_7 = (1.0725)(22,473) - \left(\dfrac{1.0725}{250,000}\right)(22,473)^2 \approx 21,936$ |
| 8 | $P_8 = (1.0725)(21,936) - \left(\dfrac{1.0725}{250,000}\right)(21,936)^2 \approx 21,462$ |
| 9 | $P_9 = (1.0725)(21,462) - \left(\dfrac{1.0725}{250,000}\right)(21,462)^2 \approx 21,042$ |
| 10 | $P_{10} = (1.0725)(21,042) - \left(\dfrac{1.0725}{250,000}\right)(21,042)^2 \approx 20,668$ |

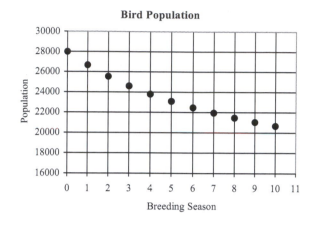

**(b)** The steady-state population is $P_{steady} = \dfrac{r \times c}{1+r} = \dfrac{0.0725 \times 250,000}{1+0.0725} = \dfrac{18,125}{1.0725} \approx 16,900$.

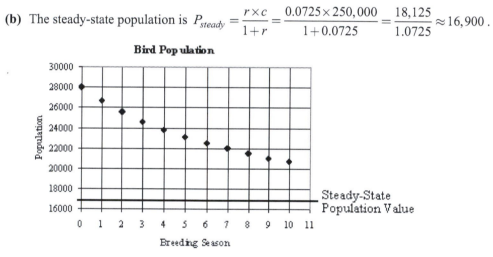

14. Calculate the population fraction for the first 10 breeding seasons. The growth parameter is $\tilde{r} = 1 + r = 1 + 0.85 = 1.85$. Verhulst's equation, $p_{m+1} = \tilde{r}(1 - p_m)p_m$, gives the population fraction after $m+1$ breeding seasons.

| Breeding Season $m$ | Population Fraction |
|---|---|
| 0 | $p_0 = \dfrac{P_0}{c} = \dfrac{450}{3000} = 0.15$ |
| 1 | $p_1 = 1.85(1 - 0.15)(0.15) = 0.235875$ |
| 2 | $p_2 = 1.85(1 - 0.235875)(0.235875) \approx 0.3334403$ |
| 3 | $p_3 = 1.85(1 - 0.3334403)(0.3334403) \approx 0.4111771$ |
| 4 | $p_4 = 1.85(1 - 0.4111771)(0.4111771) \approx 0.4479044$ |
| 5 | $p_5 = 1.85(1 - 0.4479044)(0.4479044) \approx 0.4574792$ |
| 6 | $p_6 = 1.85(1 - 0.4574792)(0.4574792) \approx 0.4591552$ |
| 7 | $p_7 = 1.85(1 - 0.4591552)(0.4591552) \approx 0.4594136$ |
| 8 | $p_8 = 1.85(1 - 0.4594136)(0.4594136) \approx 0.4594526$ |
| 9 | $p_9 = 1.85(1 - 0.4594526)(0.4594526) \approx 0.4594584$ |
| 10 | $p_{10} = 1.85(1 - 0.4594584)(0.4594584) \approx 0.4594593$ |

The steady-state population fraction is $p_{steady} = \dfrac{\tilde{r}-1}{\tilde{r}} = \dfrac{1.85-1}{1.85} = \dfrac{0.85}{1.85} = \dfrac{17}{37} \approx 0.4594595.$ Over time this population will level off to approximately 45.9% of the carrying capacity.

15. (a) Calculate the population fraction for the next 8 breeding seasons. The growth parameter is $\tilde{r} = 3.7$. Verhulst's equation, $p_{m+1} = \tilde{r}(1-p_m)p_m$, gives the population fraction after $m+1$ breeding seasons.

| Breeding Season $m$ | Population Fraction |
|---|---|
| 0 | $p_0 = 0.2$ |
| 1 | $p_1 = 3.7(1-0.2)(0.2) = 0.592$ |
| 2 | $p_2 = 3.7(1-0.592)(0.592) = 0.8936832$ |
| 3 | $p_3 = 3.7(1-0.8936832)(0.8936832) \approx 0.3515501$ |
| 4 | $p_4 = 3.7(1-0.3515501)(0.3515501) \approx 0.8434617$ |
| 5 | $p_5 = 3.7(1-0.8434617)(0.8434617) \approx 0.4885260$ |
| 6 | $p_6 = 3.7(1-0.4885260)(0.4885260) \approx 0.9245129$ |
| 7 | $p_7 = 3.7(1-0.9245129)(0.9245129) \approx 0.2582186$ |
| 8 | $p_8 = 3.7(1-0.2582186)(0.2582186) \approx 0.7087045$ |

(b) The steady-state population fraction is $p_{steady} = \dfrac{\tilde{r}-1}{\tilde{r}} = \dfrac{3.7-1}{3.7} = \dfrac{27}{37} \approx 0.7297297.$ For each breeding season calculate $p_n - p_{steady} \approx p_n - 0.7297297$. After the first breeding season, the population fraction bounces above and below the steady-state population fraction value, and at the same time, the absolute difference between the population fraction and the steady-state population fraction becomes closer to the steady-state value, then moves farther away, then moves closer again. It does not seem to settle down.

| Breeding Season $m$ | Difference Between Population Fraction and the Steady-State Population Fraction |
|---|---|
| 0 | −0.530 |
| 1 | −0.138 |
| 2 | +0.164 |
| 3 | −0.378 |
| 4 | +0.114 |
| 5 | −0.241 |
| 6 | +0.195 |
| 7 | −0.472 |
| 8 | −0.021 |

16. Calculate the population fraction for the next 4 breeding seasons for both populations. For the Green Turtles $\tilde{r} = 1.5$, $P_0 = 37$, and $c = 3000$. For the Hawksbill Sea Turtles $\tilde{r} = 1.3$, $P_0 = 56$, and $c = 3500$. Use Verhulst's equation, $p_{m+1} = \tilde{r}(1-p_m)p_m$, which gives the population fraction after $m+1$ breeding seasons.

| Breeding Season $m$ | Green Turtle Population Fraction |
|---|---|
| 0 | $p_0 = \dfrac{P_0}{c} = \dfrac{37}{3000} \approx 0.0123333$ |
| 1 | $p_1 = 1.5(1 - 0.0123333)(0.0123333) \approx 0.0182718$ |
| 2 | $p_2 = 1.5(1 - 0.0182718)(0.0182718) \approx 0.0269069$ |
| 3 | $p_3 = 1.5(1 - 0.0269069)(0.0269069) \approx 0.0392744$ |
| 4 | $p_4 = 1.5(1 - 0.0392744)(0.0392744) \approx 0.0565979$ |

| Breeding Season $m$ | Hawksbill Sea Turtle Population Fraction |
|---|---|
| 0 | $p_0 = \dfrac{P_0}{c} = \dfrac{56}{3500} = 0.016$ |
| 1 | $p_1 = 1.3(1 - 0.016)(0.016) = 0.0204672$ |
| 2 | $p_2 = 1.3(1 - 0.0204672)(0.0204672) \approx 0.0260628$ |
| 3 | $p_3 = 1.3(1 - 0.0260628)(0.0260628) \approx 0.0329986$ |
| 4 | $p_4 = 1.3(1 - 0.0329986)(0.0329986) \approx 0.0414826$ |

▲ **Green Turtles**   ■ **Hawskbill Sea Turtles**

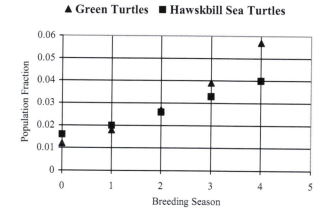

After 4 breeding seasons, the Green Turtle population is at approximately 5.7% of carrying capacity while the Hawksbill Sea Turtle population is at approximately 4.1% of carrying capacity.

# *Chapter 13*: Consumer Mathematics – Buying and Saving

## *Key Concepts*

**By the time you finish studying Section 13.1, you should be able to:**

☐ calculate simple interest (exact and ordinary).

☐ calculate the future value for an account that earns simple interest.

☐ calculate the future value for an account that earns compound interest.

☐ determine the principal needed to achieve a certain future value.

☐ calculate the effective annual rate.

☐ calculate the future value for a continuously compounded interest account.

**By the time you finish studying Section 13.2, you should be able to:**

☐ calculate the finance charge, average daily balance, daily interest rate, and new balance for a simple interest loan.

☐ calculate the monthly payment for an amortized loan using an amortization table or a monthly payment formula.

☐ chart the history of an amortized loan.

☐ calculate the interest rate for a rent-to-own transaction.

**By the time you finish studying Section 13.3, you should be able to:**

☐ use affordability guidelines to determine the maximum home price and maximum monthly housing expenses a borrower can afford.

☐ calculate high and low estimates of how much a borrower can afford to pay for monthly housing expenses.

☐ calculate the cost of the loan origination fee and the discount charge associated with a mortgage loan.

☐ calculate the annual percentage rate for a loan.

☐ calculate the down payment.

☐ determine the maximum affordable home price based on the down payment.

☐ calculate the monthly mortgage payment for a home loan given the amount financed, the interest rate, and the term of the loan using an amortization table or a monthly payment formula.

## Skills Brush-Ups

Problems in Chapter 13 will require that you be comfortable working with percents, rounding numbers to the nearest hundredth, evaluating expressions involving exponents by correctly applying rules for order of operations, and using the number $e$.

### PERCENTS:

Percent means "per 100". Therefore, 10% means 10 per 100 or 10 out of 100.
To convert a number from a percent to a decimal, divide the number by 100 and eliminate the percent symbol. In general, you can drop the percent symbol and move the decimal point two places to the left.

*Examples:* (a) $8\% = \dfrac{8}{100} = 0.08$ (b) $200\% = \dfrac{200}{100} = 2$ (c) $0.06\% = \dfrac{0.06}{100} = 0.0006$

To convert a number to a percent, multiply the number by 100%. In general, you can shift the decimal point two places to the right and add on a % symbol.

*Example:* 0.0359 is $0.0359 \times 100\% = 3.59\%$

### ROUNDING:

You will encounter many problems in this chapter for which you will round dollar amounts to the nearest cent. This is equivalent to rounding to the nearest hundredth. When rounding a number to the nearest hundredth, consider the number in the thousandths place. If it is 5 or greater, round the digit in the hundredths place up and discard the remaining digits. Otherwise, leave the digit in the hundredths place alone and discard the remaining digits.

*Examples:* (a) To round 5.379 to the nearest hundredth, we round the 7 up to an 8 since the number in the thousandths place is a 9. The result is 5.38.
(b) To round 16.024 to the nearest hundredth, we leave the 2 alone since the number in the thousandths place is less than 5. The result is 16.02.

**Note:** The only time in this chapter you will not round in this way is when calculating monthly payment values for amortized loans. In those cases, you will *always* round up.

## ORDER OF OPERATIONS:

Many calculations in this chapter involve exponents and complicated expressions with parentheses and fractions. Correctly evaluating these expressions requires you to remember the order of operations. Simplify what is in parentheses first, evaluate exponents, evaluate multiplications and divisions in order from left to right, and then evaluate additions and subtractions in order from left to right.

*Example:* To evaluate $500\left(1+\dfrac{0.2}{4}\right)^{4\times5}$, simplify inside the parentheses. This requires that you divide

0.2 by 4 first and then add 1. Simplify the exponent, raise the number in the parentheses to a power of 20, and then multiply by 500. As you are working, let your calculator keep all the digits even if you do not write all the digits down. Round your final answer to the nearest hundredth in the last step as shown next.

$$500\left(1+\frac{0.2}{4}\right)^{4\times5} = 500(1.05)^{20}$$
$$\approx 500(2.65329770514)$$
$$\approx 1326.65$$

Note: The $\approx$ symbol indicates that the number is not the exact answer and has been rounded.

## THE NUMBER e:

The number $e$ is a real number. It is also an irrational number which means it cannot be written as a fraction. The decimal expansion never terminates and never repeats. The number $e$ has many real life applications. It occurs in situations in which a quantity increases or decreases at a rate proportional to its value. You will use $e$ in Chapter 13 when you deal with continuously compounded interest and radioactive growth and decay. The value of $e$ is approximately 2.71828182846. On your calculator look for the $e^x$ button which will allow you to find any power of $e$.

*Examples:* (a) $e^2 \approx 7.389056$ (b) $e^{0.03\times5} = e^{0.15} \approx 1.161834$

## Skills Practice

1. Convert each of the following from percent notation to decimal notation.
   (a) 5% (b) 11.2% (c) 0.25% (d) 7.98%

2. Convert each of the following from decimal notation to percent notation.
   (a) 0.03 (b) 0.0975 (c) 0.0045 (d) 0.125

3. (a) What is 16% of 2427? (b) 4.5 is what percent of 15?

4. Round each of the following to the nearest hundredth.
   (a) 3.4251 (b) 10.0914 (c) 286.996 (d) 10,795.0308

5. Simplify each of the following. Round to the nearest hundredth.

   (a) $650(1.0385)^5$ (b) $950(1+0.089)^{4\times5}$ (c) $35\left(1+\dfrac{0.03}{4}\right)^{4\times6}$ (d) $10,350e^{0.045\times7}$

*Answers to Skills Practice:*
1. (a) 0.05 (b) 0.112 (c) 0.0025 (d) 0.0798
2. (a) 3% (b) 9.75% (c) 0.45% (d) 12.5%
3. (a) 388.32 (b) 30%
4. (a) 3.43 (b) 10.09 (c) 287.00 (d) 10,795.03
5. (a) 785.14 (b) 5227.35 (c) 41.87 (d) 14,182.18

# Hints for Odd-Numbered Problems

## Section 13.1

1. Use the simple interest formula.

3. Ordinary interest treats each month as if it is $\frac{1}{12}$ of a year.

5. (a) The year 2004 was a leap year.

7. The monthly payment will be the simple interest calculated for one month based on the simple interest formula.

9. Compare the expression to the compound interest formula.

11. Use the future value formula for simple interest. The future value after a compounding period becomes the principal at the beginning of the next compounding period.

13. Which future value formula is used for simple interest and which is used for compounded interest?

15. Because the bond was not held for at least five years, there is a three-month interest penalty. Subtract the last three months' interest from the value of the bond at time it is redeemed.

17. The bond was purchased for half of the face value. Interest on the bond was compounded three times in the year and a half the bond was held. Consult the table to determine the correct interest rate for each compounding period. Do not forget about the three-month interest penalty.

19. Pick a convenient dollar amount to use as the principal and calculate the future value in each case.

21. Use the appropriate future value formula and solve for the unknown principal value.

23. Use the compound interest formula. Time is unknown. Simplify the expression as much as possible and guess and test until you find a value for time that makes the equation true.

25. (a) Use the future value formula for simple interest.
    (b) Use the compound interest formula.
    (c) The principal value is $1500.

27. (a) Use the compound interest formula with a principal of $7700.
    (b) Find the future value of the investment.

29. After one year, what percentage of the purchasing power would the money retain after taking inflation into account?

31. If interest is compounded every two months, then there are six compounding periods each year.

33. (a) The effective annual rate is located in the intersection of the row labeled "8" and the column labeled "Daily".
    (b) Use the effective annual rate formula for continually compounded interest.

35. Use the annual percentage yield formula.

37. (a) The maturity value of an investment is the future value.

## Section 13.2

1. To calculate the finance charge, use the average daily balance, the daily percentage rate, and the number of days in the billing period.

3. The new balance is the sum of the previous balance, all new purchases, and the finance charge minus any payments.

5. The new balance is the sum of the previous balance, all new purchases, and the finance charge minus any payments.

7. Create a table to show how many days each balance was in effect.

9. Create a table to show how many days each balance was in effect.

11. (a) The interest is calculated for one month, so $t = \dfrac{1}{12}$.

    (b) The payment to the principal does not include the monthly interest.

13. Each month the principal is reduced by the payment to the principal. The reduced value is the new balance. Interest for the next month is calculated on the new balance.

15. (a) Use the monthly payment formula. Round the final value up.

    (b) Use the monthly payment from part (a) for each payment in the amortization schedule.

    (c) Often the final payment is slightly different from the monthly payment. The monthly payments were the same for how many months?

17. (a) Payments can be calculated using the simple interest formula.

    (b) Interest that capitalizes is compounded, in this case quarterly. Use the compound interest formula. The time period must be expressed in years. What fraction of a year is 10 months?

19. The final payment is the sum of the new balance from the previous month and the interest due on that balance.

21. Be sure to round the monthly payment up.

23. Be sure to round the monthly payment up.

25. Be sure to round the monthly payment up.

27. How much will be financed? Express the time in years.

29. What is 20% of $16,285? How much will Justin finance?

31. Treat the debt like the principal for an amortized loan. Express the debt in 1000s and use an amortization table to find the monthly payment.

33. An amortization table may be used. Find the table value given the rate and term. Use the fact that the Monthly Payment = (Amortization Table Entry)(Loan Amount in 1000s) and solve for the loan amount in 1000s.

35. An amortization table may be used. Find the table value given the rate and term. Use the fact that the Monthly Payment = (Amortization Table Entry)(Loan Amount in 1000s) and solve for the loan amount in 1000s.

37. Use the fact that the Monthly Payment = (Amortization Table Entry)(Loan Amount in 1000s), solve for the amortization table entry and determine the interest rate.

39. Given that the term is 15 years, locate the two rates between which the amortization table value falls.

41. (a) The suggested retail value is $1250. Any extra amount paid is the interest.

43. What total amount is paid to the rent-to-own store?

45. (a) Use an amortization table to determine the monthly payment.

    (b) Calculate and compare the total amount she will pay to the rent-to-own store and the future value of the loan.

## Section 13.3

1. The maximum home price and the maximum monthly expenses a family can afford is related to the family's gross annual income.

3. The maximum home price and the maximum monthly expenses Dana can afford is related to Dana's annual gross income.

5. Add their gross annual incomes before applying the affordability guidelines.

7. The low estimate is 25% of the monthly gross income while the high estimate is 38% of the monthly gross income.

9. Each point costs 1% of the principal amount of the loan.

11. Each point costs 1% of the principal amount of the loan.

13. Write an equation and solve.
$23,000 = 20% × (Purchase price).

15. Write an equation and solve.
$18,000 = 20% × (Purchase price).

17. Subtract the $2500 they will need for closing costs from the $18,000 they have available. Write an equation.

**19, 21, and 23.** How much is the down payment? Assume closing costs are paid separately. Subtract the down payment to determine the amount to be financed. Do not forget to round the monthly payment value up.

25. The steps in this problem guide you to find the annual percentage rate. See example 13.24.

27. Carry out the same steps as are given in problem 25 to determine the APR for the loan.

29. To find the taxes, convert the percent to a decimal and multiply by the assessed value. Add the taxes and insurance then divide by 12 to determine the amount to add to each monthly payment.

31. To find the taxes, convert the percent to a decimal and multiply by the assessed value. Add the taxes and insurance then divide by 12 to determine the amount to add to each monthly payment.

## Solutions to Odd-Numbered Problems in Section 13.1

1. Use the simple interest formula: $I = Prt$. Round your answers to the nearest penny.
   (a) $I = \$600(0.07)(3) = \$126.00$
   (b) $I = \$400(0.115)(5) = \$230.00$
   (c) $I = \$1235(0.0325)(8) = \$321.10$

3. When calculating ordinary interest, remember that each month is $\frac{1}{12}$ of a year. Round your answers to the nearest penny.
   (a) $I = \$800(0.0225)\left(\dfrac{4}{12}\right) = \$6.00$

   (b) $I = \$1400(0.054)\left(\dfrac{30}{12}\right) = \$189.00$

   (c) $I = \$11,500(0.0725)\left(\dfrac{26}{12}\right) \approx \$1806.46$

5. Use the future value formula for Simple Interest: $F = P(1+rt)$. Count the number of days in each term and divide by 365.
   (a) There are 3 years from January 1, 2002 to January 1, 2005 including one leap year for a total of $365 + 366 + 365 = 1096$ days. From January 1, 2005 to June 30, 2005, there are $(31 - 1) + 28 + 31 + 30 + 31 + 30 = 180$ days for a total of 1276 days.
   $$F = \$18,000\left(1 + 0.0235\left(\frac{1276}{365}\right)\right) \approx \$19,478.76.$$

   (b) There is one nonleap year from March 3, 2000 to March 3, 2001 for a total of 365 days. From March 3, 2001 to December 3, 2001 there are $(31 - 3) + 30 + 31 + 30 + 31 + 31 + 30 + 31 + 30 + 3 = 275$ days for a total of 640 days.
   $$F = \$18,000\left(1 + 0.0235\left(\frac{640}{365}\right)\right) \approx \$18,741.70.$$

**(c)** $F = \$18{,}000\left(1+0.0235\left(\dfrac{85}{365}\right)\right) \approx \$18{,}098.51.$

7. Use the simple interest formula to find the monthly interest payment in each case. To find the total interest during the interest-only payment period, multiply the monthly payment by the number of months in the interest-only period.

    **3 year Interest-Only Period:**

    Monthly Payment $= \$235{,}000(0.04375)\left(\dfrac{1}{12}\right) \approx \$856.77.$

    Total Interest $= \$856.77(36) = \$30{,}843.72.$

    **5-year Interest-Only Period:**

    Monthly Payment $= \$235{,}000(0.04625)\left(\dfrac{1}{12}\right) \approx \$905.73.$

    Total Interest $= \$905.73(60) = \$54{,}343.80.$

    **7-year Interest-Only Period:**

    Monthly Payment $= \$235{,}000(0.0525)\left(\dfrac{1}{12}\right) \approx \$1028.13.$

    Total Interest $= \$1028.13(84) = \$86{,}362.92.$

9. Compare the given expression to the compound interest formula.

    Compare the expression $F = 3420\left(1+\dfrac{0.025}{4}\right)^{(4)(3)}$ to the formula $F = P\left(1+\dfrac{r}{m}\right)^{mt}$.

    **(a)** $\$3420$
    **(b)** $2.5\%$
    **(c)** $4$
    **(d)** $\$3685.50$

11. Calculate the future value for simple interest on the balance in the account at the start of each compounding period. Each compounding period is three months long, so $t = \dfrac{3}{12} = 0.25$ years. There are four compounding periods in one year. The balance or future value at the end of a compounding period becomes the principal at the beginning of the next compounding period.

    First compounding period: $\quad F = \$1500(1+(0.12)(0.25)) = \$1545.00$

    Second compounding period: $\quad F = \$1545(1+(0.12)(0.25)) = \$1591.35$

    Third compounding period: $\quad F = \$1591.35(1+(0.12)(0.25)) \approx \$1639.09$

    Fourth compounding period: $\quad F = \$1639.09(1+(0.12)(0.25)) \approx \$1688.26$

    Therefore the future value after one year is $\$1688.26$.

13. **(a)** When finding the future value for a simple interest account, use $F = P(1+rt)$. When interest is compounded, use $F = P\left(1+\dfrac{r}{m}\right)^{mt}$. In each case, $P = \$750$, $r = 0.03$, and $t = 5$ years.

| Interest | Future Value | Future Value − Principal = Interest |
|---|---|---|
| Simple | $F = \$750(1+(0.03)(5)) = \$862.50$ | $\$862.50 - \$750.00 = \$112.50$ |
| Compounded annually: $m = 1$ | $F = \$750\left(1+\dfrac{0.03}{1}\right)^{1\times5} \approx \$869.46$ | $\$869.46 - \$750.00 = \$119.46$ |
| Compounded semiannually: $m = 2$ | $F = \$750\left(1+\dfrac{0.03}{2}\right)^{2\times5} \approx \$870.41$ | $\$870.41 - \$750.00 = \$120.41$ |
| Compounded quarterly: $m = 4$ | $F = \$750\left(1+\dfrac{0.03}{4}\right)^{4\times5} \approx \$870.89$ | $\$870.89 - \$750.00 = \$120.89$ |
| Compounded monthly: $m = 12$ | $F = \$750\left(1+\dfrac{0.03}{12}\right)^{12\times5} \approx \$871.21$ | $\$871.21 - \$750.00 = \$121.21$ |
| Compounded daily: $m = 365$ | $F = \$750\left(1+\dfrac{0.03}{365}\right)^{365\times5} \approx \$871.37$ | $\$871.37 - \$750.00 = \$121.37$ |

**(b)** Any savings account earning compounded interest exhibits geometric growth. In each compound period, the account balance is multiplied by the same amount. The simple interest account is the only account that does not exhibit geometric growth.

**15.** The bond was purchased for $2500 which was half of the face value. On May 1, 2000, the annual interest rate was 5.73%. Interest for Series EE U.S. Savings Bonds is compounded semiannually. Use the compound interest formula with $P = \$2500$, $r = 0.0573$, $m = 2$, and $t = \dfrac{1}{2}$. After six months, the bond was valued at $F = \$2500\left(1+\dfrac{0.0573}{2}\right)^{2\times\frac{1}{2}} = \$2571.625$.

Because the bond was redeemed before 5 years, there was a three month interest penalty. To calculate the penalty, determine the interest from the most recent three month period and subtract it from the current value of the bond.

In the last six months, the interest earned was $2571.625 − $2500 = $71.625, so the three month penalty is half of that value or $71.625 ÷ 2 = $35.8125. Therefore, the value of the bond was $2571.625 − $35.8125 ≈ $2535.81.

**17.** The $200 bond was purchased for $100. It sat in a drawer for a year and a half so interest was compounded three times. Because interest rates change every six months, a different interest rate was used each time interest was compounded. Consider each compounding period and keep as many digits as possible for subsequent calculations to avoid rounding error.

November 1, 1997 to May 1, 1998: $F = \$100\left(1+\dfrac{0.0559}{2}\right)^{2\times\frac{1}{2}} = \$102.795$

May 1, 1998 to November 1, 1998: $F = \$102.795\left(1+\dfrac{0.0506}{2}\right)^{2\times\frac{1}{2}} = \$105.3957135$

November 1, 1998 to May 1, 1999: $F = \$105.3957135\left(1 + \dfrac{0.0460}{2}\right)^{2 \times \frac{1}{2}} \approx \$107.8198149$

On May 1, 1999, the bond had a value of approximately $107.82, but because the bond had not been held for at least five years, there was a three-month interest penalty. In the last six months, the interest earned was $107.8198149 − $105.3957135 = $2.4241014, so the three-month penalty is half of that value or $2.4241014 ÷ 2 = $1.2120507. Therefore, the value of the bond was $107.8198149 − $1.2120507 ≈ $106.61.

19. No principal amount was given, so any convenient amount could be used. Suppose $1000 was invested. Find the future value in each case and determine which account yields the largest amount.

5% Compounded Annually: $F = \$1000\left(1 + \dfrac{0.05}{1}\right)^{1 \times 3} \approx \$1157.63$

4.95% Compounded Semiannually: $F = \$1000\left(1 + \dfrac{0.0495}{2}\right)^{2 \times 3} \approx \$1158.00$

4.9% Compounded Monthly: $F = \$1000\left(1 + \dfrac{0.049}{12}\right)^{12 \times 3} \approx \$1158.01$

The account offering 4.9% compounded monthly is the best deal.

21. Find the principal, $P$, in each case. The future value is given.
   (a) Use the future value formula for simple interest and solve for $P$.
   $$F = P(1 + rt)$$
   $$\$50,000 = P(1 + (0.0425)(20))$$
   $$\$50,000 = P(1.85)$$
   $$\dfrac{\$50,000}{1.85} = P$$
   $$\$27,027.03 \approx P$$
   Therefore, $27,027.03 would have to be invested.
   (b) Use the compound interest formula and solve for $P$.
   $$F = P\left(1 + \dfrac{r}{m}\right)^{mt}$$
   $$\$50,000 = P\left(1 + \dfrac{0.0425}{2}\right)^{2 \times 20}$$
   $$\$50,000 = P(1.02125)^{40}$$
   $$\dfrac{\$50,000}{(1.02125)^{40}} = P$$
   $$\$21,561.91 \approx P$$
   Therefore, $21,561.91 would have to be invested.

(c) Use the future value formula for continuously compounded interest and solve for $P$.

$$F = Pe^{rt}$$

$$\$50,000 = Pe^{0.0425 \times 20}$$

$$\$50,000 = Pe^{0.85}$$

$$\frac{\$50,000}{e^{0.85}} = P$$

$$\$21,370.75 \approx P$$

Therefore, $\$21,370.75$ would have to be invested.

**23.** In this case, the time is unknown. Use the compound interest formula with $P = \$1200$, $r = 0.0425$, $m = 1$, and $F = \$19,250$.

$$F = P\left(1 + \frac{r}{m}\right)^{mt}$$

$$\$19,250 = \$1200\left(1 + \frac{0.0425}{1}\right)^{1 \times t}$$

$$\$19,250 = \$1200(1.0425)^{t}$$

$$16.0416667 \approx (1.0425)^{t}$$

Guess values for $t$ until 1.0425 raised to the power of $t$ is approximately 16.0416667. It will take between 66 and 67 years.

**25.** In this case, $P = \$1500$, $r = 0.035$, and $t = 3$ years.

(a) Use the future value formula for simple interest.

$$F = P(1 + rt)$$

$$F = \$1500(1 + (0.035)(3))$$

$$F = \$1657.50$$

(b) Use the compound interest formula.

$$F = P\left(1 + \frac{r}{m}\right)^{mt}$$

$$F = \$1500\left(1 + \frac{0.035}{1}\right)^{1 \times 3}$$

$$F = \$1500(1.035)^{3}$$

$$F \approx \$1663.08$$

(c) Find the annual interest rate, $r$, if the future value is $\$1663.08$ and the account pays simple interest.

$$F = P(1 + rt)$$

$$\$1663.08 = \$1500(1 + r(3))$$

$$\frac{\$1663.08}{\$1500} = 1 + 3r$$

$$1.10872 = 1 + 3r$$

$$\frac{1.10872 - 1}{3} = r$$

$$0.03624 = r$$

A simple interest rate of 3.624% would have to be offered.

**27. (a)** Use the compound interest formula with $P = \$7700$, $r = 0.0299$, $m = 1$, and $t = 1$.

$$F = P\left(1 + \frac{r}{m}\right)^{mt}$$

$$F = \$7700\left(1 + \frac{0.0299}{1}\right)^{1 \times 1}$$

$$F = \$7700(1.0299)^1$$

$$F = \$7930.23$$

The ring will cost $7930.23 one year later.

**(b)** Find the future value of the investment. Use the compound interest formula with $P = \$7700$, $r = 0.0225$, $m = 365$, and $t = 1$.

$$F = P\left(1 + \frac{r}{m}\right)^{mt}$$

$$F = \$7700\left(1 + \frac{0.0225}{365}\right)^{365 \times 1}$$

$$F \approx \$7700(1.00006164384)^{365}$$

$$F \approx \$7700(1.02275432491)$$

$$F \approx \$7875.21$$

You cannot afford the ring. The investment did not keep up with inflation.

**29.** Inflation erodes the value of your money. The $10,000 can buy $10,000 worth of goods and services in 1976, however, due to inflation, the $10,000 from 1976 does not have the same purchasing power in 1978. It can buy only $100\% - 5.75\% = 94.25\%$ of what it used to buy for each year that the inflation rate stays at 5.75%. The $10,000 from 1976 could only buy $\$10,000(0.9425)^2 \approx \$8883.06$ worth of goods in 1978.

**31.** Use the APY Formula with $r = 0.0561$. Interest is compounded every two months or six times each year, so $m = 6$.

$$APY = \left(1 + \frac{r}{m}\right)^m - 1$$

$$= \left(1 + \frac{0.0561}{6}\right)^6 - 1$$

$$= (1.00935)^6 - 1$$

$$APY \approx 0.0574278$$

Therefore, the annual percentage yield is approximately 5.74%.

**33. (a)** Find the row containing the nominal rate of 8%. Find the column labeled "365 (Daily)". The effective annual rate is given in the intersection of the row and column and is approximately 8.32776%.

**(b)** The effective annual rate is $e^r - 1 = e^{0.0455} - 1 \approx 0.04655$ or approximately 4.66%.

**35.** Use the APY Formula in each case.

(a)

$$APY = \left(1 + \frac{r}{m}\right)^m - 1$$

$$= \left(1 + \frac{0.0337}{365}\right)^{365} - 1$$

$APY \approx 0.03427267$

The APY is approximately 3.43% in this case.

(b)

$$APY = \left(1 + \frac{r}{m}\right)^m - 1$$

$$= \left(1 + \frac{0.0334}{365}\right)^{365} - 1$$

$APY \approx 0.03396246$

The APY is approximately 3.40% in this case.

(c)

$$APY = \left(1 + \frac{r}{m}\right)^m - 1$$

$$= \left(1 + \frac{0.0330}{12}\right)^{12} - 1$$

$APY \approx 0.03350373$

The APY is approximately 3.35% in this case.

**37. (a)** Use the compound interest formula.

$$F = P\left(1 + \frac{r}{m}\right)^{mt}$$

$$F = \$500,000\left(1 + \frac{0.0183}{12}\right)^{12 \times 1}$$

$$F = \$500,000(1.001525)^{12}$$

$$F \approx \$500,000(1.01845427418)$$

$$F \approx \$509,227.14$$

The maturity value is \$509,227.14.

**(b)** Use the annual percentage yield formula.

$$APY = \left(1 + \frac{r}{m}\right)^m - 1$$

$$= \left(1 + \frac{0.0183}{12}\right)^{12} - 1$$

$$= (1.001525)^{12} - 1$$

$APY \approx 0.0184543$

The APY is approximately 1.84543%.

**(c)** Use the future value formula for simple interest.

$$F = P(1 + rt)$$
$$F = \$500,000(1 + (0.01845427)(1))$$
$$F = \$500,000(1.01845427)$$
$$F \approx \$509,227.14$$

## Solutions to Odd-Numbered Problems in Section 13.2

**1. (a)** The daily percentage rate is $\dfrac{12.9\%}{365} \approx 0.03534\%$.

Finance Charge $= Prt = \$2355\left(\dfrac{0.129}{365}\right)(30) \approx \$24.97$

**(b)** The daily percentage rate is $\dfrac{14.9\%}{365} \approx 0.04082\%$.

Finance Charge $= Prt = \$4825.80\left(\dfrac{0.149}{365}\right)(31) \approx \$61.07$

**(c)** The daily percentage rate is $\dfrac{12.68\%}{365} \approx 0.03474\%$.

From May 15 through June 14, there are $(31 - 14) + 14 = 31$ days.

Finance Charge $= Prt = \$315.42\left(\dfrac{0.1268}{365}\right)(31) \approx \$3.40$

**3.** Finance Charge $= Prt = \$275.00\left(\dfrac{0.189}{365}\right)(30) \approx \$4.27$

New Balance= Previous Balance + New Purchases + Finance Charge − Payment
$$= \$250.25 + \$245.27 + \$4.27 - \$175.00$$
$$= \$324.77$$

**5.** Finance Charge $= Prt = \$2004.78\left(\dfrac{0.0745}{365}\right)(30) \approx \$12.28$

New Balance= Previous Balance + New Purchases + Finance Charge − Payment
$$= \$1919.85 + \$378.00 + \$12.28 - \$250.00$$
$$= \$2060.13$$

**7. (a)** Daily Percentage Rate $= \left(\dfrac{9.9\%}{365}\right) \approx 0.02712\%$

**(b)** Find the balance for each day in the billing period and the number of days the balance was in effect.

| Time Period | Days | Daily Balance |
|---|---|---|
| October 11 to October 17 | 7 | Previous Balance = $5165.45 |
| October 18 to October 24 | 7 | $5165.45 − $750.00 = $4415.45 |
| October 25 to November 4 | 11 | $4415.45 + $48.90 = $4464.35 |
| November 5 to November 10 | 6 | $4464.35 + $85.64 = $4549.99 |

$$\text{Average Daily Balance} = \frac{7(\$5165.45) + 7(\$4415.45) + 11(\$4464.35) + 6(\$4549.99)}{7 + 7 + 11 + 6}$$

$$= \frac{\$143,474.09}{31}$$

$$\approx \$4628.20$$

**(c)** Finance Charge $= Prt = \$4628.20\left(\dfrac{0.099}{365}\right)(31) \approx \$38.91$

**(d)** Find the new account balance on November 11.

New Account Balance $=$ Previous Balance $+$ New Purchases $+$ Finance Charge $-$ Payment
$= \$5165.45 + \$134.54 + \$38.91 - \$750.00$
$= \$4588.90$

**9. (a)** Daily Percentage Rate $= \left(\dfrac{14.9\%}{365}\right) \approx 0.04082\%$

**(b)** Find the balance for each day in the billing period and the number of days the balance was in effect.

| Time Period | Days | Daily Balance |
|---|---|---|
| June 8 to June 19 | 12 | Previous Balance = $225.85 |
| June 20 to June 24 | 5 | $225.85 + $79.95 = $305.80 |
| June 25 to June 27 | 3 | $305.80 - $125.00 = $180.80 |
| June 28 to July 4 | 7 | $180.80 + $34.65 = $215.45 |
| July 5 to July 7 | 3 | $215.45 + $69.50 = $284.95 |

$$\text{Average Daily Balance} = \frac{12(\$225.85) + 5(\$305.80) + 3(\$180.80) + 7(\$215.45) + 3(\$284.95)}{12 + 5 + 3 + 7 + 3}$$

$$= \frac{\$7144.60}{30}$$

$$\approx \$238.15$$

**(c)** Finance Charge $= Prt = \$238.15\left(\dfrac{0.149}{365}\right)(30) \approx \$2.92$

**(d)** Find the new account balance on July 7.

New Account Balance $=$ Previous Balance $+$ New Purchases $+$ Finance Charge $-$ Payment
$= \$225.85 + \$184.10 + \$2.92 - \$125.00$
$= \$287.87$

**11.** The principal is $13,000.

**(a)** Interest $= Prt = \$13,000(0.049)\left(\dfrac{1}{12}\right) \approx \$53.08$

**(b)** Payment to Principal $=$ Monthly Payment $-$ Interest
$= \$298.79 - \$53.08$
$= \$245.71$

**(c)** New Balance $=$ Previous Balance $-$ Payment to Principal
$= \$13,000 - \$245.71$
$= \$12,754.29$

**13.** The principal is $5000. Payments of $111.23 will be made. Each month, the interest and the new balance will be calculated in the following way:

Interest $=$ (New Balance)$(0.0563)\left(\dfrac{1}{12}\right)$

New Balance = Previous Balance − Payment to Principal
Amortization Schedule: First 3 Months

| Month | Payment | Interest | Payment to Principal | *New Balance* |
|-------|---------|----------|----------------------|---------------|
| 1 | $111.23 | $23.46 | $87.77 | *$4912.23* |
| 2 | $111.23 | $23.05 | $88.18 | *$4824.05* |
| 3 | *$111.23* | *$22.63* | *$88.60* | *$4735.45* |

**15.** This is a 10-year loan with an annual interest rate of 3.37% and a principal amount of $12,350.

    **(a)** Using the monthly payment formula, we have the following:

$$PMT = \frac{P \times \left(\frac{r}{12}\right) \times \left(1+\frac{r}{12}\right)^{12t}}{\left(1+\frac{r}{12}\right)^{12t} - 1}$$

$$= \frac{\$12,350\left(\frac{0.0337}{12}\right)\left(1+\frac{0.0337}{12}\right)^{12 \times 10}}{\left(1+\frac{0.0337}{12}\right)^{12 \times 10} - 1}$$

$$\approx \$121.38$$

    **(b)** Interest = (New Balance)(0.0337)$\left(\frac{1}{12}\right)$

New Balance = Previous Balance − Payment to Principal
Amortization Schedule: First 4 Months

| Month | Payment | Interest | Payment to Principal | *New Balance* |
|-------|---------|----------|----------------------|---------------|
| 1 | $121.38 | $34.68 | $86.70 | *$12,263.30* |
| 2 | $121.38 | $34.44 | $86.94 | *$12,176.36* |
| 3 | $121.38 | $34.20 | $87.18 | *$12,089.18* |
| 4 | *$121.38* | *$33.95* | *$87.43* | *$12,001.75* |

    **(c)** The loan was paid for 119 months (9 years and 11 months) at $121.38 per month. The final payment was $120.45. Therefore, the total amount paid was 119($121.38) + 1($120.45) = $14,564.67. The principal amount of the loan was $12,350 so the total amount of interest paid was Total Amount Paid − Principal = $14,564.67 − $12,350 = $2214.67.

**17. (a)** Each interest payment is found using the simple interest formula.

    Monthly Payment = $Prt$ = $8500(0.0277)$\left(\frac{1}{12}\right)$ ≈ $19.62.

    The deferment period lasts 10 months, so Greg will pay 10($19.62) = $196.20.

    **(b)** Interest that capitalizes gets added to the principal and compounded. The compound interest formula with interest compounded quarterly gives the following:

$$P\left(1+\frac{r}{m}\right)^{mt} = \$8500\left(1+\frac{0.0277}{4}\right)^{4 \times \frac{10}{12}} = \$8500(1.006925)^{\frac{10}{3}} \approx \$8697.80.$$

    Therefore, $8697.80 − $8500.00 = $197.80 in interest accumulated.

    **(c)** Repayment would begin with a total of $8697.80 principal value, an annual interest rate of 3.37%, and a term of 10 years. Find the monthly payment using the monthly payment formula.

$$PMT = \frac{P \times \left(\dfrac{r}{12}\right) \times \left(1 + \dfrac{r}{12}\right)^{12t}}{\left(1 + \dfrac{r}{12}\right)^{12t} - 1}$$

$$= \frac{\$8697.80 \left(\dfrac{0.0337}{12}\right)\left(1 + \dfrac{0.0337}{12}\right)^{12 \times 10}}{\left(1 + \dfrac{0.0337}{12}\right)^{12 \times 10} - 1}$$

$$\approx \$85.48$$

Therefore, when the loan comes out of deferment, the monthly payment will be $85.48.

**19.** The principal amount is $600 with a 12.5% annual interest rate.

$$\text{Interest} = (\text{New Balance})(0.125)\left(\frac{1}{12}\right)$$

New Balance = Previous Balance – Payment to Principal
Amortization Schedule:

| Payment | Interest | Payment to Principal | New Balance |
|---------|----------|----------------------|-------------|
| $123.78 | $6.25 | $117.53 | *$482.47* |
| $123.78 | $5.03 | $118.75 | *$363.72* |
| $123.78 | $3.79 | $119.99 | *$243.73* |
| $123.78 | $2.54 | $121.24 | *$122.49* |
| *$123.77* | *$1.28* | *$122.49* | *$0.00* |

The last payment is $123.77.

**21. (a)** The monthly payment is $29.191515(10) ≈ $291.92.
**(b)** The monthly payment is $23.009846(18) ≈ $414.18.

**23. (a)** Use the monthly payment formula:

$$PMT = \frac{P \times \left(\dfrac{r}{12}\right) \times \left(1 + \dfrac{r}{12}\right)^{12t}}{\left(1 + \dfrac{r}{12}\right)^{12t} - 1}$$

$$= \frac{\$16,325 \left(\dfrac{0.0625}{12}\right)\left(1 + \dfrac{0.0625}{12}\right)^{12 \times 3}}{\left(1 + \dfrac{0.0625}{12}\right)^{12 \times 3} - 1}$$

$$\approx \$498.49$$

**(b)** The monthly payment is $30.535342(16.325) ≈ $498.49. They are the same.

**25. (a)** Use the monthly payment formula:

$$PMT = \frac{P \times \left(\frac{r}{12}\right) \times \left(1 + \frac{r}{12}\right)^{12t}}{\left(1 + \frac{r}{12}\right)^{12t} - 1}$$

$$= \frac{\$7350 \left(\frac{0.1015}{12}\right) \left(1 + \frac{0.1015}{12}\right)^{12 \times 4}}{\left(1 + \frac{0.1015}{12}\right)^{12 \times 4} - 1}$$

$$\approx \$186.95$$

**(b)** Use the monthly payment formula:

$$PMT = \frac{P \times \left(\frac{r}{12}\right) \times \left(1 + \frac{r}{12}\right)^{12t}}{\left(1 + \frac{r}{12}\right)^{12t} - 1}$$

$$= \frac{\$12,000 \left(\frac{0.0283}{12}\right) \left(1 + \frac{0.0283}{12}\right)^{12 \times \frac{7}{12}}}{\left(1 + \frac{0.0283}{12}\right)^{12 \times \frac{7}{12}} - 1}$$

$$\approx \$1730.50$$

**27.** You are financing $14,995 for 4 years at 6.25% annual interest. Use an amortization table to find the monthly payment. The monthly payment is $23.599820(14.995) \approx \$353.88$.

**29.** Justin makes a 20% down payment of $16,285(0.20) = \$3257$ and will finance the balance of $16,285 - \$3257 = \$13,028$ for 5 years at 11.89% annual interest. Our amortization tables do not include entries for interest rates of 11.89%, so use the monthly payment formula.

$$PMT = \frac{P \times \left(\frac{r}{12}\right) \times \left(1 + \frac{r}{12}\right)^{12t}}{\left(1 + \frac{r}{12}\right)^{12t} - 1}$$

$$= \frac{\$13,028 \left(\frac{0.1189}{12}\right) \left(1 + \frac{0.1189}{12}\right)^{12 \times 5}}{\left(1 + \frac{0.1189}{12}\right)^{12 \times 5} - 1}$$

$$\approx \$289.08$$

Justin's monthly payment will be $289.08.

**31.** An amortization table may be used to find the monthly payment. The monthly payment is $6.653025(7,368,363,360) \approx \$49,021,905,643.20$. The interest in the first month's payment is $Prt = \$7,368,363,360,000(0.07)\left(\frac{1}{12}\right) = \$42,982,119,600.00$.

**33. (a)** Use an amortization table with a 15-year term and a 6.75% annual interest rate. Solve for the loan amount in $1000s. Multiply by 1000 to find the loan amount and round to the nearest dollar.

$$\text{Monthly Payment} = (\text{Amortization Table Entry})(\text{Loan Amount in } \$1000\text{s})$$

$$\$300 = 8.849095(\text{Loan Amount in } \$1000\text{s})$$

$$\frac{\$300}{8.849095} = \text{Loan Amount in } \$1000\text{s}$$

$$\$33.9017718761 \approx \text{Loan Amount in } \$1000\text{s}$$

Therefore, the loan amount is $33.9017718761(1000) \approx $33,902.

**(b)** Use an amortization table with a 6-year term and a 8.75% annual interest rate. Solve for the loan amount in $1000s. Multiply by 1000 to find the loan amount and round to the nearest dollar.

$$\text{Monthly Payment} = (\text{Amortization Table Entry})(\text{Loan Amount in } \$1000\text{s})$$

$$\$250 = 17.901710(\text{Loan Amount in } \$1000\text{s})$$

$$\frac{\$250}{17.901710} = \text{Loan Amount in } \$1000\text{s}$$

$$\$13.9651463464 \approx \text{Loan Amount in } \$1000\text{s}$$

Therefore, the loan amount is $13.9651463464(1000) \approx $13,965.

**35.** Use an amortization table with a 15-year term and a 9.75% annual interest rate. Solve for the loan amount in $1000s. Multiply by 1000 to find the loan amount and round to the nearest dollar.

$$\text{Monthly Payment} = (\text{Amortization Table Entry})(\text{Loan Amount in } \$1000\text{s})$$

$$\$300 = 10.593627(\text{Loan Amount in } \$1000\text{s})$$

$$\frac{\$300}{10.593627} = \text{Loan Amount in } \$1000\text{s}$$

$$\$28.3189128709 \approx \text{Loan Amount in } \$1000\text{s}$$

Therefore, the maximum loan amount is $28.3189128709(1000) \approx $28,319.

**37.** Given the monthly payment and the loan amount in 1000s, solve for the amortization table entry. Then locate the entry in an amortization table for a 30-year loan.

$$\text{Monthly Payment} = (\text{Amortization Table Entry})(\text{Loan Amount in } \$1000\text{s})$$

$$\$751.27 = (\text{Amortization Table Entry})(\$100)$$

$$\frac{\$751.27}{\$100} = \text{Amortization Table Entry}$$

$$7.5127 = \text{Amortization Table Entry}$$

The approximate annual interest rate for this loan is 8.25%.

**39.** Given the monthly payment and the loan amount in 1000s, solve for the amortization table entry. Then locate the entry in an amortization table for a 15-year loan.

$$\text{Monthly Payment} = (\text{Amortization Table Entry})(\text{Loan Amount in } \$1000\text{s})$$

$$\$122.92 = (\text{Amortization Table Entry})(\$16.450)$$

$$\frac{\$122.92}{\$16.450} = \text{Amortization Table Entry}$$

$$7.47234042553 \approx \text{Amortization Table Entry}$$

The approximate annual interest rate for this loan between 4% and 4.25%.

**41. (a)** You pay a total of $65(24) = $1560 for the dining room set which is $1560 − $1250 = $310 over the cost of the set.

**(b)** Use the simple interest formula and solve for the unknown rate.

$$I = Prt$$

$$\$310 = \$1250(r)(2)$$

$$\$310 = \$2500r$$

$$\frac{\$310}{\$2500} = r$$

$$0.124 = r$$

You pay an annual interest rate of 12.4%.

**43.** You pay $50.20(24) = $1204.80 for the set which is $1204.80 – $735.98 = $468.82 over the suggested retail value of the set. Use the simple interest formula and solve for the unknown rate.

$$I = Prt$$

$$\$468.82 = \$735.98(r)(2)$$

$$\$468.82 = \$1471.96r$$

$$\frac{\$468.82}{\$1471.96} = r$$

$$0.31850050 \approx r$$

An annual rate of approximately 31.85% would yield the same amount.

**45. (a)** Her monthly payment at the rent-to-own store is $32. If she takes out a simple interest loan, use an amortization table to determine the monthly payment. The term of the loan is 4 years.

$$\text{Monthly Payment} = (\text{Amortization Table Entry})(\text{Loan Amount in } \$1000s)$$

$$= \$25.003920(0.629)$$

$$\approx \$15.73$$

Choose the loan with a monthly payment of $15.73.

**(b)** Her total cost at the rent-to-own store will be $32(30) = $960. Her total cost of the loan can be found by using the future value formula for simple interest.

$$F = P(1+rt)$$

$$= \$629(1+(0.0925)(4))$$

$$= \$861.73$$

Choose the loan with a future value of $861.73 over the total cost of $960.00 from the rent-to-own store.

## Solutions to Odd-Numbered Problems in Section 13.3

**1.** The maximum home price the family can afford is 3×(annual gross income) = 3($75,000) = $225,000.

The maximum monthly housing expenses cannot exceed $0.25 \times \left(\frac{1}{12}\right) \times (\text{annual gross income}) =$

$0.25 \times \left(\frac{1}{12}\right) \times (\$75,000) = \$1562.50.$

**3.** The maximum home price Dana can afford is 3×(annual gross income) = 3($43,550) = $130,650. Her

maximum monthly housing expenses cannot exceed $0.25 \times \left(\frac{1}{12}\right) \times (\text{annual gross income}) =$

$0.25 \times \left(\frac{1}{12}\right) \times (\$43,550) \approx \$907.29.$

5.  Together their annual gross income is \$93,000. The maximum purchase price they can afford is $3 \times$(annual gross income) = 3(\$93,000) = \$279,000. The maximum monthly housing expenses cannot exceed $0.25 \times \left(\dfrac{1}{12}\right) \times$ (annual gross income) = $0.25 \times \left(\dfrac{1}{12}\right) \times$ (\$93,000) = \$1937.50.

7.  The couple can pay a low of $0.25 \times$(monthly gross income) = 0.25(\$3650) = \$912.50 to a high of $0.38 \times$(monthly gross income) = 0.38(3650) = \$1387.

9.  One-point is 1% of the principal. The loan origination fee and the discount charge will each cost 1% of \$95,000. 1% of \$95,000 = 0.01(\$95,000) = \$950. Therefore, the loan origination fee and the discount charge will cost a total of 2(\$950) = \$1900.

11. One-point is 1% of the principal, and 1.5-points is 1.5% of the principal. The total for the two charges is 0.01(\$92,000) + 0.015(\$92,000) = \$920 + \$1380 = \$2300.

13. His down payment is \$23,000. If \$23,000 is 20% of the purchase price, then write an equation.

    $\$23,000 = 20\% \times$(Purchase Price)

    $\$23,000 = 0.20$(Purchase Price)

    $\dfrac{\$23,000}{0.20} =$ Purchase Price

    $\$115,000 =$ Purchase Price

    Therefore, based on his down payment, he could buy a home priced at \$115,000.

15. Ladonna's down payment is \$18,000. If \$18,000 is 20% of the purchase price, then write an equation.

    $\$18,000 = 20\% \times$(Purchase Price)

    $\$18,000 = 0.20$(Purchase Price)

    $\dfrac{\$18,000}{0.20} =$ Purchase Price

    $\$90,000 =$ Purchase Price

    Therefore, based on the down payment, she could buy a home priced at \$90,000. The affordability guideline states that the home should not cost more than three times the buyer's annual gross income. This assumes a standard 20% down payment.

    Maximum Home Price = $3 \times$ (Annual Gross Income)

    $\$90,000 = 3 \times$ (Annual Gross Income)

    $\dfrac{\$90,000}{3} =$ Annual Gross Income

    $\$30,000 =$ Annual Gross Income

    Ladonna should have an annual gross income of at least \$30,000.

17. Subtract the closing costs from the \$18,000 the Davis family has saved. The down payment will be \$18,000 – \$2500 = \$15,500. If \$15,500 is 10% of the purchase price, then write an equation.

    $\$15,500 = 10\% \times$(Purchase Price)

    $\$15,500 = 0.10$(Purchase Price)

    $\dfrac{\$15,500}{0.10} =$ Purchase Price

    $\$155,000 =$ Purchase Price

    The maximum price they can pay for a home is \$155,000.

**19.** The down payment is 20% of $135,000.

$$\text{Down Payment} = 20\% \text{ of } \$135,000$$
$$= 0.20(\$135,000)$$
$$= \$27,000$$

A total of $135,000 - $27,000 = $108,000 will be financed at 8% for 30 years. Use an amortization table to find the monthly payment. The monthly payment = $7.337646(108) ≈ $792.47.

**21.** The down payment is 10% of $227,500.

$$\text{Down Payment} = 10\% \text{ of } \$227,500$$
$$= 0.10(\$227,500)$$
$$= \$22,750$$

A total of $227,500 - $22,750 = $204,750 will be financed at 6.5% for 30 years. Use an amortization table to find the monthly payment. The monthly payment = $6.320680(204.75) ≈ $1294.16.

**23.** The down payment is 10% of $95,500.

$$\text{Down Payment} = 10\% \text{ of } \$95,500$$
$$= 0.10(\$95,500)$$
$$= \$9550$$

A total of $95,500 - $9550 = $85,950 will be financed at 4.6% for 30 years. Use the monthly payment formula.

$$PMT = \frac{P \times \left(\dfrac{r}{12}\right) \times \left(1 + \dfrac{r}{12}\right)^{12t}}{\left(1 + \dfrac{r}{12}\right)^{12t} - 1}$$

$$= \frac{\$85,950\left(\dfrac{0.046}{12}\right)\left(1 + \dfrac{0.046}{12}\right)^{12 \times 30}}{\left(1 + \dfrac{0.046}{12}\right)^{12 \times 30} - 1}$$

$$\approx \$440.62.$$

The monthly payment is $440.62.

**25.** The amount that will be financed is $155,000.

(a) A 2-point discount charge will cost 2% of $155,000. A 1-point origination fee will cost 1% of $155,000. The total cost of these fees is 0.02($155,000) + 0.01($155,000) = $3100 + $1550 = $4650.

(b) The revised balance is $155,000 + $4650 = $159,650.

(c) Use an amortization table to determine the monthly payment. Monthly payment = $5.216473(159.65) ≈ $832.81.

(d) Use the monthly payment formula to find the interest rate that would yield a monthly payment of $832.81 for the original balance of $155,000. First use an amortization table to narrow down the interest rate options and then use linear interpolation to find the APR.

From an amortization table we have the following:

$$\text{Monthly Payment} = (\text{Amortization Table Entry})(\text{Loan Amount in \$1000s})$$

$$\$832.81 = (\text{Amortization Table Entry})(155)$$

$$\frac{\$832.81}{155} = \text{Amortization Table Entry}$$

$$5.372968 \approx \text{Amortization Table Entry}$$

For a 30-year loan, the amortization table entry falls between 5.368216 and 5.522037 which correspond to 5% and 5.25% respectively. Use linear interpolation to obtain a good, first estimate of the correct value.

$$\frac{5.372968 - 5.368216}{5.522037 - 5.368216} \times 100\% = \frac{0.004752}{0.153821} \times 100\% \approx 0.0308931 \times 100\% \approx 3\%$$

We see that the amortization table entry we calculated is approximately 3% of the way from the lower table entry to the higher table entry. We estimate that the corresponding interest rate is itself about 3% of the way from 5% to 5.25% or $5 + 0.0308931 \times (5.25 - 5) \approx 5.0077233\%$. Use 5.0077% as the first guess at the APR.

| Guess | Monthly Payment | Compare to $832.81 |
|---|---|---|
| 5.0077% | $PMT = \dfrac{\$155{,}000 \times \left(\dfrac{0.050077}{12}\right) \times \left(1 + \dfrac{0.050077}{12}\right)^{360}}{\left(1 + \dfrac{0.050077}{12}\right)^{360} - 1} \approx \$832.81$ | *This rate is correct.* |

Therefore, the interest rate that would yield a monthly payment of $832.81 for a balance of $155,000 is 5.0077%.

**27.** The amount that will be financed is $140,000.
**Find the total cost of the additional fees:**
A 1.5-point discount charge will cost 1.5% of $140,000. A 1-point origination fee will cost 1% of $140,000. The total cost of these fees is $0.015(\$140{,}000) + 0.01(\$140{,}000) = \$2100 + \$1400 = \$3500$.
**Find the revised balance:**
The revised balance is $140,000 + $3500 = $143,500.
**Use an amortization table to determine the monthly payment:**
Monthly payment = $6.192345(143.5) \approx $888.61.
**Use the monthly payment formula:**
Find the interest rate that would yield a monthly payment of $888.61 for the original balance of $140,000. First use an amortization table to narrow down the interest rate options and then use linear interpolation to find the APR.
From an amortization table we have the following:

$$\text{Monthly Payment} = (\text{Amortization Table Entry})(\text{Loan Amount in \$1000s})$$

$$\$888.61 = (\text{Amortization Table Entry})(140)$$

$$\frac{\$888.61}{140} = \text{Amortization Table Entry}$$

$$6.34721 \approx \text{Amortization Table Entry}$$

For a 20-year loan, the amortization table entry falls between 6.326494 and 6.462236 which correspond to 4.5% and 4.75% respectively. Use linear interpolation to obtain a good, first estimate of the correct value.

$$\frac{6.34721 - 6.326494}{6.462236 - 6.326494} \times 100\% = \frac{0.020716}{0.135742} \times 100\% \approx 0.152613 \times 100\% \approx 15\%$$

We see that the amortization table entry we calculated is approximately 15% of the way from the lower table entry to the higher table entry. We estimate that the corresponding interest rate is itself about 15% of the way from 4.5% to 4.75% or $4.5 + 0.1526130453 \times (4.75 - 4.5) \approx 4.5382\%$. Use 4.5382% as the first guess at the APR.

| Guess | Monthly Payment | Compare to $888.61 |
|---|---|---|
| 4.5382% | $PMT = \dfrac{\$140,000 \times \left(\dfrac{0.045382}{12}\right) \times \left(1 + \dfrac{0.045382}{12}\right)^{240}}{\left(1 + \dfrac{0.045382}{12}\right)^{240} - 1} \approx \$888.60$ | There is a one penny difference. |

The interest rate that would yield a monthly payment of $888.61 for a balance of $140,000 is 4.5382%. This is called the annual percentage rate.

29. Use an amortization table to find the monthly payment. Monthly payment = $5.995505(115) \approx$ $689.49.

Taxes = 2.5% of Assessed Value

$$= 0.025(\$150,000)$$
$$= \$3750$$

Taxes and insurance will cost a total of $3750 + $650 = $4400 for the year. Each month add $\dfrac{\$4400}{12} \approx \$366.67$ to the monthly payment. The complete monthly payment is $689.49 + $366.67 = $1056.16.

31. Use the monthly payment formula to find the monthly payment.

$$PMT = \frac{\$75,000 \times \left(\frac{0.049}{12}\right) \times \left(1 + \frac{0.049}{12}\right)^{180}}{\left(1 + \frac{0.049}{12}\right)^{180} - 1} \approx \$589.20$$

Taxes = 2.25% of Assessed Value

$$= 0.0225(\$127,700)$$
$$= \$2873.25$$

Taxes and insurance will cost a total of $2873.25 + $740.00 = $3613.25 for the year. Each month add $\dfrac{\$3613.25}{12} \approx \$301.10$ to the monthly payment. The complete monthly payment is $589.20 + $301.10 = $890.30.

## Solutions to Chapter 13 Review Problems

1. Use the simple interest formula: $I = Prt$. Round your answers to the nearest penny.

   (a) $I = \$13,750(0.0525)\left(\dfrac{76}{365}\right) = \$150.31$

   (b) $I = \$13,750(0.0525)\left(\dfrac{4}{12}\right) \approx \$240.63$

2. Use the simple interest formula and solve for the annual interest rate, $r$.

$$I = Prt$$

$$\$131.40 = \$1500(r)(2)$$

$$\$131.40 = \$3000r$$

$$\frac{\$131.40}{\$3000} = r$$

$$0.0438 = r$$

The annual interest rate is 4.38%.

3. Use the simple interest formula. The variable, $t$, in the formula is in years. Divide $t$ by 12 to express the term in months.

$$I = Prt$$

$$\$93.10 = \$2850(0.098)\left(\frac{t}{12}\right)$$

$$\$93.10 = \$279.30\left(\frac{t}{12}\right)$$

$$\frac{\$93.10}{\$279.30} = \frac{t}{12}$$

$$\frac{(\$93.10)(12)}{\$279.30} = t$$

$$4 = t$$

Therefore, the term of the loan is 4 months.

4. **(a)** Use the future value formula for Simple Interest:

$$F = P(1 + rt)$$

$$F = \$12,500(1 + (0.0225)(3))$$

$$F = \$12,500(1.0675)$$

$$F = \$13,343.75$$

**(b)** Use the future value formula for Simple Interest:

$$F = P(1 + rt)$$

$$F = \$6890(1 + (0.0738)(4.5))$$

$$F = \$6890(1.3321)$$

$$F \approx \$9178.17$$

5. **(a)** Calculate the future value for simple interest on the balance in the account at the start of each compounding period. Each compounding period is three months long, so $t = \frac{3}{12} = 0.25$ years.

There are four compounding periods in one year. The balance or future value at the end of a compounding period becomes the principal at the beginning of the next compounding period.

First compounding period: $\quad F = \$2000(1 + (0.0275)(0.25)) = \$2013.75$

Second compounding period: $\quad F = \$2013.75(1 + (0.0275)(0.25)) \approx \$2027.59$

Third compounding period: $\quad F = \$2027.59(1 + (0.0275)(0.25)) \approx \$2041.53$

Fourth compounding period: $\quad F = \$2041.53(1 + (0.0275)(0.25)) \approx \$2055.57$

**(b)** Use the compound interest formula.

$$F = P\left(1+\frac{r}{m}\right)^{mt}$$

$$F = \$2000\left(1+\frac{0.0275}{4}\right)^{4\times 8}$$

$$F \approx \$2490.28$$

After 8 years, the future value of the account is $2490.28.

6. **(a)** Use the future value formula for simple interest.

$$F = P(1+rt)$$
$$F = \$8500(1+(0.0235)(4))$$
$$F = \$8500(1.094)$$
$$F = \$9299.00$$

**(b)** Use the compound interest formula.

$$F = P\left(1+\frac{r}{m}\right)^{mt}$$

$$F = \$8500\left(1+\frac{0.102}{1}\right)^{1\times 4}$$

$$F \approx \$12,535.61$$

**(c)** Use the compound interest formula.

$$F = P\left(1+\frac{r}{m}\right)^{mt}$$

$$F = \$8500\left(1+\frac{0.034}{3}\right)^{3\times 4}$$

$$F \approx \$9730.85$$

**(d)** Use the compound interest formula.

$$F = P\left(1+\frac{r}{m}\right)^{mt}$$

$$F = \$8500\left(1+\frac{0.0298}{12}\right)^{12\times 4}$$

$$F \approx \$9574.64$$

**(e)** Use the compound interest formula.

$$F = P\left(1+\frac{r}{m}\right)^{mt}$$

$$F = \$8500\left(1+\frac{0.0479}{365}\right)^{365\times 4}$$

$$F \approx \$10,294.95$$

7.  (a)  Use the future value formula for simple interest.
    $$F = P(1+rt)$$
    $$F = \$14,590(1+(0.0195)(2.5))$$
    $$F = \$8500(1.04875)$$
    $$F \approx \$15,301.26$$

    (b)  Use the compound interest formula.
    $$F = P\left(1+\frac{r}{m}\right)^{mt}$$
    $$F = \$14,590\left(1+\frac{0.0195}{2}\right)^{2\times2.5}$$
    $$F \approx \$15,315.27$$

    (c)  Use the compound interest formula.
    $$F = P\left(1+\frac{r}{m}\right)^{mt}$$
    $$F = \$14,590\left(1+\frac{0.0195}{4}\right)^{4\times2.5}$$
    $$F \approx \$15,317.07$$

    (d)  Use the compound interest formula.
    $$F = P\left(1+\frac{r}{m}\right)^{mt}$$
    $$F = \$14,590\left(1+\frac{0.0195}{12}\right)^{12\times2.5}$$
    $$F \approx \$15,318.28$$

    (e)  Use the compound interest formula.
    $$F = P\left(1+\frac{r}{m}\right)^{mt}$$
    $$F = \$14,590\left(1+\frac{0.0195}{365}\right)^{365\times2.5}$$
    $$F \approx \$15,318.86$$

8.  (a)  Use the future value formula for simple interest:
    $$F = P(1+rt)$$
    $$\$175 = \$150\left(1+(r)\left(\frac{6}{12}\right)\right)$$
    $$\frac{\$175}{\$150} = 1+0.5r$$
    $$\frac{\frac{\$175}{\$150}-1}{0.5} = r$$
    $$0.333333 \approx r$$
    The rate of simple interest is approximately 33.33%.

**(b)** Use the compound interest formula.

$$F = P\left(1+\frac{r}{m}\right)^{mt}$$

$$\$175 = \$150\left(1+\frac{r}{12}\right)^{12\times\frac{6}{12}}$$

$$\frac{\$175}{\$150} = \left(1+\frac{r}{12}\right)^{6}$$

$$\sqrt[6]{\frac{\$175}{\$150}} = 1+\frac{r}{12}$$

$$12\left(\sqrt[6]{\frac{\$175}{\$150}}-1\right) = r$$

$$0.3122959 \approx r$$

The rate of interest compounded monthly is approximately 31.23%.

**9.** **(a)** Use the future value formula for simple interest and solve for $P$.

$$F = P(1+rt)$$

$$\$5000 = P(1+(0.0638)(10))$$

$$\$5000 = P(1.638)$$

$$\frac{\$5000}{1.638} = P$$

$$\$3052.50 \approx P$$

**(b)** Use the compound interest formula.

$$F = P\left(1+\frac{r}{m}\right)^{mt}$$

$$\$5000 = P\left(1+\frac{0.0638}{12}\right)^{12\times10}$$

$$\frac{\$5000}{\left(1+\frac{0.0638}{12}\right)^{120}} = P$$

$$\$2646.21 \approx P$$

**(c)** Use the future value formula for continuously compounded interest.

$$F = Pe^{rt}$$

$$\$5000 = Pe^{0.0638\times10}$$

$$\$5000 = Pe^{0.638}$$

$$\frac{\$5000}{e^{0.638}} = P$$

$$\$2641.74 \approx P$$

**10.** Use the Effective Annual Rate formula.

$$EAR = \left(1 + \frac{r}{m}\right)^m - 1$$

$$= \left(1 + \frac{0.0238}{4}\right)^4 - 1$$

$EAR \approx 0.024013$

The Effective Annual Rate is approximately 2.4013%.

**11.** Use the APY Formula.

| Account paying 3.45% | Account paying 3.38% |
|---|---|
| Compounded Monthly | Compounded Daily |

$$APY = \left(1 + \frac{r}{m}\right)^m - 1 \qquad APY = \left(1 + \frac{r}{m}\right)^m - 1$$

$$APY = \left(1 + \frac{0.0345}{12}\right)^{12} - 1 \qquad APY = \left(1 + \frac{0.0338}{365}\right)^{365} - 1$$

$APY \approx 0.035051 \qquad\qquad APY \approx 0.034376$

The APY for the account paying 3.45% annual interest compounded monthly is approximately 3.5051%. The APY for the other account is approximately 3.4376%.

**12.** Use the APY Formula and use a Guess and Test strategy to find the number of compounding periods, *m*.

$$APY = \left(1 + \frac{r}{m}\right)^m - 1$$

$$0.0225 = \left(1 + \frac{0.0223}{m}\right)^m - 1$$

$$1.0225 = \left(1 + \frac{0.0223}{m}\right)^m$$

A compounding period of one month makes the equation true.

**13.** The daily percentage rate is $\dfrac{8.9\%}{365} \approx 0.024384\%$.

Use the average daily balance, the daily percentage rate, and the billing period to calculate the finance charge.

Finance Charge = $Prt$ = \$6590.00$\left(\dfrac{0.089}{365}\right)$(30) ≈ \$48.21.

**14.** The daily percentage rate is $\dfrac{5.89\%}{365} \approx 0.016136986\%$.

Use the average daily balance, the daily percentage rate, and the billing period to calculate the finance charge.

Finance Charge = $Prt$ = \$7432.52$\left(\dfrac{0.0589}{365}\right)$(31) ≈ \$37.18.

New Balance= Previous Balance + New Purchases + Finance Charge − Payment
= \$6359.00 + \$1427.00 + \$37.18 − \$250.00
= \$7573.18

**15.** On June 1, the balance is $0. The balance remained $0 for 19 days, and then on June 20, there was a purchase of $612.00.

(a) Find the balance for each day in the billing period and the number of days the balance was in effect.

| Time Period | Days | Daily Balance |
|---|---|---|
| June 1 to June 19 | 19 | Previous Balance = $0 |
| June 20 to June 30 | 11 | $0 + $612.00 = $612.00 |

$$\text{Average Daily Balance} = \frac{19(\$0) + 11(\$612.00)}{19 + 11}$$

$$= \frac{\$6732.00}{30}$$

$$= \$224.40$$

Therefore, the average daily balance was $224.40.

(b) Use the average daily balance, the daily percentage rate, and the billing period to calculate the finance charge.

$$\text{Finance Charge} = Prt = \$224.40\left(\frac{0.139}{365}\right)(30) \approx \$2.56.$$

**16.** (a) $\text{Daily Percentage Rate} = \left(\frac{6.28\%}{365}\right) \approx 0.017205\%$

(b) Find the balance for each day in the billing period and the number of days the balance was in effect.

| Time Period | Days | Daily Balance |
|---|---|---|
| September 9 to September 17 | 9 | Previous Balance = $3470.63 |
| September 18 to September 24 | 7 | $3470.63 − $1500.00 = $1970.63 |
| September 25 to October 4 | 10 | $1970.63 + $126.00 = $2096.63 |
| October 5 to October 10 | 6 | $2096.63 + $283.61 = $2380.24 |

$$\text{Average Daily Balance} = \frac{9(\$3470.63) + 7(\$1970.63) + 10(\$2096.63) + 6(\$2380.24)}{9 + 7 + 10 + 6}$$

$$= \frac{\$80,277.82}{32}$$

$$\approx \$2508.68$$

(c) $\text{Finance Charge} = Prt = \$2508.68\left(\frac{0.0628}{365}\right)(32) \approx \$13.81$

(d) Find the new account balance on October 10.

$$\text{New Account Balance} = \text{Previous Balance} + \text{New Purchases} + \text{Finance Charge} - \text{Payment}$$

$$= \$3470.63 + \$409.61 + \$13.81 - \$1500.00$$

$$= \$2394.05$$

**17.** The principal is $8500. Payments of $63.95 will be made. Each month, the interest and the new balance will be calculated in the following way:

$$\text{Interest} = (\text{New Balance})(0.0425)\left(\frac{1}{12}\right)$$

New Balance = Previous Balance − Payment to Principal

Amortization Schedule: First 3 Months

| Month | Payment | Interest | Payment to Principal | New Balance |
|---|---|---|---|---|
| 1 | $63.95 | $30.10 | $33.85 | $8466.15 |
| 2 | $63.95 | $29.98 | $33.97 | $8432.18 |
| 3 | $63.95 | $29.86 | $34.09 | $8398.09 |

**18.** The principal is $3500. Payments of $712.30 will be made. Each month, the interest and the new balance will be calculated in the following way:

$$\text{Interest} = (\text{New Balance})(0.07)\left(\frac{1}{12}\right)$$

New Balance = Previous Balance − Payment to Principal

Amortization Schedule: First 5 Months

| Month | Payment | Interest | Payment to Principal | New Balance |
|---|---|---|---|---|
| 1 | $712.30 | $20.41 | $691.89 | $2808.14 |
| 2 | $712.30 | $16.38 | $695.92 | $2112.19 |
| 3 | $712.30 | $12.32 | $699.98 | $1412.21 |
| 4 | $712.30 | $8.24 | $704.06 | $708.15 |
| 5 | $712.28 | $4.13 | $708.16 | $0.00 |

The amount of the final payment is $712.28.

**19.** Round all monthly payment values up.
   **(a)** The monthly payment is $33.333841(15) \approx $500.01.
   **(b)** The monthly payment is $10.484774(35.75) \approx $374.84.
   **(c)** The monthly payment is $4.490447(118.9) \approx $533.92.

**20. (a)** Using the monthly payment formula, we have the following:

$$PMT = \frac{P \times \left(\frac{r}{12}\right) \times \left(1+\frac{r}{12}\right)^{12t}}{\left(1+\frac{r}{12}\right)^{12t}-1}$$

$$= \frac{\$24,225\left(\frac{0.0435}{12}\right)\left(1+\frac{0.0435}{12}\right)^{12\times\frac{45}{12}}}{\left(1+\frac{0.0435}{12}\right)^{12\times\frac{45}{12}}-1}$$

$$\approx \$584.41$$

The monthly payment is $584.41.

**21.** The down payment is 15% of $19,995. Down Payment = 0.15($19,995) = $299.25. The amount financed is $19,995 − $2999.25 = $16,995.75.
   **(a)** The monthly payment is $20.637233(16.99575) \approx $350.75.

**(b)** Using the monthly payment formula, we have the following:

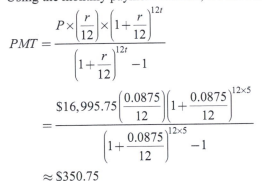

$$PMT = \frac{P \times \left(\frac{r}{12}\right) \times \left(1 + \frac{r}{12}\right)^{12t}}{\left(1 + \frac{r}{12}\right)^{12t} - 1}$$

$$= \frac{\$16,995.75 \left(\frac{0.0875}{12}\right)\left(1 + \frac{0.0875}{12}\right)^{12 \times 5}}{\left(1 + \frac{0.0875}{12}\right)^{12 \times 5} - 1}$$

$$\approx \$350.75$$

The monthly payment is $350.75.

**22.** Use an amortization table with a 4-year term and a 2.25% annual interest rate. Solve for the loan amount in $1000s. Multiply by 1000 to find the loan amount and round to the nearest dollar.

Monthly Payment = (Amortization Table Entry)(Loan Amount in $1000s)

$280.95 = 21.804406(Loan Amount in $1000s)

$$\frac{\$280.95}{21.804406} = \text{Loan Amount in } \$1000s$$

$12.885010488 ≈ Loan Amount in $1000s

The size of the loan is 1000($12.885010488) = $12,885.

**23.** Given the monthly payment and the loan amount in 1000s, solve for the amortization table entry. Then locate the entry in an amortization table for a 30-year loan.

Monthly Payment = (Amortization Table Entry)(Loan Amount in $1000s)

$2238.28 = (Amortization Table Entry)($230.45)

$$\frac{\$2238.28}{\$230.45} = \text{Amortization Table Entry}$$

$9.712649 ≈ Amortization Table Entry

The approximate annual interest rate for this loan is 11.25%.

**24. (a)** You pay a total of $75(18) = $1350 for the refrigerator which is $1350 − $1050 = $300 over the cost of the refrigerator.

**(b)** Use the simple interest formula and solve for the unknown rate.

$$I = Prt$$

$300 = $1050(r)(1.5)

$300 = $1575r

$$\frac{\$300}{\$1575} = r$$

$0.190476 ≈ r

You pay an annual interest rate of approximately 19.05%.

**25. (a)** The maximum home price the couple can afford is 3×(annual gross income) = 3($61,000) = $183,000.

**(b)** Their monthly gross income = $\left(\dfrac{1}{12}\right)$($61,000) ≈ $5083.33.

The couple can pay a low of 0.25×(monthly gross income) = 0.25($5083.33) ≈ $1270.83 to a high of 0.38×(monthly gross income) = 0.38(5083.33) ≈ $1931.67.

**26.** Jackson will have a down payment of $24,000 − $3000 = $21,000. He will finance a total of $180,000 − $21,000 = $159,000.

**(a)** Use the monthly payment formula.

$$PMT = \frac{P \times \left(\dfrac{r}{12}\right) \times \left(1 + \dfrac{r}{12}\right)^{12t}}{\left(1 + \dfrac{r}{12}\right)^{12t} - 1}$$

$$= \frac{\$159{,}000\left(\dfrac{0.061}{12}\right)\left(1 + \dfrac{0.061}{12}\right)^{12 \times 30}}{\left(1 + \dfrac{0.061}{12}\right)^{12 \times 30} - 1}$$

$$\approx \$963.54$$

The monthly payments will be approximately $963.54.

**(b)** Assume the home is valued at $180,000.

Taxes = 3.5% of Assessed Value

$$= 0.035(\$180{,}000)$$

$$= \$6300$$

Taxes and insurance will cost a total of $6300 + $650 = $6950 for the year. Each month add $\dfrac{\$6950}{12} \approx \$579.17$ to the monthly payment. The complete monthly payment is $963.54 + $579.17 = $1542.71.

**(c)** Jackson's monthly gross income is $\dfrac{\$65{,}000}{12} \approx \$5416.67$. According to the affordability guidelines, Jackson's monthly expenses cannot exceed a low of 0.25(monthly gross income) = 0.25($5416.67) ≈ $1354.17 to a high of 0.38(monthly gross income) = 0.38($5416.67) ≈ $2058.33. Yes, Jackson can afford the monthly expenses.

**27.** A 2.5-point discount charge will cost 2.5% of $78,500. A 1.5-point origination fee will cost 1.5% of $78,500. The total cost of these fees is 0.025($78,500) + 0.015($78,500) = $1962.50 + $1177.50 = $3140.00.

**28.** By paying another 0.5-point, the interest rate will be 4.5%. The amount that will be financed is $78,500.

**Find the total cost of the additional fees:**

The additional 0.5-point will cost 0.005($78,500) = $392.50 for a total of $3140.00 + $392.50 = $3532.50 in additional fees.

**Find the revised balance:**

The revised balance is $78,500.00 + $3532.50 = $82,032.50.

**Use an amortization table to determine the monthly payment:**
Monthly payment $= \$6.326494(82.0325) \approx \$518.98$.

**Use the monthly payment formula:**
Find the interest rate that would yield a monthly payment of \$518.98 for the original balance of \$78,500. First use an amortization table to narrow down the interest rate options and then use linear interpolation.

From an amortization table we have the following:

$$\text{Monthly Payment} = (\text{Amortization Table Entry})(\text{Loan Amount in } \$1000s)$$

$$\$518.98 = (\text{Amortization Table Entry})(78.5)$$

$$\frac{\$518.98}{78.5} = \text{Amortization Table Entry}$$

$$6.611210 \approx \text{Amortization Table Entry}$$

For a 20-year loan, the amortization table entry falls between 6.599557 and 6.738442 which correspond to 5% and 5.25% respectively. Use linear interpolation to obtain a good, first estimate of the correct value.

$$\frac{6.611210 - 6.599557}{6.738442 - 6.599557} \times 100\% = \frac{0.011653}{0.138885} \times 100\% \approx 0.083904 \times 100\% \approx 8\%$$

We see that the amortization table entry we calculated is approximately 8% of the way from the lower table entry to the higher table entry. We estimate that the corresponding interest rate is itself about 8% of the way from 5% to 5.25% or $5 + 0.083904 \times (5.25 - 5) \approx 5.020976\%$. Use 5.0210% as the first guess at the APR.

| Guess | Monthly Payment | Compare to $518.98 |
|---|---|---|
| 5.0210% | $PMT = \dfrac{\$78,500 \times \left(\dfrac{0.05021}{12}\right) \times \left(1 + \dfrac{0.05021}{12}\right)^{240}}{\left(1 + \dfrac{0.05021}{12}\right)^{240} - 1} \approx \$518.98$ | *This rate is correct.* |

The interest rate that would yield a monthly payment of \$518.98 for a balance of \$78,500 is 5.0210%. This is called the annual percentage rate.